Essential Concepts of Differential Equations

Volume I

Essential Concepts of Differential Equations

Volume I

Edited by **Calanthia Wright**

NY RESEARCH PRESS

New York

Published by NY Research Press,
23 West, 55th Street, Suite 816,
New York, NY 10019, USA
www.nyresearchpress.com

Essential Concepts of Differential Equations: Volume I
Edited by Calanthia Wright

International Standard Book Number: 978-1-63238-182-8 (Hardback)

Printed in China.

Contents

Preface

A differential equation is an integral part of the vast field of mathematics. It can be defined as a mathematical equation that relates some function of one or more variables with its derivatives. The mathematical theory of differential equations can be said to have developed together with the sciences where the equations had derived from and where the results found application or were needed. Differential equations arise whenever a deterministic relation concerning some constantly changing quantities and their rates of change in space and time is known or hypothesized. Such relations are extremely familiar and therefore differential equations play a fundamental role in many disciplines like physics, engineering, biology and economics. The mathematical theory behind the equations can also be viewed as a uniting principle behind various phenomena. The theory of conduction of heat is one of the examples of a phenomena governed by a differential equation, that is, the heat equation. One will find that there are many processes that, while seemingly different, are described by differential equations. Diverse problems, sometimes stemming from quite distinct scientific fields, may give rise to identical differential equations. Many fundamental laws of physics and chemistry can be formulated as differential equations. Even in fields such as biology and economics, differential equations can be used to represent the behavior of complex systems. Thus the arena of differential equations can be said to be quite a prolific one.

This book is an attempt to compile and collate all available research on the subject of differential equations under one umbrella. I am grateful to those who put their hard work, effort and expertise into these researches as well as those who were supportive in this endeavour. I also wish to thank my publisher for giving me this unmatched opportunity. I am extremely thankful to all the contributing authors who took out their precious time to interact with me and helped me understand their research perspectives in a better manner for the best output. Lastly, I wish to thank my family for their constant support.

Editor

On Uniform Exponential Stability and Exact Admissibility of Discrete Semigroups

Aftab Khan,[1,2] Gul Rahmat,[2] and Akbar Zada[3]

[1] *Shaheed Benazir Bhutto University Sheringal, Dir Upper 18000, Pakistan*
[2] *Government College University, Abdus Salam School of Mathematical Sciences (ASSMS), Lahore 54600, Pakistan*
[3] *Department of Mathematics, University of Peshawar, Peshawar 25000, Pakistan*

Correspondence should be addressed to Akbar Zada; zadababo@yahoo.com

Academic Editor: Sotiris Ntouyas

We prove that a discrete semigroup $\mathbb{T} = \{T(n) : n \in \mathbb{Z}_+\}$ of bounded linear operators acting on a complex Banach space X is uniformly exponentially stable if and only if, for each $x \in AP_0(\mathbb{Z}_+, X)$, the sequence $n \mapsto \sum_{k=0}^{n} T(n-k)x(k) : \mathbb{Z}_+ \to X$ belongs to $AP_0(\mathbb{Z}_+, X)$. Similar results for periodic discrete evolution families are also stated.

1. Introduction

The solutions of the autonomous discrete systems $x_{n+1} = Ax_n$ or $y_{n+1} = Ay_n + h_n$ lead to the idea of discrete semigroups. There are a lot of spectral criteria which characterize different types of stability (or other types of asymptotic behavior) of the solutions of above systems. For further results on asymptotic behavior of semigroups, we refer to [1].

New difficulties appear in the study of the nonautonomous systems, especially because the part of the solution generated by the forced term (h_n), that is, $\sum_{k=\nu}^{n} U(n,k)h_k$, is not a convolution in the classical sense. These difficulties may be passed by using the so-called evolution semigroups.

The evolution semigroups were exhaustively studied in [2]. Having in mind the well-known results stated in the continuous case, see for example [2, 3], we can say that this method is a very efficient one. See also [4, 5] for recent developments concerning the semigroups of evolution acting on almost periodic function spaces.

Recently, the discrete version of [6] was obtained in [7].

In this note, we study the asymptotic behavior of the discrete semigroups in terms of exact admissibility of the space of almost periodic sequences.

In this regard, we develop the theory of discrete evolution semigroups on a special space of bounded sequences. Results of this type in the continuous case may be found in [8] and the references therein. However, by contrast with the continuous case, we did not find in the existent literature papers written in the spirit of the present one referring to the discrete evolution semigroups. These results could be new and useful for people whose area of research is restricted to difference equations.

2. Definitions and Preliminary Results

Let X be a complex Banach space and $\mathscr{B}(X)$ the Banach algebra of all linear and bounded operators acting on X. The norms in X and in $\mathscr{B}(X)$ will be denoted by $\|\cdot\|$. Let $\mathbb{Z}_+$ be the set of all nonnegative integer numbers. A sequence $x : \mathbb{Z}_+ \to X$ is said to be almost periodic if for any $\epsilon > 0$ there exists an integer $l_\epsilon > 0$ such that any discrete interval of length l_ϵ contains an integer τ, such that

$$\left\|x_{n+\tau} - x_n\right\| \le \epsilon, \quad \forall n \in \mathbb{Z}_+. \tag{1}$$

The integer number τ is called ϵ-translation number of (x_n). The set of all almost periodic sequences will be denoted by $AP(\mathbb{Z}_+, X)$. For further details about almost periodic functions, we refer to the books [9, 10]. The set $l^\infty(\mathbb{Z}_+, X)$ of all bounded sequences becomes a Banach space when it is endowed with the "sup" norm denoted by $\|\cdot\|_\infty$. Clearly, $AP(\mathbb{Z}_+, X)$ is a subset of $l^\infty(\mathbb{Z}_+, X)$. Let $P_q^0(\mathbb{Z}_+, X)$ be the space of all q-periodic ($q \ge 2$ is an integer number) sequences

x with $x(0) = 0$. Denote by $\mathcal{A}_0(\mathbb{Z}_+, X)$ the set of all sequences $\{x(n)\}_{n\geq 0}$ for which there exists $n_x \in \mathbb{Z}_+$ with $n_x > 0$ and $y_x \in P_q^0(\mathbb{Z}_+, X)$ such that

$$x(n) = \begin{cases} 0, & \forall 0 \leq n < n_x, \\ y_x(n), & \text{if } n \geq n_x. \end{cases} \quad (2)$$

Let $AP_0(\mathbb{Z}_+, X) := \overline{\operatorname{span}}\{\mathcal{A}_0(\mathbb{Z}_+, X)\}$. Here the closeness is considered in the space $l^\infty(\mathbb{Z}_+, X)$.

For a bounded linear operator L, acting on X, we denote by $\sigma(L)$ the spectrum of L and by $\rho(L)$ its resolvent set. Recall that a subset $\mathbb{T} = \{T(n)\}_{n\in\mathbb{Z}_+}$ of $\mathcal{B}(X)$ is called discrete semigroup if it satisfies the following conditions:

(i) $T(0) = I$, where I is the identity operator on X.

(ii) $T(n+m) = T(n)T(m)$, for all $n, m \in \mathbb{Z}_+$.

A discrete semigroup $\mathbb{T}$ is said to be uniformly exponentially stable if there exist $N, \nu > 0$ such that

$$\|T(n)\| \leq Ne^{-\nu n} \quad \forall n \in \mathbb{Z}_+. \quad (3)$$

The spectral radius of $T(1)$ denoted by $r(T(1))$ is defined as

$$r(T(1)) := \sup\{|\lambda| : \lambda \in \sigma(T(1))\}. \quad (4)$$

It is well known that, see for example [11, page 42],

$$r(T(1)) = \lim_{n\to\infty} \|(T(1))^n\|^{1/n}. \quad (5)$$

As a consequence of (5), a discrete semigroup $\{T(n)\}_{n\in\mathbb{Z}_+}$ is uniformly exponentially stable if and only if $r(T(1)) < 1$.

Having in mind the continuous case, the "infinitesimal generator" of the discrete semigroup denoted by G is defined by $G := T(1) - I$. For discrete semigroups, the Taylor formula of order one is

$$T(n)x - x = \sum_{k=0}^{n-1} T(k)Gx, \quad \forall n \in \mathbb{Z}_+,\ n \geq 1,\ \forall x \in X. \quad (6)$$

A discrete semigroup $\mathbb{T}$ is said to be $AP_0(\mathbb{Z}_+, X)$ exact admissible, if for every $x \in AP_0(\mathbb{Z}_+, X)$ the sequence $(\sum_{k=0}^{n} T(n-k)h(k))_{n\in\mathbb{Z}_+}$ belongs with $AP_0(\mathbb{Z}_+, X)$.

The evolution semigroup $\mathbb{S} = \{S(n), n \in \mathbb{Z}_+\}$ associated with $\mathbb{T}$ on $AP_0(\mathbb{Z}_+, X)$ is defined by

$$(S(r)x)(n) = \begin{cases} T(r)x(n-r), & \forall n \geq r, \\ 0, & 0 \leq n \leq r. \end{cases} \quad (7)$$

3. Results

The following lemma shows that the associated evolution semigroup $\{S(n)\}_{n\in\mathbb{Z}_+}$ acts on $AP_0(\mathbb{Z}_+, X)$.

Lemma 1. *Let $x \in AP_0(\mathbb{Z}_+, X)$ and $\mathbb{T} = \{T(j)\}_{j\in\mathbb{Z}_+}$ be a discrete semigroup of bounded linear operators on X. The sequence $S(r)x$, given by*

$$(S(r)x)(n) = \begin{cases} T(r)x(n-r), & \forall n \geq r \\ 0, & 0 \leq n \leq r, \end{cases} \quad (8)$$

belongs to $AP_0(\mathbb{Z}_+, X)$.

Proof. First we show that $S(r)x \in \mathcal{A}_0(\mathbb{Z}_+, X)$ for any $x \in \mathcal{A}_0(\mathbb{Z}_+, X)$. Since $x \in \mathcal{A}_0(\mathbb{Z}_+, X)$ there exist $n_x \in \mathbb{Z}_+$ with $n_x > 0$, and $(y_x(n)) \in P_q^0(\mathbb{Z}_+, X)$, such that

$$x(n) = \begin{cases} 0, & \text{if } 0 \leq n < n_x \\ y_x(n), & \text{if } n \geq n_x. \end{cases} \quad (9)$$

Let $n_{S(r)x} := r + n_x$ and set $y_{S(r)x}(\cdot) = T(r)y_x(\cdot - r)$. Clearly $y_{S(r)x}$ is q-periodic sequence. It remains to show that

$$(S(r)x)(n) = \begin{cases} 0, & \text{if } 0 \leq n < n_{S(r)x} \\ y_{S(r)x}(n), & \text{if } n \geq n_{S(r)x}. \end{cases} \quad (10)$$

If $n \leq n_{S(r)x} = r + n_x$, then $n - r < n_x$ and $x(n-r) = 0$, so

$$(S(r)x)(n) = T(r)x(n-r) = 0. \quad (11)$$

If $n \geq n_{S(r)x} = r + n_x$, then $n - r \geq n_x$ and $x(n-r) = y_x(n-r)$; hence

$$\begin{aligned} (S(r)x)(n) &= T(r)x(n-r) \\ &= T(r)y_x(n-r) \\ &= y_{S(r)x}(n). \end{aligned} \quad (12)$$

Thus $S(r)x \in \mathcal{A}_0(\mathbb{Z}_+, X)$. Now, from linearity it follows that $S(r)z$ belongs to $\operatorname{span}\{\mathcal{A}_0(\mathbb{Z}_+, X)\}$ whenever $z \in \operatorname{span}\{\mathcal{A}_0(\mathbb{Z}_+, X)\}$. Let now $\epsilon > 0$, $x \in AP_0(\mathbb{Z}_+, X)$, and let $z \in \operatorname{span}\{\mathcal{A}_0(\mathbb{Z}_+, X)\}$, such that $\|x - z\|_{l^\infty(\mathbb{Z}_+,X)} < \epsilon$. Clearly $S(r)z$ belongs to $\operatorname{span}\{\mathcal{A}_0(\mathbb{Z}_+, X)\}$, and

$$\begin{aligned} \|S(r)z - S(r)x\|_{l^\infty(\mathbb{Z}_+,X)} &= \sup_{n\geq r} \|T(r)[z(n-r) - x(n-r)]\| \\ &\leq Me^{\nu r} \sup_{n\geq r} \|z(n-r) - x(n-r)\| \\ &\leq Me^{\nu r}\epsilon, \end{aligned} \quad (13)$$

that is, $S(r)x$ is in $AP_0(\mathbb{Z}_+, X)$. This completes the proof.

Lemma 2. *Let $\mathbb{T} = \{T(n)\}_{n\in\mathbb{Z}_+}$ be a discrete semigroup of bounded linear operators on X, and let $\mathbb{S} = \{S(n), n \in \mathbb{Z}_+\}$ be the evolution semigroup associated with $\mathbb{T}$ on $AP_0(\mathbb{Z}_+, X)$, having $G_{\mathbb{S}}$ as generator. Let $x, z \in AP_0(\mathbb{Z}_+, X)$. The following two statements are equivalent:*

(i) $G_{\mathbb{S}}x = -z$,

(ii) *$x(n) = \sum_{k=0}^{n} T(n-k)z(k)$, for all $n \in \mathbb{Z}_+$.*

Proof. (i) $\Rightarrow$ (ii): Using the Taylor formula (6), one has

$$S(n)x - x = \sum_{m=0}^{n-1} S(m)G_{\mathbb{S}}x = -\sum_{m=0}^{n-1} S(m)z. \quad (14)$$

Then, for every $n \in \mathbb{Z}_+$, one has

$$\begin{aligned} x(n) &= (S(n)x)(n) + \sum_{m=0}^{n-1} (S(m)z)(n) \\ &= T(n)x(0) + \sum_{m=0}^{n-1} T(m)z(n-m) \\ &= \sum_{k=0}^{n} T(n-k)z(k). \end{aligned} \tag{15}$$

(ii) ⇒ (i): For each $n \in \mathbb{Z}_+$, one has

$$\begin{aligned} (G_{\mathbb{S}}x)(n) &= (S(1)-I)x(n) \\ &= T(1)x(n-1) - x(n) \\ &= T(1)\sum_{k=0}^{n-1} T(n-1-k)z(k) - x(n) \\ &= \sum_{k=0}^{n-1} T(n-k)z(k) - \sum_{k=0}^{n} T(n-k)z(k) \\ &= -z(n). \end{aligned} \tag{16}$$

This completes the proof.

See also [12], for a variant of this lemma in other space.

The next result is the main ingredient in the proof of Theorem 5 that follows.

Theorem 3 (see [7]). *Let $\mathbb{T} = \{T(n) : n \in \mathbb{Z}_+\}$ be a discrete semigroup on X, and let μ be a real number. If*

$$\sup_{n\geq 0} \left\| \sum_{k=0}^{n} e^{i\mu k} T(n-k) f(k) \right\| < \infty, \tag{17}$$

for every $f \in P_0^q(\mathbb{Z}_+, X)$, then $T(1)$ is power bounded (i.e., $\sup_{n\in\mathbb{Z}_+}\|A^n\| < \infty$) and $e^{i\mu} \in \rho(T(1))$.

As a corollary of this theorem, we state the following.

Corollary 4 (see [7]). *Let $\mathbb{T} = \{T(n) : n \in \mathbb{Z}_+\}$ be a discrete semigroup on X. If the condition (17) holds for every $\mu \in \mathbb{R}$ and every f in $P_0^q(\mathbb{Z}_+, X)$, then the semigroup $\mathbb{T}$ is uniformly exponentially stable.*

The result of this paper reads as follows.

Theorem 5. *Let $\mathbb{T} = \{T(n)\}_{n\in\mathbb{Z}_+}$ be a discrete semigroup on X. The following four statements are equivalent:*

(i) *$\mathbb{T}$ is uniformly exponentially stable.*

(ii) *The evolution semigroup $\mathbb{S}$ associated with $\mathbb{T}$ on $AP_0(\mathbb{Z}_+, X)$ is uniformly exponentially stable.*

(iii) *The semigroup $\mathbb{T}$ is $AP_0(\mathbb{Z}_+, X)$ exact admissible.*

(iv) $\sup_{n\in\mathbb{Z}_+}\|\sum_{k=0}^{n} T(n-k)z(k)\| = M_z < \infty$, *for all $z \in AP_0(\mathbb{Z}_+, X)$.*

Proof. (i) ⇒ (ii): Let $\mathbb{T}$ be uniformly exponentially stable, and let N and ν be positive constants such that

$$\|T(n)\| \leq Ne^{-\nu n} \quad \forall n \in \mathbb{Z}_+. \tag{18}$$

Then for every f in $AP_0(\mathbb{Z}_+, X)$, one has

$$\|S(j)f\|_\infty = \sup_{n\geq j}\|T(j)f(n-j)\| \leq Ne^{-\nu j}\|f\|_\infty. \tag{19}$$

(ii) ⇒ (iii): Since $\mathbb{S}$ is uniformly exponentially stable, $1 \in \rho(S(1))$, that is, $S(1) - I$ is invertible. Then for each z in $AP_0(\mathbb{Z}_+, X)$, there exists $u \in AP_0(\mathbb{Z}_+, X)$ such that $(S(1) - I)u = -z$.

On the other hand, by Lemma 2, $u(n) = \sum_{k=0}^{n} T(k)z(n-k)$, for every $n \in \mathbb{Z}_+$; hence $\mathbb{T}$ is $AP_0(\mathbb{Z}_+, X)$ exact admissible.

(iii) ⇒ (iv) It is obvious.

(iv) ⇒ (i) Obviously, if $z \in P_q^0(\mathbb{Z}_+, X)$ and μ is a real number, then $(e^{i\mu n}z(n))_{n\in\mathbb{Z}_+}$ belongs to $AP_0(\mathbb{Z}_+, X)$. Now, we can apply Corollary 4 to finish the proof.

The following example is a concrete application of Theorem 5.

Example 6. Let X be a complex Banach space, and let A be a bounded linear operator acting on X. Consider the following two discrete Cauchy problems:

$$\begin{aligned} x_{j+1} &= Ax_j, \quad j \in \mathbb{Z}_+, \\ x_0 &= b, \end{aligned} \tag{20}$$

$$\begin{aligned} y_{j+1} &= Ay_j + f_{j+1}, \quad j \in \mathbb{Z}_+, \\ y_0 &= 0. \end{aligned} \tag{21}$$

The solutions of (20) and (21) are (resp.) given by $x_j = T(j)b$ and $y_j = \sum_{k=0}^{j} T(j-k)x(k)$. Here $T(k) := A^k$.

From Theorem 5, the following two statements are equivalent.

(1) For each $b \in X$ the solution of (20) decays exponentially, or, equivalently, there exist two positive constants K and ν such that

$$\|T(j)x\| \leq Ke^{-\nu j}\|x\| \quad \forall x \in X. \tag{22}$$

(2) For each $f \in AP_0(\mathbb{Z}_+, X)$ the solution of (21) belongs to $AP_0(\mathbb{Z}_+, X)$.

In fact, we can state a more general result concerning q-periodic discrete evolution families. To establish this result, we recall that a family $\mathcal{U} = \{U(n,m) : n \geq m \in \mathbb{Z}_+\} \subset \mathcal{B}(X)$ is said to be q-periodic discrete evolution family if it satisfies the following properties.

(i) $U(n,n) = I$ and $U(n,m)U(m,r) = U(n,r)$, for all $n,m,r \in \mathbb{Z}_+$ with $n \geq m \geq r \in \mathbb{Z}_+$, where I is the identity operator on X.

(ii) $U(n+q, m+q) = U(n,m)$, for all $n \geq m \in \mathbb{Z}_+$.

It is said to be uniformly exponentially stable if there exist the positive constants K and ν such that

$$\|U(n,m)\| \leq Ke^{-\nu(n-m)} \quad \forall m \geq n \in \mathbb{Z}_+. \tag{23}$$

Also, the family $\mathcal{U}$ is said to be $AP_0(\mathbb{Z}_+, X)$ exact admissible, if for every $z \in AP_0(\mathbb{Z}_+, X)$ the sequence $(\sum_{k=0}^{n} U(n, k)z(k))_{n\in\mathbb{Z}_+}$ belongs to $AP_0(\mathbb{Z}_+, X)$.

The discrete evolution semigroup $\mathcal{T} = \{\mathcal{T}(n), n \in \mathbb{Z}_+\}$ associated with the evolution family $\mathcal{U}$ on $AP_0(\mathbb{Z}_+, X)$ is defined by

$$(\mathcal{T}(n)z)(r) = \begin{cases} U(r, r-n)\,z(r-n), & \forall r \geq n, \\ 0, & \text{otherwise}. \end{cases} \tag{24}$$

As in Lemma 1 it can be proved that it acts on $AP_0(\mathbb{Z}_+, X)$.

Theorem 7. *Let $\mathcal{U} = \{U(n,m) : n \geq m \in \mathbb{Z}_+\}$ be a q-periodic evolution family of bounded linear operators on X. The following statements are equivalent:*

(1) *$\mathcal{U}$ is uniformly exponentially stable.*

(2) *The evolution semigroup $\mathcal{T}$ associated with $\mathcal{U}$ is uniformly exponentially stable.*

(3) *$\mathcal{U}$ is $AP_0(\mathbb{Z}_+, X)$ exact admissible.*

(4) $\sup_{n\in\mathbb{Z}_+} \| \sum_{k=0}^{n} U(n,k)h(k)\| < \infty$, *for all $h \in AP_0(\mathbb{Z}_+, X)$.*

The proofs of (1) ⇒ (2) ⇒ (3) ⇒ (4) are similar to those in the semigroup case. For the proof of (4) ⇒ (1) we use the following result from [13].

If for every $\mu \in \mathbb{R}$ and every $z \in P_q^0(\mathbb{Z}_+, X)$, one has

$$\sup_{n\in\mathbb{Z}_+} \left\| \sum_{k=0}^{n} e^{i\mu k} U(n,k)\, z(k) \right\| := M(\mu, z) < \infty, \tag{25}$$

then the family $\mathcal{U}$ is uniformly exponentially stable.

Acknowledgment

The authors would like to thank the anonymous referees for their comments and suggestions on preliminary versions of this paper, which have led to a substantial improvement in its readability. In particular, we have completed the last section of this note at the suggestions of referees.

References

[1] J. van Neerven, *The Asymptotic Behaviour of Semigroups of Linear Operators*, vol. 88 of *Operator Theory: Advances and Applications*, Birkhäuser, Basel, Switzerland, 1996.

[2] C. Chicone and Y. Latushkin, *Evolution Semigroups in Dynamical Systems and Differential Equations*, vol. 70 of *Mathematical Surveys and Monographs*, American Mathematical Society, Providence, RI, USA, 1999.

[3] S. Clark, Y. Latushkin, S. Montgomery-Smith, and T. Randolph, "Stability radius and internal versus external stability in Banach spaces: an evolution semigroup approach," *SIAM Journal on Control and Optimization*, vol. 38, no. 6, pp. 1757–1793, 2000.

[4] O. Saierli, "Spectral mapping theorem for an evolution semigroup on a space of vector-valued almost-periodic functions," *Electronic Journal of Differential Equations*, vol. 2012, no. 175, pp. 1–9, 2012.

[5] C. Buşe, D. Lassoued, T. L. Nguyen, and O. Saierli, "Exponential stability and uniform boundedness of solutions for nonautonomous periodic abstract Cauchy problems. An evolution semigroup approach," *Integral Equations and Operator Theory*, vol. 74, no. 3, pp. 345–362, 2012.

[6] C. Buşe, S. S. Dragomir, and V. Lupulescu, "Characterizations of stability for strongly continuous semigroups by boundedness of its convolutions with almost periodic functions," *International Journal of Differential Equations and Applications*, vol. 2, no. 1, pp. 103–109, 2001.

[7] A. Zada, G. Rahmat, G. Ali, and A. Tabassum, "Characterizations of stability for discrete semigroup of bounded linear operators," *International Journal of Mathematics and Soft Computing*, vol. 3, no. 3, 2013.

[8] C. Buşe and O. Jitianu, "A new theorem on exponential stability of periodic evolution families on Banach spaces," *Electronic Journal of Differential Equations*, vol. 2013, no. 14, pp. 1–10, 2003.

[9] C. Corduneanu, *Almost Periodic Oscillations and Waves*, Springer, New York, NY, USA, 2009.

[10] A. S. Besicovitch, *Almost Periodic Functions*, Dover Publications, New York, NY, USA, 1955.

[11] R. G. Douglas, *Banach Algebra Techniques in Operator Theory*, vol. 179 of *Graduate Texts in Mathematics*, Springer, New York, NY, USA, 2nd edition, 1998.

[12] C. Buşe, A. Khan, G. Rahmat, and A. Tabassum, "Uniform exponentialstability for nonautonomous system via discrete evolution semigroups," to appear in *Bulletin Mathématique de la Société des Sciences Mathématiques de Roumanie*.

[13] C. Buşe, P. Cerone, S. S. Dragomir, and A. Sofo, "Uniform stability of periodic discrete systems in Banach spaces," *Journal of Difference Equations and Applications*, vol. 11, no. 12, pp. 1081–1088, 2005.

2

Oscillations of a Class of Forced Second-Order Differential Equations with Possible Discontinuous Coefficients

Siniša Miličić, Mervan Pašić, and Darko Žubrinić

Department of Mathematics, Faculty of Electrical Engeneering and Computing, University of Zagreb, 10000 Zagreb, Croatia

Correspondence should be addressed to Mervan Pašić; mervan.pasic@gmail.com

Academic Editor: Ondrej Dosly

We study the oscillation of all solutions of a general class of forced second-order differential equations, where their second derivative is not necessarily a continuous function and the coefficients of the main equation may be discontinuous. Our main results are not included in the previously published known oscillation criteria of interval type. Many examples and consequences are presented illustrating the main results.

1. Introduction

Let $t_0 > 0$ and let $AC_{\text{loc}}([t_0,\infty),\mathbb{R})$ denote the set of all real functions absolutely continuous on every bounded interval $[a,b] \subset [t_0,\infty)$. We study the oscillatory behaviour of all solutions $x = x(t)$ of the following class of forced second-order differential equations:

$$\left(r(t)\,\Phi\left(x(t),x'(t)\right)\right)' + q(t)\,f(x(t)) = e(t), \quad \text{a.e. in } [t_0,\infty), \qquad (1)$$
$$x, r\Phi\left(x,x'\right) \in AC_{\text{loc}}\left([t_0,\infty),\mathbb{R}\right),$$

where the functions $\Phi : \mathbb{R}^2 \to \mathbb{R}$, $\Phi = \Phi(u,v)$, $f : \mathbb{R} \to \mathbb{R}$, and $f = f(u)$ satisfy some general conditions given in Section 2. A continuous function $x = x(t)$ is said to be *oscillatory* if there is a sequence $t_n \in [t_0,\infty)$, such that $x(t_n) = 0$ for all $n \in \mathbb{N}$ and $t_n \to \infty$ as $n \to \infty$. A differential equation is *oscillatory* if all its solutions are oscillatory.

The forcing term $e(t)$ is a sign-changing function (possibly discontinuous). This can be formulated by the following hypothesis: for every $T \geq t_0$ there exist two intervals (a_1,b_1) and (a_2,b_2), $T \leq a_1 < b_1 \leq a_2 < b_2$, such that

$$e(t) \geq 0, \quad t \in (a_1,b_1), \qquad e(t) \leq 0, \quad t \in (a_2,b_2). \qquad (2)$$

The coefficient $q(t)$ may be a discontinuous function on $[t_0,\infty)$ and the case $x \notin C^2((t_0,\infty),\mathbb{R})$ occurs in our main results and examples too. Two important classes of functions $\Phi(u,v)$ are included in the differential operator $(r(t)\Phi(x,x'))'$ as

$$\Phi(u,v) = \phi(u)\,v, \quad \Phi(u,v) = \frac{\phi(u)\,v}{\sqrt{1+v^2}}, \quad (u,v) \in \mathbb{R}^2. \qquad (3)$$

The first one is the classic second-order differential operator which is linear in x' and the second one is the so-called one-dimensional mean curvature differential operator; see Examples 1 and 2.

Depending on $q(t)$, we propose the following four simple models for (1):

(i) $q(t)$ is strictly positive and continuous on $[t_0,\infty)$ as

$$x'' + 4m^2 f(x) = h(\sin(mt)), \quad \text{a.e. in } [t_0,\infty), \qquad (4)$$
$$x, x' \in AC_{\text{loc}}\left([t_0,\infty),\mathbb{R}\right), \quad x \notin C^2\left((t_0,\infty),\mathbb{R}\right);$$

(ii) $q(t)$ is nonnegative and continuous on $[t_0,\infty)$ as

$$x'' + m^2\pi^2[\cos(mt)]^+ f(x) = h(\sin(mt)), \quad \text{a.e. in } [t_0,\infty), \qquad (5)$$
$$x, x' \in AC_{\text{loc}}\left([t_0,\infty),\mathbb{R}\right);$$

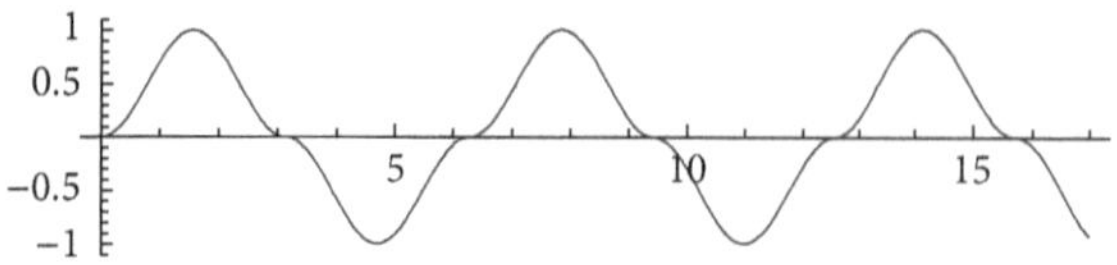

FIGURE 1: Function $x(t) = |\sin(t)| \sin(t)$ is a solution of (4).

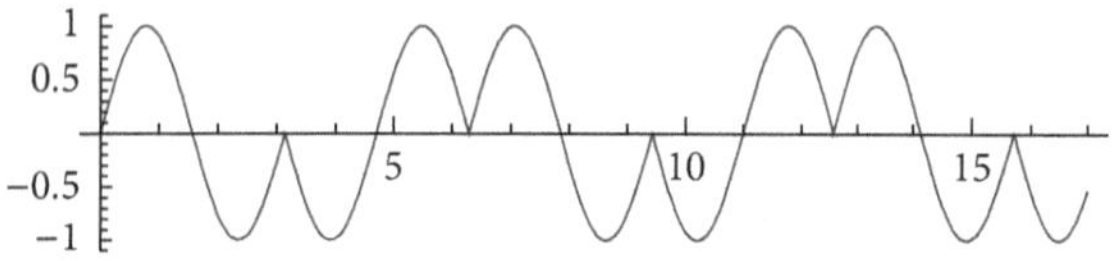

FIGURE 2: $x'(t) = 2|\sin(t)| \cos(t)$ and hence $x''(t)$ is not a continuous function.

(iii) $q(t)$ is nonnegative and discontinuous on $[t_0, \infty)$ as

$$x'' + 4m^2[\text{sign}(\cos(mt))]^+ f(x) = h(\sin(mt)), \quad \text{a.e. in } [t_0, \infty), \quad (6)$$
$$x, x' \in AC_{\text{loc}}([t_0, \infty), \mathbb{R});$$

(iv) $q(t)$ is sign changing and discontinuous on $[t_0, \infty)$ as

$$x'' + 4m^2 \,\text{sign}(\cos(mt)) f(x) = h(\sin(mt)), \quad \text{a.e. in } [t_0, \infty), \quad (7)$$
$$x, x' \in AC_{\text{loc}}([t_0, \infty), \mathbb{R}),$$

where $m \in \mathbb{N}$ and $h(s)$ is an arbitrary function such that $h(s)\, s > 0$ for all $s \neq 0$, for instance, $h(s) = s$ or $h(s) = \text{sign}(s)$. According to Corollaries 7 and 10, we will show that (4)–(7) are oscillatory provided the function $f = f(u)$ satisfies $f(u)/u \geq K \geq 1$ for all $u \neq 0$; see Examples 8–13. It is interesting that in particular for $f(u) = u$ and $h(s) = 2m^2 \,\text{sign}(s)$, (4) allows an explicit oscillatory solution $x(t) = |\sin(mt)| \sin(mt)$ as shown in Figures 1 and 2.

Moreover, as a consequence of Corollary 7, one can show that all solutions of (4) are oscillatory; for details see Example 8. The main goal of this paper is to give some sufficient conditions on functions $\Phi(u, v)$, $f(u)$ and the coefficients $r(t)$, $q(t)$, and $e(t)$ such that (1) is oscillatory; see Theorems 3 and 4. It will also cover the model equations (4)–(7) as well as some other examples presented in Section 2.

To the best of our knowledge, it seems that there are only few papers which study the oscillation of the second-order differential equations with nonsmooth (local integrable) coefficients; see [1–3]. More precisely, in [1] the author studied the interval oscillation criteria for the following second-order half-linear differential equation:

$$\left(r(t)\left|x'(t)\right|^{\sigma-1} x'(t)\right)' + q(t)\left|x(t)\right|^{\sigma-1} x(t) = 0, \quad \text{a.e. in } (0, \infty), \quad (8)$$
$$x, r\left|x'\right|^{\sigma-1} x' \in AC_{\text{loc}}((0, \infty), \mathbb{R}),$$

where $\sigma > 1$ and $1/r,\ q \in L_{\text{loc}}((0, \infty), \mathbb{R})$ such that $\int_0^\infty r^{-1/\sigma}(t)dr = \infty$. See also [2] but with the solution space $C^1((0, \infty), \mathbb{R})$ instead of $AC_{\text{loc}}((0, \infty), \mathbb{R})$, that is, x and $r|x'|^{\sigma-1}x' \in C^1((0, \infty), \mathbb{R})$.

In [3], the authors consider the following second-order differential equation:

$$\left(r(t)\, x'(t)\right)' + Q\left(t, x(t), x'(t)\right) = 0, \quad t \geq t_0, \quad (9)$$

where $r(t) > 0$ a.e. in $[t_0, \infty)$, $1/r \in L_{\text{loc}}([t_0, \infty), \mathbb{R})$, and $Q(t, y, z)$ is locally integrable function in t and continuous in (y, z). Equation (9) allows the forcing term $e(t)$ in the next sense as follows:

$$yQ(t, y, z) \geq q(t)\, y f(y) - e(t)\, y \quad \forall (t, y, z) \in [t_0, \infty) \times \mathbb{R}^2, \quad (10)$$

where $e = e(t)$ satisfies (2), but the functions $q = q(t)$ and $f = f(y)$ are smooth enough in their variables, that is, $q \in C([t_0, \infty), \mathbb{R})$ and $f \in C^1(\mathbb{R}, \mathbb{R})$.

On certain oscillation criteria for various classes of forced second-order differential equations with continuous coefficients, we refer the reader to [4–13]. Our method modifies a recently used one in [14, 15] and it contains the classic Riccati transformation of the main equation, a blow-up argument and pointwise comparison principle. The comparison principle applies to all sub- and supersolutions of a class of the generalized Riccati differential equations with nonlinear terms that are supposed to be locally integrable in the first variable and locally Lipschitz continuous in the second variable.

2. Hypotheses, Results, and Consequences

First of all, the function $\Phi(u, v)$ which appears in the second-order differential operator of (1) satisfies

$$|u|^{\gamma-2} v\Phi(u, v) \geq g(|\Phi(u, v)|) \quad \forall u, v \in \mathbb{R}, \quad (11)$$

where $\gamma \geq 2$ and $g : \mathbb{R}_+ \to \mathbb{R}_+$ is a locally Lipschitz function $g_0 : \mathbb{R}_+ \to \mathbb{R}_+$ satisfying

$$g(cs) \geq c^\gamma g_0(s) \quad \forall c > 0,\ s > 0,$$
$$g_0(s) + M_0 \geq s^2 \quad \text{for some } M_0 \geq 0 \text{ and all } s \in \mathbb{R}_+. \quad (12)$$

In most cases, $g(s) = g_0(s)g_1(s)$, $s > 0$, where $g_0(s) = s^\gamma$, $\gamma \geq 2$, and $g_1(s)$ is an arbitrary function satisfying $g_1(s) \geq 1$. Thus, for such $g(s)$ with $g_1(s) \equiv 1$, condition (11) became:

$$|u|^{\gamma-2} v\Phi(u, v) \geq |\Phi(u, v)|^\gamma \quad \forall u, v \in \mathbb{R}. \quad (13)$$

It is not difficult to check that if $g_0 \in C^1(\mathbb{R}_+)$ or $g_0(s)$ is a convex function, then it is locally Lipschitz on $\mathbb{R}_+$ too; see for instance [16, Theorem 1.3.3].

Two essential classes of the second-order differential operators $(r(t)\Phi(x, x'))'$ satisfy condition (13), as is shown in the next examples.

Example 1. We consider the second-order differential operator which is linear in x' as follows:

$$\left(r(t)\Phi\left(x, x'\right)\right)' = \alpha\left(\phi(x)x'\right)', \quad (14)$$

where $\alpha > 0$ and $0 \le \phi(u) \le 1$ for all $u \in \mathbb{R}$. Obviously, the function $\Phi(u, v) = \phi(u)v$ satisfies condition (13) in particular for $\gamma = 2$. Two usual choices for $\phi(u)$ are $\phi(u) = |\sin u|$ and $\phi(u) = |u|/(1 + |u|)$.

Example 2. We consider a quasilinear differential operator (the so-called one-dimensional prescribed mean curvature operator) as follows:

$$\left(r(t)\Phi\left(x, x'\right)\right)' = \alpha\left(\phi(x)\frac{x'}{\sqrt{1 + x'^2}}\right)', \quad (15)$$

where $\alpha > 0$ and $0 \le \phi^{\gamma-1}(u) \le |u|^{\gamma-2}$ for all $u \in \mathbb{R}$. It is not difficult to check that condition (13) is satisfied in particular for $\Phi(u, v) = \phi(u)v/(1 + v^2)^{1/2}$ and for any $\gamma > 1$. For $\phi(u)$, we can take the same choice as in the previous example.

Next, we suppose the existence of a constant K such that

$$\frac{f(u)}{u} \ge K > 0 \quad \forall u \ne 0. \quad (16)$$

In order to simplify our consideration here, in many examples we often use $f(u) = Ku$.

Condition (2) means that there exists a sequence of pairs of intervals $J_{1j} = [a_{1j}, b_{1j}]$ and $J_{2j} = [a_{2j}, b_{2j}]$, $j \in \mathbb{N}$, contained in (t_0, ∞), such that the sequences $(a_{1j})_{j\ge1}$, $(b_{1j})_{j\ge1}$, $(a_{2j})_{j\ge1}$, and $(b_{2j})_{j\ge1}$ are increasing, $a_{1j} < b_{1j} \le a_{2j} < b_{2j}$ for each j, and

$$e(t) \ge 0 \quad \text{on } J_{1j}, \quad e(t) \le 0 \quad \text{on } J_{2j} \quad \text{for each } j \in \mathbb{N}, \quad (17)$$

$$\lim_{j\to\infty} a_{1j} = \infty.$$

On the intervals J_{1j} and J_{2j}, the coefficient $r(t)$ satisfies

$$r(t) > 0 \quad \text{on } J_{ij}, r^{1-\gamma} \in L^1\left(J_{ij}\right) \quad \forall i \in \{1,2\}, \; j \in \mathbb{N}. \quad (18)$$

Let there be a real function $C = C(t)$, $C \in L^1_{\text{loc}}((t_0, \infty), \mathbb{R})$, and let there exist a sequence of positive real numbers $(\lambda_j)_{j\in\mathbb{N}}$ such that

$$C(t) \ge 0 \quad \text{on } J_{ij}, c_{ij} := \int_{J_{ij}} C(\tau)\,d\tau > 0 \quad \forall i \in \{1,2\}, \; j \in \mathbb{N},$$

$$\frac{1}{c_{ij}}C(t) \le \frac{1}{\pi}\min\left\{\left(\lambda_j r(t)\right)^{1-\gamma}, \frac{K}{M_0 + 1}\lambda_j q(t)\right\} \quad \forall t \in J_{ij}, \; i \in \{1,2\}, \; j \in \mathbb{N}, \quad (19)$$

where γ, M_0, and K are constants defined in (11), (12), and (16), respectively.

The proof of the following main result will be presented in Section 4.

Theorem 3. *Let the functions $\Phi(u, v)$, $f(u)$, $e(t)$, and $r(t)$ satisfy* (11), (12), (16), (17), *and* (18), *respectively. Let $q(t) \ge 0$ and $q(t) \not\equiv 0$ on each interval J_{ij}, $i \in \{1,2\}$, $j \in \mathbb{N}$. If* (19) *is fulfilled, then* (1) *is oscillatory.*

Condition (19) can be replaced by an equivalent one, which has a more practical value and takes a simpler form since we do not need a sequence of auxiliary parameters $(\lambda_j)_{j\in\mathbb{N}}$: let there be a real function $C = C(t)$, $C \in L^1_{\text{loc}}((t_0, \infty), \mathbb{R})$ such that

$$C(t) \ge 0 \quad \text{on } J_{ij}, c_{ij} := \int_{I_{ij}} C(\tau)\,d\tau > 0 \quad \forall i \in \{1,2\}, \; j \in \mathbb{N},$$

$$\sup_{t\in J_{ij}}\left[\frac{C(t)}{q(t)}\right]\sup_{t\in J_{ij}}\left[r(t)C(t)^{1/(\gamma-1)}\right] \le \frac{K}{M_0 + 1}\left(\frac{c_{ij}}{\pi}\right)^{\gamma/(\gamma-1)} \quad \forall i \in \{1,2\}, \; j \in \mathbb{N}, \quad (20)$$

where γ, M_0, and K are constants defined in (11), (12), and (16), respectively. Since we will show that (19) and (20) are equivalent, see page 8, the next oscillation criterion immediately follows from Theorem 3.

Theorem 4. *Let the functions $\Phi(u, v)$, $f(u)$, $e(t)$, and $r(t)$ satisfy* (11), (12), (16), (17), *and* (18), *respectively. Let $q(t) \ge 0$ and $q(t) \not\equiv 0$ on each interval J_{ij}, $i \in \{1,2\}$, $j \in \mathbb{N}$. If* (20) *is fulfilled, then* (1) *is oscillatory.*

Remark 5. Assuming that $L := \lim_{j\to\infty} a_{1j} < \infty$, we can ensure the oscillation in the point L. Note that $L = \lim_{j\to\infty} a_{2j}$ since $a_{1j} < a_{2j} < a_{1j+1}$. Thus, we can generate a one-sided (right) limit.

Now, we consider some consequences of Theorem 4, which depend on the qualitative properties of the coefficient $q(t)$.

Substituting $C(t) \equiv 1$ in (20), Theorem 4 implies the following result involving lower bounds on the lengths of intervals $|J_{ij}| = b_{ij} - a_{ij}$.

Corollary 6 ($q(t)$ is positive). *Let the functions $\Phi(u, v)$, $f(u)$, $e(t)$, and $r(t)$ satisfy* (11), (12), (16), (17), *and* (18), *respectively. Let $\inf_{t\in J_{ij}} q(t) > 0$ for each $i \in \{1,2\}$, $j \in \mathbb{N}$. If*

$$\left|J_{ij}\right| \ge \pi\left(\frac{(M_0 + 1)\sup_{t\in J_{ij}} r(t)}{K\inf_{t\in J_{ij}} q(t)}\right)^{(\gamma-1)/\gamma} \quad \forall i \in \{1,2\}, \; j \in \mathbb{N}, \quad (21)$$

then (1) *is oscillatory.*

Corollary 7 ($q(t)$ is bounded from below by a positive constant). *Let the functions* $\Phi(u,v)$, $f(u)$, $e(t)$, *and* $r(t)$ *satisfy* (11), (12), (16), (17), *and* (18), *respectively. Let there be two constants* r_0, q_0 *satisfying*

$$0 < r(t) \le r_0, \quad q(t) \ge q_0 > 0 \quad \forall i \in \{1,2\}, \ j \in \mathbb{N}. \quad (22)$$

If

$$|J_{ij}| \ge \pi \left(\frac{(M_0+1) r_0}{K q_0} \right)^{(\gamma-1)/\gamma} \quad \forall i \in \{1,2\}, \ j \in \mathbb{N}, \quad (23)$$

then (1) *is oscillatory, where* γ, M_0, *and* K *are constants defined in* (11), (12), *and* (16), *respectively.*

Example 8 (oscillation of (4)). We know that $x(t) = |\sin(mt)|\sin(mt)$ is an oscillatory solution of (4). However, according to Corollary 7, we can show that all solutions of (4) are oscillatory. Indeed, since $\Phi(u,v) \equiv v$, the conditions (11) and (12) are satisfied especially for $\gamma = 2$, $g(s) = g_0(s) = s^2$ and $M_0 = 0$. Next, $f(u) \equiv u$ implies that condition (16) is satisfied especially for $K = 1$. Since $r(t) \equiv 1$ and $q(t) \equiv 4m^2$, it is clear that conditions (18) and (22) are also satisfied in particular for $r_0 = 1$ and $q_0 = 4m^2$. Moreover, since $e(t) = h(\sin(mt))$ and $h(s)s > 0$, $s \neq 0$, we have that (17) is fulfilled for $a_{1j} = 2j\pi/m$, $b_{1j} = (2j+1)\pi/m = a_{2j}$ and $b_{2j} = (2j+2)\pi/m$. Moreover,

$$|J_{ij}| = b_i - a_i = \frac{\pi}{m} > 0, \quad i \in \{1,2\}. \quad (24)$$

Hence, we conclude that the required condition (23) is fulfilled, that is,

$$|J_{ij}| = \frac{\pi}{m} \ge \frac{\pi}{\sqrt{4m^2}} = \pi \left(\frac{(M_0+1) r_0}{q_0} \right)^{(\gamma-1)/\gamma} \ge \pi \left(\frac{(M_0+1) r_0}{K q_0} \right)^{(\gamma-1)/\gamma}. \quad (25)$$

Thus, all conditions of Corollary 7 are satisfied and hence (4) is oscillatory.

Example 9. We consider the following class of equations:

$$x'' - 2\frac{S''(t)}{S(t)} x = 2\,\mathrm{sign}(S(t)) S'^2(t), \quad \text{a.e. in } [t_0,\infty),$$
$$x, x' \in AC_{\mathrm{loc}}([t_0,\infty),\mathbb{R}), \quad x \notin C^2((t_0,\infty),\mathbb{R}), \quad (26)$$

where $S = S(t)$, $S \in C^2(\mathbb{R})$ is an oscillatory function such that the zeros t_n of the function $\mathrm{sign}(S(t))S'^2(t)$ satisfy $t_n \to \infty$, there is a $\tau_0 \in \mathbb{R}$ such that $t_{n+1} - t_n \ge \tau_0 > 0$ for all $n \in \mathbb{N}$, and $S(t) \neq 0$ on (t_n, t_{n+1}). This equation allows an explicitly given oscillatory solution $x(t) = |S(t)|S(t)$. Moreover, if there is a constant $s_0 > 0$ such that

$$-2\frac{S''(t)}{S(t)} \ge s_0, \quad t \in (t_n, t_{n+1}), \quad \tau_0 \ge \frac{\pi}{\sqrt{s_0}}, \quad (27)$$

then by Corollary 7 we conclude that (26) is oscillatory. Indeed, conditions (11), (12), and (16) are satisfied by the same reasons as in Example 8. Condition (18) is satisfied because of $r(t) \equiv 1$. Also, from (27) it follows that (22) and (23) are fulfilled in particular for $\gamma = 2$, $M_0 = 0$, $r_0 = 1$, $q_0 = s_0$, and $K \ge 1$, that is,

$$|J_{ij}| = |t_{j+1} - t_j| \ge \tau_0 \ge \frac{\pi}{\sqrt{s_0}} = \pi \left(\frac{(M_0+1) r_0}{q_0} \right)^{(\gamma-1)/\gamma} \ge \pi \left(\frac{(M_0+1) r_0}{K q_0} \right)^{(\gamma-1)/\gamma}. \quad (28)$$

Hence Corollary 7 proves this result.

As the second consequence of Theorem 3 is unlike the first one, we consider the case when the coefficient $q(t)$ is not a strictly positive function. Here by $\{q = 0\}$ we denote the set of all $t \in \mathbb{R}$ such that $q(t) = 0$.

Corollary 10 ($q(t)$ is nonnegative, but not $\equiv 0$). *Let the functions* $\Phi(u,v)$, $f(u)$, $e(t)$, *and* $r(t)$ *satisfy* (11), (12), (16), (17), *and* (18), *respectively. Let* $q(t) \ge 0$ *on each interval* J_{ij}, $i \in \{1,2\}$, $j \in \mathbb{N}$, *such that*

$$q_{ij} := \int_{J_{ij}} q(\tau)\,d\tau > 0, \quad i \in \{1,2\}, \ j \in \mathbb{N}. \quad (29)$$

If

$$\frac{q_{ij}^{\gamma}}{q(t)} \ge \pi^{\gamma} \left(\frac{(M_0+1) r(t)}{K} \right)^{\gamma-1}, \quad t \in J_{ij} \setminus \{q = 0\}, \ i \in \{1,2\}, \ j \in \mathbb{N}, \quad (30)$$

then (1) *is oscillatory.*

Proof. It suffices to show that (30) is equivalent to the existence of a real number λ_j such that

$$\lambda_j \ge \frac{\pi(M_0+1)}{K}\frac{1}{q_{ij}}, \qquad \frac{1}{q_{ij}} q(t) \le \frac{1}{\pi}\left(\lambda_j r(t)\right)^{1-\gamma}, \quad t \in J_{ij}, \ i \in \{1,2\}, \ j \in \mathbb{N}. \quad (31)$$

The claim will then follow from Theorem 3. Inequality (31) is for any $t \in J_{ij} \setminus \{q = 0\}$ equivalent to

$$\frac{\pi(M_0+1)}{K}\frac{1}{q_{ij}} \le \lambda_j \le \left(\frac{q_{ij}}{\pi q(t)} \right)^{1/(\gamma-1)} \frac{1}{r(t)}, \quad (32)$$

that is, to

$$\frac{\pi(M_0+1)}{K}\frac{1}{q_{ij}} \le \left(\frac{q_{ij}}{\pi q(t)} \right)^{1/(\gamma-1)} \frac{1}{r(t)}. \quad (33)$$

This inequality is easily seen to be equivalent to (30) for any $t \in J_{ij} \setminus \{q = 0\}$. Note that if $t \in \{q = 0\}$, then the second inequality in (31) is trivially satisfied.

The following three examples are simple consequences of Corollary 10.

Example 11 (oscillation of (5)). Equation (5) as well as equation

$$x'' + m^2\pi^2[\sin(mt)]^+ f(x) = h(\cos(mt)), \quad \text{a.e. in } [t_0,\infty), \quad x, x' \in AC_{\mathrm{loc}}([t_0,\infty),\mathbb{R}), \tag{34}$$

are oscillatory, where $f(u)$ satisfies (16) with $K \geq 1$ and $h(s)s > 0$, $s \neq 0$. Indeed, in (5) we have $r(t) \equiv 1$, $q(t) = m^2\pi^2[\cos(mt)]$ and $e(t) = h(\sin(mt))$, and so, (17) and (18) are fulfilled, and $q(t) \geq 0$, $q(t) \not\equiv 0$ on each interval J_{ij}, $i \in \{1,2\}$, $j \in \mathbb{N}$, where $J_{1j} = [a_{1j}, b_{1j}]$, $J_{2j} = [a_{2j}, b_{2j}]$, and

$$a_{1j} = \frac{2j\pi}{m}, \quad b_{1j} = \frac{(2j+1/2)\pi}{m}, \quad a_{2j} = \frac{(2j+3/2)\pi}{m}, \quad b_{2j} = \frac{(2j+2)\pi}{m}. \tag{35}$$

Moreover,

$$q_{ij} = \int_{J_{ij}} q(\tau)\,d\tau = m^2\pi^2 \int_{a_{ij}}^{b_{ij}} [\cos(m\tau)]^+ d\tau = m\pi^2, \quad i \in \{1,2\}, \; j \in \mathbb{N}, \tag{36}$$

and since $\gamma = 2$, $r(t) \equiv 1$, $K \geq 1$, and $M_0 = 0$, for $\lambda_j = 1/(m\pi)$ we have

$$\begin{aligned} \lambda_j &= \frac{1}{m\pi} \geq \frac{1}{K}\frac{1}{m\pi} = \frac{\pi(M_0+1)}{K}\frac{1}{m\pi^2} = \frac{\pi(M_0+1)}{K}\frac{1}{q_{ij}}, \quad i \in \{1,2\}, \; j \in \mathbb{N}, \\ \frac{1}{q_{ij}} q(t) &= \frac{1}{m\pi^2} m^2\pi^2 [\cos(mt)]^+ \leq m = \frac{1}{\pi}\frac{1}{(1/m\pi)} = \frac{1}{\pi}\frac{1}{(\lambda_j r(t))^{\gamma-1}}, \quad t \in J_{ij}, \; i \in \{1,2\}, \; j \in \mathbb{N}. \end{aligned} \tag{37}$$

Thus, the required condition (30) is fulfilled in this case and hence we may apply Corollary 10 to (5) to verify that this equation is oscillatory. Analogously, we can show that (34) is oscillatory too.

In the next example, the coefficient $q(t)$ is a discontinuous function on $[t_0, \infty)$.

Example 12 (oscillation of (6)). If $f(u)$ satisfies (16) with $K \geq 1$, then (6) is oscillatory since the required condition (30) is satisfied especially for $q(t) = 4m^2 \operatorname{sign}(\cos(mt))$ on J_{ij}, $\lambda_j = 1/(2m)$, and J_{ij} as in (35), $\gamma = 2$, $r(t) \equiv 1$, $K \geq 1$, and $M_0 = 0$. Indeed,

$$\begin{aligned} q_{ij} &= \int_{J_{ij}} q(\tau)\,d\tau = 4m^2 \int_{J_{ij}} [\operatorname{sign}(\cos(m\tau))]^+ d\tau = 4m^2 \frac{\pi}{2m} = 2m\pi, \quad i \in \{1,2\}, \; j \in \mathbb{N}, \\ \lambda_j &= \frac{1}{2m} = \frac{\pi}{2m\pi} = \pi(M_0+1)\frac{1}{q_{ij}} \geq \frac{\pi(M_0+1)}{K}\frac{1}{q_{ij}}, \quad i \in \{1,2\}, \; j \in \mathbb{N}, \\ \frac{1}{q_i} q(t) &= \frac{1}{2m\pi} 4m^2 [\operatorname{sign}(\cos(mt))]^+ \leq \frac{2m}{\pi} = \frac{1}{\pi}\frac{1}{\sqrt{1/(4m^2)}} = \frac{1}{\pi}(\lambda_j r(t))^{\gamma-1}, \quad t \in J_{ij}, \; i \in \{1,2\}, \; j \in \mathbb{N}, \end{aligned} \tag{38}$$

which shows that (30) is satisfied.

At the end of this section, we consider the case when $q(t)$ changes sign on $[t_0, \infty)$ but with the help of Corollary 10 since $q(t)$ is strictly positive on all intervals J_{ij}.

Example 13 (oscillation of (7)). If $f(u)$ satisfies (16) with $K \geq 1$, then model equation (7) is oscillatory. In fact, let J_{ij} be as in (35). If $h(s)s > 0$, $s \neq 0$, it is clear that $e(t) = h(\sin(mt))$ satisfies (2), that is, $e(t) \geq 0$ on J_{1j} and $e(t) \leq 0$ on J_{2j}. On the other hand, we have $q(t) = 4m^2 \operatorname{sign}(\cos(mt))$ on J_{ij} and so $q(t) \geq 0$, $q(t) \not\equiv 0$ on J_{ij} for all $i \in \{1,2\}$, $j \in \mathbb{N}$. Hence, the proof of the fact that all assumptions of Corollary 7 are fulfilled is the same as in the preceding example.

3. Proofs of Theorems 3 and 4

Let $AC_{\mathrm{loc}}([a,b),\mathbb{R})$ denote the set of all real functions which are absolutely continuous on every interval $[a, b-\varepsilon]$, where $\varepsilon \in (0, b-a)$. For arbitrary numbers $T_0 < T^*$ and functions $G = G(t,u)$ and $b = b(t)$, we consider the ordinary differential equation

$$w' = G(t,w) + b(t), \quad \text{a.e. in } [T_0, T^*), \tag{39}$$

which generalizes the classic Riccati equation $w' = a(t)w^2 + b(t)$, where $a(t)$ is an arbitrary function. We associate to (39) the corresponding sub- and supersolutions $\underline{w}, \overline{w} \in AC_{\mathrm{loc}}([T_0,T^*),\mathbb{R})$ which satisfy, respectively,

$$\underline{w}' \leq G(t,\underline{w}) + b(t), \quad \overline{w}' \geq G(t,\overline{w}) + b(t) \quad \text{a.e. in } [T_0, T^*). \tag{40}$$

We are interested in studying the following property:

$$\underline{w}(T_0) \leq \overline{w}(T_0) \quad \text{implies } \underline{w}(t) \leq \overline{w}(t) \quad \forall t \in [T_0, T^*). \tag{41}$$

In this way, we introduce the following definition.

Definition 14. We say that *comparison principle holds for* (39) *on an interval* $[T_0, T^*)$, $T \le T_0 < T^*$, if property (41) is fulfilled for all sub- and supersolutions $\underline{w}, \overline{w} \in AC_{\text{loc}}([T_0, T^*), \mathbb{R})$ of (39) on $[T_0, T^*)$.

Now, we are able to state the next result.

Lemma 15. *Let $T \le T_0 < T^*$. Let $G = G(t,u)$ and $b = b(t)$ be two arbitrary function. For any $M > 0$, let there be a function $L = L(t) > 0$ depending on T_0, T^*, M such that*

$$\begin{gathered} L(t) > 0 \quad \text{on } [T_0, T^*), \ L \in L^1(T_0, T^*), \\ |G(t,u_1) - G(t,u_2)| \le L(t)|u_1 - u_2| \\ \forall t \in [T_0, T^*), \ u_1, u_2 \in [-M, M]. \end{gathered} \tag{42}$$

Then comparison principle (41) *holds for* (39) *on an interval* $[T_0, T^*)$, $T \le T_0 < T^*$.

Proof. It is enough to use the idea of the proof of [14, Lemma 4.1]. Hence, this proof is left to the reader.

Corollary 16. *Let $\lambda > 0$, and let $g_0 : \mathbb{R}_+ \to \mathbb{R}_+$ be a locally Lipschitz function on $\mathbb{R}_+$. Let K be an arbitrary real number and $q(t)$ an arbitrary function. Let $T \le T_0 < T^*$. If $r(t) > 0$ on $[T_0, T^*)$ and $r^{-1+\gamma} \in L^1(T_0, T^*)$, then comparison principle* (41) *holds for the Riccati differential equation*

$$w' = \frac{1}{(\lambda r(t))^{\gamma-1}} g_0(|w|) + K\lambda q(t), \quad \text{a.e. in } [T_0, T^*). \tag{43}$$

Proof. It is clear that (43) is a particular case of (39) in particular for

$$G(t,u) = \frac{1}{(\lambda r(t))^{\gamma-1}} g_0(|u|), \qquad b(t) = K\lambda q(t). \tag{44}$$

Since g_0 is a locally Lipschitz function on $\mathbb{R}_+$, for every $M > 0$ there is an $L_0 > 0$ depending on M such that

$$|g_0(s_1) - g_0(s_2)| \le L_0 |s_1 - s_2|, \quad \forall s_1, s_1 \in [-M, M]. \tag{45}$$

Hence, for any $M > 0$ and all $u_1, u_2 \in [-M, M]$, we obtain:

$$\begin{aligned} |G(t,u_1) - G(t,u_2)| &= \frac{1}{(\lambda r(t))^{\gamma-1}} |g_0(|u_1|) - g_0(|u_2|)| \\ &\le \frac{L_0}{(\lambda r(t))^{\gamma-1}} ||u_1| - |u_2|| \\ &\le \frac{L_0}{(\lambda r(t))^{\gamma-1}} |u_1 - u_2|. \end{aligned} \tag{46}$$

Thus, according to assumption $r^{-1+\gamma} \in L^1(T_0, T^*)$, the required condition (42) is fulfilled in particular for $L(t) = L_0(\lambda r(t))^{-\gamma+1}$ and so Lemma 15 proves this corollary.

Before we give the proof of Theorem 3, we state and prove the next two propositions. In the first one, by a nonoscillatory solution $x(t)$ of the main equation (1), we get the existence of a supersolution $\overline{w}(t)$ of the Riccati differential equation (43) on the interval (a_1, b_1) or (a_2, b_2). In the second one, we construct two subsolutions $\underline{w}_1(t)$ and $\underline{w}_2(t)$ of (43) which blow up on intervals $(a_1, T_1^*) \subseteq (a_1, b_1)$ and $(a_2, T_2^*) \subseteq (a_2, b_2)$, respectively.

Proposition 17. *Let $\Phi(u,v)$ and $f(u)$ satisfy* (11), (12), *and* (16), *respectively, and let $e(t)$ satisfy* (2). *Let $r(t) > 0$, $q(t) \ge 0$ and $q(t) \not\equiv 0$ on $[a_1, b_1] \cup [a_2, b_2]$. Let $x = x(t)$ be a nonoscillatory solution of* (1) *and let, for some $T \ge t_0$ and $\lambda > 0$, the function $\overline{w} = \overline{w}(t)$ be defined by*

$$\overline{w}(t) = -\frac{\lambda r(t)\Phi(x(t), x'(t))}{x(t)}, \quad t \ge T. \tag{47}$$

Then $\overline{w} \in AC_{\text{loc}}([T, \infty), \mathbb{R})$ and $\overline{w}(t)$ satisfies the inequality

$$\overline{w}' \ge (\lambda r(t))^{1-\gamma} g_0(|\overline{w}|) + K\lambda q(t) \quad \text{a.e. in } J, \tag{48}$$

where $J = (a_1, b_1)$ if $x(t) < 0$ and $J = (a_2, b_2)$ if $x(t) > 0$.

Proof. Since $x = x(t)$ is a nonoscillatory solution of (1), there is a $T \ge t_0$ such that $x(t) \ne 0$ on $[T, \infty)$. Hence, $\overline{w}(t)$ is well defined by (47).

Because of (2), we have $e(t)/x(t) \le 0$ for all $t \in J$, where $J = (a_1, b_1)$ if $x(t) < 0$ and $J = (a_2, b_2)$ if $x(t) > 0$ and since $\lambda > 0$ we have

$$-\lambda \frac{e(t)}{x(t)} \ge 0 \quad \text{in } J. \tag{49}$$

Next, since x and $r(t)\Phi(x(t), x'(t))$ are from $AC_{\text{loc}}([t_0, \infty), \mathbb{R})$, we can take the first derivative of $\overline{w}(t)$ for almost everywhere in J, which together with $r(t) > 0$, $q(t) \ge 0$ and $q(t) \not\equiv 0$ on J gives

$$\begin{aligned} \overline{w}'(t) &= -\frac{\lambda\left(r(t)\Phi(x(t), x'(t))\right)'}{x(t)} \\ &\quad + \frac{\lambda r(t)\Phi(x(t), x'(t))}{x^2(t)} x'(t) \\ &= \frac{\lambda r(t)}{|x(t)|^\gamma} \Phi(x(t), x'(t)) x'(t) |x(t)|^{\gamma-2} \\ &\quad + \lambda q(t) \frac{f(x(t))}{x(t)} - \lambda \frac{e(t)}{x(t)} \end{aligned}$$

$$\geq \frac{\lambda r(t)}{|x(t)|^{\gamma}} g\left(\left|\Phi\left(x(t), x'(t)\right)\right|\right) + K\lambda q(t) - \lambda \frac{e(t)}{x(t)}$$
$$= \frac{\lambda r(t)}{|x(t)|^{\gamma}} g\left(\frac{|x(t)|}{\lambda r(t)} |\overline{w}(t)|\right) + K\lambda q(t) - \lambda \frac{e(t)}{x(t)}$$
$$\geq \frac{\lambda r(t)}{|x(t)|^{\gamma}} \frac{|x(t)|^{\gamma}}{(\lambda r(t))^{\gamma}} g_0(|\overline{w}(t)|) + K\lambda q(t) - \lambda \frac{e(t)}{x(t)} \quad \text{a.e. in } J, \tag{50}$$

that is,

$$\overline{w}' \geq (\lambda r(t))^{1-\gamma} g_0(|\overline{w}|) + K\lambda q(t) - \lambda \frac{e(t)}{x(t)} \quad \text{a.e. in } J. \tag{51}$$

Hence, (49) and (51) prove the desired inequality (48).

Proposition 18. *Let (19) be satisfied. Let R_1 and R_2 be two arbitrary real numbers. Then there are two points $T_1^* \in (a_1, b_1)$ and $T_2^* \in (a_2, b_2)$ and two continuous functions $\underline{w}_1(t)$ and $\underline{w}_2(t)$, such that*

$$\underline{w}_i(a_i) \leq R_i, \quad \lim_{t \to T_i^*} \underline{w}_i(t) = \infty, \; i \in \{1, 2\},$$
$$\underline{w}_i' \leq (\lambda r(t))^{1-\gamma} g_0(|\underline{w}_i|) + K\lambda q(t) \tag{52}$$
$$\text{for a.e. } t \in (a_i, T_i^*), \; i \in \{1, 2\}.$$

Proof. Since the function $y(s) = \tan(s)$ is a bijection from the interval $(-\pi/2, \pi/2)$ into $\mathbb{R}$, we observe that there are two $s_1, s_2 \in (-\pi/2, \pi/2)$ such that

$$\tan(s_i) \leq R_i, \quad i \in \{1, 2\}. \tag{53}$$

Next, we define two functions $V_1(t)$ and $V_2(t)$ by

$$V_i(t) = s_i + \frac{\pi}{c_i} \int_{a_i}^{t} C(\tau) \, d\tau, \quad t \in [a_i, b_i], \; i \in \{1, 2\}, \tag{54}$$

where the real numbers c_1, c_2 and the function $C(t)$ are defined in (19). From (19) and (54), one can immediately conclude that

$$V_i(a_i) = s_i < \frac{\pi}{2}, \qquad V_i(b_i) = s_i + \pi > \frac{\pi}{2}, \; i \in \{1, 2\}. \tag{55}$$

Since $C \in L^1_{\text{loc}}((t_0, \infty), \mathbb{R})$, it follows that $C \in L^1((t_0, \infty), \mathbb{R})$, which implies that $V_i \in AC([a_i, b_i], \mathbb{R})$, $i \in \{1, 2\}$. Therefore, exploiting the fact that in (55) we have $V_i \in C([a_i, b_i], \mathbb{R})$, we obtain the existence of two points $T_1^* \in (a_1, b_1)$ and $T_2^* \in (a_2, b_2)$ such that

$$V_i(T_i^*) = \frac{\pi}{2}, \qquad -\frac{\pi}{2} < V_i(t) < \frac{\pi}{2} \quad \forall t \in [a_i, T_i^*), \; i \in \{1, 2\}. \tag{56}$$

Now, we are able to define the following two functions $\underline{\omega}_1(t)$ and $\underline{\omega}_2(t)$ by

$$\underline{\omega}_i(t) = \tan(V_i(t)), \quad t \in [a_i, T_i^*), \; i \in \{1, 2\}. \tag{57}$$

That are well defined because of (56). Moreover, from (53), (54), and (56) we obtain

$$\underline{\omega}_i(a_i) = \tan(s_i) \leq R_i,$$
$$\lim_{t \to T_m^*} \underline{\omega}_i(t) = \tan(V_i(T_i^*)) = \tan\left(\frac{\pi}{2}\right) = \infty. \tag{58}$$

Also, since $V_i \in AC([a_i, b_i], \mathbb{R})$, we can take the first derivative of $\underline{\omega}_i(t)$ and hence from (19) and (57) we obtain

$$\underline{w}_i'(t) = \frac{\pi}{c_i} C(t) \frac{1}{\cos^2(V_i(t))}$$
$$= \frac{\pi}{c_i} C(t) \left(|\underline{\omega}_i(t)|^2 + 1\right)$$
$$\leq \frac{\pi}{c_i} C(t) \left[g_0(|\underline{\omega}_i(t)|) + M_0 + 1\right]$$
$$\leq \frac{\pi}{c_i} C(t) g_0(|\underline{\omega}_i(t)|) + \frac{\pi}{c_i} C(t) (M_0 + 1)$$
$$\leq (\lambda r(t))^{1-\gamma} g_0(|\underline{\omega}_i(t)|) + K\lambda q(t), \quad \text{a.e. in } (a_i, T_i^*), \tag{59}$$

which together with (58) proves this proposition.

Now we are able to present the proof of Theorem 3 based on Lemma 15 and Propositions 17 and 18.

Proof of Theorem 3. Assuming the contrary, then there is a nonoscillatory solution $x(t)$ and $T \geq t_0$ such that $x(t) \neq 0$ for all $t \geq T$, $x(t)$ and $r(t)\Phi(x, x')$ are from $AC_{\text{loc}}([t_0, \infty), \mathbb{R})$. In order to simplify notation, let every $j \in \mathbb{N}$ be fixed and omitted in the notation. For instance, instead of (a_{1j}, b_{1j}) and (a_{2j}, b_{2j}) we write (a_1, b_1) and (a_2, b_2), respectively, and so on.

On one hand, we observe by Proposition 17 that the function $\overline{w}(t)$ defined by (47) is a supersolution of the following Riccati differential equation:

$$w' = (\lambda r(t))^{1-\gamma} g_0(|w|) + K\lambda q(t) \quad \text{a.e. in } J, \tag{60}$$

where $\overline{w} \in AC_{\text{loc}}(J, \mathbb{R})$ and $J = (a_1, b_1)$ if $x(t) < 0$ and $J = (a_2, b_2)$ if $x(t) > 0$; see the proof of Proposition 17. Let for instance $x(t) < 0$ and thus $J = (a_1, b_1)$.

On the other hand, by Proposition 18 there are a number $T_1^* \in (a_1, b_1)$ and subsolutions $\underline{w}_1(t)$, $\underline{w}_1 \in AC([a_1, T_1^*), \mathbb{R})$ of the Riccati differential equation (60) such that

$$\underline{w}_1(a_1) \leq \overline{w}(a_1), \qquad \lim_{t \to T_1^*} \underline{w}_1(t) = \infty, \tag{61}$$

(in the case when $J = (a_2, b_2)$, then we work with $T_2^* \in (a_2, b_2)$ and the other subsolution $\underline{w}_2(t)$, $\underline{w}_2 \in AC([a_2, T_2^*), \mathbb{R})$). Hence by Corollary 16 and (61) we conclude that $\underline{w}_1(t) \leq \overline{w}(t)$ for all $t \in [a_1, T_1^*)$ and therefore,

$$\infty = \lim_{t \to T_1^*} \underline{w}_1(t) \leq \lim_{t \to T_1^*} \overline{w}(t), \tag{62}$$

which contradicts $\overline{w} \in AC_{\text{loc}}([T, \infty), \mathbb{R})$. Thus, the assumption that $x(t)$ is nonoscillatory is not possible and hence every solution of (1) is oscillatory.

Proof of Theorem 4. There is only one difference between Theorems 3 and 4, and it is the difference between conditions (19) and (20). Hence, Theorem 4 immediately follows from Theorem 3 provided that we prove the equivalence between (19) and (20).

First of all, we need the following two lemmas.

Lemma 19. *Let a, b, and c be positive real numbers, and* $\gamma > 1$. *Then the existence of a positive real number* λ *such that*

$$c \le \min\left\{\frac{a}{\lambda^{\gamma-1}}, b\lambda\right\} \tag{63}$$

is equivalent to

$$c \le a^{1/\gamma} b^{1/\gamma'}. \tag{64}$$

Here $\gamma' := \gamma/(\gamma-1)$ *is the conjugate exponent of* γ.

Proof. Let us define $g(\lambda) = \min\{a/\lambda^{\gamma-1}, b\lambda\}$. Since $\lambda \mapsto a/\lambda^{\gamma-1}$ is decreasing and $\lambda \mapsto b\lambda$ is increasing, there exists a unique point of maximum of g. It is achieved at $\lambda = \lambda_0$ which is a solution of $a/\lambda_0^{\gamma-1} = b\lambda_0$, hence $\lambda_0 = (a/b)^{1/\gamma}$.

If there is a positive real number λ such that $c \le g(\lambda)$, then $c \le g(\lambda_0) = b\lambda_0 = b(a/b)^{1/\gamma} = a^{1/\gamma}b^{1/\gamma'}$. Conversely, if $c \le a^{1/\gamma}b^{1/\gamma'}$, then $c \le g(\lambda_0)$, and the equivalence in the lemma is proved.

The preceding lemma is a special case of the following more general statement.

Lemma 20. *Assume that* $a(t)$, $b(t)$, *and* $c(t)$ *are positive real functions defined on a subset* $J \subseteq \mathbb{R}$. *Then the existence of a positive real number* λ *such that for all* $t \in J$,

$$c(t) \le \min\left\{\frac{a(t)}{\lambda^{\gamma-1}}, b(t)\lambda\right\}, \tag{65}$$

is equivalent to

$$\sup_{t\in J}\frac{c(t)}{b(t)} \le \left(\inf_{t\in J}\frac{a(t)}{c(t)}\right)^{1/(\gamma-1)}. \tag{66}$$

Proof. The condition that there exists $\lambda > 0$ such that (65) is true for all $t \in J$ is equivalent with the following two inequalities, which have to hold simultaneously:

$$c(t) \le \frac{a(t)}{\lambda^{\gamma-1}}, \qquad c(t) \le b(t)\lambda, \tag{67}$$

that is, with

$$\frac{c(t)}{b(t)} \le \lambda \le \left(\frac{a(t)}{c(t)}\right)^{1/(\gamma-1)} \qquad \forall t \in J. \tag{68}$$

Taking the supremum of the left-hand side and the infimum of the right-hand side, we obtain (66). Conversely, if (66) is satisfied, then, since the left-hand side in (66) cannot be equal to zero, while the right-hand side is less than ∞, then inequality (66) defines a nonempty closed interval in $\mathbb{R}$ (possibly reducing to just one point), in which we can choose any positive λ. Hence, (67) is satisfied for all $t \in J$, and therefore (65) as well.

Remark 21. If $a = a(t)$, $b = b(t)$, and $c = c(t)$, appearing in Lemma 20, are constant functions, then condition (66) is equivalent to $c/b = (a/c)^{1/(\gamma-1)}$, that is, to (64).

Proof of Equivalence of (19) *and* (20). We will use Lemma 20. Let us fix J_{ij}, where $i \in \{1,2\}$ and $j \in \mathbb{N}$. We see that that the inequality appearing in the second line of condition (19) is of the form (65), where

$$\begin{gathered} c(t) := \frac{1}{c_{ij}}C(t), \qquad a(t) := \frac{1}{\pi}r(t)^{1-\gamma}, \\ b(t) := \frac{Kq(t)}{\pi(M_0+1)}. \end{gathered} \tag{69}$$

Condition (66) is equivalent to

$$\begin{aligned} \sup_{t\in J_{ij}}\frac{C(t)}{q(t)} &\le \frac{Kc_{ij}^{\gamma'}}{\pi^{\gamma'}(M_0+1)}\inf_{t\in J_{ij}}\frac{1}{r(t)\,C(t)^{\gamma-1}} \\ &= \frac{Kc_{ij}^{\gamma'}}{(M_0+1)\sup_{t\in J_{ij}}r(t)\,C(t)^{\gamma-1}}\left(\frac{c_{ij}}{\pi}\right)^{\gamma'}. \end{aligned} \tag{70}$$

This inequality is equivalent to the corresponding one in (19), and the claim follows from Lemma 19.

Proof of Corollary 6. Substituting $\inf_{t\in J_{ij}}q(t)$ instead of $q(t)$, and $\sup_{t\in J_{ij}}r(t)$ instead of $r(t)$ in the inequality appearing in the second line of (20), after a short computation we obtain a stronger inequality than (20) as

$$\begin{aligned} &\frac{1}{c_{ij}}\sup_{t\in J_{ij}}C(t) \\ &\quad\le \frac{1}{\pi}\left(\frac{K\inf_{t\in J_{ij}}q(t)}{(M_0+1)\sup_{t\in J_{ij}}r(t)}\right)^{(\gamma-1)/\gamma} \qquad \forall i \in \{1,2\},\ j \in \mathbb{N}. \end{aligned} \tag{71}$$

Substituting $C(t) \equiv 1$ we obtain that the left-hand side of (71) is equal to $|J_{ij}|^{-1}$, and the resulting inequality is equivalent to (21). Since (21) implies (20), the claim is proved.

Remark 22. Here we show that the choice of $C(t) \equiv 1$ in the proof of Corollary 6 is the best possible. To prove this, note that on the left-hand side of (71) we have the expression depending on an auxiliary function $C(t)$, and this function is absent on the right-hand side. Therefore, it has sense to try to find a function $C(t)$, $t \in J_{ij}$, such that the value of $Q(C) := (1/c_{ij})\sup_{t\in J_{ij}}C(t)$ is minimal (note that c_{ij} depends on the function $C = C(t)$ as well). It is easy to see that the minimum is achieved for $C(t) \equiv 1$ (or any positive constant). Indeed, since $Q(C) = Q(\mu c)$ for any $\mu > 0$, by taking $\mu = (\sup_{t\in J_{ij}}C(t))^{-1}$ it suffices to assume that $\sup_{t\in J_{ij}}C(t) = 1$. The value of $Q(C)$ is minimal if $c_{ij} = \int_{J_{ij}}C(\tau)\,d\tau$ is maximal possible, and since $0 \le C(t) \le 1$, it is clear that the maximum of $c_{ij} = c_{ij}(C)$ is achieved when $C(t) \equiv 1$.

Proof of Corollary 7. It is clear that the condition (23) implies (21). Furthermore, since the lengths $|J_{ij}|$ are uniformly bounded from below by a positive constant, see (23), then obviously $a_{1j} \to \infty$ as $j \to \infty$. Thus, the claim follows immediately from Corollary 6.

Proof of Corollary 10. Let $C(t) := q(t)$. Then the required condition (19) is clearly satisfied because of assumption (30). Hence, this corollary immediately follows from Theorem 3.

4. An Extension of Condition (12) to the Case of $\gamma > 1$

In this section, we consider the oscillation of (1) in the case when $\gamma \in (1,2)$ is allowed in the assumption (12). In this way, the assumption (11) is slightly modified by a real number $p > 1$ such that

$$|u|^{(p-1)\gamma-p} v\Phi(u,v) \geq g(|\Phi(u,v)|), \tag{72}$$

where (12) is supposed, that is, $g(cs) \geq c^\gamma g_0(s)$ for every $c, s > 0$ and for a locally Lipschitz function $g_0 : [0,\infty) \to [0,\infty)$ for which there exists a constant M_0 such that $g_0(s) + M_0 \geq s^p$ for all $s \in [0,\infty)$.

For the function f, we assume that there exists a constant $K > 0$ such that

$$f(u)\operatorname{sign}(u) \geq K|u|^{p-1} \quad \text{for} \quad \text{every } u \in \mathbb{R}. \tag{73}$$

We will need the following two lemmas.

Lemma 23. *Let $a < b$, and let $f : [a,b] \to [0,\infty)$ be a measurable function. Then,*

$$\int_a^b f(t)\,dt \geq 1 \tag{74}$$

if and only if there exists a function $g \in L_1([a,b],[0,\infty))$, $\|g\|_1 = 1$, such that

$$f(t) \geq g(t), \quad \forall t \in [a,b]. \tag{75}$$

Proof. We assume (74). If $\int_a^b f(t)dt$ is finite, we may choose $g(x) := f(x)/\|f\|_1$.

For $\int_a^b f(t)dt = \infty$, we may define for every natural number n a function

$$f_n(t) = \begin{cases} f(t), & f(t) \leq n, \\ n, & f(t) > n. \end{cases} \tag{76}$$

Since $f_n(t) \leq n$, we have $f_n \in L_1([a,b],\mathbb{R})$. Obviously, $\lim_n \int_a^b f_n(t)dt$ is also infinite. This implies that there exists an n_0 such that $\int_a^b f_{n_0}(t)dt \geq 1$. We define $g(t) := f_{n_0}(t)/\|f_{n_0}\|_1$.

Now, we assume (75). Integrating over the whole interval, we get

$$\int_a^b f(t)\,dt \geq \int_a^b g(t)\,dt = 1. \tag{77}$$

Lemma 24. *Let $p > 1$. There exists a unique function $y = y(s)$ that satisfies the following Cauchy problem:*

$$y'(s) = 1 + |y(s)|^p, \quad s \in \mathbb{R}, \tag{78}$$

$$y(0) = 0. \tag{79}$$

Furthermore, there exists a finite real number S^* such that

$$\lim_{s \to \pm S^*} y(s) = \pm\infty, \tag{80}$$

and $y \in C^1((-S^*,S^*),\mathbb{R})$ is injective and monotonous. We have $S^* = (\pi/p)(1/(\sin(\pi/p)))$ and y is odd.

We will denote by $\tan_p$ the solution of (78)-(79).

Proof. Obviously every solution to (78) is injective and monotonous on a connected set, since $y' = 1 + |y|^p \geq 1 > 0$.

Let $f(x) = 1/(1+|x|^p)$. Obviously $1 \geq f(x) > 0$ for all $x \in \mathbb{R}$ and $C_p > |f'(x)| > 0$ for some constants $C_p \in \mathbb{R}$. We may assume $C_p > 1$. Hence, $z' = f(t)$, $z(0) = 0$ has (by [17, Theorem 3.1]) a locally unique solution on $(-1/C_p, 1/C_p)$. By the same argument z can be extended to the whole of $\mathbb{R}$, interval by interval of type $((n-1)/C_p, (n+1)/C_p)$ for all $n \in \mathbb{Z}$.

We know from [18] that the for the integral of the left-hand side we have

$$\int_0^\infty f(x)\,dx = \int_0^\infty \frac{dx}{1+x^p} = \frac{\pi}{p}\frac{1}{\sin(\pi/p)} = S^*, \tag{81}$$

so $\lim_{t\to\pm\infty} z(t) = \pm S^*$.

Since $z'(t) \neq 0$, z is injective, monotonous, and of class $C^1(\mathbb{R},(-S^*,S^*))$ and so we define $y(s) = z^{-1}(s)$. It immediately follows that $y \in C^1((-S^*,S^*),\mathbb{R})$ and that y is injective and monotonous. Also, by continuity of z, we have $\lim_{s\to\pm S^*} y(s) = \lim_{w\to\pm\infty} y(z(w)) = \lim_{w\to\pm\infty} w = \pm\infty$.

Differentiating, it also follows that $y' = 1 + |y|^p$ and by definition of y, $y(0) = z^{-1}(0) = 0$. So, y is a solution of (78)-(79). Since any other such solution must have z as its inverse, y is unique on its domain.

Proposition 25. *Let $a < b$ and $\gamma > p > 1$. Assume (72) and (73) and let $q(t) \geq 0$ and $e(t) \leq 0$ (or $e(t) \geq 0$) for all $t \in [a,b]$. If there exists a constant $\lambda > 0$ such that*

$$\frac{p}{2\pi}\sin\left(\frac{\pi}{p}\right)\int_a^b \min\left\{\frac{p-1}{r^{\gamma-1}(t)\,\lambda^{\gamma-1}}, \frac{K\lambda q(t)}{1+M_0}\right\} dt \geq 1, \tag{82}$$

then for any solution $x(t)$ of (1) with $x(a) \geq 0$ (or $x(a) \leq 0$) there exists a number $t^* \in [a,b]$ such that $x(t^*) = 0$.

Proof. We assume that $x(t) \neq 0$ for all $t \in [a,b]$. Then we may define the function

$$\overline{\omega}(t) := -\frac{\lambda r(t)\,\Phi\left(x(t), x'(t)\right)}{\operatorname{sign}(x(t))\,|x(t)|^{p-1}}. \tag{83}$$

This function is well defined on $[a,b]$ and its derivative is

$$\overline{w}'(t) = -\frac{\lambda\left(r(t)\Phi\left(x(t),x'(t)\right)\right)'}{\operatorname{sign}(x(t))|x(t)|^{p-1}} + \frac{(p-1)\lambda r(t)\Phi\left(x(t),x'(t)\right)}{|x(t)|^{p}}x'(t). \tag{84}$$

Using the fact that $x(t)$ is a solution of (1) and omitting t from x and x' for clarity, we get

$$\begin{aligned}
\overline{w}'(t) &= \frac{\lambda(q(t)f(x)-e(t))}{\operatorname{sign}(x)|x|^{p-1}} + \frac{(p-1)\lambda r(t)\Phi(x,x')}{x^{p}}x' \\
&= \frac{(p-1)\lambda r(t)\Phi(x,x')}{|x|^{p}}x' + \lambda q(t)\frac{f(x)}{\operatorname{sign}(x)|x|^{p-1}} - \lambda\frac{e(t)}{\operatorname{sign}(x)|x|^{p-1}} \\
&= \frac{(p-1)\lambda r(t)}{|x|^{\gamma(p-1)}}\Phi(x,x')x'|x|^{(p-1)\gamma-p} + \lambda q(t)\frac{f(x)}{\operatorname{sign}(x)|x|^{p-1}} - \lambda\frac{e(t)}{\operatorname{sign}(x)|x|^{p-1}} \\
&\overset{(72)}{\geq} \frac{(p-1)\lambda r(t)}{|x|^{\gamma(p-1)}}g\left(\left|\Phi(x,x')\right|\right) + \lambda q(t)\frac{f(x)}{\operatorname{sign}(x)|x|^{p-1}} - \lambda\frac{e(t)}{\operatorname{sign}(x)|x|^{p-1}} \\
&= \frac{(p-1)\lambda r(t)}{|x|^{\gamma(p-1)}}g\left(\frac{|x|^{p-1}}{\lambda r(t)}|\overline{\omega}(t)|\right) + \lambda q(t)\frac{f(x)}{\operatorname{sign}(x)|x|^{p-1}} - \lambda\frac{e(t)}{\operatorname{sign}(x)|x|^{p-1}} \\
&\overset{(12)}{\geq} \frac{(p-1)\lambda r(t)}{|x|^{\gamma(p-1)}}\left(\frac{|x|^{p-1}}{\lambda r(t)}\right)^{\gamma}g_0(|\overline{\omega}(t)|) + \lambda q(t)\frac{f(x)}{\operatorname{sign}(x)|x|^{p-1}} - \lambda\frac{e(t)}{\operatorname{sign}(x)|x|^{p-1}} \\
&= \frac{p-1}{(\lambda r(t))^{\gamma-1}}g_0(|\overline{\omega}(t)|) + \lambda q(t)\frac{f(x)}{\operatorname{sign}(x)|x|^{p-1}} - \lambda\frac{e(t)}{\operatorname{sign}(x)|x|^{p-1}} \\
&\overset{(73)}{\geq} \frac{p-1}{(\lambda r(t))^{\gamma-1}}g_0(|\overline{\omega}(t)|) + \lambda q(t)K - \lambda\frac{e(t)}{\operatorname{sign}(x)|x|^{p-1}}.
\end{aligned} \tag{85}$$

Since $e(t)/(\operatorname{sign}(x)|x|^{p-1}) \leq 0$, we have

$$\overline{w}'(t) \geq \frac{p-1}{(\lambda r(t))^{\gamma-1}}g_0(|\overline{\omega}(t)|) + \lambda K q(t). \tag{86}$$

We apply Lemma 23 to (82) to obtain a function $C(t)$ such that

$$\int_a^b C(t)\,dt = 1, \tag{87}$$

and by (75)

$$0 \leq C(t) \leq \frac{p}{2\pi}\sin\left(\frac{\pi}{p}\right)\min\left\{\frac{p-1}{r^{\gamma-1}(t)\lambda^{\gamma-1}}, \frac{K\lambda q(t)}{1+M_0}\right\}. \tag{88}$$

Let $s_0 \in (-(\pi/p)(1/\sin(\pi/p)), (\pi/p)(1/\sin(\pi/p)))$ be such that $\tan_p(s_0) \leq \overline{\omega}(a)$. We define the function

$$V(t) = s_0 + \frac{2\pi}{p}\frac{1}{\sin(\pi/p)}\int_a^t C(\tau)\,d\tau. \tag{89}$$

Note that $(-(\pi/p)(1/\sin(\pi/p))) < s_0 = V(a) < (\pi/p)(1/\sin(\pi/p))$ and $V(b) = s_0 + (2\pi/p)(1/\sin(\pi/p)) > (\pi/p)(1/\sin(\pi/p))$. Since V is continuous, there exists a $T^* \in (a,b)$ such that $V(t) < (\pi/p)(1/\sin(\pi/p))$ for $t \in [a,T^*)$ and $V(T^*) = (\pi/p)(1/\sin(\pi/p))$.

We define

$$\underline{\omega}(t) := \tan_p(V(t)),\quad t \in [a,T^*), \tag{90}$$

and note that by Lemma 24, $\lim_{t\to T^*}\underline{\omega}(t) = +\infty$, where $\tan_p(s) = y(s)$ and $y(s)$ is determined by Lemma 24.

The derivative of $\underline{\omega}$ then satisfies

$$\begin{aligned}
\underline{\omega}'(t) &= \tan_p'(V(t))V'(t) \\
&= \left(1+\tan_p^p(V(t))\right)V'(t) \\
&= \frac{2\pi}{p}\frac{1}{\sin(\pi/p)}(1+\underline{\omega}^p(t))C(t) \\
&\leq (1+M_0+g_0(|\underline{\omega}(t)|))\min\left\{\frac{p-1}{r^{\gamma-1}(t)\lambda^{\gamma-1}}, \frac{K\lambda q(t)}{1+M_0}\right\} \\
&\leq \frac{p-1}{r^{\gamma-1}(t)\lambda^{\gamma-1}}g_0(|\underline{\omega}(t)|) + K\lambda q(t).
\end{aligned} \tag{91}$$

Thus, we may apply the comparison principle of Lemma 15 to $\underline{\omega}(a) \leq \overline{\omega}(a)$ and their respective differential inequalities to conclude that $\underline{\omega}(t) \leq \overline{\omega}(t)$ for all $t \in [a,b]$. But, since $\underline{\omega} \to +\infty$ as $t \to T^* < b$ and $\overline{\omega}(t)$ is well defined on the whole segment $[a,b]$, we arrive at a contradiction. Hence, $x(t) > 0$ cannot hold for all $t \in [a,b]$, and since $x(t)$ is continuous, there exists a $t^* \in [a,b]$ such that $x(t^*) = 0$.

Theorem 26. *Let $\gamma > p > 1$. Assume (72) and (73) and let for any $T > t_0$ there exist $T \leq a_1 < b_1 \leq a_2 < b_2$ such that $g(t) \geq 0$ on $[a_1,b_1] \cup [a_2,b_2]$ and $e(t) \leq 0$ on $[a_1,b_1]$ and $e(t) \geq 0$ on*

$[a_2, b_2]$. *If for both* $i \in \{1,2\}$ *there exist constants* $\lambda_i > 0$ *such that*

$$\frac{p}{2\pi}\sin(\pi/p)\int_{a_i}^{b_i}\min\left\{\frac{p-1}{r^{\gamma-1}(t)\,\lambda_i^{\gamma-1}},\frac{K\lambda_i q(t)}{1+M_0}\right\}dt \geq 1, \tag{92}$$

then (1) *is oscillatory.*

Proof. For (1) to be oscillatory it is sufficient that for every $n \in \mathbb{N}$ there exists a $t_n > n$ such that $x(t_n) = 0$.

Let $n \in \mathbb{N}$ and let $n < a_1 < b_1 \leq a_2 < b_2$ as assumed in the theorem. It is enough to show that every solution $x(t)$ of (1) has a zero on $[a_1, b_2]$. If $x(a_1) \geq 0$, we apply Proposition 25 to the segment $[a_1, b_1]$ to obtain the sought zero. If $x(a_2) \leq 0$ we again apply Proposition 25 to the segment $[a_2, b_2]$. Suppose $x(a_1) < 0$ and $x(a_2) > 0$. Then, since x is continuous, there exists a number $t_1 \in [a_1, a_2]$ such that $x(t_1) = 0$.

Acknowledgment

This work is supported by the Ministry of Science of the Republic of Croatia under Grant no. 036-0361621-1291.

References

[1] Q. Kong, "Nonoscillation and oscillation of second order half-linear differential equations," *Journal of Mathematical Analysis and Applications*, vol. 332, no. 1, pp. 512–522, 2007.

[2] Q.-R. Wang and Q.-G. Yang, "Interval criteria for oscillation of second-order half-linear differential equations," *Journal of Mathematical Analysis and Applications*, vol. 291, no. 1, pp. 224–236, 2004.

[3] Q. Long and Q.-R. Wang, "New oscillation criteria of second-order nonlinear differential equations," *Applied Mathematics and Computation*, vol. 212, no. 2, pp. 357–365, 2009.

[4] Z. Zheng and F. Meng, "Oscillation criteria for forced second-order quasi-linear differential equations," *Mathematical and Computer Modelling*, vol. 45, no. 1-2, pp. 215–220, 2007.

[5] D. Çakmak and A. Tiryaki, "Oscillation criteria for certain forced second-order nonlinear differential equations," *Applied Mathematics Letters*, vol. 17, no. 3, pp. 275–279, 2004.

[6] F. Jiang and F. Meng, "New oscillation criteria for a class of second-order nonlinear forced differential equations," *Journal of Mathematical Analysis and Applications*, vol. 336, no. 2, pp. 1476–1485, 2007.

[7] Q. Yang, "Interval oscillation criteria for a forced second order nonlinear ordinary differential equations with oscillatory potential," *Applied Mathematics and Computation*, vol. 135, no. 1, pp. 49–64, 2003.

[8] A. Tiryaki and B. Ayanlar, "Oscillation theorems for certain nonlinear differential equations of second order," *Computers & Mathematics with Applications*, vol. 47, no. 1, pp. 149–159, 2004.

[9] J. Jaroš, T. Kusano, and N. Yoshida, "Generalized Picone's formula and forced oscillations in quasilinear differential equations of the second order," *Archivum Mathematicum*, vol. 38, no. 1, pp. 53–59, 2002.

[10] R. P. Agarwal, S. R. Grace, and D. O'Regan, *Oscillation Theory for Second Order Linear, Half-Linear, Superlinear and Sublinear Dynamic Equations*, Kluwer Academic Publishers, Dordrecht, The Netherlands, 2002.

[11] S. Jing, "A new oscillation criterion for forced second-order quasilinear differential equations," *Discrete Dynamics in Nature and Society*, vol. 2011, Article ID 428976, 8 pages, 2011.

[12] W.-T. Li and S. S. Cheng, "An oscillation criterion for nonhomogeneous half-linear differential equations," *Applied Mathematics Letters*, vol. 15, no. 3, pp. 259–263, 2002.

[13] J. S. W. Wong, "Oscillation criteria for a forced second-order linear differential equation," *Journal of Mathematical Analysis and Applications*, vol. 231, no. 1, pp. 235–240, 1999.

[14] M. Pašić, "Fite-Wintner-Leighton type oscillation criteria for second-order differential equations with nonlinear damping," *Abstract and Applied Analysis*, vol. 2013, Article ID 852180, 10 pages, 2013.

[15] M. Pašić, "New interval oscillation criteria for forced second-order differential equations with nonlinear damping," *International Journal of Mathematical Analysis*, vol. 7, no. 25, pp. 1239–1255, 2013.

[16] L. Gasiński and N. S. Papageorgiou, *Nonsmooth Critical Point Theory and Nonlinear Boundary Value Problems*, vol. 8 of *Series in Mathematical Analysis and Applications, Edited by R. P. Agarwal and D. O'Regan*, Chapman & Hall/CRC, Boca Raton, Fla, USA, 2005.

[17] S. Lang, *Real and Functional Analysis*, vol. 142 of *Graduate Texts in Mathematics*, Springer-Verlag, New York, NY, USA, 3rd edition, 1993.

[18] I. N. Bronshtein, K. A. Semendyayev, G. Musiol, and H. Muehlig, *Handbook of Mathematics*, Springer, Berlin, Germany, 5th edition, 2007.

Unboundedness of Solutions of Timoshenko Beam Equations with Damping and Forcing Terms

Kusuo Kobayashi and Norio Yoshida

Department of Mathematics, University of Toyama, Toyama 930-8555, Japan

Correspondence should be addressed to Norio Yoshida; nori@sci.u-toyama.ac.jp

Academic Editor: Jaroslav Jaros

Timoshenko beam equations with external damping and internal damping terms and forcing terms are investigated, and boundary conditions (end conditions) to be considered are hinged ends (pinned ends), hinged-sliding ends, and sliding ends. Unboundedness of solutions of boundary value problems for Timoshenko beam equations is studied, and it is shown that the magnitude of the displacement of the beam grows up to ∞ as $t \to \infty$ under some assumptions on the forcing term. Our approach is to reduce the multidimensional problems to one-dimensional problems for fourth-order ordinary differential inequalities.

1. Introduction

The most fundamental beam equations are of the following form:

$$\rho A\frac{\partial^2 u}{\partial t^2} + EI\frac{\partial^4 u}{\partial x^4} = 0 \quad (0 < x < L,\ t > 0) \tag{1}$$

with the length L, the mass density ρ, the cross-sectional area A, the modulus of elasticity E, and the flexural rigidity EI (see [1, page 416]). Taking account of the rotary inertia and the deflection due to shear, we obtain the following fourth-order beam equation for the transverse vibrations of prismatic beams on elastic foundations:

$$\frac{\rho^2 I}{k_1 G}\frac{\partial^4 u}{\partial t^4} + \rho A\frac{\partial^2 u}{\partial t^2} - \rho\left(I + \frac{EI}{k_1 G}\right)\frac{\partial^4 u}{\partial t^2\partial x^2} + EI\frac{\partial^4 u}{\partial x^4} = q(x,t) \tag{2}$$

(see [1, page 433] and Wang and Stephens [2, page 150]).

Dividing the above equation by $\rho^2 I/(k_1 G)$, letting $\alpha = \rho A/EI$, $\beta = k_1 G/\rho$, $\gamma = E/\rho$, $f(x,t) = (k_1 G/\rho^2 I)\,q(x,t)$, and taking into account the nonlinear term $\varphi(u)$, the external damping term

$$\delta\frac{\partial u}{\partial t}, \tag{3}$$

and the internal damping terms

$$-\epsilon\frac{\partial^3 u}{\partial t\partial x^2}, \qquad \sigma\frac{\partial^5 u}{\partial t\partial x^4}, \tag{4}$$

we obtain the Timoshenko beam equation

$$\begin{aligned}\frac{\partial^4 u}{\partial t^4} &+ \alpha\beta\gamma\frac{\partial^2 u}{\partial t^2} - (\beta+\gamma)\frac{\partial^4 u}{\partial t^2\partial x^2} + \beta\gamma\frac{\partial^4 u}{\partial x^4} + \delta\frac{\partial u}{\partial t} \\ &- \epsilon\frac{\partial^3 u}{\partial t\partial x^2} + \sigma\frac{\partial^5 u}{\partial t\partial x^4} + \varphi(u) \\ &= f(x,t), \quad (x,t) \in (0,L)\times(0,\infty),\end{aligned} \tag{5}$$

where α, β, γ, δ, ϵ, σ, and L are positive constants.

Let $J = (0,L)$, $\overline{J} = [0,L]$ and we assume throughout this paper that

(H1) $\varphi(\xi)$ is a real-valued continuous function in $\mathbb{R}$;

(H2) $\varphi(\xi) > 0$ and $\varphi(-\xi) = -\varphi(\xi)$ for $\xi > 0$;

(H3) $\varphi(\xi)$ is a nondecreasing function in $\mathbb{R}$;

(H4) $f(x,t)$ is a real-valued continuous function on $\overline{J} \times [0,\infty)$.

Definition 1. By a *solution* of (5), one means a function $u \in C(\overline{J} \times [0,\infty))$ such that the partial derivatives

$$\frac{\partial^4 u}{\partial t^4}, \quad \frac{\partial^4 u}{\partial t^2 \partial x^2}, \quad \frac{\partial^4 u}{\partial x^4}, \quad \frac{\partial^5 u}{\partial t \partial x^4} \tag{6}$$

exist and are continuous on $\overline{J} \times [0,\infty)$.

Oscillations of beam equations have been treated by numerous authors; see, for example, Feireisl and Herrmann [3], Herrmann [4], Kopáčková [5], Kusano and Yoshida [6], Yoshida [7–10], and the references therein. In particular, we mention the paper [4] by Herrmann which deals with the Euler-Bernoulli beam equations that is similar to (5). We note that the oscillation of (5) was studied by Yoshida [10]. We refer to Ball [11], Fitzgibbon [12], and Narazaki [13] for stability and existence results for beam equations.

However, there appears to be no known unboundedness results for beam equations. The objective of this paper is to provide unboundedness results for (5) by reducing the multi-dimensional problems to one-dimensional problems for ordinary differential inequalities of fourth-order.

In Section 2 we treat the hinged ends and reduce unboundedness problem for (5) to that for ordinary differential inequalities. Sections 3 and 4 are devoted to the hinged-sliding ends and sliding ends, respectively. In Section 5, we study fourth-order differential inequalities, and we derive unboundedness results for (5) in Section 6.

2. Hinged Ends

In this section, we treat the case where the ends of the beam are hinged, so that solutions $u = u(x,t)$ are required to satisfy the boundary condition

$$u(0,t) = \frac{\partial^2 u}{\partial x^2}(0,t) = u(L,t) = \frac{\partial^2 u}{\partial x^2}(L,t) = 0. \tag{HE}$$

Theorem 2. *Every solution u of (5) satisfying (HE) is unbounded on $\overline{J} \times [0,\infty)$ if for any constant $\widetilde{M} > 0$, all solutions $y(t)$ of the fourth-order differential inequalities*

$$\begin{aligned} y^{(4)}(t) &+ \left(\alpha\beta\gamma + (\beta+\gamma)\left(\frac{\pi}{L}\right)^2\right) y''(t) \\ &+ \left(\delta + \epsilon\left(\frac{\pi}{L}\right)^2 + \sigma\left(\frac{\pi}{L}\right)^4\right) y'(t) \\ &+ \beta\gamma\left(\frac{\pi}{L}\right)^4 y(t) - \frac{2L}{\pi}\varphi(\widetilde{M}) \le F(t), \end{aligned} \tag{7}$$

$$\begin{aligned} y^{(4)}(t) &+ \left(\alpha\beta\gamma + (\beta+\gamma)\left(\frac{\pi}{L}\right)^2\right) y''(t) \\ &+ \left(\delta + \epsilon\left(\frac{\pi}{L}\right)^2 + \sigma\left(\frac{\pi}{L}\right)^4\right) y'(t) \\ &+ \beta\gamma\left(\frac{\pi}{L}\right)^4 y(t) - \frac{2L}{\pi}\varphi(\widetilde{M}) \le -F(t) \end{aligned} \tag{8}$$

are not bounded from below, where

$$F(t) = \int_0^L f(x,t) \sin\frac{\pi}{L}x\,dx. \tag{9}$$

Proof. Suppose to the contrary that there exists a solution u of the boundary value problem (5), (HE) which is bounded on $\overline{J} \times [0,\infty)$. Then, there exists a constant $M > 0$ such that

$$|u(x,t)| \le M, \quad (x,t) \in \overline{J} \times [0,\infty), \tag{10}$$

that is,

$$-M \le u(x,t) \le M, \quad (x,t) \in \overline{J} \times [0,\infty). \tag{11}$$

First we consider the case where $-M \le u(x,t)$ on $\overline{J} \times [0,\infty)$. It follows from the hypotheses (H2) and (H3) that

$$\varphi(u) \ge \varphi(-M) = -\varphi(M) \tag{12}$$

and therefore we see from (5) that

$$\begin{aligned} &\frac{\partial^4 u}{\partial t^4} + \alpha\beta\gamma\frac{\partial^2 u}{\partial t^2} - (\beta+\gamma)\frac{\partial^4 u}{\partial t^2 \partial x^2} + \beta\gamma\frac{\partial^4 u}{\partial x^4} \\ &\quad + \delta\frac{\partial u}{\partial t} - \epsilon\frac{\partial^3 u}{\partial t \partial x^2} + \sigma\frac{\partial^5 u}{\partial t \partial x^4} - \varphi(M) \le f(x,t) \end{aligned} \tag{13}$$

on $\overline{J} \times [0,\infty)$. Multiplying (13) by $\psi(x) = \sin(\pi x/L)$ and then integrating over $[0,L]$, we derive

$$\begin{aligned} &\frac{d^4}{dt^4}\int_0^L u\psi(x)\,dx + \alpha\beta\gamma\frac{d^2}{dt^2}\int_0^L u\psi(x)\,dx \\ &\quad - (\beta+\gamma)\frac{d^2}{dt^2}\int_0^L \frac{\partial^2 u}{\partial x^2}\psi(x)\,dx + \beta\gamma\int_0^L \frac{\partial^4 u}{\partial x^4}\psi(x)\,dx \\ &\quad + \delta\frac{d}{dt}\int_0^L u\psi(x)\,dx - \epsilon\frac{d}{dt}\int_0^L \frac{\partial^2 u}{\partial x^2}\psi(x)\,dx \\ &\quad + \sigma\frac{d}{dt}\int_0^L \frac{\partial^4 u}{\partial x^4}\psi(x)\,dx - \frac{2L}{\pi}\varphi(M) \\ &\le \int_0^L f(x,t)\psi(x)\,dx, \quad t \ge 0. \end{aligned} \tag{14}$$

Integrating by parts and using (HE), we obtain

$$\begin{aligned} \int_0^L \frac{\partial^2 u}{\partial x^2}\psi(x)\,dx &= -\left(\frac{\pi}{L}\right)^2 \int_0^L u\psi(x)\,dx, \\ \int_0^L \frac{\partial^4 u}{\partial x^4}\psi(x)\,dx &= \left(\frac{\pi}{L}\right)^4 \int_0^L u\psi(x)\,dx. \end{aligned} \tag{15}$$

Combining (14) and (15) yields

$$\begin{aligned} U^{(4)}(t) &+ \left(\alpha\beta\gamma + (\beta+\gamma)\left(\frac{\pi}{L}\right)^2\right) U''(t) \\ &+ \left(\delta + \epsilon\left(\frac{\pi}{L}\right)^2 + \sigma\left(\frac{\pi}{L}\right)^4\right) U'(t) \\ &+ \beta\gamma\left(\frac{\pi}{L}\right)^4 U(t) - \frac{2L}{\pi}\varphi(M) \le F(t), \end{aligned} \tag{16}$$

where $U(t) = \int_0^L u\psi(x)dx$. It is easy to check that

$$U(t) = \int_0^L u\psi(x)\,dx \geq -\frac{2L}{\pi}M, \tag{17}$$

that is, $U(t)$ is bounded from below. Hence, we conclude that $U(t)$ is a solution of (7) which is bounded from below. This contradicts the hypothesis. In the case where $u(x,t) \leq M$ on $\overline{J} \times [0,\infty)$, $v := -u$ satisfies

$$\begin{aligned}\frac{\partial^4 v}{\partial t^4} &+ \alpha\beta\gamma\frac{\partial^2 v}{\partial t^2} - (\beta+\gamma)\frac{\partial^4 v}{\partial t^2\partial x^2} + \beta\gamma\frac{\partial^4 v}{\partial x^4} \\ &+ \delta\frac{\partial v}{\partial t} - \epsilon\frac{\partial^3 v}{\partial t\partial x^2} + \sigma\frac{\partial^5 v}{\partial t\partial x^4} + \varphi(v) = -f(x,t)\end{aligned} \tag{18}$$

in view of the hypothesis (H2). It is readily verified that $v \geq -M$ and $\varphi(v) \geq \varphi(-M) = -\varphi(M)$. By the same arguments as in the case where $u(x,t) \geq -M$, we conclude that $V(t) = \int_0^L v\psi(x)dx$ is a solution of (8) which is bounded from below in light of $V(t) \geq -(2L/\pi)M$. This contradicts the hypothesis and the proof is complete.

3. Hinged-Sliding Ends

In this section, we deal with the case of hinged-sliding ends, for which the boundary condition takes the form

$$u(0,t) = \frac{\partial^2 u}{\partial x^2}(0,t) = \frac{\partial u}{\partial x}(L,t) = \frac{\partial^3 u}{\partial x^3}(L,t) = 0. \tag{HSE}$$

Theorem 3. *Every solution u of (5) satisfying (HSE) is unbounded on $\overline{J} \times [0,\infty)$ if for any constant $\widetilde{M} > 0$, all solutions $y(t)$ of the fourth-order differential inequalities*

$$\begin{aligned}y^{(4)}(t) &+ \left(\alpha\beta\gamma + (\beta+\gamma)\left(\frac{\pi}{2L}\right)^2\right)y''(t) \\ &+ \left(\delta + \epsilon\left(\frac{\pi}{2L}\right)^2 + \sigma\left(\frac{\pi}{2L}\right)^4\right)y'(t) \\ &+ \beta\gamma\left(\frac{\pi}{2L}\right)^4 y(t) - \frac{2L}{\pi}\varphi(\widetilde{M}) \leq G(t),\end{aligned} \tag{19}$$

$$\begin{aligned}y^{(4)}(t) &+ \left(\alpha\beta\gamma + (\beta+\gamma)\left(\frac{\pi}{2L}\right)^2\right)y''(t) \\ &+ \left(\delta + \epsilon\left(\frac{\pi}{2L}\right)^2 + \sigma\left(\frac{\pi}{2L}\right)^4\right)y'(t) \\ &+ \beta\gamma\left(\frac{\pi}{2L}\right)^4 y(t) - \frac{2L}{\pi}\varphi(\widetilde{M}) \leq -G(t)\end{aligned} \tag{20}$$

are not bounded from below, where

$$G(t) = \int_0^L f(x,t)\sin\frac{\pi}{2L}x\,dx. \tag{21}$$

Proof. Suppose that there is a solution u of the boundary value problem (5), (HSE) which is bounded on $\overline{J} \times [0,\infty)$. Let $|u(x,t)| \leq M$ on $\overline{J} \times [0,\infty)$ for some $M > 0$. First, consider the case where $-M \leq u(x,t)$ on $\overline{J} \times [0,\infty)$. Proceeding as in the proof of Theorem 2, we find that the inequality (13) holds. Multiplying (13) by $\omega(x) = \sin(\pi x/(2L))$ and then integrating over $[0,L]$, we obtain the inequality (14) with $\psi(x)$ replaced by $\omega(x)$. Integration by parts yields

$$\begin{aligned}\int_0^L \frac{\partial^2 u}{\partial x^2}\omega(x)\,dx &= -\left(\frac{\pi}{2L}\right)^2\int_0^L u\omega(x)\,dx, \\ \int_0^L \frac{\partial^4 u}{\partial x^4}\omega(x)\,dx &= \left(\frac{\pi}{2L}\right)^4\int_0^L u\omega(x)\,dx.\end{aligned} \tag{22}$$

As in the proof of Theorem 2, we observe that $\int_0^L u\omega(x)dx$ is a solution of (19) which is bounded from below. This contradicts the hypothesis. The case where $u(x,t) \leq M$ on $\overline{J} \times [0,\infty)$ can be treated similarly, and we find that $\int_0^L (-u)\omega(x)dx$ is a solution of (20) which is bounded from below. The contradiction establishes the theorem.

4. Sliding Ends

We study the case of sliding ends for which the boundary condition takes the form

$$\frac{\partial u}{\partial x}(0,t) = \frac{\partial^3 u}{\partial x^3}(0,t) = \frac{\partial u}{\partial x}(L,t) = \frac{\partial^3 u}{\partial x^3}(L,t) = 0. \tag{SE}$$

Theorem 4. *Every solution u of (5) satisfying (SE) is unbounded on $\overline{J} \times [0,\infty)$ if for any constant $\widetilde{M} > 0$, all solutions $y(t)$ of the fourth-order differential inequalities*

$$y^{(4)}(t) + \alpha\beta\gamma y''(t) + \delta y'(t) - L\varphi(\widetilde{M}) \leq H(t), \tag{23}$$

$$y^{(4)}(t) + \alpha\beta\gamma y''(t) + \delta y'(t) - L\varphi(\widetilde{M}) \leq -H(t) \tag{24}$$

are not bounded from below, where

$$H(t) = \int_0^L f(x,t)\,dx. \tag{25}$$

Proof. Suppose that the boundary value problem (5), (SE) has a solution u which is bounded on $\overline{J} \times [0,\infty)$. There exists a constant $M > 0$ such that $|u(x,t)| \leq M$ on $\overline{J} \times [0,\infty)$. First, consider the case where $-M \leq u(x,t)$ on $\overline{J} \times [0,\infty)$. Arguing as in the proof of Theorem 2, we see that the inequality (13) holds. Integrating (13) over $[0,L]$, we obtain the inequality (14) with $-((2L)/\pi)\varphi(M)$ replaced by $-L\varphi(M)$ and $\psi(x) \equiv 1$. It is easy to see that

$$\begin{aligned}\int_0^L \frac{\partial^2 u}{\partial x^2}dx &= \frac{\partial u}{\partial x}(L,t) - \frac{\partial u}{\partial x}(0,t) = 0, \\ \int_0^L \frac{\partial^4 u}{\partial x^4}dx &= \frac{\partial^3 u}{\partial x^3}(L,t) - \frac{\partial^3 u}{\partial x^3}(0,t) = 0.\end{aligned} \tag{26}$$

Hence, we have the inequality

$$U^{(4)}(t) + \alpha\beta\gamma U''(t) + \delta U'(t) - L\varphi(M) \leq H(t), \quad t \geq 0, \tag{27}$$

where $U(t) = \int_0^L u dx$. Therefore, we conclude that $U(t)$ is a solution of (23) which is bounded from below. This contradicts the hypothesis. The case where $u(x,t) \le M$ on $\bar{J} \times [0,\infty)$ can be treated analogously, and we observe that $\int_0^L (-u)dx$ is a solution of (24) which is bounded from below. This is a contradiction and the proof is complete.

5. Fourth-Order Ordinary Differential Inequalities

We deal with the ordinary differential inequality of the fourth order

$$y^{(4)}(t) + ky''(t) + \ell y'(t) + my(t) - \mu \le g(t), \quad t \ge 0, \tag{28}$$

and derive sufficient condition for every solution $y(t)$ of (28) to be unbounded from below. It is assumed that k, ℓ, m, and μ are nonnegative constants, and $g(t)$ is a continuous function on $[0,\infty)$.

Theorem 5. *Every solution $y(t)$ of (28) is not bounded from below if*

$$\liminf_{t\to\infty} \frac{1}{t}\int_0^t \left(1 - \frac{\xi}{t}\right)^3 g(\xi)\, d\xi = -\infty. \tag{29}$$

Proof. Assume, on the contrary, that there exists a solution $y(t)$ of (28) which is bounded from below. Let $y(t) \ge -M$ on $[0,\infty)$ for some constant $M > 0$. Then we obtain from (28) the inequality

$$y^{(4)}(t) + ky''(t) + \ell y'(t) - Mm - \mu \le g(t), \quad t \ge 0. \tag{30}$$

Integrating (30) over $[0,t]$, we get

$$y^{(3)}(t) + ky'(t) + \ell y(t) - Mmt - \mu t \le c_1 + \int_0^t g(\xi)\, d\xi, \tag{31}$$

and hence

$$y^{(3)}(t) + ky'(t) - M\ell - Mmt - \mu t \le c_1 + \int_0^t g(\xi)\, d\xi, \tag{32}$$

where $c_1 = y^{(3)}(0) + ky'(0) + \ell y(0)$. Integration of (32) over $[0,t]$ yields

$$\begin{aligned} y''(t) &+ ky(t) - M\ell t - \frac{Mm}{2}t^2 - \frac{\mu}{2}t^2 \\ &\le c_0 + c_1 t + \int_0^t ds_1 \int_0^{s_1} g(\xi)\, d\xi, \end{aligned} \tag{33}$$

that is,

$$\begin{aligned} y''(t) &- Mk - M\ell t - \frac{Mm}{2}t^2 - \frac{\mu}{2}t^2 \\ &\le c_0 + c_1 t + \int_0^t ds_1 \int_0^{s_1} g(\xi)\, d\xi, \end{aligned} \tag{34}$$

where $c_0 = y''(0) + ky(0)$. Integrating (34) over $[0,t]$ twice, we obtain

$$\begin{aligned} y(t) &- \frac{Mk}{2}t^2 - \frac{M\ell}{6}t^3 - \frac{Mm}{24}t^4 - \frac{\mu}{24}t^4 \\ &\le d_0 + d_1 t + d_2 t^2 + d_3 t^3 \\ &\quad + \int_0^t ds_3 \int_0^{s_3} ds_2 \int_0^{s_2} ds_1 \int_0^{s_1} g(\xi)\, d\xi, \end{aligned} \tag{35}$$

and therefore

$$\begin{aligned} &- M - \frac{Mk}{2}t^2 - \frac{M\ell}{6}t^3 - \frac{Mm}{24}t^4 - \frac{\mu}{24}t^4 \\ &\le d_0 + d_1 t + d_2 t^2 + d_3 t^3 + \frac{1}{6}\int_0^t (t-\xi)^3 g(\xi)\, d\xi, \end{aligned} \tag{36}$$

in view of the identity

$$\int_0^t ds_3 \int_0^{s_3} ds_2 \int_0^{s_2} ds_1 \int_0^{s_1} g(\xi)\, d\xi = \frac{1}{6}\int_0^t (t-\xi)^3 g(\xi)\, d\xi, \tag{37}$$

where $d_0 = y(0)$, $d_1 = y'(0)$, $d_2 = c_0/2$, and $d_3 = c_1/6$. Dividing (36) by t^4 yields

$$\begin{aligned} &-\frac{M}{t^4} - \frac{Mk}{2t^2} - \frac{M\ell}{6t} - \frac{Mm}{24} - \frac{\mu}{24} - \left(\frac{d_0}{t^4} + \frac{d_1}{t^3} + \frac{d_2}{t^2} + \frac{d_3}{t}\right) \\ &\le \frac{1}{6t}\int_0^t \left(1 - \frac{\xi}{t}\right)^3 g(\xi)\, d\xi. \end{aligned} \tag{38}$$

The left hand side of (38) is bounded from below, whereas the right hand side of (38) is not bounded from below from the condition (29). This is a contradiction and the proof is complete.

6. Unboundedness Results for Timoshenko Beam Equations

Combining Theorems 2–4 with Theorem 5, we present unboundedness results for the three types of boundary value problems for (5) under consideration.

Theorem 6. *Every solution u of (5) satisfying (HE) is unbounded on $\bar{J} \times [0,\infty)$ if*

$$\liminf_{t\to\infty} \frac{1}{t}\int_0^t \left(1 - \frac{\xi}{t}\right)^3 F(\xi)\, d\xi = -\infty, \tag{39}$$

$$\limsup_{t\to\infty} \frac{1}{t}\int_0^t \left(1 - \frac{\xi}{t}\right)^3 F(\xi)\, d\xi = \infty. \tag{40}$$

Proof. The hypothesis (39) implies that every solution $y(t)$ of (7) is not bounded from below via Theorem 5. Since the hypothesis (40) implies that

$$\begin{aligned} &\liminf_{t\to\infty} \frac{1}{t}\int_0^t \left(1 - \frac{\xi}{t}\right)^3 (-F(\xi))\, d\xi \\ &= -\limsup_{t\to\infty} \frac{1}{t}\int_0^t \left(1 - \frac{\xi}{t}\right)^3 F(\xi)\, d\xi = -\infty, \end{aligned} \tag{41}$$

we observe that every solution $y(t)$ of (8) is not bounded from below. The conclusion follows from Theorem 2. The proof is complete. □

We combine Theorems 3 and 4 with Theorem 5 to establish the following two theorems.

Theorem 7. *Every solution u of (5) satisfying (HSE) is unbounded on $\overline{J} \times [0, \infty)$ if*

$$\liminf_{t\to\infty} \frac{1}{t}\int_0^t \left(1-\frac{\xi}{t}\right)^3 G(\xi)\,d\xi = -\infty, \qquad \limsup_{t\to\infty} \frac{1}{t}\int_0^t \left(1-\frac{\xi}{t}\right)^3 G(\xi)\,d\xi = \infty. \tag{42}$$

Theorem 8. *Every solution u of (5) satisfying (SE) is unbounded on $\overline{J} \times [0, \infty)$ if*

$$\liminf_{t\to\infty} \frac{1}{t}\int_0^t \left(1-\frac{\xi}{t}\right)^3 H(\xi)\,d\xi = -\infty, \qquad \limsup_{t\to\infty} \frac{1}{t}\int_0^t \left(1-\frac{\xi}{t}\right)^3 H(\xi)\,d\xi = \infty. \tag{43}$$

Remark 9. If a solution u of (5) is unbounded on $\overline{J} \times [0, \infty)$, then there exists a sequence $\{(x_n, t_n)\} \subset \overline{J} \times [0, \infty)$ such that

$$\lim_{n\to\infty} u(x_n, t_n) = \infty. \tag{44}$$

Remark 10. In the case where

$$\varphi(\xi) = p|\xi|^{r-1}\xi, \tag{45}$$

$p > 0$, and $r > 0$ being constants, we see that $\varphi(\xi)$ satisfies the hypotheses (H1)–(H3).

Example 11. We consider the Timoshenko beam equation

$$\frac{\partial^4 u}{\partial t^4} + 3\frac{\partial^2 u}{\partial t^2} - 6\frac{\partial^4 u}{\partial t^2 \partial x^2} + 9\frac{\partial^4 u}{\partial x^4} + \frac{1}{3}\frac{\partial u}{\partial t} - \frac{1}{3}\frac{\partial^3 u}{\partial t \partial x^2} + \frac{1}{3}\frac{\partial^5 u}{\partial t \partial x^4} + u = f(x,t), \quad (x,t) \in (0,\pi) \times (0,\infty) \tag{46}$$

with the hinged ends (HE), where

$$f(x,t) = (\sin x)\left[2t^5 \sin t + t^5 \sin\left(t+\frac{\pi}{2}\right) + 5t^4 \sin t - 70t^4 \sin\left(t+\frac{\pi}{2}\right) + 60t^3 \sin t + 240t^2 \sin\left(t+\frac{\pi}{2}\right) + 120t \sin t\right]. \tag{47}$$

Here $L = \pi$, $\alpha = 1/3$, $\beta = \gamma = 3$, $\delta = \epsilon = \sigma = 1/3$, and $\varphi(u) = u$. An easy computation shows that

$$F(t) = \frac{\pi}{2}\left[2t^5 \sin t + t^5 \sin\left(t+\frac{\pi}{2}\right) + 5t^4 \sin t - 70t^4 \sin\left(t+\frac{\pi}{2}\right) + 60t^3 \sin t + 240t^2 \sin\left(t+\frac{\pi}{2}\right) + 120t \sin t\right]. \tag{48}$$

Since

$$\frac{1}{t}\int_0^t \left(1-\frac{\xi}{t}\right)^3 \xi^5 \sin(\xi+\theta_1)\,d\xi = 6t \sin(\xi+\theta_1) + B_1(t,\theta_1), \qquad \frac{1}{t}\int_0^t \left(1-\frac{\xi}{t}\right)^3 \xi^N \sin(\xi+\theta_2)\,d\xi = B_2(t,\theta_2) \quad (N = 1,2,3,4), \tag{49}$$

we observe that

$$\frac{1}{t}\int_0^t \left(1-\frac{\xi}{t}\right)^3 F(\xi)\,d\xi = 3\pi t\left(2\sin t + \sin\left(t+\frac{\pi}{2}\right)\right) + B_3(t), \tag{50}$$

where θ_i $(i = 1, 2)$ are constants, $B_i(t, \theta_i)$ $(i = 1, 2)$ and $B_3(t)$ are bounded functions on $[0, \infty)$. Hence, the conditions (39) and (40) are satisfied. It follows from Theorem 6 that every solution u of the problem (46), (HE) is unbounded on $[0, \pi] \times [0, \infty)$. For example, $u = (\sin x)t^5 \sin t$ is such a solution.

References

[1] S. Timoshenko, D. H. Young, and W. Weaver Jr., *Vibration Problems in Engineering*, John Wiley & Sons, New York, NY, USA, 4th edition, 1974.

[2] T. M. Wang and J. E. Stephens, "Natural frequencies of Timoshenko beams on pasternak foundations," *Journal of Sound and Vibration*, vol. 51, no. 2, pp. 149–155, 1977.

[3] E. Feireisl and L. Herrmann, "Oscillations of a non-linearly damped extensible beam," *Applications of Mathematics*, vol. 37, no. 6, pp. 469–478, 1992.

[4] L. Herrmann, "Vibration of the Euler-Bernoulli beam with allowance for dampings," in *Proceedings of the World Congress on Engineering*, vol. 2, London, UK, 2008.

[5] M. Kopáčková, "On periodic solution of a nonlinear beam equation," *Aplikace Matematiky*, vol. 28, no. 2, pp. 108–115, 1983.

[6] T. Kusano and N. Yoshida, "Forced oscillations of Timoshenko beams," *Quarterly of Applied Mathematics*, vol. 43, no. 2, pp. 167–177, 1985.

[7] N. Yoshida, "Forced oscillations of extensible beams," *SIAM Journal on Mathematical Analysis*, vol. 16, no. 2, pp. 211–220, 1985.

[8] N. Yoshida, "Forced oscillations of nonlinear extensible beams," in *Proceedings of the 10th International Conference on Nonlinear Oscillations (Varna, 1984)*, pp. 814–817, Sofia, Bulgaria, 1985.

[9] N. Yoshida, "On the zeros of solutions of beam equations," *Annali di Matematica Pura ed Applicata*, vol. 151, pp. 389–398, 1988.

[10] N. Yoshida, "Vibrations of Timoshenko beams with damping and forcingterms," in *Proceedings of the IEEE International Conference on Industrial Engineering and Engineering Management of (IEEM '12)*, 2012.

[11] J. M. Ball, "Stability theory for an extensible beam," *Journal of Differential Equations*, vol. 14, pp. 399–418, 1973.

[12] W. E. Fitzgibbon, "Global existence and boundedness of solutions to the extensible beam equation," *SIAM Journal on Mathematical Analysis*, vol. 13, no. 5, pp. 739–745, 1982.

[13] T. Narazaki, "On the time global solutions of perturbed beam equations," *Proceedings of the Faculty of Science of Tokai University*, vol. 16, pp. 51–71, 1981.

4

Entropy Solutions for Nonlinear Elliptic Anisotropic Homogeneous Neumann Problem

B. K. Bonzi, S. Ouaro, and F. D. Y. Zongo

Laboratoire d'Analyse Mathématique des Equations (LAME), UFR, Sciences Exactes et Appliquées, Université de Ouagadougou, 03 BP 7021 Ouaga 03, Ouagadougou, Burkina Faso

Correspondence should be addressed to S. Ouaro; ouaro@yahoo.fr

Academic Editor: Elena Braverman

We prove the existence and uniqueness of entropy solution for nonlinear anisotropic elliptic equations with Neumann homogeneous boundary value condition for L^1-data. We prove first, by using minimization techniques, the existence and uniqueness of weak solution when the data is bounded, and by approximation methods, we prove a result of existence and uniqueness of entropy solution.

1. Introduction

Let Ω be an open bounded domain of $\mathbb{R}^N$ $(N \geq 3)$ with smooth boundary. Our aim is to prove the existence and uniqueness of entropy solution for the anisotropic nonlinear elliptic problem of the form

$$-\sum_{i=1}^{N} \frac{\partial}{\partial x_i} a_i\left(x, \frac{\partial}{\partial x_i} u\right) + |u|^{p_M(x)-2} u = f \quad \text{in } \Omega,$$
$$\sum_{i=1}^{N} a_i\left(x, \frac{\partial}{\partial x_i} u\right) \nu_i = 0 \quad \text{on } \partial\Omega, \tag{1}$$

where the right-hand side $f \in L^1(\Omega)$ and ν_i, $i \in \{1, \ldots, N\}$ are the components of the outer normal unit vector.

For the rest of the functions involved in (1), we are going to enumerate their properties after we make some notations.

For any $\Omega \subset \mathbb{R}^N$, we set

$$C_+\left(\overline{\Omega}\right) = \left\{h \in C\left(\overline{\Omega}\right) : \inf_{x\in\Omega} h(x) > 1\right\}, \tag{2}$$

and we denote

$$h^+ = \sup_{x\in\Omega} h(x), \qquad h^- = \inf_{x\in\Omega} h(x). \tag{3}$$

For the exponents, $\vec{p}(\cdot) : \overline{\Omega} \to \mathbb{R}^N$, $\vec{p}(\cdot) = (p_1(\cdot), \ldots, p_N(\cdot))$ with $p_i \in C_+(\overline{\Omega})$ for every $i \in \{1, \ldots, N\}$ and for all $x \in \overline{\Omega}$, we put $p_M(x) = \max\{p_1(x), \ldots, p_N(x)\}$ and $p_m(x) = \min\{p_1(x), \ldots, p_N(x)\}$. Now, we can give the properties of the rest of the functions involved in (1).

We assume that for $i = 1, \ldots, N$, the function $a_i : \Omega \times \mathbb{R} \to \mathbb{R}$ is Carathéodory and satisfies the following conditions: $a_i(x, \xi)$ is the continuous derivative with respect to ξ of the mapping $A_i : \Omega \times \mathbb{R} \to \mathbb{R}$, $A_i = A_i(x, \xi)$, that is, $a_i(x, \xi) = (\partial/\partial\xi) A_i(x, \xi)$ such that the following equality and inequalities holds

$$A_i(x, 0) = 0, \tag{4}$$

for almost every $x \in \Omega$.

There exists a positive constant C_1 such that

$$\left|a_i(x, \xi)\right| \leq C_1 \left(j_i(x) + |\xi|^{p_i(x)-1}\right), \tag{5}$$

for almost every $x \in \Omega$ and for every $\xi \in \mathbb{R}$, where j_i is a nonnegative function in $L^{p_i'(\cdot)}(\Omega)$, with $1/p_i(x) + 1/p_i'(x) = 1$.

There exists a positive constant C_2 such that

$$(a_i(x, \xi) - a_i(x, \eta)) \cdot (\xi - \eta) \geq \begin{cases} C_2 |\xi - \eta|^{p_i(x)} & \text{if } |\xi - \eta| \geq 1, \\ C_2 |\xi - \eta|^{p_i^-} & \text{if } |\xi - \eta| < 1, \end{cases} \tag{6}$$

for almost every $x \in \Omega$ and for every $\xi, \eta \in \mathbb{R}$, with $\xi \neq \eta$ and

$$|\xi|^{p_i(x)} \leq a_i(x,\xi) \cdot \xi \leq p_i(x) A_i(x,\xi), \tag{7}$$

for almost every $x \in \Omega$ and for every $\xi \in \mathbb{R}$.

We also assume that the variable exponents $p_i(\cdot) : \overline{\Omega} \to [2, N)$ are continuous functions for all $i = 1, \ldots, N$ such that

$$\frac{\overline{p}(N-1)}{N(\overline{p}-1)} < p_i^- < \frac{\overline{p}(N-1)}{N-\overline{p}}, \qquad \sum_{i=1}^{N} \frac{1}{p_i^-} > 1,$$
$$\frac{p_i^+ - p_i^- - 1}{p_i^-} < \frac{\overline{p}-N}{\overline{p}(N-1)}, \tag{8}$$

where $1/\overline{p} = (1/N)\sum_{i=1}^{N}(1/p_i^-)$.

We introduce the numbers

$$q = \frac{N(\overline{p}-1)}{N-1}, \qquad q^* = \frac{Nq}{N-q} = \frac{N(\overline{p}-1)}{N-\overline{p}}. \tag{9}$$

A prototype example, that is, covered by our assumptions is the following anisotropic equation:

Set $A_i(x,\xi) = (1/p_i(x))|\xi|^{p_i(x)}$, $a_i(x,\xi) = |\xi|^{p_i(x)-2}\xi$ where $p_i(x) \geq 2$. Then, we get the following equation.

$$-\sum_{i=1}^{N} \frac{\partial}{\partial x_i}\left(\left|\frac{\partial}{\partial x_i}u\right|^{p_i(x)-2} \frac{\partial}{\partial x_i}u\right) + |u|^{p_M(x)-2} = f. \tag{10}$$

Actually, one of the topics from the field of PDEs that continuously gained interest is the one concerning the Sobolev space with variable exponents, $W^{1,p(\cdot)}(\Omega)$ or $W_0^{1,p(\cdot)}(\Omega)$ depending on the boundary condition (see [1–23]). In that context, problems involving the $p(\cdot)$-Laplace operator

$$\Delta_{p(x)} u = \operatorname{div}\left(|\nabla u|^{p(x)-2} \nabla u\right) \tag{11}$$

or the more general operator

$$\operatorname{div} a(x, \nabla u) \tag{12}$$

were intensively studied (see [13]). At the same time, some authors was interested by PDEs involving anisotropic Sobolev spaces with variable exponent $W^{1,\vec{p}(\cdot)}$ when the boundary condition is the homogeneous Dirichlet boundary condition (see [15, 16, 18, 20, 24–26]). In that context, the authors have considered the anisotropic $p(\cdot)$-Laplace operator

$$\Delta_{\vec{p}(x)} u = \sum_{i=1}^{N} \partial_{x_i}\left(\left|\partial_{x_i} u\right|^{p_i(x)-2} \partial_{x_i} u\right) \tag{13}$$

or the more general variable exponent anisotropic operator

$$\sum_{i=1}^{N} \partial_{x_i} a\left(x, \partial_{x_i} u\right). \tag{14}$$

When the homogeneous Dirichlet boundary condition is replaced by the Neumann boundary condition, one has to work with the anisotropic variable exponent Sobolev space $W^{1,\vec{p}(\cdot)}(\Omega)$ instead of $W_0^{1,\vec{p}(\cdot)}(\Omega)$. The main difficulty which appears is that the famous Poincaré inequality does not apply and then it is very difficult to get a priori estimates which are necessary for the proof of the existence result of entropy solutions. Sometimes one can use the Wirtinger inequality which does not apply, in some problems like (1). The first systematic study of anisotropic Neumann problem was done by Fan (see [11]). In a second time, Boureanu and Rădulescu studied an anisotropic nonhomogeneous Neumann problem with obstacle (see [2]). In the two papers, the authors were interested by the existence and multiplicity results of weak solution even if in [2], Boureanu and Rădulescu have showed some conditions under which we can get uniqueness of weak solution. In this paper, we are interested to the existence and uniqueness of entropy solution. For the proof of the existence of entropy solution of (1), we follow [27] and derive a priori estimates for the approximated solutions u_n and the partial derivatives $\partial u_n/\partial x_i$ in the Marcinkiewicz spaces $\mathcal{M}^{\overline{p}}$ and $\mathcal{M}^{\overline{p}_i}$, respectively (see Section 2 or [27, 28] for definition and properties of Marcinkiewicz spaces).

The study of anisotropic problems are motivated, for example, by their applications to the mathematical analysis of a system of nonlinear partial differential equations arising in a population dynamics model describing the spread of an epidemic disease through a heterogeneous habitat.

The paper is organized as follows. In Section 2, we introduce some notations/functional spaces. In Section 3, we prove for the problem (1), the existence and uniqueness of weak solution when the data is bounded, and the existence and uniqueness of entropy solution when the data is in $L^1(\Omega)$.

2. Preliminaries

In this section, we define Lebesgue, Sobolev, and anisotropic spaces with variable exponent and give some of their properties (see [29] for more details about Lebesgue and Sobolev spaces with variable exponent). Roughly speaking, anistropic Lebesgue and Sobolev spaces are functional spaces of Lebesgue's and Sobolev's type in which different space directions have different roles.

Given a measurable function $p(\cdot) : \Omega \to [1, \infty)$, we define the Lebesgue space with variable exponent $L^{p(\cdot)}(\Omega)$ as the set of all measurable functions $u : \Omega \to \mathbb{R}$ for which the convex modular

$$\rho_{p(\cdot)}(u) := \int_{\Omega} |u|^{p(x)} dx \tag{15}$$

is finite. If the exponent is bounded, that is, if $p_+ < \infty$, then the expression

$$|u|_{p(\cdot)} := \inf\left\{\lambda > 0 : \rho_{p(\cdot)}\left(\frac{u}{\lambda}\right) \leq 1\right\} \tag{16}$$

defines a norm in $L^{p(\cdot)}(\Omega)$, called the Luxembourg norm. The space $(L^{p(\cdot)}(\Omega), |\cdot|_{p(\cdot)})$ is a separable Banach space. Moreover, if $p_- > 1$, then $L^{p(\cdot)}(\Omega)$ is uniformly convex, hence reflexive, and its dual space is isomorphic to $L^{p'(\cdot)}(\Omega)$, where $1/p(x) + 1/p'(x) = 1$. Finally, we have the Hölder type inequality.

Proposition 1 (generalized Hölder inequality, see [10]). (i) *For any* $u \in L^{p(\cdot)}(\Omega)$ *and* $v \in L^{p'(\cdot)}(\Omega)$, *we have*

$$\left|\int_\Omega u\, v\, dx\right| \le \left(\frac{1}{p_-} + \frac{1}{p'_-}\right) |u|_{p(\cdot)} |v|_{p'(\cdot)}. \tag{17}$$

(ii) *If* $p_1, p_2 \in \mathscr{C}_+(\overline{\Omega})$, $p_1(x) \le p_2(x)$ *for any* $x \in \overline{\Omega}$, *then* $L^{p_2(x)}(\Omega) \hookrightarrow L^{p_1(x)}(\Omega)$ *and the imbedding is continuous.*

Moreover, the application $\rho_{p(\cdot)} : L^{p(\cdot)}(\Omega) \to \mathbb{R}$ called the $p(\cdot)$-modular of the $L^{p(\cdot)}(\Omega)$ space is very useful in handling these Lebesgue spaces with variable exponent. Indeed we have the following properties (see [10]). If $u \in L^{p(\cdot)}(\Omega)$ and $p < \infty$ then

$$|u|_{p(\cdot)} < 1 \Longrightarrow |u|_{p(\cdot)}^{p^+} \le \rho_{p(\cdot)}(u) \le |u|_{p(\cdot)}^{p^-}, \tag{18}$$

$$|u|_{p(\cdot)} > 1 \Longrightarrow |u|_{p(\cdot)}^{p^-} \le \rho_{p(\cdot)}(u) \le |u|_{p(\cdot)}^{p^+}, \tag{19}$$

$$|u|_{p(\cdot)} < 1 \,(= 1; > 1) \Longrightarrow \rho_{p(\cdot)}(u) < 1 \,(= 1; > 1), \tag{20}$$

$$|u|_{p(\cdot)} \longrightarrow 0 \,(\longrightarrow \infty) \Longleftrightarrow \rho_{p(\cdot)}(u) \longrightarrow 0 \,(\longrightarrow \infty). \tag{21}$$

If, in addition, $(u_n)_n \subset L^{p(\cdot)}(\Omega)$, then $\lim_{n\to\infty}|u_n - u|_{p(\cdot)} = 0 \Leftrightarrow \lim_{n\to\infty}\rho_{p(\cdot)}(u_n - u) = 0 \Leftrightarrow (u_n)_n$ converges to u in measure and $\lim_{n\to\infty}\rho_{p(\cdot)}(u_n) = \rho_{p(\cdot)}(u)$.

Now, let us introduce the definition of the isotropic Sobolev space with variable exponent, $W^{1,p(\cdot)}(\Omega)$.

We set

$$W^{1,p(\cdot)}(\Omega) := \left\{u \in L^{p(\cdot)}(\Omega) : |\nabla u| \in L^{p(\cdot)}(\Omega)\right\}, \tag{22}$$

which is a Banach space equipped with the norm

$$\|u\|_{1,p(\cdot)} := |u|_{p(\cdot)} + |\nabla u|_{p(\cdot)}. \tag{23}$$

Now, we present a natural generalization of the variable exponent Sobolev space $W^{1,p(\cdot)}(\Omega)$ that will enable us to study the problem (1) with sufficient accuracy.

The *anisotropic variable exponent Sobolev space* $W^{1,\vec{p}(\cdot)}(\Omega)$ is defined as follows:

$$W^{1,\vec{p}(\cdot)}(\Omega) = \left\{u \in L^{p_M(\cdot)}(\Omega); \frac{\partial u}{\partial x_i} \in L^{p_i(\cdot)}(\Omega),\ \forall i \in \{1,\ldots,N\}\right\}. \tag{24}$$

Endowed with the norm

$$\|u\|_{\vec{p}(\cdot)} := |u|_{p_M(\cdot)} + \sum_{i=1}^{N}\left|\frac{\partial}{\partial x_i}u\right|_{p_i(\cdot)}, \tag{25}$$

the space $(W^{1,\vec{p}(\cdot)}(\Omega), \|\cdot\|_{\vec{p}(\cdot)})$ is a reflexive Banach space (see [11, Theorems 2.1 and 2.2]).

We have the following result.

Theorem 2 (see [11, Corollary 2.1]). *Let* $\Omega \subset \mathbb{R}^N$ $(N \ge 3)$ *be a bounded open set and for all* $i \in \{1,\ldots,N\}$, $p_i \in L^\infty(\Omega)$, $p_i(x) \ge 1$ *a.e. in* Ω. *Then, for any* $r \in L^\infty(\Omega)$ *with* $r(x) \ge 1$ *a.e. in* Ω *such that*

$$\operatorname*{ess\,inf}_{x\in\Omega}(p_M(x) - r(x)) > 0, \tag{26}$$

we have the compact embedding

$$W^{1,\vec{p}(\cdot)}(\Omega) \hookrightarrow L^{r(\cdot)}(\Omega). \tag{27}$$

Next, we define

$$\mathscr{T}^{1,\vec{p}(\cdot)}(\Omega) = \left\{u : \Omega \to \mathbb{R}; T_k(u) \in W^{1,\vec{p}(\cdot)}(\Omega), \forall k > 0\right\}. \tag{28}$$

Finally, in this paper, we will use the Marcinkiewicz spaces $\mathscr{M}^q(\Omega)$ $(1 < q < \infty)$ with constant exponent. Note that the Marcinkiewicz spaces $\mathscr{M}^{q(\cdot)}(\Omega)$ in the variable exponent setting were introduced for the first time by Sanchon and Urbano (see [23]).

Marcinkiewicz spaces $\mathscr{M}^q(\Omega)$ $(1 < q < \infty)$ contain the measurable functions $h : \Omega \to \mathbb{R}$ for which the distribution function

$$\lambda_h(\gamma) = \left|\{x \in \Omega : |h(x)| > \gamma\}\right|, \quad \gamma \ge 0 \tag{29}$$

satisfies an estimate of the form

$$\lambda_h(\gamma) \le C\gamma^{-q}, \quad \text{for some finite constant } C > 0. \tag{30}$$

The space $\mathscr{M}^q(\Omega)$ is a Banach space under the norm

$$\|h\|^*_{\mathscr{M}^q(\Omega)} = \sup_{t>0} t^{1/q}\left(\frac{1}{t}\int_0^t h^*(s)\, ds\right), \tag{31}$$

where h^* denotes the nonincreasing rearrangement of h:

$$h^*(t) = \inf\{\gamma > 0 : \lambda_h(\gamma) \le t\}. \tag{32}$$

We will use the following pseudonorm

$$\|h\|_{\mathscr{M}^q(\Omega)} = \inf\{C : \lambda_h(\gamma) \le C\gamma^{-q},\ \forall \gamma > 0\}, \tag{33}$$

which is equivalent to the norm $\|h\|^*_{\mathscr{M}^q(\Omega)}$ (see [27]).

We need the following Lemma (see [28, Lemma A.2]).

Lemma 3. *Let* $1 \le q < p < +\infty$. *Then, for every measurable function* u *on* Ω, *we have*

(i) $((p-1)^p/p^{p+1})\|u\|^p_{\mathscr{M}^p(\Omega)} \le \sup_{\lambda>0}\{\lambda^p \text{ meas}[x \in \Omega : |u(x)| > \lambda]\} \le \|u\|^p_{\mathscr{M}^p(\Omega)}$.

Moreover,

(ii) $\int_K |u|^q dx \le (p/(p-q))(p/q)^{q/p}\|u\|^q_{\mathscr{M}^p(\Omega)}(\text{meas}(K))^{p-q/p}$, *for every measurable subset* $K \subset \Omega$.

In particular, $\mathscr{M}^p(\Omega) \subset L^q_{\text{loc}}(\Omega)$ *with continuous injection and* $u \in \mathscr{M}^p(\Omega)$ *implies* $|u|^q \in \mathscr{M}^{p/q}(\Omega)$.

The following result is due to Troisi (see [30]).

Theorem 4. *Let $p_1, p_2, \ldots, p_N \in [1, +\infty)$; $g \in W^{1,(p_1,p_2,\ldots,p_N)}(\Omega)$ and let*

$$q = \overline{p}^* \quad \text{if } \overline{p}^* < N, \qquad q \in [1, +\infty) \quad \text{if } \overline{p}^* \geq N. \tag{34}$$

Then, there exists a constant $C > 0$ depending on $N, p_1, p_2, \ldots, p_N$ if $\overline{p} < N$ and also on q and $\text{meas}(\Omega)$ if $\overline{p} \geq N$ such that

$$\|g\|_{L^q(\Omega)} \leq C \prod_{i=1}^{N} \left[\|g\|_{L^{p_M}(\Omega)} + \left\| \frac{\partial g}{\partial x_i} \right\|_{L^{p_i}(\Omega)} \right]^{1/N}, \tag{35}$$

where $p_M = \max\{p_1, p_2, \ldots, p_N\}$ and $1/\overline{p} = (1/N) \sum_{i=1}^{N} (1/p_i)$.

We will use through the paper, the truncation function T_γ at height ($\gamma > 0$), that is

$$T_\gamma(s) = \begin{cases} s & \text{if } |s| \leq \gamma, \\ \gamma \operatorname{sign}(s) & \text{if } |s| > \gamma. \end{cases} \tag{36}$$

We need the following lemma.

Lemma 5. *Let g be a nonnegative function in $W^{1,\vec{p}(\cdot)}(\Omega)$. Assume $\overline{p} < N$ and there exists a constant $C > 0$ such that*

$$\int_\Omega |T_\gamma(g)|^{p_M^-} dx + \sum_{i=1}^{N} \int_{\{|g| \leq \gamma\}} \left| \frac{\partial g}{\partial x_i} \right|^{p_i^-} dx \leq C(\gamma + 1), \quad \forall \gamma > 0. \tag{37}$$

Then, there exists a constant D, depending on C, such that

$$\|g\|_{\mathcal{M}^{\tilde{p}}(\Omega)} \leq D, \tag{38}$$

where $\tilde{p} = N(\overline{p} - 1)/(N - \overline{p})$.

Proof. Consider the following

Step 1 ($\|T_\gamma(g)\|_{L^{p_M^-}(\Omega)} \leq 1$). Then, obviously we have $\|g\|_{\mathcal{M}^{\tilde{p}}(\Omega)} \leq D$, for some positive constant D. Indeed, since $1 < \tilde{p} \leq \overline{p} \leq p_M^-$, according to Proposition 1 there exists a positive constant C such that

$$\|T_\gamma(g)\|_{L^{\tilde{p}}(\Omega)} \leq C \|T_\gamma(g)\|_{L^{p_M^-}(\Omega)} \leq C. \tag{39}$$

It follows that there exists a positive constant D such that

$$\|g\|_{\mathcal{M}^{\tilde{p}}(\Omega)} \leq D. \tag{40}$$

Step 2 ($\|T_\gamma(g)\|_{L^{p_M^-}(\Omega)} > 1$). We get from (37)

$$\|T_\gamma(g)\|_{L^{p_M^-}(\Omega)}^{p_M^-} + \left\| \frac{\partial T_\gamma(g)}{\partial x_i} \right\|_{L^{p_i^-}(\Omega)}^{p_i^-} \leq C(\gamma + 1). \tag{41}$$

Not also that

$$\begin{aligned} &\left(\|T_\gamma(g)\|_{L^{p_M^-}(\Omega)} + \left\| \frac{\partial T_\gamma(g)}{\partial x_i} \right\|_{L^{p_i^-}(\Omega)} \right)^{p_i^-} \\ &\quad \leq 2^{(p_i^- - 1)} \left(\|T_\gamma(g)\|_{L^{p_M^-}(\Omega)}^{p_i^-} + \left\| \frac{\partial T_\gamma(g)}{\partial x_i} \right\|_{L^{p_i^-}(\Omega)}^{p_i^-} \right) \\ &\quad \leq 2^{(p_i^- - 1)} \left(\|T_\gamma(g)\|_{L^{p_M^-}(\Omega)}^{p_M^-} + \left\| \frac{\partial T_\gamma(g)}{\partial x_i} \right\|_{L^{p_i^-}(\Omega)}^{p_i^-} \right). \end{aligned} \tag{42}$$

Therefore, by using (35), we obtain for $\gamma > 1$,

$$\begin{aligned} \|T_\gamma(g)\|_{L^q(\Omega)} &\leq C \prod_{i=1}^{N} \left[2^{(p_i^- - 1)/Np_i^-} \gamma^{1/Np_i^-} \right] \\ &\leq D\gamma^{\sum_{i=1}^{N} (1/Np_i^-)} = D\gamma^{1/\overline{p}}. \end{aligned} \tag{43}$$

It follows that

$$\int_{\{|g| > \gamma\}} |T_\gamma(g)|^q dx \leq D\gamma^{q/\overline{p}} \tag{44}$$

which is equivalent to

$$\gamma^q \text{meas}(\{|g| > \gamma\}) \leq D\gamma^{q/\overline{p}}. \tag{45}$$

Therefore,

$$\text{meas}(\{|g| > \gamma\}) \leq D\gamma^{-q(\overline{p}-1)/\overline{p}}. \tag{46}$$

Since, $q = \overline{p}^* = N\overline{p}/(N - \overline{p})$ we get

$$\text{meas}(\{|g| > \gamma\}) \leq D\gamma^{-N(\overline{p}-1)/(N-\overline{p})} \tag{47}$$

which implies that $\|g\|_{\mathcal{M}^{\tilde{p}}(\Omega)} \leq D$.

For $0 < \gamma \leq 1$ we have

$$\text{meas}(\{|g| > \gamma\}) \leq \text{meas}(\Omega) \leq \text{meas}(\Omega)\gamma^{-\tilde{p}}. \tag{48}$$

So,

$$\|g\|_{\mathcal{M}^{\tilde{p}}(\Omega)} \leq D. \tag{49}$$

We need the following well-known results.

Theorem 6 (see [31, Theorem 6.2.1]). *Let X be a reflexive Banach space and let $f : M \subset X \to \mathbb{R}$ be Gateaux differentiable over the closed set M. Then, the following are equivalent.*

(i) *f is convex over M.*

(ii) *We have*

$$f(u) - f(v) \geq \langle f'(v), u - v \rangle_{X^* \times X} \quad \forall u, v \in M, \tag{50}$$

where X^* denotes the dual of the space X.

(iii) *The first Gateaux derivative is monotone, that is,*

$$\left\langle f'(u) - f'(v), u - v\right\rangle_{X^* \times X} \geq 0 \quad \forall u, v \in M. \tag{51}$$

(iv) *The second Gateaux derivative of f exists and it is positive, that is,*

$$\left\langle f''(u) \circ v, v\right\rangle_{X^* \times X} \geq 0 \quad \forall v \in M. \tag{52}$$

Theorem 7 (see [32, Theorem 1.2]). *Suppose X is a reflexive Banach space with norm $\|\cdot\|_X$, and let $M \subset X$ be a weakly closed subset of X. Suppose $\Psi : M \subset X \to \mathbb{R} \cup \{\infty\}$ is coercive and (sequentially) weakly lower semicontinuous on M with respect to X, that is, suppose the following conditions are fulfilled.*

(i) $\Psi(u) \to \infty$ *as* $\|u\|_X \to \infty$, $u \in M$.

(ii) *For any $u \in M$, any subsequence (u_m) in M such that $u_m \rightharpoonup u$ weakly in X there holds*

$$\Psi(u) \leq \liminf_{m \to \infty} \Psi(u_m). \tag{53}$$

Then, Ψ is bounded from below and attains its infinimun in M.

3. Main Results

In the sequel, we denote $W^{1,\vec{p}(\cdot)}(\Omega) = E$ and $\|\cdot\|_{W^{1,\vec{p}(\cdot)}(\Omega)} = \|\cdot\|_E$.

3.1. Weak Solutions. Let us define first the notion of weak solution.

Definition 8. Let $u : \Omega \to \mathbb{R}$ be a measurable function, we say that u is a weak solution of problem (1) if u belongs to $W^{1,\vec{p}(\cdot)}(\Omega)$ and satisfies the following equation:

$$\int_\Omega \sum_{i=1}^N a_i\left(x, \frac{\partial u}{\partial x_i}\right)\frac{\partial v}{\partial x_i}dx + \int_\Omega |u|^{p_M(x)-2}uv\,dx - \int_\Omega f(x)\,v\,dx = 0, \tag{54}$$

for every $v \in W^{1,\vec{p}(\cdot)}(\Omega)$.

We associate to problem (1) the energy functional $I : E \to \mathbb{R}$, defined by

$$I(u) = \int_\Omega \sum_{i=1}^N A_i\left(x, \frac{\partial u}{\partial x_i}\right)dx + \int_\Omega \frac{1}{p_M(x)}|u|^{p_M(x)}\,dx - \int_\Omega f(x)\,u\,dx. \tag{55}$$

To simplify our writing, we denote by $\Lambda : E \to \mathbb{R}$ the functional

$$\Lambda(u) = \int_\Omega \sum_{i=1}^N A_i\left(x, \frac{\partial u}{\partial x_i}\right)dx. \tag{56}$$

We recall the following result (see [15, Lemma 3.4]).

Lemma 9. *The functional Λ is well-defined on E. In addition, Λ is of class $\mathscr{C}^1(E, \mathbb{R})$ and*

$$\left\langle \Lambda'(u), v\right\rangle = \int_\Omega \sum_{i=1}^N a_i\left(x, \frac{\partial u}{\partial x_i}\right)\frac{\partial v}{\partial x_i}dx, \tag{57}$$

for all $u, v \in E$.

Due to Lemma 9, a standard calculus leads to the facts that I is well-defined on E and $I \in \mathscr{C}^1(E, \mathbb{R})$ with the derivative given by

$$\left\langle I'(u), v\right\rangle = \int_\Omega \sum_{i=1}^N a_i\left(x, \frac{\partial u}{\partial x_i}\right)\frac{\partial v}{\partial x_i}dx + \int_\Omega |u|^{p_M(x)-2}uv\,dx - \int_\Omega f(x)\,v\,dx \tag{58}$$

for all $u, v \in E$. Obviously, the weak solutions of (1) are the critical points of I; so by means of Theorem 7, we intend to prove the existence of critical points in order to deduce the existence of weak solutions.

Theorem 10. *Assume* (4)–(8) *and $f \in L^\infty(\Omega)$. Then, there exists a unique weak solution of problem* (1).

Let us start the proof by establishing some useful lemmas.

Lemma 11. *If hypotheses* (4)–(8) *are fulfilled, then the functional I is coercive.*

Proof. Let $u \in E$ be such that $\|u\|_E \to \infty$. Using (7), we deduce that

$$\Lambda(u) \geq \frac{1}{p_M^+}\sum_{i=1}^N \int_\Omega \left|\frac{\partial u}{\partial x_i}\right|^{p_i(x)} dx. \tag{59}$$

We make the following notations:

$$\mathcal{I}_1 = \left\{i \in \{1, \dots, N\} : \left|\frac{\partial u}{\partial x_i}\right|_{L^{p_i(\cdot)}(\Omega)} \leq 1\right\}, \quad \mathcal{I}_2 = \left\{i \in \{1, \dots, N\} : \left|\frac{\partial u}{\partial x_i}\right|_{L^{p_i(\cdot)}(\Omega)} > 1\right\}. \tag{60}$$

We then have

$$\Lambda(u) \geq \frac{1}{p_M^+}\sum_{i \in \mathcal{I}_1} \int_\Omega \left|\frac{\partial u}{\partial x_i}\right|^{p_i(x)} dx + \frac{1}{p_M^+}\sum_{i \in \mathcal{I}_2} \int_\Omega \left|\frac{\partial u}{\partial x_i}\right|^{p_i(x)} dx. \tag{61}$$

Using (19), (20), and (21), we have

$$\begin{aligned}\Lambda(u) &\geq \frac{1}{p_M^+}\sum_{i\in\mathcal{I}_1}\left|\frac{\partial u}{\partial x_i}\right|_{p_i(\cdot)}^{p_i^+} + \frac{1}{p_M^+}\sum_{i\in\mathcal{I}_2}\left|\frac{\partial u}{\partial x_i}\right|_{p_i(\cdot)}^{p_i^-} \\ &\geq \frac{1}{p_M^+}\sum_{i\in\mathcal{I}_2}\left|\frac{\partial u}{\partial x_i}\right|_{p_i(\cdot)}^{p_i^-} \\ &\geq \frac{1}{p_M^+}\sum_{i\in\mathcal{I}_2}\left|\frac{\partial u}{\partial x_i}\right|_{p_i(\cdot)}^{p_m^-} \\ &\geq \frac{1}{p_M^+}\left(\sum_{i=1}^{N}\left|\frac{\partial u}{\partial x_i}\right|_{p_i(\cdot)}^{p_m^-} - \sum_{i\in\mathcal{I}_1}\left|\frac{\partial u}{\partial x_i}\right|_{p_i(\cdot)}^{p_m^-}\right) \\ &\geq \frac{1}{p_M^+}\left(\sum_{i=1}^{N}\left|\frac{\partial u}{\partial x_i}\right|_{p_i(\cdot)}^{p_m^-} - N\right).\end{aligned} \tag{62}$$

By the generalized mean inequality or the Jensen's inequality applied to the convex function $z : \mathbb{R}^+ \to \mathbb{R}^+$, $z(t) = t^{p_m^-}$, $p_m^- > 1$, we get

$$\sum_{i=1}^{N}\left|\frac{\partial u}{\partial x_i}\right|_{p_i(\cdot)}^{p_m^-} \geq \frac{1}{N^{p_m^- - 1}}\left(\sum_{i=1}^{N}\left|\frac{\partial u}{\partial x_i}\right|_{p_i(\cdot)}\right)^{p_m^-}, \tag{63}$$

thus,

$$\Lambda(u) \geq \frac{1}{p_M^+}\left[\frac{1}{N^{p_m^- - 1}}\left(\sum_{i=1}^{N}\left|\frac{\partial u}{\partial x_i}\right|_{p_i(\cdot)}\right)^{p_m^-} - N\right]. \tag{64}$$

Case 1 ($|u|_{p_M(\cdot)} \geq 1$). We have

$$\begin{aligned}I(u) &\geq \frac{1}{p_M^+}\left[\frac{1}{N^{p_m^- - 1}}\left(\sum_{i=1}^{N}\left|\frac{\partial u}{\partial x_i}\right|_{p_i(\cdot)}\right)^{p_m^-} - N\right] \\ &\quad + \frac{1}{p_M^+}|u|_{p_M(\cdot)}^{p_M^-} - \|f\|_{L^\infty(\Omega)}\|u\|_{L^1(\Omega)} \\ &\geq \frac{1}{p_M^+}\left[\frac{1}{N^{p_m^- - 1}}\left(\sum_{i=1}^{N}\left|\frac{\partial u}{\partial x_i}\right|_{p_i(\cdot)}\right)^{p_m^-} + |u|_{p_M(\cdot)}^{p_m^-}\right] \\ &\quad - C\|f\|_{L^\infty(\Omega)}\|u\|_E - \frac{N}{p_M^+} \\ &\geq \frac{1}{2^{p_m^- - 1}p_M^+}\min\left(1, \frac{1}{N^{p_m^- - 1}}\right) \\ &\quad \cdot \left[\sum_{i=1}^{N}\left|\frac{\partial u}{\partial x_i}\right|_{p_i(\cdot)} + |u|_{p_M(\cdot)}\right]^{p_m^-} \\ &\quad - C\|f\|_{L^\infty(\Omega)}\|u\|_E - \frac{N}{p_M^+}.\end{aligned} \tag{65}$$

Therefore,

$$\begin{aligned}I(u) &\geq \frac{1}{2^{p_m^- - 1}p_M^+}\min\left(1, \frac{1}{N^{p_m^- - 1}}\right)\|u\|_E^{p_m^-} \\ &\quad - C\|f\|_{L^\infty(\Omega)}\|u\|_E - \frac{N}{p_M^+}.\end{aligned} \tag{66}$$

Case 2 ($|u|_{p_M(\cdot)} < 1$). Then $|u|_{p_M(\cdot)}^{p_m^-} - 1 < 0$, and we get

$$\begin{aligned}I(u) &\geq \frac{1}{p_M^+}\left[\frac{1}{N^{p_m^- - 1}}\left(\sum_{i=1}^{N}\left|\frac{\partial u}{\partial x_i}\right|_{p_i(\cdot)}\right)^{p_m^-} + |u|_{p_M(\cdot)}^{p_m^-} - N - 1\right] \\ &\quad - \|f\|_{L^\infty(\Omega)}\|u\|_{L^1(\Omega)} \\ &\geq \frac{1}{p_M^+}\left[\frac{1}{N^{p_m^- - 1}}\left(\sum_{i=1}^{N}\left|\frac{\partial u}{\partial x_i}\right|_{p_i(\cdot)}\right)^{p_m^-} + |u|_{p_M(\cdot)}^{p_m^-}\right] \\ &\quad - C\|f\|_{L^\infty(\Omega)}\|u\|_E - \frac{N+1}{p_M^+} \\ &\geq \frac{1}{2^{p_m^- - 1}p_M^+}\min\left(1, \frac{1}{N^{p_m^- - 1}}\right)\left[\sum_{i=1}^{N}\left|\frac{\partial u}{\partial x_i}\right|_{p_i(\cdot)} + |u|_{p_M(\cdot)}\right]^{p_m^-} \\ &\quad - C\|f\|_{L^\infty(\Omega)}\|u\|_E - \frac{N+1}{p_M^+}.\end{aligned} \tag{67}$$

So, we obtain

$$\begin{aligned}I(u) &\geq \frac{1}{2^{p_m^- - 1}p_M^+}\min\left(1, \frac{1}{N^{p_m^- - 1}}\right)\|u\|_E^{p_m^-} \\ &\quad - C\|f\|_{L^\infty(\Omega)}\|u\|_E - \frac{N+1}{p_M^+}.\end{aligned} \tag{68}$$

Then, letting $\|u\|_E$ goes to infinity in (66) and (68), we conclude that $I(u)$ reaches infinity. Thus, I is coercive.

Lemma 12. *The functional I is weakly lower semicontinuous.*

Proof. By [33, Corollary III.8], it is enough to show that I is lower semicontinuous. To this aim, fix $u \in E$ and $\epsilon > 0$. Since for every $i \in \{1, \ldots, N\}$, $a_i(x, \cdot)$ is monotone, Theorem 6 yields

$$\begin{aligned}&A_i\left(x, \frac{\partial v}{\partial x_i}\right) - A_i\left(x, \frac{\partial u}{\partial x_i}\right) \geq a_i\left(x, \frac{\partial u}{\partial x_i}\right)\left(\frac{\partial v}{\partial x_i} - \frac{\partial u}{\partial x_i}\right) \\ &\Longrightarrow \sum_{i=1}^{N}\int_\omega A_i\left(x, \frac{\partial v}{\partial x_i}\right)dx \geq \sum_{i=1}^{N}\int_\omega A_i\left(x, \frac{\partial u}{\partial x_i}\right)dx \\ &\quad + \sum_{i=1}^{N}\int_\omega a_i\left(x, \frac{\partial u}{\partial x_i}\right)\left(\frac{\partial v}{\partial x_i} - \frac{\partial u}{\partial x_i}\right)dx\end{aligned}$$

$$\Longrightarrow I(v) \geq I(u) + \sum_{i=1}^{N} \int_{\omega} a_i\left(x, \frac{\partial u}{\partial x_i}\right)\left(\frac{\partial v}{\partial x_i} - \frac{\partial u}{\partial x_i}\right) dx + \int_{\omega} \frac{1}{p_M(x)}\left(|v|^{p_M(x)} - |u|^{p_M(x)}\right) dx + \int_{\omega} f(x)(u-v)\, dx. \tag{69}$$

Since the map $t \mapsto t^{p_M(x)}$, $t > 0$ is convex, again by Theorem 6, we have

$$|v|^{p_M(x)} - |u|^{p_M(x)} \geq p_M(x)|u|^{p_M(x)-2}u(v-u), \tag{70}$$

then (69) becomes

$$I(v) \geq I(u) + \sum_{i=1}^{N} \int_{\Omega} a_i\left(x, \frac{\partial u}{\partial x_i}\right)\left(\frac{\partial v}{\partial x_i} - \frac{\partial u}{\partial x_i}\right) dx + \int_{\Omega} |u|^{p_M(x)-2}u(v-u)\, dx + \int_{\Omega} f(x)(u-v)\, dx. \tag{71}$$

Consider the second term in the right-hand side of (71). By (5) and Hölder type inequality, we have

$$\begin{aligned}
&\sum_{i=1}^{N} \int_{\Omega} a_i\left(x, \frac{\partial u}{\partial x_i}\right)\left(\frac{\partial v}{\partial x_i} - \frac{\partial u}{\partial x_i}\right) dx \\
&\geq -\sum_{i=1}^{N} \int_{\Omega} \left|a_i\left(x, \frac{\partial u}{\partial x_i}\right)\right| \left|\frac{\partial v}{\partial x_i} - \frac{\partial u}{\partial x_i}\right| dx \\
&\geq -\max\{C_1, \ldots, C_N\} \\
&\quad \cdot \sum_{i=1}^{N} \int_{\Omega} \left(j_i + \left|\frac{\partial u}{\partial x_i}\right|^{p_i(x)-1}\right) \left|\frac{\partial v}{\partial x_i} - \frac{\partial u}{\partial x_i}\right| dx \\
&\geq -K \sum_{i=1}^{N} \int_{\Omega} j_i(x) \left|\frac{\partial v}{\partial x_i} - \frac{\partial u}{\partial x_i}\right| dx \\
&\quad - K \sum_{i=1}^{N} \int_{\Omega} \left|\frac{\partial u}{\partial x_i}\right|^{p_i(x)-1} \left|\frac{\partial v}{\partial x_i} - \frac{\partial u}{\partial x_i}\right| dx \\
&\geq -K' \sum_{i=1}^{N} |j_i|_{p_i'(\cdot)} \left|\frac{\partial v}{\partial x_i} - \frac{\partial u}{\partial x_i}\right|_{p_i(\cdot)} \\
&\quad - K' \sum_{i=1}^{N} \left|\left|\frac{\partial u}{\partial x_i}\right|^{p_i(x)-1}\right|_{p_i'(\cdot)} \left|\frac{\partial v}{\partial x_i} - \frac{\partial u}{\partial x_i}\right|_{p_i(\cdot)} \\
&\geq -K' \max_i \left\{ |j_i|_{p_i'(\cdot)}, \left|\left|\frac{\partial u}{\partial x_i}\right|^{p_i(x)-1}\right|_{p_i'(\cdot)} \right\} \\
&\quad \cdot \sum_{i=1}^{N} \left|\frac{\partial v}{\partial x_i} - \frac{\partial u}{\partial x_i}\right|_{p_i(\cdot)} \geq -C_1 \sum_{i=1}^{N} \left|\frac{\partial u}{\partial x_i} - \frac{\partial v}{\partial x_i}\right|_{p_i(\cdot)}.
\end{aligned} \tag{72}$$

For the fourth term in the right-hand side of (71), we have

$$\begin{aligned}
\int_{\Omega} f(x)(u-v)\, dx &\geq -\int_{\Omega} |f(x)|\, |u-v|\, dx \\
&\geq \|f\|_{L^{\infty}(\Omega)} \|u-v\|_{L^1(\Omega)} \\
&\geq -C_2 \|u-v\|_E.
\end{aligned} \tag{73}$$

The third term in the right-hand side of (71) gives by using Hölder type inequality

$$\begin{aligned}
&\int_{\Omega} |u|^{p_M(x)-2}u(v-u)\, dx \\
&\geq -\int_{\Omega} |u|^{p_M(x)-1} |u-v|\, dx \\
&\geq -C' \left||u|^{p_M(x)-1}\right|_{p_M'(\cdot)} |u-v|_{p_M(\cdot)} \\
&\geq -C_3 |u-v|_{p_M(\cdot)}.
\end{aligned} \tag{74}$$

Gathering these inequalities, it follows that

$$I(v) \geq I(u) - C\|u-v\|_E \geq I(u) - \epsilon, \tag{75}$$

for every $v \in E$ such that $\|u-v\|_E < \epsilon/C$. Thus, I is lower semicontinuous.

Proof of Theorem 10. Consider the following

Step 1. Existence of weak solutions. The proof follows directly from Lemmas 11 and 12 and Theorem 7.

Step 2. Uniqueness of weak solution. Let $u, v \in E$ be two weak solutions of problem (1). Choosing a test function in (54), $\varphi = v-u$ for the weak solution u and $\varphi = u-v$ for the weak solution v, we get

$$\int_{\Omega} \sum_{i=1}^{N} a_i\left(x, \frac{\partial u}{\partial x_i}\right) \frac{\partial (v-u)}{\partial x_i}\, dx + \int_{\Omega} |u|^{p_M(x)-2}u(v-u)\, dx - \int_{\Omega} f(x)(v-u)\, dx = 0, \tag{76}$$

$$\int_{\Omega} \sum_{i=1}^{N} a_i\left(x, \frac{\partial v}{\partial x_i}\right) \frac{\partial (u-v)}{\partial x_i}\, dx + \int_{\Omega} |v|^{p_M(x)-2}v(u-v)\, dx - \int_{\Omega} f(x)(u-v)\, dx = 0. \tag{77}$$

Summing up (76) and (77), we obtain

$$\int_{\Omega} \sum_{i=1}^{N} \left(a_i\left(x, \frac{\partial u}{\partial x_i}\right) - a_i\left(x, \frac{\partial v}{\partial x_i}\right)\right) \frac{\partial (u-v)}{\partial x_i}\, dx + \int_{\Omega} \frac{1}{p_M(x)}\left(|u|^{p_M(x)-2}u - |v|^{p_M(x)-2}v\right)(u-v)\, dx = 0. \tag{78}$$

Thus, by the monotonicity of the functions $a_i(x,\cdot)$ and $t \mapsto |t|^{p_M(x)-2}t$, we deduce that $u = v$ almost everywhere.

3.2. Entropy Solutions. First of all, we define a space in which we will look for entropy solutions. We define the space $\mathcal{T}^{1,\vec{p}(\cdot)}(\Omega)$ as the set of every measurable function $u : \Omega \to \mathbb{R}$ which satisfies for every $k > 0$, $T_k(u) \in W^{1,\vec{p}(\cdot)}(\Omega)$.

Lemma 13 (see [34, 35]). *Let $u \in \mathcal{T}^{1,\vec{p}(\cdot)}(\Omega)$. Then, there exists a unique measurable function $v_i : \Omega \to \mathbb{R}$ such that*

$$v_i \chi_{\{|u|<k\}} = \frac{\partial T_k(u)}{\partial x_i} \quad \text{for a.e. } x \in \Omega,\ \forall k > 0,\ i \in \{1,\dots,N\}, \tag{79}$$

where χ_A denotes the characteristic function of a measurable set A. The functions v_i are called the weak partial gradients of u and are still denoted $\partial u/\partial x_i$. Moreover, if u belongs to $W^{1,\vec{p}(\cdot)}(\Omega)$, then $v_i \in L^{p_i(\cdot)}(\Omega)$ and coincides with the standard distributional gradient of u, that is, $v_i = \partial u/\partial x_i$.

Definition 14. We define the space $\mathcal{T}^{1,\vec{p}(\cdot)}_{\mathcal{H}}(\Omega)$ as the set of function $u \in \mathcal{T}^{1,\vec{p}(\cdot)}(\Omega)$ such that there exists a sequence $(u_n)_n \subset W^{1,\vec{p}(\cdot)}(\Omega)$ satisfying

(a) $u_n \to u$ a.e. in Ω,

(b) $\partial T_k(u_n)/\partial x_i \to \partial T_k(u)/\partial x_i$ in $L^1(\Omega)$, for all $k > 0$.

Definition 15. A measurable function u is an entropy solution of (1) if $u \in \mathcal{T}^{1,\vec{p}(\cdot)}_{\mathcal{H}}(\Omega)$ and for every $k > 0$,

$$\sum_{i=1}^{N} \int_\Omega a_i\left(x, \frac{\partial u}{\partial x_i}\right) \frac{\partial}{\partial x_i} T_k(u-\varphi)\,dx + \int_\Omega |u|^{p_M(x)-2} u T_k(u-\varphi)\,dx \le \int_\Omega f(x) T_k(u-\varphi)\,dx, \tag{80}$$

for all $\varphi \in W^{1,\vec{p}(\cdot)}(\Omega) \cap L^\infty(\Omega)$.

Our main result in this section is the following.

Theorem 16. *Assume (4)–(8) and $f \in L^1(\Omega)$. Then, there exists a unique entropy solution u to problem (1).*

Proof. The proof of this Theorem will be done in three steps.

Step 1 (a priori estimates).

Lemma 17. *Assume (4)–(8) and $f \in L^1(\Omega)$. Let u be an entropy solution of (1). If there exists a positive constant M such that*

$$\sum_{i=1}^{N} \int_{\{|u|>t\}} t^{q_i(x)}dx \le M, \quad \forall t > 0, \tag{81}$$

then

$$\sum_{i=1}^{N} \int_{\{|(\partial/\partial x_i)u|^{\alpha_i(\cdot)}>t\}} t^{q_i(x)}dx \le \|f\|_1 + M \quad \forall t > 0, \tag{82}$$

where $\alpha_i(\cdot) = p_i(\cdot)/(q_i(\cdot)+1)$, for all $i = 1,\dots,N$.

Proof. Take $\varphi = 0$ in (80), we have

$$\sum_{i=1}^{N} \int_\Omega a_i\left(x, \frac{\partial}{\partial x_i} T_t(u)\right) \cdot \frac{\partial}{\partial x_i} T_t(u)\,dx + \int_\Omega |u|^{p_M(x)-2} u T_t(u)\,dx \le \int_\Omega f(x) T_t(u)\,dx. \tag{83}$$

Since the second term in the previous inequality is nonnegative, it follows that

$$\sum_{i=1}^{N} \int_\Omega a_i\left(x, \frac{\partial}{\partial x_i} T_t(u)\right) \cdot \frac{\partial}{\partial x_i} T_t(u)\,dx \le \int_\Omega f(x) T_t(u)\,dx. \tag{84}$$

According to (7), we deduce that

$$\sum_{i=1}^{N} \int_\Omega \left|\frac{\partial}{\partial x_i} T_t(u)\right|^{p_i(x)} dx \le t\|f\|_1, \quad \forall t > 0. \tag{85}$$

Therefore, defining $\psi := T_t(u)/t$, we have for all $t > 0$,

$$\sum_{i=1}^{N} \int_\Omega t^{p_i(x)-1} \left|\frac{\partial}{\partial x_i}\psi\right|^{p_i(x)} dx = \sum_{i=1}^{N} \frac{1}{t} \int_\Omega \left|\frac{\partial}{\partial x_i} T_t(u)\right|^{p_i(x)} dx \le \|f\|_1. \tag{86}$$

From the previous inequality, the definition of $\alpha_i(\cdot)$ and (81), we have

$$\begin{aligned}
&\sum_{i=1}^{N} \int_{\{|(\partial/\partial x_i)u|^{\alpha_i(\cdot)}>t\}} t^{q_i(x)}dx \\
&\le \sum_{i=1}^{N} \int_{\{|(\partial/\partial x_i)u|^{\alpha_i(\cdot)}>t\}\cap\{|u|\le t\}} t^{q_i(x)}dx + \sum_{i=1}^{N} \int_{\{|u|>t\}} t^{q_i(x)}dx \\
&\le \sum_{i=1}^{N} \int_{\{|u|\le t\}} t^{q_i(x)} \left(\frac{|(\partial/\partial x_i)u|^{\alpha_i(x)}}{t}\right)^{p_i(x)/\alpha_i(x)} dx + M \\
&\le \sum_{i=1}^{N} \int_{\{|(\partial/\partial x_i)u|^{\alpha_i(\cdot)}>t;\ |u|\le t\}} t^{q_i(x)-(p_i(x)/\alpha_i(x))} \left|\frac{\partial u}{\partial x_i}\right|^{p_i(x)} dx + M \\
&\le \sum_{i=1}^{N} \frac{1}{t} \int_{\{|(\partial/\partial x_i)u|^{\alpha_i(\cdot)}>t;\ |u|\le t\}} \left|\frac{\partial u}{\partial x_i}\right|^{p_i(x)} dx + M \le \|f\|_1 + M.
\end{aligned} \tag{87}$$

Lemma 18. *Assume* (4)–(8) *and* $f \in L^1(\Omega)$. *Let* u *be an entropy solution of* (1), *then*

$$\frac{1}{h}\sum_{i=1}^{N}\int_{\{|u|\le h\}}\left|\frac{\partial}{\partial x_i}T_h(u)\right|^{p_i(x)}dx \le M \tag{88}$$

for every $h > 0$, *with* M *a positive constant. Moreover, we have*

$$\left\| |u|^{p_M(x)-2}u \right\|_1 = \left\| |u|^{p_M(x)-1} \right\|_1 \le \|f\|_1 \tag{89}$$

and there exists a constant $D > 0$ *which depends on* f *and* Ω *such that*

$$\text{meas}\,\{|u| > h\} \le \frac{D}{h^{p_M^- -1}}, \quad \forall h > 0. \tag{90}$$

Proof. Taking $\varphi = 0$ in the entropy inequality (80) and using (7), we obtain

$$\begin{gathered}\sum_{i=1}^{N}\int_{\{|u|\le h\}}\left|\frac{\partial}{\partial x_i}T_h(u)\right|^{p_i(x)}dx \le h\|f\|_1 \le Mh,\\ \int_\Omega |u|^{p_M(x)-2}uT_h(u)\,dx \le h\|f\|_1,\end{gathered} \tag{91}$$

for all $h > 0$. This yields

$$\int_{\{|u|>h\}} |u|^{p_M(x)-2}uT_h(u)\,dx \le h\|f\|_1. \tag{92}$$

As $uT_h(u)\chi_{\{|u|>h\}} = h|u|\chi_{\{|u|>h\}}$, we get from the previous inequality by using Fatou's lemma

$$\int_\Omega |u|^{p_M(x)-2}\,|u|\,dx \le \|f\|_1. \tag{93}$$

Now, since $|T_h(u)| \le |u|$ we have

$$\int_\Omega |T_h(u)|^{p_M(x)-1}dx \le \int_\Omega |u|^{p_M(x)-1}dx \le \|f\|_1. \tag{94}$$

We deduce that

$$\int_\Omega |T_h(u)|^{p_M^- -1}\,dx \le D(f,\Omega). \tag{95}$$

Indeed,

$$\begin{aligned}&\int_\Omega |T_h(u)|^{p_M^- -1}dx\\ &\quad\le \int_{\{|T_h(u)|\le 1\}} |T_h(u)|^{p_M^- -1}dx\\ &\quad\quad + \int_{\{|T_h(u)|>1\}} |T_h(u)|^{p_M^- -1}dx\\ &\quad\le \text{meas}(\Omega) + \int_\Omega |T_h(u)|^{p_M(x)-1}dx\\ &\quad\le \text{meas}(\Omega) + \|f\|_1.\end{aligned} \tag{96}$$

From aforementioned, we get

$$\int_{\{|u|>h\}} |T_h(u)|^{p_M^- -1}dx \le D(f,\Omega). \tag{97}$$

Therefore,

$$h^{p_M^- -1}\,\text{meas}\,\{|u| > h\} \le D(f,\Omega) \tag{98}$$

which implies

$$\text{meas}\,\{|u| > h\} \le \frac{D(f,\Omega)}{h^{p_M^- -1}}. \tag{99}$$

Lemma 19. *If* u *is an entropy solution of* (1) *then there exists a constant* $C > 0$ *such that*

$$\int_\Omega |T_k(u)|^{p_M^-}dx + \sum_{i=1}^{N}\int_{\{|u|\le k\}}\left|\frac{\partial u}{\partial x_i}\right|^{p_i^-}dx \le C(k+1), \quad \forall k > 0. \tag{100}$$

Proof. Taking $\varphi = 0$ in the entropy inequality (80) and using (7), we get

$$\begin{aligned}&\int_\Omega |u|^{p_M(x)-2}uT_k(u)\,dx\\ &\quad + \sum_{i=1}^{N}\int_{\{|u|\le k\}}\left|\frac{\partial}{\partial x_i}T_k(u)\right|^{p_i(x)}dx \le k\|f\|_1.\end{aligned} \tag{101}$$

Note that

$$\begin{aligned}&\sum_{i=1}^{N}\int_{\{|u|\le k\}}\left|\frac{\partial}{\partial x_i}T_k(u)\right|^{p_i^-}dx\\ &\quad= \sum_{i=1}^{N}\int_{\{|u|\le k, |\partial u/\partial x_i|\le 1\}}\left|\frac{\partial}{\partial x_i}T_k(u)\right|^{p_i^-}dx\\ &\quad\quad+ \sum_{i=1}^{N}\int_{\{|u|\le k, |\partial u/\partial x_i|> 1\}}\left|\frac{\partial}{\partial x_i}T_k(u)\right|^{p_i^-}dx\\ &\quad\le N\,\text{meas}(\Omega) + \sum_{i=1}^{N}\int_{\{|u|\le k, |\partial u/\partial x_i|> 1\}}\left|\frac{\partial}{\partial x_i}T_k(u)\right|^{p_i(x)}dx\\ &\quad\le N\,\text{meas}(\Omega) + \sum_{i=1}^{N}\int_{\{|u|\le k\}}\left|\frac{\partial}{\partial x_i}T_k(u)\right|^{p_i(x)}dx,\\ &\int_\Omega |T_k(u)|^{p_M^-}dx\\ &\quad\le \int_{\{|T_k(u)|\le 1\}}|T_k(u)|^{p_M^-}dx + \int_{\{|T_k(u)|> 1\}}|T_k(u)|^{p_M^-}\,dx\\ &\quad\le \text{meas}(\Omega) + \int_\Omega |T_k(u)|^{p_M(x)}dx\\ &\quad\le \text{meas}(\Omega) + \int_\Omega |u|^{p_M(x)-2}uT_k(u)\,dx.\end{aligned} \tag{102}$$

Therefore, we deduce according to (101) that

$$\int_{\Omega} |T_k(u)|^{p_M^-} dx + \sum_{i=1}^{N} \int_{\{|u|\le k\}} \left|\frac{\partial u}{\partial x_i}\right|^{p_i^-} dx \le (N+1)\operatorname{meas}(\Omega) + k\|f\|_1, \quad \forall k > 0. \tag{103}$$

Lemma 20. *If u is an entropy solution of (1) then*

$$\rho_{p'(\cdot)}\left(\left|\frac{\partial}{\partial x_i}u\right|^{p_i(x)-1} \chi_F\right) \le C, \quad \forall i = 1,\dots,N, \tag{104}$$

where $F = \{h < |u| \le h+t\}$, $h > 0$, $t > 0$.

Proof. Taking $\varphi = T_h(u)$ as a test function in the entropy inequality (80), we get

$$\sum_{i=1}^{N} \int_{\Omega} a_i\left(x, \frac{\partial}{\partial x_i}u\right) \cdot \frac{\partial}{\partial x_i} T_t(u - T_h(u))\, dx + \int_{\Omega} |u|^{p_M(x)-2} u T_t(u - T_h(u))\, dx \le \int_{\Omega} f(x) T_t(u - T_h(u))\, dx. \tag{105}$$

It follows by using (7) that

$$\int_F \left|\frac{\partial}{\partial x_i}u\right|^{p_i(x)} dx \le t\|f\|_1. \tag{106}$$

Therefore,

$$\rho_{p'(\cdot)}\left(\left|\frac{\partial}{\partial x_i}u\right|^{p_i(x)-1} \chi_F\right) \le C, \quad \forall i = 1,\dots,N. \tag{107}$$

Lemma 21. *If u is an entropy solution of (1) then*

$$\lim_{h\to+\infty} \int_{\Omega} |f|\, \chi_{\{|u|>h-t\}} dx = 0, \tag{108}$$

where $h > 0$, $t > 0$.

Proof. By Lemma 18, we deduce that

$$\lim_{h\to+\infty} |f|\, \chi_{\{|u|>h-t\}} = 0 \tag{109}$$

and as $f \in L^1(\Omega)$, it follows by using the Lebesgue dominated convergence theorem that

$$\lim_{h\to+\infty} \int_{\Omega} |f|\, \chi_{\{|u|>h-t\}} dx = 0. \tag{110}$$

The proof of the following lemma can be found in [1].

Lemma 22. *Assume (4)–(8) and $f \in L^1(\Omega)$. Let u be an entropy solution of (1), then*

$$\operatorname{meas}\left\{\left|\frac{\partial}{\partial x_i}u\right| > h\right\} \le \frac{D'}{h^{1/(p_M^-)'}}, \quad \forall h \ge 1,\ \forall i = 1,\dots,N, \tag{111}$$

where D' is a positive constant which depends on f and p_M^-.

Step 2 (uniqueness of entropy solution). The proof of the uniqueness of entropy solutions follows the same techniques by Ouaro [20] (see also [35]). Indeed, let $h > 0$ and u, v be two entropy solutions of (1). We write the entropy inequality (54) corresponding to the solution u, with $T_h(v)$ as test function, and to the solution v, with $T_h(u)$ as test function. Upon addition, we get

$$\begin{aligned}
&\int_{\{|u-T_h(v)|\le t\}} \sum_{i=1}^{N} a_i\left(x, \frac{\partial}{\partial x_i}u\right) \cdot \frac{\partial}{\partial x_i}(u - T_h(v))\, dx \\
&+ \int_{\{|v-T_h(u)|\le t\}} \sum_{i=1}^{N} a_i\left(x, \frac{\partial}{\partial x_i}v\right) \cdot \frac{\partial}{\partial x_i}(v - T_h(u))\, dx \\
&+ \int_{\Omega} |u|^{p_M(x)-2} u T_t(u - T_h(v))\, dx \\
&+ \int_{\Omega} |v|^{p_M(x)-2} v T_t(v - T_h(u))\, dx \\
&\le \int_{\Omega} f(x)\left(T_t(u - T_h(v)) + T_t(v - T_h(u))\right) dx.
\end{aligned} \tag{112}$$

Define

$$\begin{aligned}
E_1 &:= \{|u-v| \le t, |v| \le h\}, \\
E_2 &:= E_1 \cap \{|u| \le h\}, \\
E_3 &:= E_1 \cap \{|u| > h\}.
\end{aligned} \tag{113}$$

We start with the first integral in (112). By (7), we have

$$\begin{aligned}
&\int_{\{|u-T_h v|\le t\}} \sum_{i=1}^{N} a_i\left(x, \frac{\partial}{\partial x_i}u\right) \cdot \frac{\partial}{\partial x_i}(u - T_h(v))\, dx \\
&\ge \int_{E_2} \sum_{i=1}^{N} a_i\left(x, \frac{\partial}{\partial x_i}u\right) \cdot \frac{\partial}{\partial x_i}(u - v)\, dx \\
&- \int_{E_3} \sum_{i=1}^{N} a_i\left(x, \frac{\partial}{\partial x_i}u\right) \cdot \frac{\partial}{\partial x_i} v\, dx.
\end{aligned} \tag{114}$$

Using (5) and Proposition 1, we estimate the last integral in (114) as follows:

$$\begin{aligned}\left|\int_{E_3}\sum_{i=1}^{N}a_i\left(x,\frac{\partial}{\partial x_i}u\right)\cdot\frac{\partial}{\partial x_i}v\,dx\right| &\le C_1\int_{E_3}\sum_{i=1}^{N}\left(j_i(x)+\left|\frac{\partial}{\partial x_i}u\right|^{p_i(x)-1}\right)\left|\frac{\partial}{\partial x_i}v\right|dx \\ &\le C\sum_{i=1}^{N}\left(|j|_{p_i'(\cdot)}+\left\|\left|\frac{\partial}{\partial x_i}u\right|^{p_i(x)-1}\right|_{p_i'(\cdot),\{h<|u|\le h+t\}}\right) \\ &\quad\cdot\left|\frac{\partial}{\partial x_i}v\right|_{p_i(\cdot),\{h-t<|v|\le h\}},\end{aligned} \tag{115}$$

where

$$\begin{aligned}\left\|\left|\frac{\partial}{\partial x_i}u\right|^{p_i(x)-1}\right|_{p_i'(\cdot),\{h<|u|\le h+t\}} = \left\|\left|\frac{\partial}{\partial x_i}u\right|^{p_i(x)-1}\right\|_{L^{p_i'(\cdot)}(\{h<|u|\le h+t\})}.\end{aligned} \tag{116}$$

For all $i = 1,\dots,N$, the quantity $(|j_i|_{p_i'(\cdot)} + \||(\partial/\partial x_i)u|^{p_i(x)-1}|_{p_i'(\cdot),\{h<|u|\le h+t\}})$ is finite according to relations (18), (19) and Lemma 20. The quantity $|(\partial/\partial x_i)v|_{p_i(\cdot),\{h-t<|v|\le h\}}$ converges to zero as h goes to infinity according to Lemma 21. Then, the last expression in (115) converges to zero as h tends to infinity. Therefore, from (114), we obtain

$$\begin{aligned}\int_{\{|u-T_h v|\le t\}}\sum_{i=1}^{N}a_i\left(x,\frac{\partial}{\partial x_i}u\right)\cdot\frac{\partial}{\partial x_i}(u-T_h(v))\,dx \ge I_h+\int_{E_2}\sum_{i=1}^{N}a_i\left(x,\frac{\partial}{\partial x_i}u\right)\cdot\frac{\partial}{\partial x_i}(u-v)\,dx,\end{aligned} \tag{117}$$

where I_h converges to zero as h tends to infinity. We may adopt the same procedure to treat the second term in (112) to obtain

$$\begin{aligned}\int_{\{|v-T_h(u)|\le t\}}\sum_{i=1}^{N}a_i\left(x,\frac{\partial}{\partial x_i}v\right)\cdot\frac{\partial}{\partial x_i}(v-T_h(u))\,dx \ge J_h-\int_{E_2}\sum_{i=1}^{N}a_i\left(x,\frac{\partial}{\partial x_i}v\right)\cdot\frac{\partial}{\partial x_i}(u-v)\,dx,\end{aligned} \tag{118}$$

where J_h converges to zero as h tends to infinity.

For the two other terms in the left-hand side of (112), we denote

$$K_h=\int_\Omega|u|^{p_M(x)-2}uT_t(u-T_h(v))\,dx+\int_\Omega|v|^{p_M(x)-2}vT_t(v-T_h(u))\,dx. \tag{119}$$

We have $|u|^{p_M(x)-2}uT_t(u-T_h(v))\longrightarrow|u|^{p_M(x)-2}uT_t(u-v)$ a.e. as h goes to infinity and

$$|u|^{p_M(x)-2}uT_t(u-T_h(v))\,|\le t|u|^{p_M(x)-2}u\in L^1(\Omega). \tag{120}$$

Then, by the Lebesgue dominated convergence theorem, we obtain

$$\int_\Omega|u|^{p_M(x)-2}uT_t(u-T_h(v))\,dx\longrightarrow\int_\Omega|u|^{p_M(x)-2}uT_t(u-v)\,dx,\quad\text{as }h\longrightarrow\infty. \tag{121}$$

In the same way, we get

$$\int_\Omega|v|^{p_M(x)-2}vT_t(v-T_h(u))\,dx\longrightarrow\int_\Omega|v|^{p_M(x)-2}vT_t(v-u)\,dx,\quad\text{as }h\longrightarrow\infty. \tag{122}$$

Therefore,

$$\lim_{h\to\infty}K_h=\int_\Omega\left(|u|^{p_M(x)-2}u-|v|^{p_M(x)-2}v\right)T_t(u-v)\,dx. \tag{123}$$

Furthermore, consider the right-hand side of inequality (112). We have

$$\lim_{h\to\infty}\int_\Omega f(x)\left(T_t(u-T_h(v))+T_t(v-T_h(u))\right)dx=0. \tag{124}$$

Indeed,

$$\begin{aligned}&f(x)\left(T_t(u-T_h(v))+T_t(v-T_h(u))\right)\\&\quad\longrightarrow f(x)\left(T_t(u-v)+T_t(v-u)\right)=0\\&\qquad\text{a.e. in }\Omega\quad\text{as }h\longrightarrow\infty,\\&\left|f(x)\left(T_t(u-T_h(v))+T_t(v-T_h(u))\right)\right|\\&\quad\le 2t\,|f(x)|\in L^1(\Omega),\end{aligned} \tag{125}$$

so that we are able to apply the Lebesgue dominated convergence theorem. Then, we deduce from relations (112)–(124) after passing to the limit as $h\to\infty$ in (112) the following:

$$\begin{aligned}&\sum_{i=1}^{N}\int_{\{|u-v|\le t\}}\left(a_i\left(x,\frac{\partial}{\partial x_i}u\right)-a_i\left(x,\frac{\partial}{\partial x_i}v\right)\right)\cdot\frac{\partial}{\partial x_i}(u-v)\\&\quad+\int_\Omega\left(|u|^{p_M(x)-2}u-|v|^{p_M(x)-2}v\right)T_t(u-v)\le 0.\end{aligned} \tag{126}$$

Using (6) and as $t\mapsto|t|^{p_M(x)-2}t$ is monotone, we deduce from (126) that

$$\int_\Omega\left(|u|^{p_M(x)-2}u-|v|^{p_M(x)-2}v\right)T_t(u-v)\,dx\le 0. \tag{127}$$

Since $p_M^->1$, the following relation is true for any $\xi,\eta\in\mathbb{R}$, $\xi\ne\eta$ (cf. [12])

$$\left(|\xi|^{p_M(x)-2}\xi-|\eta|^{p_M(x)-2}\eta\right)(\xi-\eta)>0. \tag{128}$$

Therefore, from (127), we get that $(|u|^{p_M(x)-2}u - |v|^{p_M(x)-2}v)\, T_t(u-v) = 0$ a.e. in Ω, which means that for all $t \in \mathbb{R}^+$, there exists $\Omega_t \subset \Omega$ with meas $(\Omega_t) = 0$ such that for all $x \in \Omega \setminus \Omega_t$,

$$\left(|u|^{p_M(x)-2}u - |v|^{p_M(x)-2}v\right) T_t(u-v) = 0. \tag{129}$$

Therefore,

$$\left(|u|^{p_M(x)-2}u - |v|^{p_M(x)-2}v\right)(u-v) = 0, \quad \forall x \in \Omega \setminus \bigcup_{t\in\mathbb{N}^*} \Omega_t. \tag{130}$$

Now, using (128) and (130), we obtain

$$u_1 = u_2 \quad \text{a.e. in } \Omega. \tag{131}$$

Step 3 (Existence of entropy solutions). Let $(f_n)_{n\in\mathbb{N}^*}$ be a sequence of bounded functions, strongly converging to $f \in L^1(\Omega)$ and such that

$$\|f_n\|_1 \le \|f\|_1, \quad \forall n \in \mathbb{N}^*. \tag{132}$$

We consider the problem

$$\begin{gathered} -\sum_{i=1}^N \frac{\partial}{\partial x_i} a_i\left(x, \frac{\partial}{\partial x_i}u_n\right) + |u_n|^{p_M(x)-2}u_n = f_n \quad \text{in } \Omega, \\ \sum_{i=1}^N a_i\left(x, \frac{\partial}{\partial x_i}u_n\right)\nu_i = 0 \quad \text{on } \partial\Omega. \end{gathered} \tag{133}$$

It follows from Theorem 10 that problem (133) admits a unique weak solution $u_n \in W^{1,\vec{p}(\cdot)}(\Omega)$ which satisfies

$$\begin{aligned} \sum_{i=1}^N \int_\Omega a_i\left(x, \frac{\partial u_n}{\partial x_i}\right)\frac{\partial}{\partial x_i}\varphi dx + \int_\Omega |u_n|^{p_M(x)-2}u_n\varphi dx \\ = \int_\Omega f_n(x)\varphi dx, \end{aligned} \tag{134}$$

for all $\varphi \in W^{1,\vec{p}(\cdot)}(\Omega)$.

Our interest is to prove that these approximated solutions u_n tend, as n goes to infinity, to a measurable function u which is an entropy solution of the problem (1). We announce the following important lemma, useful to get some convergence results.

Lemma 23. *If u_n is a weak solution of* (126) *then there exist some constants $C_1, C_2 > 0$ such that*

(i) $\|u_n\|_{\mathcal{M}^{\tilde{p}}(\Omega)} \le C_1$,

(ii) $\|\partial u_n/\partial x_i\|_{\mathcal{M}^{p_i^- q/\bar{p}}(\Omega)} \le C_2$, *for all $i = 1, \ldots, N$.*

Proof. (i) is a consequence of Lemmas 19 and 5 by using $T_k(u_n)$ for all $k > 0$ as a test function in (134).

(ii) We first use $T_\gamma(u_n)$ for all $\gamma > 0$ as a test function in (134) to get

$$\sum_{i=1}^N \int_{\{|u|\le\gamma\}} \left|\frac{\partial u}{\partial x_i}\right|^{p_i^-} dx \le C(\gamma+1). \tag{135}$$

Then, let $\lambda_{|\partial u_n/\partial x_i|}(\alpha) = \text{meas}\{x \in \Omega : |\partial u_n/\partial x_i| > \alpha\}$ for all $i = 1, \ldots, N$, we have for any $\alpha > 1, \gamma > 0$,

$$\begin{aligned} \lambda_{|\partial u_n/\partial x_i|}(\alpha) &\le \text{meas}\left\{x \in \Omega : \left|\frac{\partial u_n}{\partial x_i}\right| > \alpha, |u_n| \le \gamma\right\} \\ &\quad + \text{meas}\left\{x \in \Omega : \left|\frac{\partial u_n}{\partial x_i}\right| > \alpha, |u_n| > \gamma\right\} \\ &\le \int_{\{|\partial u_n/\partial x_i|>\alpha, |u_n|\le\gamma\}} \left(\frac{1}{\alpha}\left|\frac{\partial u_n}{\partial x_i}\right|\right)^{p_i^-} dx + \lambda_{|u_n|}(\gamma) \\ &\le \frac{1}{\alpha^{p_i^-}} \int_{\{|u_n|\le\gamma\}} \left|\frac{\partial u_n}{\partial x_i}\right|^{p_i^-} dx + \lambda_{|u_n|}(\gamma). \end{aligned} \tag{136}$$

Using (135) and (i), we get

$$\lambda_{|\partial u_n/\partial x_i|}(\alpha) \le C\left(\frac{\gamma}{\alpha^{p_i^-}} + \gamma^{-\tilde{p}}\right), \tag{137}$$

from which we deduce (ii).

By lemmas 3 and 23, it follows that $(u_n)_{n\in\mathbb{N}^*}$ is uniformly bounded in $L^{s_0}(\Omega)$ for some $1 \le s_0 < \tilde{p}$, and in the same way, $(|\partial u_n/\partial x_i|)_{n\in\mathbb{N}^*}$ is uniformly bounded in $L^{s_i}(\Omega)$ for some $1 \le s_i < \tilde{p}_i$. From this, we get that the sequence $(u_n)_{n\in\mathbb{N}^*}$ is uniformly bounded in $W^{1,s}(\Omega)$, where $s = \min(s_0, s_1, \ldots, s_N)$. Consequently, we can extract a subsequence, still denoted (u_n) satisfying

$$\begin{gathered} u_n \longrightarrow u \quad \text{a.e. in } \Omega, \text{ in } L^s(\Omega), \\ u_n \rightharpoonup u \quad \text{in } W^{1,s}(\Omega), \\ \left|\frac{\partial u_n}{\partial x_i} - \frac{\partial u}{\partial x_i}\right| \rightharpoonup \mathcal{H}_i(x) \quad \text{in } L^s(\Omega), \ \forall i = 1, \ldots, N. \end{gathered} \tag{138}$$

By the same way as in the proof of [16, Lemma 3.5] (see also [27]), we prove that

$$\mathcal{H}_i(x) = 0 \quad \text{a.e. } x \in \Omega \ \forall i = 1, \ldots, N. \tag{139}$$

We deduce from (139) that

$$\begin{aligned} &a_i\left(x, \frac{\partial u_n}{\partial x_i}\right) \\ &\quad \longrightarrow a_i\left(x, \frac{\partial u}{\partial x_i}\right) \ \text{a.e. in } \Omega, \text{in } L^1(\Omega), \ \forall i = 1, \ldots, N. \end{aligned} \tag{140}$$

In order to pass to the limit in relation (134), we need also the following convergence results which can be proved by the same way as in [1].

Proposition 24. *Assume* (4)–(8), $f \in L^1(\Omega)$ *and* (132). *Let* $u_n \in W^{1,\vec{p}(\cdot)}(\Omega)$ *be the solution of* (133). *The sequence* $(u_n)_{n\in\mathbb{N}}$ *is Cauchy in measure. In particular, there exists a measurable function* u *and a subsequence still denoted by* u_n *such that* $u_n \to u$ *in measure.*

Proposition 25. *Assume* (4)–(8), $f \in L^1(\Omega)$ *and* (132). *Let* $u_n \in W^{1,\vec{p}(\cdot)}(\Omega)$ *be the solution of* (133). *The following assertions hold.*

(i) *For all* $i = 1, \ldots, N$, $\partial u_n/\partial x_i$ *converges in measure to the weak partial gradient of* u.

(ii) *For all* $i = 1, \ldots, N$ *and all* $k > 0$, $a_i(x, \partial T_k(u_n)/\partial x_i)$ *converges to* $a_i(x, \partial T_k(u_n)/\partial x_i)$ *in* $L^1(\Omega)$ *strongly and in* $L^{p_i'(\cdot)}(\Omega)$ *weakly.*

We can now pass to the limit in (134). To this end, let $\varphi \in W^{1,\vec{p}(\cdot)}(\Omega) \cap L^\infty(\Omega)$. For any $k > 0$, choose $T_k(u_n - \varphi)$ as a test function in (134), we get

$$\sum_{i=1}^{N} \int_\Omega a_i\left(x, \frac{\partial u_n}{\partial x_i}\right) \frac{\partial}{\partial x_i} T_k(u_n - \varphi)\, dx + \int_\Omega |u_n|^{p_M(x)-2} u_n T_k(u_n - \varphi)\, dx = \int_\Omega f_n(x) T_k(u_n - \varphi)\, dx. \tag{141}$$

For the right-hand side of (141), the convergence is obvious since f_n converges strongly to f in $L^1(\Omega)$, and $T_k(u_n - \varphi)$ converges weakly-$*$ to $T_k(u - \varphi)$ in $L^\infty(\Omega)$ and a.e in Ω.

For the second term of (141), we have

$$\int_\Omega |u_n|^{p_M(x)-2} u_n T_k(u_n - \varphi)\, dx = \int_\Omega \left(|u_n|^{p_M(x)-2} u_n - |\varphi|^{p_M(x)-2}\varphi\right) T_k(u_n - \varphi)\, dx + \int_\Omega |\varphi|^{p_M(x)-2}\varphi T_k(u_n - \varphi)\, dx. \tag{142}$$

The quantity $(|u_n|^{p_M(x)-2}u_n - |\varphi|^{p_M(x)-2}\varphi)T_k(u_n - \varphi)$ is nonnegative and since for all $x \in \Omega, s \mapsto |s|^{p_M(x)-2}s$ is continuous; we get

$$\left(|u_n|^{p_M(x)-2} u_n - |\varphi|^{p_M(x)-2}\varphi\right) T_k(u_n - \varphi) \longrightarrow \left(|u|^{p_M(x)-2} u - |\varphi|^{p_M(x)-2}\varphi\right) T_k(u - \varphi) \quad \text{a.e. in } \Omega. \tag{143}$$

Then, it follows by Fatou's Lemma that

$$\liminf_{n\to+\infty} \int_\Omega \left(|u_n|^{p_M(x)-2} u_n - |\varphi|^{p_M(x)-2}\varphi\right) T_k(u_n - \varphi)\, dx \geq \int_\Omega \left(|u|^{p_M(x)-2} u - |\varphi|^{p_M(x)-2}\varphi\right) T_k(u - \varphi)\, dx. \tag{144}$$

Let us show that $|\varphi|^{p_M(x)-2}\varphi \in L^1(\Omega)$.

We have

$$\int_\Omega \left||\varphi|^{p_M(x)-2}\varphi\right| dx = \int_\Omega |\varphi|^{p_M(x)-1} dx \leq \int_\Omega \left(\|\varphi\|_\infty\right)^{p_M(x)-1} dx. \tag{145}$$

If $\|\varphi\|_\infty \leq 1$, then $\int_\Omega ||\varphi|^{p_M(x)-2}\varphi| dx \leq \text{meas}(\Omega) < +\infty$.

If $\|\varphi\|_\infty > 1$, then

$$\int_\Omega \left||\varphi|^{p_M(x)-2}\varphi\right| dx \leq \int_\Omega \left(\|\varphi\|_\infty\right)^{p_M^+-1} dx = \left(\|\varphi\|_\infty\right)^{p_M^+-1} \text{meas}(\Omega) < +\infty. \tag{146}$$

Hence, $|\varphi|^{p_M(x)-2}\varphi \in L^1(\Omega)$.

Since $T_k(u_n - \varphi)$ converges weakly-$*$ to $T_k(u - \varphi)$ in $L^\infty(\Omega)$ and $|\varphi|^{p_M(x)-2}\varphi \in L^1(\Omega)$, it follows that

$$\lim_{n\to+\infty} \int_\Omega |\varphi|^{p_M(x)-2}\varphi T_k(u_n - \varphi)\, dx = \int_\Omega |\varphi|^{p_M(x)-2}\varphi T_k(u - \varphi)\, dx. \tag{147}$$

For the first term of (141), we write it as follows:

$$\sum_{i=1}^{N} \int_{\{|u_n - \varphi| \leq k\}} a_i\left(x, \frac{\partial u_n}{\partial x_i}\right) \frac{\partial}{\partial x_i} u_n dx - \sum_{i=1}^{N} \int_{\{|u_n - \varphi| \leq k\}} a_i\left(x, \frac{\partial u_n}{\partial x_i}\right) \frac{\partial}{\partial x_i} \varphi dx. \tag{148}$$

The first term of (148) is nonnegative by (7), then by Fatou's Lemma and (138), we get

$$\sum_{i=1}^{N} \int_{\{|u - \varphi| \leq k\}} a_i\left(x, \frac{\partial u}{\partial x_i}\right) \frac{\partial}{\partial x_i} u dx \leq \liminf_{n\to\infty} \sum_{i=1}^{N} \int_{\{|u_n - \varphi| \leq k\}} a_i\left(x, \frac{\partial u_n}{\partial x_i}\right) \frac{\partial}{\partial x_i} u_n dx. \tag{149}$$

According to Proposition 25, the second term of (148) converges to

$$\sum_{i=1}^{N} \int_{\{|u - \varphi| \leq k\}} a_i\left(x, \frac{\partial u}{\partial x_i}\right) \frac{\partial}{\partial x_i} \varphi dx. \tag{150}$$

Combining the previous convergence results, we obtain

$$\sum_{i=1}^{N} \int_\Omega a_i\left(x, \frac{\partial u}{\partial x_i}\right) \frac{\partial}{\partial x_i} T_k(u - \varphi)\, dx + \int_\Omega |u|^{p_M(x)-2} u T_k(u - \varphi)\, dx \leq \int_\Omega f(x) T_k(u - \varphi)\, dx. \tag{151}$$

References

[1] B. K. Bonzi and S. Ouaro, "Entropy solutions for a doubly nonlinear elliptic problem with variable exponent," *Journal of Mathematical Analysis and Applications*, vol. 370, no. 2, pp. 392–405, 2010.

[2] M.-M. Boureanu and V. D. Rădulescu, "Anisotropic Neumann problems in Sobolev spaces with variable exponent," *Nonlinear Analysis: Theory, Methods & Applications*, vol. 75, no. 12, pp. 4471–4482, 2012.

[3] Y. Chen, S. Levine, and M. Rao, "Variable exponent, linear growth functionals in image restoration," *SIAM Journal on Applied Mathematics*, vol. 66, no. 4, pp. 1383–1406, 2006.

[4] L. Diening, *Theoretical and numerical results for electrorheological fluids [Ph.D. thesis]*, University of Frieburg, Freiburg, Germany, 2002.

[5] L. Diening, "Riesz potential and Sobolev embeddings on generalized Lebesgue and Sobolev spaces $L^{p(.)}$ and $W^{1,p(.)}$," *Mathematische Nachrichten*, vol. 268, pp. 31–43, 2004.

[6] D. E. Edmunds and J. Rákosník, "Density of smooth functions in $W^{k,p(x)}(\Omega)$," *Proceedings of the Royal Society of London. Series A*, vol. 437, no. 1899, pp. 229–236, 1992.

[7] D. E. Edmunds and J. Rákosník, "Sobolev embeddings with variable exponent," *Studia Mathematica*, vol. 143, no. 3, pp. 267–293, 2000.

[8] D. E. Edmunds and J. Rákosník, "Sobolev embeddings with variable exponent. II," *Mathematische Nachrichten*, vol. 246/247, pp. 53–67, 2002.

[9] A. El Hamidi, "Existence results to elliptic systems with nonstandard growth conditions," *Journal of Mathematical Analysis and Applications*, vol. 300, no. 1, pp. 30–42, 2004.

[10] X. Fan and D. Zhao, "On the spaces $L^{p(.)}(\Omega)$ and $W^{1,p(.)}(\Omega)$," *Journal of Mathematical Analysis and Applications*, vol. 263, no. 2, pp. 424–446, 2001.

[11] X. Fan, "Anisotropic variable exponent Sobolev spaces and $\vec{p}(.)$-Laplacian equations," *Complex Variables and Elliptic Equations*, vol. 55, pp. 1–20, 2010.

[12] X.-L. Fan and Q.-H. Zhang, "Existence of solutions for $p(x)$-Laplacian Dirichlet problem," *Nonlinear Analysis: Theory, Methods & Applications*, vol. 52, no. 8, pp. 1843–1852, 2003.

[13] P. Harjulehto, P. Hästö, Út V. Lê, and M. Nuortio, "Overview of differential equations with non-standard growth," *Nonlinear Analysis: Theory, Methods & Applications*, vol. 72, no. 12, pp. 4551–4574, 2010.

[14] P. Harjulehto, P. Hästö, M. Koskenoja, and S. Varonen, "The Dirichlet energy integral and variable exponent Sobolev spaces with zero boundary values," *Potential Analysis*, vol. 25, no. 3, pp. 205–222, 2006.

[15] B. Koné, S. Ouaro, and S. Traoré, "Weak solutions for anisotropic nonlinear elliptic equations with variable exponents," *Electronic Journal of Differential Equations*, vol. 2009, no. 144, pp. 1–11, 2009.

[16] B. Koné, S. Ouaro, and S. Soma, "Weak solutions for anisotropic nonlinear elliptic problem with variable exponent and measure data," *International Journal of Evolution Equations*, vol. 5, no. 3, p. 327–350 (2011), 2010.

[17] O. Kováčik and J. Rákosník, "On spaces $L^{p(x)}$ and $W^{1,p(x)}$," *Czechoslovak Mathematical Journal*, vol. 41(116), no. 4, pp. 592–618, 1991.

[18] M. Mihăilescu, P. Pucci, and V. Rădulescu, "Eigenvalue problems for anisotropic quasilinear elliptic equations with variable exponent," *Journal of Mathematical Analysis and Applications*, vol. 340, no. 1, pp. 687–698, 2008.

[19] M. Mihăilescu and V. Rădulescu, "A multiplicity result for a nonlinear degenerate problem arising in the theory of electrorheological fluids," *Proceedings of the Royal Society of London. Series A*, vol. 462, no. 2073, pp. 2625–2641, 2006.

[20] S. Ouaro, "Well-posedness results for anisotropic nonlinear elliptic equations with variable exponent and L^1-data," *Cubo*, vol. 12, no. 1, pp. 133–148, 2010.

[21] K. R. Rajagopal and M. Ruzicka;, "Mathematical modelling of electrorheological fluids," *Continuum Mechanics and Thermodynamics*, vol. 13, pp. 59–78, 2001.

[22] M. Ruzicka, *Electrorheological Fluids: Modeling and Mathematical Theory*, vol. 1748 of *Lecture Notes in Mathematics*, Springer, Berlin, Germany, 2000.

[23] M. Sanchón and J. M. Urbano, "Entropy solutions for the $p(x)$-Laplace equation," *Transactions of the American Mathematical Society*, vol. 361, no. 12, pp. 6387–6405, 2009.

[24] D. Stancu-Dumitru, "Multiplicity of solutions for anisotropic quasilinear elliptic equations with variable exponents," *Bulletin of the Belgian Mathematical Society*, vol. 17, no. 5, pp. 875–889, 2010.

[25] D. Stancu-Dumitru, "Two nontrivial solutions for a class of anisotropic variable exponent problems," *Taiwanese Journal of Mathematics*, vol. 16, no. 4, pp. 1205–1219, 2012.

[26] D. Stancu-Dumitru, "Multiplicity of solutions for a nonlinear degenerate problem in anisotropic variable exponent spaces," *Bulletin of the Malaysian Mathematical Sciences Society*, vol. 36, no. 21, pp. 117–130, 2013.

[27] M. Bendahmane and K. H. Karlsen, "Anisotropic nonlinear elliptic systems with measure data and anisotropic harmonic maps into spheres," *Electronic Journal of Differential Equations*, vol. 2006, no. 46, pp. 1–30, 2006.

[28] P. Benilan, H. Brezis, and M. G. Crandall, "A semilinear equation in $L^1(\mathbb{R})^N$," *Annali della Scuola Normale Superiore di Pisa*, vol. 2, no. 4, pp. 523–555, 1975.

[29] L. Diening, P. Harjulehto, P. Hästö, and M. Ruzicka, *Lebesgue and Sobolev Spaces with Variable Exponents*, vol. 2017 of *Lecture Notes in Mathematics*, Springer, Heidelberg, Germany, 2011.

[30] M. Troisi, "Teoremi di inclusione per spazi di Sobolev non isotropi," *Ricerche di Matematica*, vol. 18, pp. 3–24, 1969.

[31] A. J. Kurdila and M. Zabarankin, *Convex Functional Analysis*, Systems & Control: Foundations & Applications, Birkhäuser, Basel, Switzerland, 2005.

[32] M. Struwe, *Variational Methods: Applications to Nonlinear Partial Dierential Equations and Hamiltonian Systems*, vol. 34 of *Results in Mathematics and Related Areas (3)*, Springer, Berlin, Germany, 2nd edition, 1996.

[33] H. Brezis, *Analyse Fonctionnelle: Théorie et Applications*, Collection Mathématiques Appliquées pour la Maîtrise, Masson, Paris, Farnce, 1983.

[34] A. Alvino, L. Boccardo, V. Ferone, L. Orsina, and G. Trombetti, "Existence results for nonlinear elliptic equations with degenerate coercivity," *Annali di Matematica Pura ed Applicata*, vol. 182, no. 1, pp. 53–79, 2003.

[35] P. Bénilan, L. Boccardo, T. Gallouët, R. Gariepy, M. Pierre, and J. L. Vázquez, "An L^1-theory of existence and uniqueness of solutions of nonlinear elliptic equations," *Annali della Scuola Normale Superiore di Pisa*, vol. 22, no. 2, pp. 241–273, 1995.

5

Nonlocal Problems for Fractional Differential Equations via Resolvent Operators

Zhenbin Fan[1] and Gisèle Mophou[2]

[1] *Department of Mathematics, Changshu Institute of Technology, Suzhou, Jiangsu 215500, China*
[2] *Laboratoire CEREGMIA, Université des Antilles et de la Guyane, Campus Fouillole, Guadeloupe (FWI) 97159 Pointe-à-Pitre, France*

Correspondence should be addressed to Gisèle Mophou; gmophou@univ-ag.fr

Academic Editor: Sotiris Ntouyas

We discuss the continuity of analytic resolvent in the uniform operator topology and then obtain the compactness of Cauchy operator by means of the analytic resolvent method. Based on this result, we derive the existence of mild solutions for nonlocal fractional differential equations when the nonlocal item is assumed to be Lipschitz continuous and neither Lipschitz nor compact, respectively. An example is also given to illustrate our theory.

1. Introduction

In this paper, we are concerned with the existence of mild solutions for a fractional differential equation with nonlocal conditions of the form

$$D^{\alpha}u(t) = Au(t) + J_t^{1-\alpha} f(t, u(t)), \quad 0 < t \le b,$$
$$u(0) = u_0 - g(u), \tag{1}$$

where D^{α} is the Caputo fractional derivative of order α with $0 < \alpha < 1$, $A : D(A) \subset X \to X$ is the infinitesimal generator of a resolvent $S_{\alpha}(t)$, $t \ge 0$, X is a real Banach space endowed with the norm $\|\cdot\|$, and f and g are appropriate continuous functions to be specified later.

The theory of fractional differential equations has received much attention over the past twenty years, since they are important in describing the natural models such as diffusion processes, stochastic processes, finance, and hydrology. Many notions associated with resolvent are developed such as integral resolvent, solution operators, α-resolvent operator functions, (a,k)-regularized resolvent, and α-order fractional semigroups. All of these notions play a central role in the study of Volterra equations, especially the fractional differential equations. Concerning the literature, we refer the reader to the books [1, 2], the recent papers [3–20], and the references therein.

On the other hand, abstract differential equations with nonlocal conditions have also been studied extensively in the literature, since it is demonstrated that the nonlocal problems have better effects in applications than the classical ones. It was Byszewski and Lakshmikantham [21] who first studied the existence and uniqueness of mild solutions for nonlocal differential equations. And the main difficulty in dealing with the nonlocal problem is how to get the compactness of solution operator at zero, especially when the nonlocal item is only assumed to be Lipschitz continuous or continuous. Many authors developed different techniques and methods to solve this problem. For more details on this topic, we refer to [10, 11, 22–33] and references therein.

In this paper, we combine the above two directions and study the nonlocal fractional differential equation (1) governed by operator A generating an analytic resolvent. A standard approach in deriving the mild solution of (1) is to define the solution operator Q. Then, conditions are given such that some fixed point theorems such as Browder's and Schauder's fixed point theorems can be applied to get a fixed point for solution operator Q, which gives rise to a mild solution of (1). The key step of using this approach is to prove the compactness of Cauchy operator G associated with solution operator Q. When the operator A generates a compact semigroup, it is well known that the Cauchy operator G is also compact. However, to the best of our knowledge, it is unknown when A generates a compact resolvent. The main difficulty of this problem lies in the fact that there is no property of semigroups for resolvent.

To this end, we will first discuss the continuity of resolvent in the uniform operator topology in this paper. In fact, we prove that the compact analytic resolvent is continuous in the uniform operator topology. Based on this result, we can prove the compactness of Cauchy operator. As a consequence, we obtain the existence of mild solutions for (1) when the nonlocal item is Lipschitz continuous. At the same time, we also derive the existence of mild solutions for (1) without the Lipschitz or compact assumption on the nonlocal item g by using the techniques developed in [24, 30]. Actually, we only assume that g is continuous on $C([0,b],X)$ and g is completely determined on $[\delta,b]$ for some small $\delta > 0$ or g is continuous on $C([0,b],X)$ with $L^1([0,b],X)$ topology (see Corollaries 17–19).

This paper has four sections. In Section 2, we recall some definitions on Caputo fractional derivatives, analytic resolvent, and mild solutions to (1). In Section 3, we prove the compactness of Cauchy operator. Finally, in Section 4 we establish the existence of mild solutions of (1) when the nonlocal item satisfies different conditions. An example is also given in this section.

2. Preliminaries

Throughout this paper, let $b > 0$ be fixed, and let $\mathbb{N}$, $\mathbb{R}$, and $\mathbb{R}_+$ be the set of positive integers, real numbers, and nonnegative real numbers, respectively. We denote by X the Banach space with the norm $\|\cdot\|$, $C([0,b],X)$ the space of all X-valued continuous functions on $[0,b]$ with the norm $\|u\| = \sup\{\|u(t)\|, t \in [0,b]\}$, and $L^p([0,b],X)$ the space of X-valued Bochner integrable functions on $[0,b]$ with the norm $\|f\|_{L^p} = (\int_0^b \|f(t)\|^p dt)^{1/p}$, where $1 \le p < \infty$. Also, we denote by $\mathscr{L}(X)$ the space of bounded linear operators from X into X endowed with the norm of operators.

Now, let us recall some basic definitions and results on fractional derivative and fractional differential equations.

Definition 1 (see [1]). The fractional order integral of the function $f \in L^1([0,b],X)$ of order $\alpha > 0$ is defined by

$$J_t^\alpha f(t) = \frac{1}{\Gamma(\alpha)} \int_0^t (t-s)^{\alpha-1} f(s)\,ds, \tag{2}$$

where Γ is the Gamma function.

Definition 2 (see [1]). The Riemann-Liouville fractional order derivative of order α of a function $f \in L^1([0,b],X)$ given on the interval $[0,b]$ is defined by

$$D_L^\alpha f(t) = \frac{1}{\Gamma(n-\alpha)} \frac{d^n}{dt^n} \int_0^t (t-s)^{n-\alpha-1} f(s)\,ds, \tag{3}$$

where $\alpha \in (n-1,n]$, $n \in \mathbb{N}$.

Definition 3 (see [1]). The Caputo fractional order derivative of order α of a function $f \in L^1([0,b],X)$ given on the interval $[0,b]$ is defined by

$$D^\alpha f(t) = \frac{1}{\Gamma(n-\alpha)} \int_0^t (t-s)^{n-\alpha-1} f^{(n)}(s)\,ds, \tag{4}$$

where $\alpha \in (n-1,n]$, $n \in \mathbb{N}$.

In the remainder of this paper, we always suppose that $0 < \alpha < 1$ and A is a closed and densely defined linear operator on X.

Definition 4. A family $\{S_\alpha(t)\}_{t\ge0} \subseteq \mathscr{L}(X)$ of bounded linear operators in X is called a resolvent (or solution operator) generating by A if the following conditions are satisfied:

(S1) $S_\alpha(t)$ is strong continuous on $\mathbb{R}_+$ and $S_\alpha(0) = I$;

(S2) $S_\alpha(t)D(A) \subseteq D(A)$ and $AS_\alpha(t)x = S_\alpha(t)Ax$ for all $x \in D(A)$ and $t \ge 0$;

(S3) the resolvent equation holds

$$S_\alpha(t)x = x + \int_0^t g_\alpha(t-s) AS_\alpha(s)x\,ds \quad \forall x \in D(A),\ t \ge 0. \tag{5}$$

For $\omega, \theta \in \mathbb{R}$, let

$$\sum(\omega,\theta) := \{\lambda \in \mathbb{C} : |\arg(\lambda-\omega)| < \theta\}. \tag{6}$$

Definition 5. A resolvent $S_\alpha(t)$ is called analytic, if the function $S_\alpha(\cdot) : \mathbb{R}_+ \to \mathscr{L}(X)$ admits analytic extension to a sector $\sum(0,\theta_0)$ for some $0 < \theta_0 \le \pi/2$. An analytic resolvent $S_\alpha(t)$ is said to be of analyticity type (ω_0,θ_0) if for each $\theta < \theta_0$ and $\omega > \omega_0$ there is $M_1 = M_1(\omega,\theta)$ such that $\|S(z)\| \le M_1 e^{\omega \operatorname{Re} z}$ for $z \in \sum(0,\theta)$, where $\operatorname{Re} z$ denotes the real part of z.

Definition 6. A resolvent $S_\alpha(t)$ is called compact for $t > 0$ if for every $t > 0$, $S_\alpha(t)$ is a compact operator.

According to Proposition 1.2 in [2], we can give the following definition of mild solutions for (1).

Definition 7. A function $u \in C([0,b],X)$ is called a mild solution of fractional evolution equation (1) if it satisfies

$$u(t) = S_\alpha(t)[u_0 - g(u)] + \int_0^t S_\alpha(t-s) f(s,u(s))\,ds, \quad 0 \le t \le b \tag{7}$$

for every $u_0 \in X$.

Next, we introduce the Hausdorff measure of noncompactness $\beta(\cdot)$ defined on each bounded subset Ω of Banach space Y by

$$\beta(\Omega) = \inf\{\varepsilon > 0;\ \Omega \text{ has a finite } \varepsilon\text{-net in } Y\}. \tag{8}$$

Some basic properties of $\beta(\cdot)$ are given in the following lemma.

Lemma 8 (see [34]). *Let Y be a real Banach space and let $B, C \subseteq Y$ be bounded; then the following properties are satisfied:*

(1) *B is precompact if and only if $\beta(B) = 0$;*

(2) *$\beta(B) = \beta(\overline{B}) = \beta(\text{conv } B)$, where $\overline{B}$ and conv B mean the closure and convex hull of B, respectively;*

(3) *$\beta(B) \leq \beta(C)$ when $B \subseteq C$;*

(4) *$\beta(B + C) \leq \beta(B) + \beta(C)$, where $B + C = \{x + y; x \in B, y \in C\}$;*

(5) *$\beta(B \cup C) \leq \max\{\beta(B), \beta(C)\}$;*

(6) *$\beta(\lambda B) = |\lambda|\beta(B)$ for any $\lambda \in \mathbb{R}$;*

(7) *if the map $Q : D(Q) \subseteq Y \to Z$ is Lipschitz continuous with constant k, then $\beta(QB) \leq k\beta(B)$ for any bounded subset $B \subseteq D(Q)$, where Z is a Banach space.*

The map $Q : W \subseteq Y \to Y$ is said to be a β-contraction if there exists a positive constant $k < 1$ such that $\beta(QC) \leq k\beta(C)$ for any bounded closed subset $C \subseteq W$, where Y is a Banach space.

Lemma 9 ([34], Darbo-Sadovskii). *If $W \subseteq Y$ is bounded closed and convex, the continuous map $Q : W \to W$ is a β-contraction, then the map Q has at least one fixed point in W.*

3. Compactness of Cauchy Operators

Let Cauchy operator $G : C([0, b], X) \to C([0, b], X)$ be defined by

$$(Gf)(t) = \int_0^t S_\alpha(t - s) f(s)\, ds, \quad t \in [0, b]. \tag{9}$$

If $S_\alpha(t)$ is a compact C_0-semigroup, it is well known that G is compact. However, it is unknown in case of compact resolvent. The main difficulty is that the resolvent does not have the property of semigroups. Thus, it seems to be more complicated to prove the compactness of Cauchy operator. Here, we will first discuss the continuity of resolvent in the uniform operator topology. Then, we can give the positive answer to the above problem.

In the remainder of this paper, we always assume that $M = \sup_{t\in[0,b]} \|S_\alpha(t)\| < +\infty$.

Lemma 10. *Suppose $S_\alpha(t)$ is a compact analytic resolvent of analyticity type (ω_0, θ_0). Then the following are hold:*

(i) *$\lim_{h\to 0} \|S_\alpha(t + h) - S_\alpha(t)\| = 0$ for $t > 0$;*

(ii) *$\lim_{h\to 0^+} \|S_\alpha(t + h) - S_\alpha(h)S_\alpha(t)\| = 0$ for $t > 0$;*

(iii) *$\lim_{h\to 0^+} \|S_\alpha(t) - S_\alpha(h)S_\alpha(t - h)\| = 0$ for $t > 0$.*

Proof. (i) Let $S_\alpha(t)$ be an analytic resolvent of analyticity type (ω_0, θ_0), and let $t > 0$ be given. Then, by means of Cauchy integral formula, we have

$$S'_\alpha(t) = \frac{1}{2\pi i} \int_{|z-t|=t\sin\theta} \frac{S_\alpha(z)}{(z - t)^2} dz. \tag{10}$$

Thus, for any $\omega > \omega_0$, $0 < \theta < \theta_0$, there exists a constant $M_1 = M_1(\omega, \theta)$ such that

$$\begin{aligned}\left\|S'_\alpha(t)\right\| &\leq M_1 (2\pi)^{-1} \int_{-\pi}^{\pi} e^{\omega(t+t\sin\theta\cos\gamma)} (t\sin\theta)^{-1} d\gamma \\ &\leq M_1 e^{2\omega t} (t\sin\theta)^{-1}.\end{aligned} \tag{11}$$

Now, let $\|x\| \leq 1$ and $|h| < t$. It follows from (11) that there exists a constant M' such that

$$\begin{aligned}&\left\|S_\alpha(t + h)x - S_\alpha(t)x\right\| \\ &\quad = \left\|\int_t^{t+h} S'_\alpha(s)x\, ds\right\| \\ &\quad \leq M' \|x\| \left|\int_t^{t+h} s^{-1}\, ds\right| \\ &\quad \leq M' |\ln(t + h) - \ln t| \longrightarrow 0 \quad \text{as } h \longrightarrow 0,\end{aligned} \tag{12}$$

which implies that $S_\alpha(t)$ is continuous in the uniform operator topology for $t > 0$, that is,

$$\lim_{h\to 0} \left\|S_\alpha(t + h) - S_\alpha(t)\right\| = 0 \quad \text{for } t > 0. \tag{13}$$

(ii) Let $\|x\| \leq 1$, $t > 0$, and $\varepsilon > 0$ be given. Since $S_\alpha(t)$ is compact, the set $W_t := \{S_\alpha(t)x : \|x\| \leq 1\}$ is also compact. Thus, there exists a finite family $\{S_\alpha(t)x_1, S_\alpha(t)x_2, \ldots, S_\alpha(t)x_m\} \subset W_t$ such that for any x with $\|x\| \leq 1$, there exists x_i $(1 \leq i \leq m)$ such that

$$\left\|S_\alpha(t)x - S_\alpha(t)x_i\right\| \leq \frac{\varepsilon}{3(M + 1)}. \tag{14}$$

From the strong continuity of $S_\alpha(t)$, $t \geq 0$, there exists $0 < h_1 < \min\{t, b\}$ such that

$$\left\|S_\alpha(t)x_i - S_\alpha(h)S_\alpha(t)x_i\right\| \leq \frac{\varepsilon}{3}, \quad \forall 0 \leq h \leq h_1,\ 1 \leq i \leq m. \tag{15}$$

On the other hand, from (i), there exists $0 < h_2 < \min\{t, b\}$ such that

$$\left\|S_\alpha(t + h)x - S_\alpha(t)x\right\| \leq \frac{\varepsilon}{3}, \quad \forall 0 \leq h \leq h_2,\ \|x\| \leq 1. \tag{16}$$

Thus, for $0 \le h \le \min\{h_1, h_2\}$ and $\|x\| \le 1$, it follows from (14)–(16) that

$$\begin{aligned}
&\|S_\alpha(t+h)x - S_\alpha(h)S_\alpha(t)x\| \\
&\quad\le \|S_\alpha(t+h)x - S_\alpha(t)x\| \\
&\qquad+ \|S_\alpha(t)x - S_\alpha(t)x_i\| \\
&\qquad+ \|S_\alpha(t)x_i - S_\alpha(h)S_\alpha(t)x_i\| \\
&\qquad+ \|S_\alpha(h)S_\alpha(t)x_i - S_\alpha(h)S_\alpha(t)x\| \\
&\quad\le \|S_\alpha(t+h)x - S_\alpha(t)x\| \\
&\qquad+ (M+1)\|S_\alpha(t)x - S_\alpha(t)x_i\| \\
&\qquad+ \|S_\alpha(t)x_i - S_\alpha(h)S_\alpha(t)x_i\| \\
&\quad\le \frac{\varepsilon}{3} + \frac{\varepsilon}{3} + \frac{\varepsilon}{3} \\
&\quad\le \varepsilon,
\end{aligned} \tag{17}$$

which implies that $\lim_{h\to 0^+}\|S_\alpha(t+h) - S_\alpha(h)S_\alpha(t)\| = 0$ for all $t > 0$.

(iii) Let $t > 0$ and $0 < h < \min\{t, b\}$. Then, there exists $M > 0$ such that

$$\begin{aligned}
&\|S_\alpha(t) - S_\alpha(h)S_\alpha(t-h)\| \\
&\quad\le \|S_\alpha(t) - S_\alpha(t+h)\| + \|S_\alpha(t+h) - S_\alpha(h)S_\alpha(t)\| \\
&\qquad+ \|S_\alpha(h)S_\alpha(t) - S_\alpha(h)S_\alpha(t-h)\| \\
&\quad\le \|S_\alpha(t) - S_\alpha(t+h)\| + \|S_\alpha(t+h) - S_\alpha(h)S_\alpha(t)\| \\
&\qquad+ M\|S_\alpha(t) - S_\alpha(t-h)\|.
\end{aligned} \tag{18}$$

It follows from (i) and (ii) that

$$\lim_{h\to 0^+}\|S_\alpha(t) - S_\alpha(h)S_\alpha(t-h)\| = 0 \quad \text{for } t > 0. \tag{19}$$

This completes the proof.

Lemma 11. *Suppose $S_\alpha(t)$ is a compact analytic resolvent of analyticity type (ω_0, θ_0). Then Cauchy operator G defined by (9) is a compact operator.*

Proof. We will show $G : C([0,b], X) \to C([0,b], X)$ is compact by using the Arzela-Ascoli theorem. Let $D_l = \{f \in C([0,b], X) : \|f\| \le l\}$ be any bounded subset of $C([0,b], X)$.

First, we claim that the set GD_l is equicontinuous on $C([0,b], X)$. In fact, let $0 \le t_1 \le t_2 \le b$ and $f \in D_l$; then we have

$$\begin{aligned}
I &:= \|(Gf)(t_2) - (Gf)(t_1)\| \\
&= \left\|\int_0^{t_2} S_\alpha(t_2-s)f(s)\,ds - \int_0^{t_1} S_\alpha(t_1-s)f(s)\,ds\right\| \\
&\le \int_0^{t_1}\|S_\alpha(t_2-s) - S_\alpha(t_1-s)\|\,\|f(s)\|\,ds \\
&\quad+ \int_{t_1}^{t_2}\|S_\alpha(t_2-s)f(s)\|\,ds \\
&\le l\int_0^{t_1}\|S_\alpha(t_2-s) - S_\alpha(t_1-s)\|\,ds + Ml(t_2-t_1).
\end{aligned} \tag{20}$$

If $t_1 = 0$, it is easy to see that

$$\lim_{t_2\to 0} I = 0, \quad \text{uniformly for } f \in D_l. \tag{21}$$

If $0 < t_1 < b$, for $0 < \delta < t_1$, we have

$$\begin{aligned}
I &\le l\int_0^{t_1-\delta}\|S_\alpha(t_2-s) - S_\alpha(t_1-s)\|\,ds \\
&\quad+ l\int_{t_1-\delta}^{t_1}\|S_\alpha(t_2-s) - S_\alpha(t_1-s)\|\,ds + Ml(t_2-t_1) \\
&\le bl\sup_{s\in[0,t_1-\delta]}\|S_\alpha(t_2-s) - S_\alpha(t_1-s)\| \\
&\quad+ 2Ml\delta + Ml(t_2-t_1).
\end{aligned} \tag{22}$$

Note that from Lemma 10(i), we know $S_\alpha(t)$ is operator norm continuous uniformly for $t \in [\delta, b]$. Combining this and the arbitrariness of δ with the above estimation on I, we can conclude that

$$\lim_{|t_1-t_2|\to 0} I = 0, \quad \text{uniformly for } f \in D_l. \tag{23}$$

Thus, GD_l is equicontinuous on $C([0,b], X)$.

Next, we will show that the set $\{(Gf)(t) : f \in D_l\}$ is precompact in X for every $t \in [0,b]$. It is easy to see that the set $\{(Gf)(0) : f \in D_l\}$ is precompact in X. Now, let $0 < t \le b$ be given and $0 < \varepsilon < t$. Then

$$\left\{S_\alpha(\varepsilon)\int_0^{t-\varepsilon} S_\alpha(t-s-\varepsilon)f(s)\,ds : f \in D_l\right\} \tag{24}$$

is precompact since $S_\alpha(\varepsilon)$ is compact. Moreover, for arbitrary $\varepsilon < \delta < t$, we have

$$\begin{aligned}&\left\|S_\alpha(\varepsilon)\int_0^{t-\varepsilon} S_\alpha(t-s-\varepsilon) f(s)\,ds - \int_0^{t-\varepsilon} S_\alpha(t-s) f(s)\,ds\right\| \\ &\leq l\int_0^{t-\varepsilon} \|S_\alpha(\varepsilon) S_\alpha(t-s-\varepsilon) - S_\alpha(t-s)\|\,ds \\ &\leq l\int_0^{t-\delta} \|S_\alpha(\varepsilon) S_\alpha(t-s-\varepsilon) - S_\alpha(t-s)\|\,ds \\ &\quad + l\int_{t-\delta}^{t-\varepsilon} \|S_\alpha(\varepsilon) S_\alpha(t-s-\varepsilon) - S_\alpha(t-s)\|\,ds \\ &\leq l\int_0^{t-\delta} \|S_\alpha(\varepsilon) S_\alpha(t-s-\varepsilon) - S_\alpha(t-s)\|\,ds + l\delta\left(M^2 + M\right).\end{aligned} \tag{25}$$

From Lemma 10(iii), we know

$$\begin{aligned}S_\alpha(\varepsilon) S_\alpha(t-s-\varepsilon) - S_\alpha(t-s) \longrightarrow 0, \\ \text{as} \quad \varepsilon \longrightarrow 0 \text{ for } s \in [0, t-\delta].\end{aligned} \tag{26}$$

Then, it follows from the Lebesgue dominated convergence theorem and the arbitrariness of δ that

$$\begin{aligned}\lim_{\varepsilon\to 0}&\left\|S_\alpha(\varepsilon)\int_0^{t-\varepsilon} S_\alpha(t-s-\varepsilon) f(s)\,ds\right. \\ &\left.-\int_0^{t-\varepsilon} S_\alpha(t-s) f(s)\,ds\right\| = 0.\end{aligned} \tag{27}$$

On the other hand,

$$\begin{aligned}&\left\|S_\alpha(\varepsilon)\int_0^{t-\varepsilon} S_\alpha(t-s-\varepsilon) f(s)\,ds - \int_0^{t} S_\alpha(t-s) f(s)\,ds\right\| \\ &\leq \left\|S_\alpha(\varepsilon)\int_0^{t-\varepsilon} S_\alpha(t-s-\varepsilon) f(s)\,ds\right. \\ &\qquad\left.-\int_0^{t-\varepsilon} S_\alpha(t-s) f(s)\,ds\right\| \\ &\quad + \left\|\int_0^{t-\varepsilon} S_\alpha(t-s) f(s)\,ds - \int_0^{t} S_\alpha(t-s) f(s)\,ds\right\| \\ &\leq \left\|S_\alpha(\varepsilon)\int_0^{t-\varepsilon} S_\alpha(t-s-\varepsilon) f(s)\,ds\right. \\ &\qquad\left.-\int_0^{t-\varepsilon} S_\alpha(t-s) f(s)\,ds\right\| + l\varepsilon M.\end{aligned} \tag{28}$$

Thus,

$$\begin{aligned}\lim_{\varepsilon\to 0}&\left\|S_\alpha(\varepsilon)\int_0^{t-\varepsilon} S_\alpha(t-s-\varepsilon) f(s)\,ds\right. \\ &\left.-\int_0^{t} S_\alpha(t-s) f(s)\,ds\right\| = 0,\end{aligned} \tag{29}$$

which implies that $\{(Gf)(t) : f \in D_l\}$ is precompact in X by using the total boundedness, and therefore G is compact in view of Arzela-Ascoli theorem.

4. Nonlocal Problems

In this section, we always assume that $u_0 \in X$ and that the operator A generates a compact analytic resolvent $S_\alpha(t)$ of analyticity type (ω_0, θ_0), and we will prove the existence of mild solutions of (1) when the nonlocal item g is assumed to be Lipschitz continuous and neither Lipschitz nor compact, respectively.

Let r be a fixed positive real number and

$$W_r := \{u \in C([0,b], X) : \|u\| \leq r\}. \tag{30}$$

Clearly, W_r is a bounded closed and convex set. We make the following assumptions.

(H1) $f : [0,b] \times X \to X$ is continuous.

(H2) $g : C([0,b], X) \to X$ is Lipschitz continuous with Lipschitz constant k such that $Mk < 1$.

Under these assumptions, we can prove the first main result in this paper.

Theorem 12. *Assume that conditions (H1) and (H2) are satisfied. Then the nonlocal problem* (1) *has at least one mild solution provided that*

$$M\left[\|u_0\| + \sup_{u\in W_r}\|g(u)\| + b\sup_{s\in[0,b],u\in W_r}\|f(s,u(s))\|\right] \leq r. \tag{31}$$

Proof. We consider the solution operator $Q : C([0,b], X) \to C([0,b], X)$ defined by

$$\begin{aligned}(Qu)(t) &= S_\alpha(t)\left[u_0 - g(u)\right] + \int_0^t S_\alpha(t-s) f(s,u(s))\,ds \\ &:= (Q_1u)(t) + (Q_2u)(t), \quad t \in [0,b].\end{aligned} \tag{32}$$

It is easy to see that the fixed point of Q is the mild solution of nonlocal Cauchy problem (1). Subsequently, we will prove that Q has a fixed point by using Lemma 9 (Darbo-Sadovskii's fixed point theorem).

Firstly, we prove that the mapping Q is continuous on $C([0,b], X)$. For this purpose, let $\{u_n\}_{n\geq 1}$ be a sequence in $C([0,b], X)$ with $\lim_{n\to\infty} u_n = u$ in $C([0,b], X)$. Then

$$\begin{aligned}&\|Qu_n - Qu\| \\ &\leq M\left[\|g(u_n) - g(u)\| + b\sup_{s\in[0,b]}\|f(s,u_n(s)) - f(s,u(s))\|\right].\end{aligned} \tag{33}$$

By the continuity of f and g, we deduce that Q is continuous on $C([0,b], X)$.

Secondly, we claim that $QW_r \subseteq W_r$, where W_r is defined by (30). In fact, for any $u \in W_r$, by (31), we have

$$\begin{aligned}\|Qu\| &\le M\left(\|u_0\| + \|g(u)\| + \int_0^b \|f(s,u(s))\|\,ds\right)\\ &\le M\left(\|u_0\| + \sup_{u\in W_r}\|g(u)\| + b\sup_{s\in[0,b],u\in W_r}\|f(s,u(s))\|\right)\\ &\le r,\end{aligned} \tag{34}$$

which implies that Q maps W_r into itself.

Now, according to Lemma 9, it remains to prove that Q is a β-contraction in W_r. From condition (H2), we get that $Q_1 : W_r \to C([0,b],X)$ is Lipschitz continuous with constant Mk. In fact, for $u, v \in W_r$, by (H2), we have

$$\|Q_1u - Q_1v\| \le M\|g(u) - g(v)\| \le Mk\|u - v\|. \tag{35}$$

Thus, it follows from Lemma 8-(30) that $\beta(Q_1W_r) \le Mk\beta(W_r)$.

For operator $Q_2 : W_r \to C([0,b],X)$, we have $Q_2u = Gf_u$, where $f_u(\cdot) = f(\cdot,u(\cdot))$ is continuous on $[0,b]$ and G is the Cauchy operator defined by (9). Thus, in view of Lemma 11, we know that Q_2 is compact on $C([0,b],X)$, and hence $\beta(Q_2W_r) = 0$. Consequently,

$$\beta(QW_r) \le \beta(Q_1W_r) + \beta(Q_2W_r) \le Mk\beta(W_r). \tag{36}$$

Since $Mk < 1$, the mapping Q is a β-contraction on W_r. By Darbo-Sadovskii's fixed point theorem, the operator Q has a fixed point in W_r, which is just the mild solution of nonlocal Cauchy problem (1).

Now, we give the following technical condition on function g.

(H) $g : C([0,b],X) \to X$ is continuous, and the set $g(\overline{\text{conv}}QW_r)$ is precompact, where $\overline{\text{conv}}B$ denotes the convex closed hull of set $B \subseteq C([0,T],X)$, and Q is given by (32).

Remark 13. It is easy to see that condition (H) is weaker than the compactness and convexity of g. The same hypothesis can be seen from [24, 30], where the authors considered the existence of mild solutions for semilinear nonlocal problems of integer order when A is a linear, densely defined operator on X which generates a C_0-semigroup. After the proof of our main results, we will give some special types of nonlocal item g which is neither Lipschitz nor compact, but satisfies the condition (H) in the next Corollaries.

Theorem 14. *Assume that conditions (H1) and (H) are satisfied. Then the nonlocal problem* (1) *has at least one mild solution provided that* (31) *holds.*

Proof. We will prove that Q has a fixed point by using Schauder's fixed point theorem. According to the proof of Theorem 12, we have proven that $Q : W_r \to W_r$ is continuous. Next, we will prove that there exists a set $W \subseteq W_r$ such that $Q : W \to W$ is compact.

For this purpose, let $0 < t \le b$ and $0 < \delta < t$. It is easy to see that the set

$$\{S_\alpha(\delta)S_\alpha(t-\delta)[u_0 - g(u)] : u \in W_r\} \tag{37}$$

is precompact since $S_\alpha(\delta)$ is compact. On the other hand, by Lemma 10(iii), we obtain

$$\begin{aligned}\lim_{\delta\to 0}\|S_\alpha(t)[u_0 - g(u)]\\ -S_\alpha(\delta)S_\alpha(t-\delta)[u_0 - g(u)]\| = 0, \quad \forall u \in W_r,\end{aligned} \tag{38}$$

which implies that $Q_1W_r(t)$ is precompact in X by using the total boundedness. Next, we claim Q_1W_r is equicontinuous on $[\eta,b]$ for any small positive number η. In fact, for $u \in W_r$ and $\eta \le t_1 \le t_2 \le b$, we have

$$\begin{aligned}&\|(Q_1u)(t_2) - (Q_1u)(t_1)\|\\ &\quad\le \|S_\alpha(t_2) - S_\alpha(t_1)\|(\|u_0\| + \|g(u)\|).\end{aligned} \tag{39}$$

By Lemma 10(i), $S_\alpha(t)$ is operator norm continuous for $t > 0$. Thus $S_\alpha(t)$ is operator norm continuous uniformly for $t \in [\eta,b]$, and hence

$$\begin{aligned}\lim_{|t_2-t_1|\to 0}\|(Q_1u)(t_2) - (Q_1u)(t_1)\| = 0,\\ \text{uniformly for } u \in W_r,\end{aligned} \tag{40}$$

that is, Q_1W_r is equicontinuous on $[\eta,b]$. Note that Q_2 is compact by Lemma 11. Therefore, we have proven that $QW_r(t)$ is precompact for every $t \in (0,b]$ and QW_r is equicontinuous on $[\eta,b]$ for any small positive number η.

Now, let $W = \overline{\text{conv}}QW_r$, we get that W is a bounded closed and convex subset of $C([0,b],X)$ and $QW \subseteq W$. It is easy to see that $QW(t)$ is precompact in X for every $t \in (0,b]$ and QW is equicontinuous on $[\eta,b]$ for any small positive number η. Moreover, we have that $g(W) = g(\overline{\text{conv}}QW_r)$ is precompact due to condition (H).

Thus, we can now claim that $Q : W \to W$ is compact. In fact, it is easy to see that $Q_1W(0) = u_0 - g(W)$ is precompact since $g(W) = g(\overline{\text{conv}}QW_r)$ is precompact. It remains to prove that Q_1W is equicontinuous on $[0,b]$. To this end, let $u \in W$ and $0 \le t_1 < t_2 \le b$; then we have

$$\|(Q_1u)(t_2) - (Q_1u)(t_1)\| \le \|S_\alpha(t_2) - S_\alpha(t_1)\|\|u_0 - g(u)\|. \tag{41}$$

In view of the compactness of $g(W)$ and the strong continuity of $S_\alpha(t)$ on $[0,b]$, we obtain the equicontinuous of Q_1W on $[0,b]$. Thus, $Q_1 : W \to C([0,b],X)$ is compact by Arzela-Ascoli theorem, and hence $Q : W \to W$ is also compact. Now, Schauder's fixed point theorem implies that Q has a fixed point on W, which gives rise to a mild solution of nonlocal problem (1).

The following theorem is a direct consequence of Theorem 14.

Theorem 15. *Assume that conditions (H1) and (H) are satisfied for each $r > 0$. If*

$$\frac{\|g(u)\|}{\|u\|} \longrightarrow 0, \quad \|u\| \longrightarrow \infty,$$
$$\frac{\|f(t,x)\|}{\|x\|} \longrightarrow 0, \quad \|x\| \longrightarrow \infty \tag{42}$$

uniformly for $t \in [0,b]$, then the nonlocal problem (1) *has at least one mild solution.*

Remark 16. It is easy to see that if there exist constants L_1, $L_2 > 0$ and $\gamma_1, \gamma_2 \in [0,1)$ such that

$$\|g(u)\| \le L_1(1+\|u\|)^{\gamma_1},$$
$$\|f(t,x)\| \le L_2(1+\|x\|)^{\gamma_2} \tag{43}$$

for $t \in [0,b]$, then conditions (42) are satisfied.

Next, we will give special types of nonlocal item g which is neither Lipschitz nor compact, but satisfies condition (H).

We give the following assumptions.

(H3) $g : C([0,b],X) \to X$ is a continuous mapping which maps W_r into a bounded set, and there is a $\delta = \delta(r) \in (0,b)$ such that $g(u) = g(v)$ for any $u, v \in W_r$ with $u(s) = v(s)$, $s \in [\delta, b]$.

(H4) $g : (C([0,b],X), \|\cdot\|_{L^1}) \to X$ is continuous.

Corollary 17. *Assume that conditions (H1) and (H3) are satisfied. Then the nonlocal problem* (1) *has at least one mild solution on $[0,b]$ provided that* (31) *holds.*

Proof. Let

$$(QW_r)_\delta = \{u \in C([0,b],X); u(t) = v(t) \text{ for } t \in [\delta,b],$$
$$u(t) = u(\delta) \text{ for } t \in [0,\delta), \text{where } v \in QW_r\}. \tag{44}$$

From the proof of Theorem 14, we know that $(QW_r)_\delta$ is precompact in $C([0,b],X)$. Moreover, by condition (H3), $g(\overline{\text{conv}}QW_r) = g(\overline{\text{conv}}(QW_r)_\delta)$ is also precompact in $C([0,b],X)$. Thus, all the hypotheses in Theorem 14 are satisfied. Therefore, there is at least one mild solution of nonlocal problem (1).

Corollary 18. *Let condition (H1) be satisfied. Suppose that $g(u) = \sum_{j=1}^{p} c_j u(t_j)$, where c_j are given positive constants, and $0 < t_1 < t_2 < \cdots < t_p \le b$. Then the nonlocal problem* (1) *has at least one mild solution on $[0,b]$ provided that*

$$M\left[\|u_0\| + \sum_{j=1}^{p} c_j r + b \sup_{s\in[0,b],u\in W_r} \|f(s,u(s))\|\right] \le r. \tag{45}$$

Proof. It is easy to see that the mapping g with $g(u) = \sum_{j=1}^{p} c_j u(t_j)$ satisfies condition (H3). And all the conditions in Corollary 17 are satisfied. So the conclusion holds.

Corollary 19. *Assume that conditions (H1) and (H4) are satisfied. Then the nonlocal problem* (1) *has at least one mild solution on $[0,b]$ provided that* (31) *holds.*

Proof. According to Theorem 14, it is sufficient to prove that the hypothesis (H) is satisfied. For arbitrary $\epsilon > 0$, there exists $0 < \delta < b$ such that $\int_0^\delta \|u(s)\| ds < \epsilon$ for all $u \in QW_r$. Let

$$(QW_r)_\delta = \{u \in C([0,b],X); u(t) = v(t) \text{ for } t \in [\delta,b],$$
$$u(t) = u(\delta) \text{ for } t \in [0,\delta), \text{where } v \in QW_r\}. \tag{46}$$

From the proof of Theorem 14, we know that $(QW_r)_\delta$ is precompact in $C([0,b],X)$, which implies that $(QW_r)_\delta$ is precompact in $L^1([0,b],X)$. Thus, QW_r is precompact in $L^1([0,b],X)$ as it has an ϵ-net $(QW_r)_\delta$. By condition $g : (C([0,b],X), \|\cdot\|_{L^1}) \to X$ is continuous and $\overline{\text{conv}}QW_r \subseteq (L)\overline{\text{conv}}QW_r$, it follows that condition (H) is satisfied, where $(L)\overline{\text{conv}}B$ denotes the convex and closed hull of B in $L^1([0,b],X)$. Therefore, the nonlocal problem (1) has at least one mild solution on $[0,b]$.

Finally, we give a simple example to illustrate our theory.

Example 20. Consider the following fractional partial differential heat equation in $\mathbb{R}$:

$$D_t^\alpha w(t,x) = \frac{\partial^2 w(t,x)}{\partial x^2} + J_t^{1-\alpha} F(t,w(t,x)), \quad 0 < x < 1,\ 0 < t < 1,$$
$$w(t,0) = w(t,1), \quad w_x'(t,0) = w_x'(t,1), \quad 0 \le t \le 1,$$
$$w(0,x) = w_0(x) - \sum_{j=1}^{p} c_j w(t_j,x), \quad 0 < x < 1,\ 0 < t_j \le 1, \tag{47}$$

where $0 < \alpha < 1$ and D_t^α denotes the Caputo fractional derivative.

We introduce the abstract frame as follows. Let $X = \{u \in C([0,1],\mathbb{R}) : u(0) = u(1)\}$. Let A be the linear operator in X defined by $D(A) = \{u \in X : u', u'' \in X\}$ and for $u \in D(A)$, $Au = u''$. Then, it is well known that A generates a compact C_0 semigroup $T(t)$ for $t > 0$ on X. From the subordination principle [6, Theorems 3.1 and 3.3], A also generates a compact resolvent $S_\alpha(t)$ for $t > 0$.

Assume that $f : [0,1] \times X \to X$ is a continuous function defined by $f(t,z)(x) = F(t,z(x))$, $0 \le x \le 1$ and $g : C([0,1],X) \to X$ is also a continuous function defined by $g(u)(x) = w_0(x) - \sum_{j=1}^{p} c_j u(t_j)(x)$, $0 \le t, x \le 1$, where $u(t)(x) := w(t,x)$.

Under these assumptions, the fractional partial differential heat equation (47) can be reformulated as the abstract problem (1). If the inequality

$$M\left[\|w_0\| + \sum_{j=1}^{p} c_j r + \sup_{s\in[0,1],u\in W_r} \|f(s,u(s))\|\right] \le r \quad (48)$$

holds for some constant $r > 0$, there exists at least one mild solution for fractional equation (47) in view of Corollary 18.

Acknowledgments

This work was completed when the first author visited the Pennsylvania State University. The author is grateful to Professor Alberto Bressan and the Department of Mathematics for their hospitality and providing good working conditions. The work was supported by the NSF of China (11001034) and Jiangsu Overseas Research & Training Program for University Prominent Young & Middle-aged Teachers and Presidents.

References

[1] I. Podlubny, *Fractional Differential Equations*, vol. 198 of *Mathematics in Science and Engineering*, Academic Press, San Diego, Calif, USA, 1999.

[2] J. Prüss, *Evolutionary Integral Equations and Applications*, vol. 87 of *Monographs in Mathematics*, Birkhäuser, Basel, Switzerland, 1993.

[3] R. P. Agarwal, Y. Zhou, and Y. He, "Existence of fractional neutral functional differential equations," *Computers & Mathematics with Applications*, vol. 59, no. 3, pp. 1095–1100, 2010.

[4] R. P. Agarwal, M. Belmekki, and M. Benchohra, "Existence results for semilinear functional differential inclusions involving Riemann-Liouville fractional derivative," *Dynamics of Continuous, Discrete & Impulsive Systems*, vol. 17, no. 3, pp. 347–361, 2010.

[5] R. P. Agarwal, V. Lakshmikantham, and J. J. Nieto, "On the concept of solution for fractional differential equations with uncertainty," *Nonlinear Analysis*, vol. 72, no. 6, pp. 2859–2862, 2010.

[6] E. Bajlekova, *Fractional evolution equations in Banach spaces [Ph.D. thesis]*, University Press Facilities, Eindhoven University of Technology, 2001.

[7] M. Benchohra, J. Henderson, S. K. Ntouyas, and A. Ouahab, "Existence results for fractional order functional differential equations with infinite delay," *Journal of Mathematical Analysis and Applications*, vol. 338, no. 2, pp. 1340–1350, 2008.

[8] M. Benchohra and B. A. Slimani, "Existence and uniqueness of solutions to impulsive fractional differential equations," *Electronic Journal of Differential Equations*, no. 10, pp. 1–11, 2009.

[9] M. Fečkan, J. Wang, and Y. Zhou, "Controllability of fractional functional evolution equations of Sobolev type via characteristic solution operators," *Journal of Optimization Theory and Applications*, vol. 156, no. 1, pp. 79–95, 2013.

[10] E. Hernández, D. O'Regan, and K. Balachandran, "On recent developments in the theory of abstract differential equations with fractional derivatives," *Nonlinear Analysis*, vol. 73, no. 10, pp. 3462–3471, 2010.

[11] E. Hernández, D. O'Regan, and K. Balachandran, "Existence results for abstract fractional differential equations with nonlocal conditions via resolvent operators," *Indagationes Mathematicae*, vol. 24, no. 1, pp. 68–82, 2013.

[12] C. Lizama, "Regularized solutions for abstract Volterra equations," *Journal of Mathematical Analysis and Applications*, vol. 243, no. 2, pp. 278–292, 2000.

[13] C. Lizama, "An operator theoretical approach to a class of fractional order differential equations," *Applied Mathematics Letters*, vol. 24, no. 2, pp. 184–190, 2011.

[14] C. Lizama and G. M. N'Guérékata, "Bounded mild solutions for semilinear integro differential equations in Banach spaces," *Integral Equations and Operator Theory*, vol. 68, no. 2, pp. 207–227, 2010.

[15] M. Li, C. Chen, and F.-B. Li, "On fractional powers of generators of fractional resolvent families," *Journal of Functional Analysis*, vol. 259, no. 10, pp. 2702–2726, 2010.

[16] K. Li, J. Peng, and J. Jia, "Cauchy problems for fractional differential equations with Riemann-Liouville fractional derivatives," *Journal of Functional Analysis*, vol. 263, no. 2, pp. 476–510, 2012.

[17] G. M. Mophou and G. M. N'Guérékata, "Existence of mild solutions of some semilinear neutral fractional functional evolution equations with infinite delay," *Applied Mathematics and Computation*, vol. 216, no. 1, pp. 61–69, 2010.

[18] G. M. Mophou and G. M. N'Guérékata, "A note on a semilinear fractional differential equation of neutral type with infinite delay," *Advances in Difference Equations*, vol. 2010, Article ID 674630, 8 pages, 2010.

[19] J. Wang, Z. Fan, and Y. Zhou, "Nonlocal controllability of semilinear dynamic systems with fractional derivative in Banach spaces," *Journal of Optimization Theory and Applications*, vol. 154, no. 1, pp. 292–302, 2012.

[20] J. Wang, Y. Zhou, and M. Fečkan, "Abstract Cauchy problem for fractional differential equations," *Nonlinear Dynamics*, vol. 71, no. 4, pp. 685–700, 2013.

[21] L. Byszewski and V. Lakshmikantham, "Theorem about the existence and uniqueness of a solution of a nonlocal abstract Cauchy problem in a Banach space," *Applicable Analysis*, vol. 40, no. 1, pp. 11–19, 1991.

[22] L. Chen and Z. Fan, "On mild solutions to fractional differential equations with nonlocal conditions," *Electronic Journal of Qualitative Theory of Differential Equations*, no. 53, pp. 1–13, 2011.

[23] Z. Fan, "Existence of nondensely defined evolution equations with nonlocal conditions," *Nonlinear Analysis*, vol. 70, no. 11, pp. 3829–3836, 2009.

[24] Z. Fan, "Impulsive problems for semilinear differential equations with nonlocal conditions," *Nonlinear Analysis*, vol. 72, no. 2, pp. 1104–1109, 2010.

[25] Z. Fan and G. Li, "Existence results for semilinear differential equations with nonlocal and impulsive conditions," *Journal of Functional Analysis*, vol. 258, no. 5, pp. 1709–1727, 2010.

[26] E. Hernández, J. S. dos Santos, and K. A. G. Azevedo, "Existence of solutions for a class of abstract differential equations with nonlocal conditions," *Nonlinear Analysis*, vol. 74, no. 7, pp. 2624–2634, 2011.

[27] J. Liang, J. H. Liu, and T.-J. Xiao, "Nonlocal problems for integrodifferential equations," *Dynamics of Continuous, Discrete & Impulsive Systems*, vol. 15, no. 6, pp. 815–824, 2008.

[28] J. Liang, J. H. Liu, and T.-J. Xiao, "Nonlocal impulsive problems for nonlinear differential equations in Banach spaces," *Mathematical and Computer Modelling*, vol. 49, no. 3-4, pp. 798–804, 2009.

[29] J. Liang and Z. Fan, "Nonlocal impulsive Cauchy problems for evolution equations," *Advances in Difference Equations*, vol. 2011, Article ID 784161, 17 pages, 2011.

[30] X. M. Xue, "Existence of semilinear differential equations with nonlocal initial conditions," *Acta Mathematica Sinica*, vol. 23, no. 6, pp. 983–988, 2007.

[31] Y. Zhou and F. Jiao, "Nonlocal Cauchy problem for fractional evolution equations," *Nonlinear Analysis*, vol. 11, no. 5, pp. 4465–4475, 2010.

[32] L. Zhu and G. Li, "Existence results of semilinear differential equations with nonlocal initial conditions in Banach spaces," *Nonlinear Analysis*, vol. 74, no. 15, pp. 5133–5140, 2011.

[33] L. Zhu, Q. Huang, and G. Li, "Existence and asymptotic properties of solutions of nonlinear multivalued differential inclusions with nonlocal conditions," *Journal of Mathematical Analysis and Applications*, vol. 390, no. 2, pp. 523–534, 2012.

[34] J. Banaś and K. Goebel, *Measures of Noncompactness in Banach Spaces*, vol. 60 of *Lecture Notes in Pure and Applied Mathematics*, Marcel Dekker, New York, NY, USA, 1980.

6

Positive Periodic Solutions of Cooperative Systems with Delays and Feedback Controls

Tursuneli Niyaz[1] and Ahmadjan Muhammadhaji[2]

[1] *College of Applied Mathematics, Xinjiang University of Finance and Economics, Ürümqi 830012, China*
[2] *College of Mathematical Sciences, Xinjiang Normal University, Urumqi 830054, China*

Correspondence should be addressed to Ahmadjan Muhammadhaji; ahmadjanm@gmail.com

Academic Editor: Norio Yoshida

This paper studies a class of periodic n species cooperative Lotka-Volterra systems with continuous time delays and feedback controls. Based on the continuation theorem of the coincidence degree theory developed by Gaines and Mawhin, some new sufficient conditions on the existence of positive periodic solutions are established.

1. Introduction

Mathematical ecological system has become one of the most important topics in the study of modern applied mathematics. Its dynamical behavior includes persistence, permanence and extinction of species, global stability of systems, the existence of positive periodic solutions, positive almost periodic solutions, and strictly positive solutions. The existence of positive periodic solutions has already become one of the most interesting subjects for scholars. In the recent years, the application of fixed-point theorems to the existence of positive periodic solutions in mathematical ecology has been studied extensively, for example, Brouwer's fixed point theorem [1–4], Schauder's fixed-point theorem [5–8], Krasnoselskii's fixed-point theorem [9–14], Horn's fixed-point theorem [15, 16], and Mawhins continuation theorem [17–36], and so forth. In particular, Mawhins continuation theorem is a powerful tool for studying the existence of periodic solutions of periodic high-dimensional time-delayed problems. When dealing with a time-delayed problem, it is very convenient and the result is relatively simple [30]. Recently, a considerable number of mathematical models with delays have been proposed in the study of population dynamics. One of the most celebrated models for population is the Lotka-Volterra system. Subsequently, a lot of the literature related to the study of the existence of positive periodic solutions for various Lotka-Volterra-type population dynamical systems with delays by using the method of continuation theorem was published and extensive research results were obtained [17–21, 24–34].

On the other hand, in some situations, people may wish to change the position of the existing periodic solution but to keep its stability. This is of significance in the control of ecology balance. One of the methods for the realization of it is to alter the system structurally by introducing some feedback control variables so as to get a population stabilizing at another periodic solution. The realization of the feedback control mechanism might be implemented by means of some biological control scheme or by harvesting procedure [21]. In fact, during the last decade, the existence of positive periodic solutions for the population dynamics with feedback control has been studied extensively [8, 14, 17–21, 24, 29]. To the best of our knowledge, studies on the existence of positive periodic solutions for cooperative systems with delays and feedback controls are fairly rare.

In [21], the authors studied the following neutral Lotka-Volterra system with feedback controls:

$$\dot{y}_i(t) = y_i(t)\Bigg[r_i(t) - \sum_{j=1}^{n} a_{ij} y_j(t) - \sum_{j=1}^{n} b_{ij} y_j\left(t - \tau_{ij}(t)\right) - \sum_{j=1}^{n} c_{ij} \dot{y}_j\left(t - \gamma_{ij}(t)\right)$$

$$-f_i(t)\,u_i(t) - e_i(t)\,u_i(t-\sigma_i(t))\Big]$$

$$\dot{u}_i(t) = -\alpha_i(t)\,u_i(t) + \beta_i(t)\,y_i(t) + \gamma_i(t)\,y_i(t)\,(t-\delta_i(t)), \quad i = 1,2,\ldots,n. \tag{1}$$

By using Mawhin's continuation theorem, the sufficient conditions on the existence of positive periodic solutions are established. In [24], the authors considered the following delay differential system with feedback control:

$$\frac{dx}{dt} = F(t, x(t-\tau_1(t)), \ldots, x(t-\tau_n(t)), u(t-\delta(t))) \tag{2}$$
$$\frac{du}{dt} = -\eta(t) + a(t)\,x(t-\sigma(t)).$$

A set of natural and easily verifiable sufficient conditions of the existence of positive periodic solutions are established, by using Mawhin's continuation theorem. In [29], the authors considered the following single-species periodic logistic systems with feedback regulation and infinite distributed delay:

$$\dot{N}(t) = r(t)\,N(t) \times \left(1 - \frac{1}{K(t)}\int_0^{\infty} H(s)\,N(t-s)\,ds - c(t)\,u(t)\right)$$
$$\dot{u}(t) = -a(t)\,u(t) + b(t)\int_0^{\infty} H(s)\,N^2(t-s)\,ds. \tag{3}$$

The sufficient conditions for the existence of positive periodic solutions are established, based on Mawhin's continuation theorem.

Motivated by the above works, in this paper, we investigate the following n species periodic Lotka-Volterra-type cooperative systems with continuous time delays and feedback controls:

$$\dot{x}_i(t) = x_i(t)\Bigg[r_i(t) - \sum_{l=1}^{m} a_{iil}(t)\,x_i(t-\tau_{iil}(t)) + \sum_{j\neq i}^{n}\sum_{l=1}^{m} a_{ijl}(t)\,x_j(t-\tau_{ijl}(t))\Bigg] - d_i(t)\,u_i(t) - e_i(t)\,u_i(t-\varepsilon_i(t)) \tag{4}$$
$$\dot{u}_i(t) = -b_i(t)\,u_i(t) + \beta_i(t)\,x_i(t) + \gamma_i(t)\,x_i(t-\sigma_i(t)), \quad i = 1,2,\ldots,n.$$

By using the technique of coincidence degree developed by Gaines and Mawhin in [36], we will establish some new sufficient conditions which guarantee that the system has at least one positive periodic solution.

2. Preliminaries

In system (4), we have that $x_i(t)$ $(i = 1,2,\ldots n)$ represent the density of n cooperative species x_i $(i = 1,2,\ldots n)$ at time t, respectively; $r_i(t)$ $(i = 1,2,\ldots n)$ represent the intrinsic growth rate of species x_i $(i = 1,2,\ldots n)$ at time t, respectively; $a_{iil}(t)$ $(i = 1,2,\ldots,n,\ l = 1,2,\ldots,m)$ represent the intrapatch restriction density of species x_i $(i = 1,2,\ldots n)$ at time t, respectively; $a_{ijl}(t)$ $(l = 1,2,\ldots,m,\ i\neq j,\ i,j = 1,2,\ldots,n)$ represent the cooperative coefficients between n species x_i $(i = 1,2,\ldots n)$ at time t, respectively. $u_i(t)$ $(i = 1,2,\ldots n)$ represent the indirect feedback control variables [21] at time t, respectively. $\beta_i(t), e_i(t), b_i(t), d_i(t)$, and $\gamma_i(t)$ $(i = 1,2,\ldots n)$ represent the feedback control coefficients at time t, respectively. In this paper, we always assume that

(H1) $\tau_{ijl}(t)$ $(l = 1,2,\ldots,m,\ i,j = 1,2,\ldots,n)$, $\sigma_i(t), \varepsilon_i(t), r_i(t)$ $(i = 1,2,\ldots n)$ are continuous ω-periodic functions with $\tau'_{ijl}(t) < 1$ and $\int_0^{\omega} r_i(t)dt > 0$. $a_{ijl}(t)$ $(i,j = 1,2,\ l = 1,2,\ldots,m)$, $\beta_i(t)$, $e_i(t)$, $b_i(t)$, $d_i(t)$, and $\gamma_i(t)$ $(i = 1,2,\ldots n)$ are continuous, positive ω-periodic functions.

From the viewpoint of mathematical biology, in this paper for system (4) we only consider the solution with the following initial conditions:

$$x_i(t) = \phi_i(t), \quad \forall t \in [-\sigma, 0],\ i = 1,2,\ldots n,$$
$$u_i(t) = \psi_i(t), \quad \forall t \in [-\sigma, 0],\ i = 1,2,\ldots n, \tag{5}$$

where $\phi_i(t), \psi_i(t)$ $(i = 1,2,\ldots n)$ are nonnegative continuous functions defined on $[-\sigma, 0]$ satisfying $\phi_i(0) > 0$, $\psi_i(0) > 0$ $(i = 1,2,\ldots n)$ with $\sigma = \max_{t\in[0,\omega]}\{\tau_{ijl}(t), \sigma_i(t), \varepsilon_i(t)\ (i,j = 1,2,\ l = 1,2,\ldots,m)\}$.

In this paper, for any ω-periodic continuous function $f(t)$ we denote

$$f^L = \min_{t\in[0,\omega]} f(t), \qquad f^M = \max_{t\in[0,\omega]} f(t),$$
$$\overline{f} = \frac{1}{\omega}\int_0^{\omega} f(t)\,dt. \tag{6}$$

In order to obtain the existence of positive ω-periodic solutions of system (4), we will use the continuation theorem developed by Gaines and Mawhin in [36]. For the reader's convenience, we will introduce the continuation theorem in the following.

Let X and Z be two normed vector spaces. Let L : $\mathrm{Dom}\, L \subset X \to Z$ be a linear operator and let $N : X \to Z$ be a continuous operator. The operator L is called a Fredholm operator of index zero, if $\dim\mathrm{Ker}\, L = \mathrm{codim}\,\mathrm{Im}\, L < \infty$ and $\mathrm{Im}\, L$ is a closed set in Z. If L is a Fredholm operator of index zero, then there exist continuous projectors $P : X \to X$ and $Q : Z \to Z$ such that $\mathrm{Im}\, P = \mathrm{Ker}\, L$ and $\mathrm{Im}\, L = \mathrm{Ker}\, Q = \mathrm{Im}\,(I - Q)$. It follows that $L \mid \mathrm{Dom}\, L \cap \mathrm{Ker}\, P$: $\mathrm{Dom}\, L \cap \mathrm{Ker}\, P \to \mathrm{Im}\, L$ is invertible and its inverse is denoted by K_P; denote by $J : \mathrm{Im}\, Q \to \mathrm{Ker}\, L$ an isomorphism of $\mathrm{Im}\, Q$ onto $\mathrm{Ker}\, L$. Let Ω be a bounded open subset of X; we say that the operator N is L-compact on $\overline{\Omega}$, where $\overline{\Omega}$ denotes the closure of Ω in X, if $QN(\overline{\Omega})$ is bounded and $K_P(I-Q)N : \overline{\Omega} \to X$ is compact.

Lemma 1 (see [35]). *Suppose $\tau \in C^1(R,R)$ with $\tau(t+\omega) \equiv \tau(t)$ and $\tau'(t) < 1$, $\forall t \in [0,\omega]$. Then the function $t - \tau(t)$ has a unique inverse function $\mu(t)$ satisfying $\mu \in C(R,R)$, $\mu(u+\omega) = \mu(u) + \omega, \forall u \in R$.*

Lemma 2 (see [36]). *Let L be a Fredholm operator of index zero and let N be L-compact on $\overline{\Omega}$. If*

(a) *for each $\lambda \in (0,1)$ and $x \in \partial\Omega \cap \operatorname{Dom} L$, $Lx \neq \lambda Nx$;*

(b) *for each $x \in \partial\Omega \cap \operatorname{Ker} L$, $QNx \neq 0$;*

(c) $\deg\{JQN, \Omega \cap \operatorname{Ker} L, 0\} \neq 0$,

then the operator equation $Lx = Nx$ has at least one solution lying in $\operatorname{Dom} L \cap \overline{\Omega}$.

3. Main Results

In order to obtain the existence of positive periodic solutions of system (4), firstly, we introduce the following lemma.

Lemma 3. *Suppose that $(x_1^*(t), x_2^*(t), \ldots, x_n^*(t), u_1^*(t), u_2^*(t), \ldots, u_n^*(t))$ is an ω-periodic solution of (4) and (5); then $(x_1^*(t), x_2^*(t), \ldots, x_n^*(t), u_1^*(t), u_2^*(t), \ldots, u_n^*(t))$ satisfies the system*

$$\dot{x}_i(t) = x_i(t)\left(r_i(t) - \sum_{j=1}^{2}(-1)^{i+j}\sum_{l=1}^{m} a_{ijl}(t)\, x_j\left(t - \tau_{ijl}(t)\right)\right) - d_i(t)\, u_i(t) - e_i(t)\, u_i\left(t - \varepsilon_i(t)\right),$$

$$u_i(t) = \int_t^{t+\omega} \left[\beta_i(s)\, x_i(s) + \gamma_i(s)\, x_i\left(s - \sigma_i(s)\right)\right] G_i(t,s)\, ds, \quad i = 1, 2, \ldots, n, \tag{7}$$

where

$$G_i(t,s) = \frac{\exp\left\{\int_t^s b_i(\theta)\, d\theta\right\}}{\exp\left\{\int_0^\omega b_i(\theta)\, d\theta\right\} - 1}, \quad i = 1, 2, \ldots, n. \tag{8}$$

The converse is also true.

Proof. By (4), (5), and the variation constants formula in ordinary differential equations, we have

$$u_i(t) = \left(\int_0^t \left(\beta_i(s)\, x_i(s) + \gamma_i(s)\, x_i\left(s - \sigma_i(s)\right)\right) \times \exp\left\{\int_0^s b_i(\theta)\, d\theta\right\} ds + \psi_i(0)\right) \times \exp\left\{-\int_0^t b_i(\theta)\, d\theta\right\}, \quad i = 1, 2, \ldots, n. \tag{9}$$

From (9), we obtain

$$u_i(t+\omega) = \left(\int_0^{t+\omega} \left(\beta_i(s)\, x_i^*(s) + \gamma_i(s)\, x_i^*\left(s - \sigma_i(s)\right)\right) \times \exp\left\{\int_0^s b_i(\theta)\, d\theta\right\} ds + \psi_i(0)\right) \times \exp\left\{-\int_0^{t+\omega} b_i(\theta)\, d\theta\right\}, \quad i = 1, 2, \ldots, n,$$

$$u_i^*(t) = \left(\int_0^t \left(\beta_i(s)\, x_i^*(s) + \gamma_i(s)\, x_i^*\left(s - \sigma_i(s)\right)\right) \times \exp\left\{\int_0^s b_i(\theta)\, d\theta\right\} ds + \psi_i(0)\right) \times \exp\left\{-\int_0^t b_i(\theta)\, d\theta\right\}, \quad i = 1, 2, \ldots, n. \tag{10}$$

Considering that $(x_1^*(t), x_2^*(t), \ldots, x_n^*(t), u_1^*(t), u_2^*(t), \ldots, u_n^*(t))$ is an ω-periodic solution of system (4) and (5), we obtain

$$\left(\int_0^t \left(\beta_i(s)\, x_i^*(s) + \gamma_i(s)\, x_i^*\left(s - \sigma_i(s)\right)\right) \times \exp\left\{\int_0^s b_i(\theta)\, d\theta\right\} ds + \psi_i(0)\right) \exp\left\{-\int_0^t b_i(\theta)\, d\theta\right\}$$
$$= \left(\int_0^{t+\omega} \left(\beta_i(s)\, x_i^*(s) + \gamma_i(s)\, x_i^*\left(s - \sigma_i(s)\right)\right) \times \exp\left\{\int_0^s b_i(\theta)\, d\theta\right\} ds + \psi_i(0)\right) \times \exp\left\{-\int_0^{t+\omega} b_i(\theta)\, d\theta\right\}, \quad i = 1, 2, \ldots, n. \tag{11}$$

Then

$$\left(\int_0^t \left(\beta_i(s)\, x_i^*(s) + \gamma_i(s)\, x_i^*\left(s - \sigma_i(s)\right)\right) \times \exp\left\{\int_0^s b_i(\theta)\, d\theta\right\} ds + \psi_i(0)\right) \exp\left\{\int_0^\omega b_i(\theta)\, d\theta\right\}$$
$$= \left(\int_0^{t+\omega} \left(\beta_i(s)\, x_i^*(s) + \gamma_i(s)\, x_i^*\left(s - \sigma_i(s)\right)\right) \times \exp\left\{\int_0^s b_i(\theta)\, d\theta\right\} ds + \psi_i(0)\right)$$
$$= \int_0^t \left(\beta_i(s)\, x_i^*(s) + \gamma_i(s)\, x_i^*\left(s - \sigma_i(s)\right)\right) \times \exp\left\{\int_0^s b_i(\theta)\, d\theta\right\} ds + \psi_i(0)$$

$$+\int_t^{t+\omega}(\beta_i(s)x_i^*(s)+\gamma_i(s)x_i^*(s-\sigma_i(s)))\times\exp\left\{\int_0^s b_i(\theta)\,d\theta\right\}ds,\quad i=1,2,\ldots,n,\tag{12}$$

which implies

$$\begin{aligned}&\left(\int_0^t(\beta_i(s)x_i^*(s)+\gamma_i(s)x_i^*(s-\sigma_i(s)))\times\exp\left\{\int_0^s b_i(\theta)\,d\theta\right\}ds+\psi_i(0)\right)\left(e^{\bar{b}\omega}-1\right)\\&=\int_t^{t+\omega}(\beta_i(s)x_i^*(s)+\gamma_i(s)x_i^*(s-\sigma_i(s)))\times\exp\left\{\int_0^s b_i(\theta)\,d\theta\right\}ds,\quad i=1,2,\ldots,n.\end{aligned}\tag{13}$$

That is,

$$\begin{aligned}&\int_0^t(\beta_i(s)x_i^*(s)+\gamma_i(i)x_i^*(s-\sigma_i(s)))\times\exp\left\{\int_0^s b_i(\theta)\,d\theta\right\}ds+\psi_i(0)\\&=u_i^*(t)\exp\left\{\int_0^t b_i(\theta)\,d\theta\right\}\\&=\frac{1}{e^{\bar{b}\omega}-1}\int_t^{t+\omega}(\beta_i(s)x_i^*(s)+\gamma_i(s)x_i^*(s-\sigma_i(s)))\times\exp\left\{\int_0^s b_i(\theta)\,d\theta\right\}ds,\quad i=1,2,\ldots,n.\end{aligned}\tag{14}$$

Hence

$$u_i^*(t)=\int_t^{t+\omega}(\beta_i(s)x_i^*(s)+\gamma_i(s)x_i^*(s-\sigma_i(s)))\times\frac{\exp\left\{\int_t^s b_i(\theta)\,d\theta\right\}}{\exp\left\{\int_0^\omega b_i(\theta)\,d\theta\right\}-1}ds,\quad i=1,2,\ldots,n.\tag{15}$$

On the other hand, assume that $(x_1^*(t),x_2^*(t),\ldots,x_n^*(t),u_1^*(t),u_2^*(t),\ldots,u_n^*(t))$ is an ω-periodic solution of system (7), then

$$u_i^*(t)=\int_t^{t+\omega}(\beta_i(s)x_i^*(s)+\gamma_i(s)x_i^*(s-\sigma_i(s)))\times\frac{\exp\left\{\int_t^s b_i(\theta)\,d\theta\right\}}{\exp\left\{\int_0^\omega b_i(\theta)\,d\theta\right\}-1}ds,\quad i=1,2,\ldots,n.\tag{16}$$

By a direct calculation, we have

$$\begin{aligned}\frac{du^*(t)}{dt}&=-b_i(t)\int_t^{t+\omega}(\beta_i(s)x_i^*(s)+\gamma_i(s)x_i^*(s-\sigma_i(s)))\times\frac{\exp\left\{\int_t^s b_i(\theta)\,d\theta\right\}}{\exp\left\{\int_0^\omega b_i(\theta)\,d\theta\right\}-1}ds\\&\quad+(\beta_i(t+\omega)x_i^*(t+\omega)+\gamma_i(t+\omega)x_i^*(t+\omega-\sigma_i(t+\omega)))\times\frac{\exp\left\{\int_t^{t+\omega} b_i(\theta)\,d\theta\right\}}{\exp\left\{\int_0^\omega b_i(\theta)\,d\theta\right\}-1}\\&\quad-(\beta_i(t)x_i^*(t)+\gamma_i(t)x_i^*(t-\sigma_i(t)))\times\frac{1}{\exp\left\{\int_0^\omega b_i(\theta)\,d\theta\right\}-1}\\&=-b_i(t)u_i(t)+\beta_i(t)x_i^*(t)+\gamma_i(t)x_i^*(t-\sigma_i(t)),\quad i=1,2,\ldots,n.\end{aligned}\tag{17}$$

This completes the proof.

It is easy to see that system (7) is equivalent to the following system:

$$\begin{aligned}\dot{x}_i(t)&=x_i(t)\left(r_i(t)-\sum_{j=1}^{2}(-1)^{i+j}\sum_{l=1}^{m}a_{ijl}(t)x_j\left(t-\tau_{ijl}(t)\right)\right)-d_i(t)u_i(t)-e_i(t)u_i(t-\varepsilon_i(t)),\\u_i(t)&=\int_t^{t+\omega}K(x_i)G_i(t,s)\,ds,\quad i=1,2,\ldots,n,\end{aligned}\tag{18}$$

where

$$\begin{aligned}G_i(t,s)&=\frac{\exp\left\{\int_t^s b_i(\theta)\,d\theta\right\}}{\exp\left\{\int_0^\omega b_i(\theta)\,d\theta\right\}-1},\\K(x_i)&=\beta_i(s)x_i(s)+\gamma_i(s)x_i(s-\sigma_i(s)),\quad i=1,2,\ldots,n.\end{aligned}\tag{19}$$

It is clear that in order to prove that systems (4) and (5) have at least one ω-periodic solution, we only need to prove that system (18) has at least one ω-periodic solution.

Now, for convenience of statements we denote the functions

$$a_{ij}(t)=\sum_{l=1}^{m}a_{ijl}(t),\quad i,j=1,2,\ldots,n.\tag{20}$$

The following theorem is about the existence of positive periodic solutions of system (4).

Theorem 4. *Suppose that assumption (H1) holds and there exists a constant $\theta_i > 0$, $\zeta_i > 0$, $i = 1, 2, \ldots, n$, such that*

$$\min_{t\in[0,\omega]} \left\{ \sum_{l=1}^{m} \left[\delta_{iil}(t)\theta_i - \sum_{j\neq i}^{n} \delta_{jil}(t)\theta_j \right] \right\} =: \zeta_i > 0, \quad i = 1, 2, \ldots, n, \tag{21}$$

where

$$\delta_{ijl}(t) = \frac{a_{ijl}(\varphi_{ijl}(t))}{1 - \tau'_{ijl}(\varphi_{ijl}(t))}, \quad i, j = 1, 2, \ldots, n, \; l = 1, 2, \ldots, m, \tag{22}$$

and the algebraic equation

$$\overline{r}_i - \overline{H}_i - \overline{a}_{ii}v_i + \sum_{j\neq i}^{n} \overline{a}_{ij}v_j = 0, \quad i = 1, 2, \ldots, n, \tag{23}$$

where

$$H_i = \int_0^{\omega} (\beta_i(s) + \gamma_i(s)) \times (d_i(t)G_i(t,s) + e_i(t)G_i(t - \varepsilon_i(t), s))\, ds, \quad i = 1, 2, \ldots, n, \tag{24}$$

has a unique positive solution. Then system (4) has at least one positive ω-periodic solution.

Proof. For system (18) we introduce new variables $y_i(t)$ $(i = 1, 2, \ldots, n)$ such that

$$x_i(t) = \exp\{y_i(t)\}, \quad i = 1, 2, \ldots, n. \tag{25}$$

Then system (18) is rewritten in the following form:

$$\begin{aligned} \dot{y}_i(t) &= r_i(t) - \sum_{j=1}^{2} (-1)^{i+j} \sum_{l=1}^{m} a_{ijl}(t)\exp\{y_j(t - \tau_{ijl}(t))\} \\ &\quad - \exp\{-y_i(t)\} d_i(t) U_i(t), \\ &\quad - \exp\{-y_i(t)\} e_i(t) U_i(t - \varepsilon_i(t)) \end{aligned} \tag{26}$$

$$U_i(t) = \int_t^{t+\omega} K(e^{y_i}) G_i(t,s)\, ds, \quad i = 1, 2,$$

where

$$K(e^{y_i}) = \beta_i(s)\exp\{y_i(s)\} + \gamma_i(s)\exp\{y_i(s - \sigma_i(s))\}, \quad i = 1, 2. \tag{27}$$

In order to apply Lemma 2 to system (26), we introduce the normed vector spaces X and Z as follows. Let $C(R, R^n)$ denote the space of all continuous functions $y(t) = (y_1(t), y_2(t), \ldots, y_n(t)) : R \to R^n$. We take

$$\begin{aligned} X &= Z \\ &= \{y(t) \in C(R, R^n) : y(t) \text{ is an } \omega\text{-periodic function}\}, \end{aligned} \tag{28}$$

with norm

$$\|y\| = \sum_{i=1}^{n} \max_{t\in[0,\omega]} |y_i(t)|. \tag{29}$$

It is obvious that X and Z are the Banach spaces. We define a linear operator $L : \operatorname{Dom} L \subset X \to Z$ and a continuous operator $N : X \to Z$ as follows:

$$\begin{aligned} Ly(t) &= \dot{y}(t), \\ Ny(t) &= (Ny_1(t), Ny_2(t), \ldots, Ny_n(t)), \end{aligned} \tag{30}$$

where

$$\begin{aligned} Ny_i(t) &= r_i(t) - \sum_{l=1}^{m} a_{iil}(t)\exp\{y_i(t - \tau_{iil}(t))\} \\ &\quad + \sum_{j\neq i}^{n} \sum_{l=1}^{m} a_{ijl}(t)\exp\{y_j(t - \tau_{ijl}(t))\} \\ &\quad - \exp\{-y_i(t)\} d_i(t) u_i(t) - \exp\{-y_i(t)\} \\ &\quad \times e_i(t) u_i(t - \varepsilon_i(t)). \end{aligned} \tag{31}$$

Further, we define continuous projectors $P : X \to X$ and $Q : Z \to Z$ as follows:

$$Py(t) = \frac{1}{\omega}\int_0^{\omega} y(t)\, dt, \qquad Qv(t) = \frac{1}{\omega}\int_0^{\omega} v(t)\, dt. \tag{32}$$

We easily see $\operatorname{Im} L = \{v \in Z : \int_0^{\omega} v(t)dt = 0\}$ and $\operatorname{Ker} L = R^n$. It is obvious that $\operatorname{Im} L$ is closed in Z and $\dim\operatorname{Ker} L = n$. Since for any $v \in Z$ there are unique $v_1 \in R^n$ and $v_2 \in \operatorname{Im} L$ with

$$v_1 = \frac{1}{\omega}\int_0^{\omega} v(t)\, dt, \qquad v_2(t) = v(t) - v_1, \tag{33}$$

such that $v(t) = v_1 + v_2(t)$, we have $\operatorname{codim}\operatorname{Im} L = n$. Therefore, L is a Fredholm mapping of index zero. Furthermore, the generalized inverse (to L) $K_p : \operatorname{Im} L \to \operatorname{Ker} P \cap \operatorname{Dom} L$ is given in the following form:

$$K_p v(t) = \int_0^t v(s)\, ds - \frac{1}{\omega}\int_0^{\omega}\int_0^t v(s)\, ds\, dt. \tag{34}$$

For convenience, we denote $F(t) = (F_1(t), F_2(t), \ldots, F_n(t))$ as follows:

$$\begin{aligned} F_i(t) &= r_i(t) - \sum_{l=1}^{m} a_{iil}(t)\exp\{y_i(t - \tau_{iil}(t))\} \\ &\quad + \sum_{j\neq i}^{n} \sum_{l=1}^{m} a_{ijl}(t)\exp\{y_j(t - \tau_{ijl}(t))\} \\ &\quad - \exp\{-y_i(t)\} d_i(t) u_i(t) - \exp\{-y_i(t)\} \\ &\quad \times e_i(t) u_i(t - \varepsilon_i(t)). \end{aligned} \tag{35}$$

Thus, we have

$$QNy(t) = \frac{1}{\omega}\int_0^{\omega} F(t)\,dt,$$

$$\begin{aligned} K_p(I-Q)Nu(t) &= K_pINu(t) - K_pQNu(t) \\ &= \int_0^t F(s)\,ds - \frac{1}{\omega}\int_0^{\omega}\int_0^t F(s)\,ds\,dt \\ &\quad + \left(\frac{1}{2} - \frac{t}{\omega}\right)\int_0^{\omega} F(s)\,ds. \end{aligned} \tag{36}$$

From formulas (36), we easily see that QN and $K_p(I-Q)N$ are continuous operators. Furthermore, it can be verified that $\overline{K_p(I-Q)N(\overline{\Omega})}$ is compact for any open bounded set $\Omega \subset X$ by using the Arzela-Ascoli theorem and $QN(\overline{\Omega})$ is bounded. Therefore, N is L-compact on $\overline{\Omega}$ for any open bounded subset $\Omega \subset X$.

Now, we reach the position to search for an appropriate open bounded subset Ω for the application of the continuation theorem (Lemma 2) to system (26).

Corresponding to the operator equation $Ly(t) = \lambda Ny(t)$ with parameter $\lambda \in (0,1)$, we have

$$\dot{y}_i(t) = \lambda F_i(t), \quad i = 1, 2, \ldots, n, \tag{37}$$

where $F_i(t)$ $(i = 1, 2, \ldots, n)$ are given in (35).

Assume that $y(t) = (y_1(t), y_2(t), \ldots, y_n(t)) \in X$ is a solution of system (37) for some parameter $\lambda \in (0,1)$. By integrating system (37) over the interval $[0,\omega]$, we obtain

$$\begin{aligned} \int_0^{\omega} \Bigg[& r_i(t) - \sum_{l=1}^{m} a_{iil}(t)\exp\{y_i(t-\tau_{iil}(t))\} \\ & + \sum_{j\neq i}^{n}\sum_{l=1}^{m} a_{ijl}(t)\exp\{y_j(t-\tau_{ijl}(t))\} \\ & - e_i(t)\exp\{-y_i(t)\}\int_{t-\varepsilon_i(t)}^{t-\varepsilon_i(t)+\omega} K(e^{y_i})G_i(t-\varepsilon_i(t),s)\,ds \\ & - d_i(t)\exp\{-y_i(t)\}\int_t^{t+\omega} K(e^{y_i})G_i(t,s)\,ds \Bigg] dt = 0, \\ & i = 1, 2, \ldots, n. \end{aligned} \tag{38}$$

Consequently,

$$\begin{aligned} \overline{r}_i\omega &+ \int_0^{\omega}\left[\sum_{j\neq i}^{n}\sum_{l=1}^{m} a_{ijl}(t)\exp\{y_j(t-\tau_{ijl}(t))\}\right] dt \\ &= \int_0^{\omega}\left[\sum_{l=1}^{m} a_{iil}(t)\exp\{y_i(t-\tau_{iil}(t))\}\right] dt \\ &\quad + \int_0^{\omega} e_i(t)\exp\{-y_i(t)\} \\ &\quad \times \int_{t-\varepsilon_i(t)}^{t-\varepsilon_i(t)+\omega} K(e^{y_i})G_i(t-\varepsilon_i(t),s)\,ds\,dt \\ &\quad + \int_0^{\omega} d_i(t)\exp\{-y_i(t)\}\int_t^{t+\omega} K(e^{y_i})G_i(t,s)\,ds\,dt, \\ &\qquad i = 1, 2, \ldots, n. \end{aligned} \tag{39}$$

From the continuity of $y(t) = (y_1(t), y_2(t), \ldots, y_n(t))$, there exist constants $\xi_i, \eta_i \in [0,\omega]$ $(i = 1, 2, \ldots, n)$ such that

$$y_i(\xi_i) = \max_{t\in[0,\omega]} y_i(t), \qquad y_i(\eta_i) = \min_{t\in[0,\omega]} y_i(t) \qquad i = 1, 2, \ldots, n. \tag{40}$$

By (39) and (40) we obtain

$$\begin{aligned} \overline{r}_i\omega \le & \int_0^{\omega} a_{ii}(t)\exp\{y_i(\xi_i)\}\,dt \\ & + \int_0^{\omega} A_i(t)\exp\{y_i(\xi_i)\}\,dt, \quad i = 1, 2, \ldots, n, \end{aligned} \tag{41}$$

where

$$\begin{aligned} A_i(t) = & \, e_i(t)\int_{t-\varepsilon_i(t)}^{t-\varepsilon_i(t)+\omega} G_i(t-\varepsilon_i(t),s)(\beta_i(s)+\gamma_i(s))\,ds \\ & + d_i(t)\int_t^{t+\omega} G_i(t,s)(\beta_i(s)+\gamma_i(s))\,ds, \\ & i = 1, 2, \ldots, n. \end{aligned} \tag{42}$$

Therefore, we further have

$$y_i(\xi_i) \ge \ln\left(\frac{\overline{r}_i}{\overline{a}_{ii} + \overline{A}_i}\right), \quad i = 1, 2, \ldots, n. \tag{43}$$

Let $s_{ijl}(t) = t - \tau_{ijl}(t)$ $(i, j = 1, 2, \ldots, n,\ l = 1, 2, \ldots, m)$; then from Lemma 1 and (H1) we get that function $s_{ijl}(t)$ has a unique ω periodic inverse function $\varphi_{ijl}(t)$; then, for every $i, j = 1, 2, \ldots, n,\ l = 1, 2, \ldots, m$, we have

$$\begin{aligned} & \int_0^{\omega} a_{ijl}(t)\exp\{y_i(t-\tau_{ijl}(t))\}\,dt \\ & = \int_{-\tau_{ijl}(0)}^{\omega-\tau_{ijl}(\omega)} \frac{a_{ijl}(\varphi_{ijl}(t))}{1-\tau'_{ijl}(\varphi_{ijl}(t))}\exp\{y_i(t)\}\,dt. \end{aligned} \tag{44}$$

One can see that

$$\frac{a_{ijl}(\varphi_{ijl}(t))}{1-\tau'_{ijl}(\varphi_{ijl}(t))} =: \delta_{ijl}(t), \qquad i, j = 1, 2, \ldots, n,\ l = 1, 2, \ldots, m, \tag{45}$$

are ω periodic functions. Then, for every $i, j = 1, 2, \ldots, n,\ l = 1, 2, \ldots, m$, we have

$$\int_0^{\omega} a_{ijl}(t)\exp\{y_i(t-\tau_{ijl}(t))\}\,dt = \int_0^{\omega} \delta_{ijl}(t)\exp\{y_i(t)\}\,dt. \tag{46}$$

From (39) and (46) we further obtain

$$\sum_{i=1}^{n}\left(\int_0^{\omega}\left(\sum_{l=1}^{m}\delta_{iil}(t)\exp\{y_i(t)\} + d_i(t)\exp\{-y_i(t)\}\int_t^{t+\omega}K(e^{y_i})G_i(t,s)\,ds + e_i(t)\exp\{-y_i(t)\} \times\int_{t-\varepsilon_i(t)}^{t-\varepsilon_i(t)+\omega}K(e^{y_i})G_i(t-\varepsilon_i(t),s)\,ds\right)dt\right) = \sum_{i=1}^{n}\left(\overline{r}_i\omega + \int_0^{\omega}\left[\sum_{j\neq i}^{n}\sum_{l=1}^{m}\delta_{ijl}(t)\exp\{y_j(t)\}\right]dt\right). \quad (47)$$

From the above equality we have

$$\sum_{i=1}^{n}\left(\int_0^{\omega}\left[\sum_{l=1}^{m}\delta_{iil}(t) - \sum_{j\neq i}^{n}\sum_{l=1}^{m}\delta_{jil}(t)\right]\exp\{y_i(t)\}\,dt\right) \leq \sum_{i=1}^{n}\overline{r}_i\omega,$$
$$\int_0^{\omega}\left[\sum_{l=1}^{m}\delta_{iil}(t) - \sum_{j\neq i}^{n}\sum_{l=1}^{m}\delta_{jil}(t)\right]\exp\{y_i(t)\}\,dt \leq \sum_{i=1}^{n}\overline{r}_i\omega, \quad i=1,2,\ldots,n. \quad (48)$$

From the assumptions of Theorem 4 and (48) we can obtain

$$\zeta_i\int_0^{\omega}\exp\{y_i(t)\}\,dt \leq \sum_{i=1}^{n}\overline{r}_i\omega, \quad i=1,2,\ldots,n. \quad (49)$$

Consequently,

$$\int_0^{\omega}\exp\{y_i(t)\}\,dt \leq \frac{\sum_{i=1}^{n}\overline{r}_i\omega}{\zeta_i}, \quad i=1,2,\ldots,n. \quad (50)$$

From (40) and (50), we further obtain

$$y_i(\eta_i) \leq \ln\left(\frac{\sum_{i=1}^{n}\overline{r}_i}{\zeta_i}\right), \quad i=1,2,\ldots,n. \quad (51)$$

On the other hand, directly from system (26) we have

$$\begin{aligned}\int_0^{\omega}|\dot{y}_i(t)|\,dt &\leq \int_0^{\omega}|r_i(t)|\,dt + \int_0^{\omega}\left(\sum_{l=1}^{m}a_{iil}(t)\exp\{y_i(t-\tau_{iil}(t))\} + \sum_{j\neq i}^{n}\sum_{l=1}^{m}a_{ijl}(t)\exp\{y_j(t-\tau_{ijl}(t))\}\right)dt \\ &\leq \int_0^{\omega}|r_i(t)|\,dt + \int_0^{\omega}\sum_{l=1}^{m}\left(\delta_{iil}(t) + \sum_{j\neq i}^{n}\delta_{jil}(t)\right)\exp\{y_i(t)\}\,dt \\ &\leq \int_0^{\omega}|r_i(t)|\,dt + D_i^M\int_0^{\omega}\exp\{y_i(t)\}\,dt \\ &\leq |\overline{r}_i|\,\omega + \frac{D_i^M\sum_{i=1}^{n}\overline{r}_i\omega}{\zeta_i}, \quad i=1,2,\ldots,n,\end{aligned} \quad (52)$$

where

$$D_i(t) = \sum_{l=1}^{m}\left(\delta_{iil}(t) + \sum_{j\neq i}^{n}\delta_{jil}(t)\right), \quad i=1,2,\ldots,n. \quad (53)$$

From (43), (51), and (52), we have for any $t\in[0,\omega]$

$$y_i(t) \leq y_i(\eta_i) + \int_0^{\omega}|\dot{y}_i(t)|\,dt \leq \ln\left(\frac{\sum_{i=1}^{n}\overline{r}_i}{\zeta_i}\right) + |\overline{r}_i|\,\omega + \frac{D_i^M\sum_{i=1}^{n}\overline{r}_i\omega}{\zeta_i} =: M_i, \quad i=1,2,\ldots,n, \quad (54)$$

$$y_i(t) \geq y_i(\xi_i) - \int_0^{\omega}|\dot{y}_i(t)|\,dt \geq \ln\left(\frac{\overline{r}_i}{\overline{a}_{ii}+\overline{A}_i}\right) - |\overline{r}_i|\,\omega - \frac{D_i^M\sum_{i=1}^{n}\overline{r}_i\omega}{\zeta_i} =: N_i, \quad i=1,2,\ldots,n. \quad (55)$$

Therefore, from (54) and (55), we have

$$\max_{t\in[0,\omega]}|y_i(t)| \leq \max\{|M_i|,|N_i|\} =: B_i, \quad i=1,2,\ldots,n. \quad (56)$$

It can be seen that the constants B_i $(i = 1, 2, \ldots, n)$ are independent of parameter $\lambda \in (0, 1)$. For any $y = (y_1, y_2, \ldots, y_n) \in R^n$, from (31) we can obtain

$$QNy = (QNy_1, QNy_2, \ldots, QNy_n)$$

$$QNy_i = \overline{r}_i - \overline{H}_i - \overline{a}_{ii} \exp\{y_i\} + \sum_{j \neq i}^{n} \overline{a}_{ij} \exp\{y_j\}, \quad i = 1, 2, \ldots, n, \tag{57}$$

$$H_i(t) = \int_0^{\omega} (\beta_i(s) + \gamma_i(s))(d_i(t) G_i(t, s)) + e_i(t) G_i(t - \varepsilon_i(t), s)\, ds, \quad i = 1, 2, \ldots, n.$$

We consider the following algebraic equation:

$$\overline{r}_i - \overline{H}_i - \overline{a}_{ii} v_i + \sum_{j \neq i}^{n} \overline{a}_{ij} v_j = 0, \quad i = 1, 2, \ldots, n. \tag{58}$$

From the assumption of Theorem 4, the equation has a unique positive solution $v^* = (v_1^*, v_2^*, \ldots, v_n^*)$. Hence, the equation $QNy = 0$ has a unique solution $y^* = (y_1^*, y_2^*, \ldots, y_n^*) = (\ln v_1^*, \ln v_2^*, \ldots, \ln v_n^*) \in R^n$.

Choosing constant $B > 0$ large enough such that $|y_1^*| + |y_2^*| + \cdots + |y_n^*| < B$ and $B > B_1 + B_2 + \cdots + B_n$, we define a bounded open set $\Omega \subset X$ as follows:

$$\Omega = \{y \in X : \|y\| < B\}. \tag{59}$$

It is clear that Ω satisfies conditions (a) and (b) of Lemma 2. On the other hand, by directly calculating we can obtain

$$\deg\{JQN, \Omega \cap \operatorname{Ker} L, (0, 0, \ldots, 0)\} = \operatorname{sgn} \begin{vmatrix} f_{y_1}^1 & f_{y_2}^1 & \cdots & f_{y_n}^1 \\ f_{y_1}^2 & f_{y_2}^2 & \cdots & f_{y_n}^2 \\ \cdots & \cdots & \cdots & \cdots \\ f_{y_1}^n & f_{y_2}^n & \cdots & f_{y_n}^n \end{vmatrix}, \tag{60}$$

where

$$f_{y_j}^i = -\overline{a}_{ij} \exp\{y_j^*\}, \quad i = j$$
$$f_{y_j}^i = \overline{a}_{ij} \exp\{y_j^*\}, \quad i \neq j\ i, j = 1, 2, \ldots, n. \tag{61}$$

From the assumption of Theorem 4, we have

$$\begin{vmatrix} f_{y_1}^1 & f_{y_2}^1 & \cdots & f_{y_n}^1 \\ f_{y_1}^2 & f_{y_2}^2 & \cdots & f_{y_n}^2 \\ \cdots & \cdots & \cdots & \cdots \\ f_{y_1}^n & f_{y_2}^n & \cdots & f_{y_n}^n \end{vmatrix} \neq 0. \tag{62}$$

From this, we finally have

$$\deg\{JQN, \Omega \cap \operatorname{Ker} L, (0, 0, \ldots, 0)\} \neq 0. \tag{63}$$

This shows that Ω satisfies condition (c) of Lemma 2. Therefore, system (26) has an ω-periodic solution $y^*(t) = (y_1^*(t), y_2^*(t), \ldots, y_n^*(t)) \in \overline{\Omega}$. Finally, we have system that (4) has a positive ω-periodic solution. This completes the proof.

Acknowledgments

This work was supported by the Natural Science Foundation of Xinjiang (Grants no. 2012223177, 200721104) and the National Natural Science Foundation of China (Grants no. 10661010, 11261061).

References

[1] X. Song and L. Chen, "Persistence and global stability for nonautonomous predator-prey system with diffusion and time delay," *Computers and Mathematics with Applications*, vol. 35, no. 6, pp. 33–40, 1998.

[2] Y. Pinghua and X. Rui, "Global attractivity of the periodic Lotka-Volterra system," *Journal of Mathematical Analysis and Applications*, vol. 233, no. 1, pp. 221–232, 1999.

[3] F. Wei, Y. Lin, L. Que, Y. Chen, Y. Wu, and Y. Xue, "Periodic solution and global stability for a nonautonomous competitive Lotka-Volterra diffusion system," *Applied Mathematics and Computation*, vol. 216, no. 10, pp. 3097–3104, 2010.

[4] R. Tan, Z. Liu, and R. A. Cheke, "Periodicity and stability in a single-species model governed by impulsive differential equation," *Applied Mathematical Modelling*, vol. 36, no. 3, pp. 1085–1094, 2012.

[5] K. Deimling, *Nonlinear Functional Analysis*, Springer, Beijing, China, 1988.

[6] Y. Xiao, W. Wang, and J. Chen, "Permanence and periodic solution of nonautonomous delay Lotka-Volterra diffusion system," *Systems Science and Mathematical Sciences*, vol. 12, no. 4, pp. 344–349, 1999.

[7] T. Zhidong, "The almost periodic Kolmogorov competitive systems," *Nonlinear Analysis: Theory, Methods and Applications*, vol. 42, no. 7, pp. 1221–1230, 2000.

[8] F. Chen, "On the periodic solutions of periodic multi-species Kolmogorov type competitive system with delays and feedback controls," *Applied Mathematics and Computation*, vol. 180, no. 1, pp. 366–373, 2006.

[9] K. Deimling, *Nonlinear Functional Analysis*, Springer, New York, NY, USA, 1985.

[10] M. A. Krasnoselskii, *Positive Solutions of Operator Equations*, P. Noordhoff, Groningen, The Netherlands, 1964.

[11] D. J. Guo and V. Lakshmikantham, *Nonlinear Problems in Abstract Cones*, vol. 5, Academic Press, New York, NY, USA, 1988.

[12] J. Yan, "Global positive periodic solutions of periodic n-species competition systems," *Journal of Mathematical Analysis and Applications*, vol. 356, no. 1, pp. 288–294, 2009.

[13] W. Yan and J. Yan, "Periodicity and asymptotic stability of a predator-prey system with infinite delays," *Computers and Mathematics with Applications*, vol. 60, no. 5, pp. 1465–1472, 2010.

[14] J. Yan and G. Liu, "Periodicity and stability for a Lotka-Volterra type competition system with feedback controls and deviating arguments," *Nonlinear Analysis: Theory, Methods and Applications*, vol. 74, no. 9, pp. 2916–2928, 2011.

[15] W. A. Horn, "Some fixed point theorems for compact maps and flows in Banach spaces," *Transactions of the American Mathematical Society*, vol. 149, pp. 391–404, 1970.

[16] Z. D. Teng and L. S. Chen, "The positive periodic solutions of periodic Kolmogorov type systems with delays," *Acta Mathematicae Applicatae Sinica*, vol. 22, no. 3, pp. 446–456, 1999 (Chinese).

[17] H.-F. Huo and W.-T. Li, "Positive periodic solutions of a class of delay differential system with feedback control," *Applied Mathematics and Computation*, vol. 148, no. 1, pp. 35–46, 2004.

[18] P. Weng, "Existence and global stability of positive periodic solution of periodic integrodifferential systems with feedback controls," *Computers and Mathematics with Applications*, vol. 40, no. 6-7, pp. 747–759, 2000.

[19] F. Yang and D. Jiang, "Existence and global attractivity of positive periodic solution of a logistic growth system with feedback control and deviating arguments," *Annals of Differential Equations*, vol. 17, no. 4, pp. 377–384, 2001.

[20] Y. Li, P. Liu, and L. Zhu, "Positive periodic solutions of a class of functional differential systems with feedback controls," *Nonlinear Analysis: Theory, Methods and Applications*, vol. 57, no. 5-6, pp. 655–666, 2004.

[21] F. Chen, "Positive periodic solutions of neutral Lotka-Volterra system with feedback control," *Applied Mathematics and Computation*, vol. 162, no. 3, pp. 1279–1302, 2005.

[22] D. Hu and Z. Zhang, "Four positive periodic solutions to a Lotka-Volterra cooperative system with harvesting terms," *Nonlinear Analysis: Real World Applications*, vol. 11, no. 2, pp. 1115–1121, 2010.

[23] K. Zhao and Y. Ye, "Four positive periodic solutions to a periodic Lotka-Volterra predatory-prey system with harvesting terms," *Nonlinear Analysis: Real World Applications*, vol. 11, no. 4, pp. 2448–2455, 2010.

[24] H.-F. Huo and W.-T. Li, "Positive periodic solutions of a class of delay differential system with feedback control," *Applied Mathematics and Computation*, vol. 148, no. 1, pp. 35–46, 2004.

[25] H. Fang and Y. Xiao, "Existence of multiple periodic solutions for delay Lotka-Volterra competition patch systems with harvesting," *Applied Mathematical Modelling*, vol. 33, no. 2, pp. 1086–1096, 2009.

[26] G. Lin and Y. Hong, "Periodic solutions to non autonomous predator prey system with delays," *Nonlinear Analysis: Real World Applications*, vol. 10, no. 3, pp. 1589–1600, 2009.

[27] S. Lu, "On the existence of positive periodic solutions to a Lotka Volterra cooperative population model with multiple delays," *Nonlinear Analysis: Theory, Methods and Applications*, vol. 68, no. 6, pp. 1746–1753, 2008.

[28] J. Shen and J. Li, "Existence and global attractivity of positive periodic solutions for impulsive predator-prey model with dispersion and time delays," *Nonlinear Analysis: Real World Applications*, vol. 10, no. 1, pp. 227–243, 2009.

[29] F. Yin and Y. Li, "Positive periodic solutions of a single species model with feedback regulation and distributed time delay," *Applied Mathematics and Computation*, vol. 153, no. 2, pp. 475–484, 2004.

[30] A. Muhammadhaji and Z. Teng, "Positive periodic solutions of n-species Lotka-Volterra cooperative systems with delays," *Vietnam Journal of Mathematics*, vol. 40, no. 4, pp. 453–467, 2012.

[31] A. Muhammadhaji and Z. Teng, "Global attractivity of a periodic delayed N-species model of facultative mutualism," *Discrete Dynamics in Nature and Society*, vol. 2013, Article ID 580185, 11 pages, 2013.

[32] M. Kouche, N. Tatar, and S. Liu, "Permanence and existence of a positive periodic solution to a periodic stage-structured system with infinite delay," *Applied Mathematics and Computation*, vol. 202, no. 2, pp. 620–638, 2008.

[33] L. Zhang, H.-X. Li, and X.-B. Zhang, "Periodic solutions of competition Lotka-Volterra dynamic system on time scales," *Computers and Mathematics with Applications*, vol. 57, no. 7, pp. 1204–1211, 2009.

[34] S. Lu, "On the existence of positive periodic solutions for neutral functional differential equation with multiple deviating arguments," *Journal of Mathematical Analysis and Applications*, vol. 280, no. 2, pp. 321–333, 2003.

[35] Z. Zhang, J. Wu, and Z. Wang, "Periodic solutions of nonautonomous stage-structured cooperative system," *Computers and Mathematics with Applications*, vol. 47, no. 4-5, pp. 699–706, 2004.

[36] R. E. Gaines and J. L. Mawhin, *Coincidence Degree, and Nonlinear Differential Equations*, vol. 568 of *Lecture Notes in Mathematics*, Springer, Berlin, Germany, 1977.

Analysis of Mixed Elliptic and Parabolic Boundary Layers with Corners

Gung-Min Gie,[1] Chang-Yeol Jung,[2] and Roger Temam[3]

[1] *Department of Mathematics, University of California, Riverside, 900 University Avenue, Riverside, CA 92521, USA*
[2] *Ulsan National Institute of Science and Technology, San 194, Banyeon-ri, Eonyang-eup, Ulju Gun, Ulsan 689-798, Republic of Korea*
[3] *The Institute for Scientific Computing and Applied Mathematics, Indiana University, 831 East Third Street, Bloomington, IN 47405, USA*

Correspondence should be addressed to Gung-Min Gie; ggie@math.ucr.edu

Academic Editor: Norio Yoshida

We study the asymptotic behavior at small diffusivity of the solutions, u^ε, to a convection-diffusion equation in a rectangular domain Ω. The diffusive equation is supplemented with a Dirichlet boundary condition, which is smooth along the edges and continuous at the corners. To resolve the discrepancy, on $\partial\Omega$, between u^ε and the corresponding limit solution, u^0, we propose asymptotic expansions of u^ε at any arbitrary, but fixed, order. In order to manage some singular effects near the four corners of Ω, the so-called elliptic and ordinary corner correctors are added in the asymptotic expansions as well as the parabolic and classical boundary layer functions. Then, performing the energy estimates on the difference of u^ε and the proposed expansions, the validity of our asymptotic expansions is established in suitable Sobolev spaces.

1. Introduction

We consider a singularly perturbed convection-diffusion equation in a rectangular domain $\Omega = (0,1) \times (0,1)$:

$$L_\varepsilon u^\varepsilon := -\varepsilon\Delta u^\varepsilon - \partial_x u^\varepsilon = f \quad \text{in } \Omega, \qquad u^\varepsilon = g \quad \text{on } \partial\Omega. \tag{1}$$

Here ε is a small but strictly positive diffusivity parameter, and $f = f(x,y)$ is a given smooth data with $\|D^\alpha f\|_{L^2(\Omega)} \leq \kappa_\alpha$, independent of ε, for some α's as needed in the analysis below. The function $g = g(x,y)$ is assumed to be continuous on $\partial\Omega$ and smooth on each edge of $\partial\Omega$. Namely, defining the restriction of g to the edges of $\partial\Omega$ as follows:

$$g|_{x=0} = g_1(y), \quad g|_{x=1} = g_2(y), \qquad g|_{y=0} = g_3(x), \quad g|_{y=1} = g_4(x), \tag{2}$$

we assume that

$$g_1(0) = g_3(0), \quad g_1(1) = g_4(0), \qquad g_2(0) = g_3(1), \quad g_2(1) = g_4(1), \qquad g_i \text{ is smooth for } 1 \leq i \leq 4. \tag{3}$$

If these compatibility conditions were not satisfied, some additional considerations would be necessary, which we do not address here.

In what follows, we study the asymptotic behavior of the solutions to (1) at small diffusivity ε.

In a very nice related earlier work, [1], the asymptotic behavior of the solutions of a convection-diffusion equation similar to (1) was discussed. More precisely, in a rectangular domain $\widetilde{\Omega} = (0,l_0) \times (0,l_1)$, the authors considered

$$-\varepsilon'\Delta u^{\varepsilon'} - b\partial_x u^{\varepsilon'} + cu^{\varepsilon'} = \widetilde{f} \quad \text{in } \widetilde{\Omega}, \qquad u^{\varepsilon'} = \widetilde{g} \quad \text{on } \partial\widetilde{\Omega}, \tag{4}$$

where $b > 0$, $c \geq 0$ are constants and $\tilde{f}$ is a given smooth function in $\tilde{\Omega}$. The function $\tilde{g}$, satisfying the analogue version of (3) in $\tilde{\Omega}$, is assumed to be continuous and piecewise smooth on $\partial\tilde{\Omega}$. By constructing, in [1], asymptotic expansions of $u^{\varepsilon'}$ with respect to a small diffusivity ε', the boundary layers of (4) were analyzed in a systematic way. The validity of their asymptotic expansions was established using the maximum principle.

Using a simple change of variables which maps $\tilde{\Omega}$ to Ω, and setting $u^{\varepsilon'} = u^{\varepsilon}e^{\lambda x}$ with a suitable λ, one can transform (4) to (1) where the two diffusivity parameters ε and ε', respectively, in (1) and (4) are compatible; that is, ε/ε' is of order one. Hence, via this transformation, our analysis of (1) in this paper is applicable to (4) as well. Our motivations for conducting the present study appear below. In particular we significantly simplify the proofs of [1] and make the study suitable for more general equations or systems, in particular, those which do not satisfy the maximum principle.

In the boundary layer analysis associated with the convection-diffusion equation (1) (or (4)), one of the most important and difficult points is to resolve any possible singularity near the four corners of the rectangular domain Ω. Towards this end, we simplify and improve the method used in [1]. More precisely, concerning the asymptotic expansion of u^{ε}, we introduce the so-called elliptic boundary layer corrector, appearing in Section 3.2, near the inflow boundary at $x = 1$ only, while in [1], it was used near both the inflow and outflow boundaries at $x = 1$ and $x = 0$. This simplification in the construction of the asymptotic expansions is mainly based on an observation that the corner singularities at $(0,0)$ and $(0,1)$ (where the flows go out) can be handled by the ordinary corner corrector, appearing in Section 3.4 (which is much easier to analyze than the elliptic corrector). Using the energy estimates rather than the maximum principle as in [1], we prove the validity of our proposed asymptotic expansions in suitable Sobolev spaces. Here we make use of the Hardy inequality (see, e.g., [2]) in the estimates. As we said above, our energy estimation approach can be easily extended to some higher order equations or systems which do not admit a maximum principle.

This paper is organized as follows. In Section 2, we introduce a formal expansion of u^{ε} as the sum of the outer and inner expansions. The outer expansion (outside of the boundary layers), which is easy to obtain, appears in Section 2. Then, in Section 3, by performing the boundary layer analysis in a systematic way, we construct the inner expansion (inside the boundary layers). In Section 4, we state and prove the main convergence result, Theorem 9, concerning the difference between u^{ε} and the proposed asymptotic expansion at a given order. In addition, in the appendix, we prove Lemma 1 that contains some delicate pointwise estimates on the parabolic boundary layer correctors introduced in Section 3.1. The study of the elliptic corrector and its approximation also appears in the appendix.

Throughout this paper, we use the notation conventions below.

Notation 1. $\kappa := \kappa(f, g, \Omega, n)$ is a constant depending on the data, but independent of ε; here $n \geq 0$ is the fixed (but arbitrary) order of the asymptotic expansion, as it appears in (6).

Notation 2. An e.s.t. is a function (or a constant) whose norm in all Sobolev spaces H^s (and thus spaces C^s) is exponentially small with, for each s, a bound of the form $c_1 e^{-c_2/\varepsilon^{\gamma}}$, $c_1, c_2, \gamma > 0$, c_i, γ depending possibly on s.

Notation 3. For a fixed $\varepsilon > 0$, we define the energy norm of $H^1(\Omega)$,

$$\|\cdot\|_{\varepsilon} := \sqrt{\varepsilon}\, \|\cdot\|_{H^1(\Omega)} + \|\cdot\|_{L^2(\Omega)}. \tag{5}$$

2. Asymptotic Expansions

To study the asymptotic behavior of u^{ε}, solution of (1), we propose an asymptotic expansion u^{ε} of the following type:

$$u^{\varepsilon} \cong \sum_{j=0}^{n} \varepsilon^{j} \left(u^{j} + \Theta^{j}\right). \tag{6}$$

Here, at each order of ε^j, $j \geq 0$, u^j corresponds to the outer expansion (outside of the boundary layer) of u^{ε}. To balance the discrepancy on the boundary $\partial\Omega$ of u^{ε} and of the u^j, $j \geq 0$, we introduce the correctors Θ^j, $j \geq 0$, which will contribute mainly inside of the boundary layer: Θ^j will be itself the sum of several boundary layer functions as we will see below. The asymptotic expansion (6) is made at order $n \geq 0$. As we will see below, the expansion itself depends on n, but n, which is set at the beginning of the study, can be chosen arbitrary large.

To determine the asymptotic expansion (6), we start with the outer expansion for u^{ε}, $u^{\varepsilon} \cong \sum_{j=0}^{n} \varepsilon^j u^j$. Inserting this expansion into (1), we formally find the equations for the u^j:

$$\begin{aligned} -\partial_x u^0 &= f, \\ -\partial_x u^j &= \Delta u^{j-1}, \quad 1 \leq j \leq n. \end{aligned} \tag{7}$$

We supplement these equations with the following inflow boundary conditions (see (1)–(3)):

$$\begin{aligned} u^0 &= g_2(y) \quad \text{at } x = 1, \\ u^j &= 0 \quad \text{at } x = 1, \ 1 \leq j \leq n. \end{aligned} \tag{8}$$

While (7) are "natural", the choice of the boundary conditions in (8) is not obvious; it will be eventually justified by the convergence theorem below. Then, by integrating the equations in x, we find the smooth outer solutions u^j in the following form:

$$u^0 = \int_x^1 f(x_1, y)\, dx_1 + g_2(y), \tag{9}$$

$$u^j = \int_x^1 \Delta u^{j-1}(x_1, y)\, dx_1, \quad 1 \leq j \leq n. \tag{10}$$

Under this construction of the outer expansion, we notice that $u^\varepsilon \cong \sum_{j=0}^{n} \varepsilon^j u^j$ satisfies the boundary condition $(1)_2$ along the edge $x = 1$ only, and not on the three other sides of $\partial\Omega$, $y = 0$, $y = 1$, and $x = 0$ (in general). Hence, we expect boundary layers to occur near those edges, and the so-called correctors (corresponding to the Θ^j) will be necessary to account for these discrepancies.

In Section 3, performing the boundary layer analysis, we define, for each $0 \le j \le n$, the corrector Θ^j appearing in (11) below. Then, in Section 4, using (6), (9), (10), and (11), we prove our main result which is stated as Theorem 9.

3. Construction of the Correctors: Boundary Layers Analysis

We want the boundary value of u^ε to match that of its approximation. In general, the boundary values of the diffusive solution u^ε and the outer expansion $\sum_{j=0}^{n} \varepsilon^j u^j$ match only at $x = 1$, but not on the other sides of $\partial\Omega$, $y = 0$, $y = 1$, and $x = 0$. Hence, we expect that some boundary layers will occur near those three edges. To resolve this inconsistency, we will define a number of boundary layer functions on the sides and corners of $\partial\Omega$, following an analysis reminiscent of the theory of the Prandtl boundary layer in [3–6].

For the moment, comparing u^ε with a finite sum $\sum_{j=0}^{n} \varepsilon^j u^j$, we infer from the definition of the u^j in (7)–(10) that the values of these functions coincide on the segment $x = 1$. In general, they will not coincide anywhere else on $\partial\Omega$. First, to correct the inconsistency on the sides $y = 0, 1$, we will define the so-called parabolic correctors $\varphi^j = \varphi_l^j + \varphi_u^j$. Then, to correct the inconsistency at $x = 0$, we will introduce the (classical) corrector θ^j. However, this will not be enough because additional inconsistencies appear at the corners of Ω, and, as we shall see below, it will be necessary to introduce additional correctors ξ^j, ζ^j, and η^j to handle these inconsistencies.

In summary, in the subsections below, we will define the Θ^j in the form,

$$\Theta^j = \varphi^j + \xi^j + \theta^j + \zeta^j + \eta^j, \quad 0 \le j \le n, \tag{11}$$

where the role and location of each corrector are explained, respectively, in Box 1 and Figure 1. The stretched variables of the corrector functions, which will be introduced and used in the following subsections, are indicated in the following equation:

$$\varphi_l^j = \varphi_l^j\left(x, \frac{y}{\sqrt{\varepsilon}}\right), \qquad \varphi_u^j = \varphi_u^j\left(x, \frac{(1-y)}{\sqrt{\varepsilon}}\right),$$
$$\xi_l^j = \xi_l^j\left(\frac{(1-x)}{\varepsilon}, \frac{y}{\varepsilon}\right), \qquad \xi_u^j = \xi_u^j\left(\frac{(1-x)}{\varepsilon}, \frac{(1-y)}{\varepsilon}\right),$$
$$\theta^j = \theta^j\left(\frac{x}{\varepsilon}, y\right),$$
$$\zeta_l^j = \zeta_l^j\left(\frac{x}{\varepsilon}, \frac{y}{\sqrt{\varepsilon}}\right), \qquad \zeta_u^j = \zeta_u^j\left(\frac{x}{\varepsilon}, \frac{(1-y)}{\varepsilon}\right),$$
$$\eta_l^j = \eta_l^j\left(\frac{x}{\varepsilon}, y\right), \qquad \eta_u^j = \eta_u^j\left(\frac{x}{\varepsilon}, y\right). \tag{12}$$

3.1. Parabolic Boundary Layers (PBL). At the bottom (or top) boundary, that is, at $y = 0$ (or $y = 1$), to balance the discrepancy of u^ε and the outer expansion $\sum_{j=0}^{n} \varepsilon^j u^j$, (see the u^j as defined in (9) and (10)), we construct below the parabolic boundary layer correctors φ^j, $0 \le j \le n$.

We formally insert the asymptotic expansion $u^\varepsilon \cong \sum_{j=0}^{n} \varepsilon^j \varphi^j$ into (1). Using the arguments similar to those of the Prandtl theory, we see that the thickness of the boundary layer near $y = 0$ or $y = 1$ should be of order $\varepsilon^{1/2}$. Then, by collecting the terms at the same order of ε^j, $0 \le j \le n$, we find the equations of the parabolic boundary layer correctors:

$$-\varepsilon\partial_y^2\varphi^j - \partial_x\varphi^j = \partial_x^2\varphi^{j-1} \quad \text{in } \Omega, \ 0 \le j \le n, \tag{13}$$

where we set $\varphi^{-1} = 0$.

Along the boundaries $y = 0$, $y = 1$, and $x = 1$, comparing $(1)_2$ and the boundary values of the functions in (9) and (10), we find the boundary conditions of φ^j, $0 \le j \le n$, which read.

$$\begin{aligned}
&\varphi^j = 0 \quad \text{at } x = 1, \ 0 \le j \le n,\\
&\varphi^0 = g_3(x) - u^0(x,0) \quad \text{at } y = 0,\\
&\varphi^j = -u^j(x,0) \quad \text{at } y = 0, \ 1 \le j \le n,\\
&\varphi^0 = g_4(x) - u^0(x,1) \quad \text{at } y = 1,\\
&\varphi^j = -u^j(x,1) \quad \text{at } y = 1, \ 1 \le j \le n.
\end{aligned} \tag{14}$$

We will show below how to actually construct φ^j. For the moment, comparing the functions u^ε and $\sum_{j=0}^{n} \varepsilon^j(u^j - \varphi^j)$ for some $n \ge 0$, we see that the boundary values of these functions coincide at $x = 1$ and $y = 0, 1$. However this has been achieved at the price of introducing some new inconsistencies. Indeed we notice, in (14), that, for each $j \ge 0$, the boundary conditions at $y = 0$ and $y = 1$ are not consistent with that at $x = 1$. That is, in general, for any $i \ge 1$,

$$\partial_x^i u^0(1,0) - \partial_x^i g_3(1) \ne 0, \qquad \partial_x^i u^0(1,1) - \partial_x^i g_4(1) \ne 0,$$
$$\partial_x^i u^j(1,0) \ne 0, \qquad \partial_x^i u^j(1,1) \ne 0, \quad j \ge 1. \tag{15}$$

Due to this inconsistency of the boundary data, if we define φ^j, $0 \le j \le n$, as a solution of (13) with (14), some derivatives of φ^j, $0 \le j \le n$ get singular at the two corners $(1,k)$, $k = 0, 1$. Therefore, to define smooth (smoother) correctors φ^j, $0 \le j \le n$, we will modify the boundary conditions (14) as introduced in (23).

$\varphi^j = \varphi_l^j + \varphi_u^j$ is the parabolic boundary layer corrector near $y = 0$ and $y = 1$,
$\xi^j = \xi_l^j + \xi_u^j$ is the elliptic corrector which resolves the compatibility issues, in the construction of φ^j, at the corners $(1, 0)$ and $(1, 1)$,
θ^j is the ordinary boundary layer corrector near $x = 0$,
$\zeta^j = \zeta_l^j + \zeta_u^j$ is the ordinary corner layer corrector which manages the effect of φ^j along $x = 0$,
$\eta^j = \eta_l^j + \eta_u^j$ is the supplementary corrector that manages the effects of θ^j and ζ^j along the edges $y = 0$ and $y = 1$.

Box 1: The different correctors.

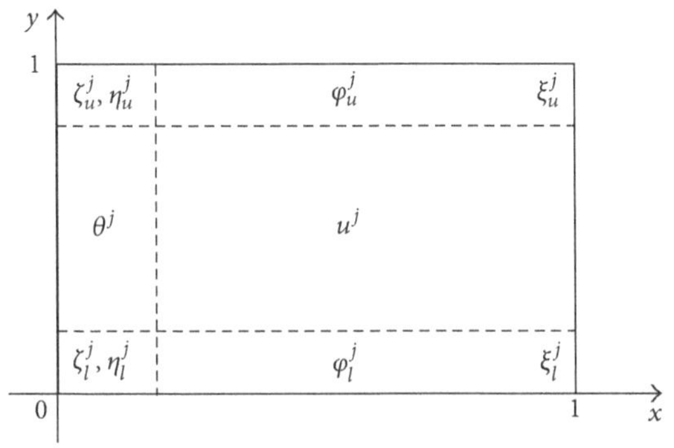

FIGURE 1: Location of the outer solution u^j and the boundary layer correctors.

3.1.1. On Modification of Boundary Conditions for Smooth Correctors at $y = 0, 1$. Equation (13) with initial and boundary conditions (14) is a parabolic problem (heat equation), in which $-x$ is the (positive) time like variable so that the boundary conditions at $x = 1$ are the analogue of the initial conditions for an evolution problem, and the possible inconsistencies in (15) correspond to the (analogue of) the compatibility conditions between the initial and boundary values for a parabolic equation; see [7–9] and the references therein. Due to the lack of consistency at the corners $(1, 0)$ and $(1, 1)$, the correctors constructed as in (13) and (14) might get singular derivatives at those corners. We will overcome this difficulty by considering some corner functions at $x = 1$ and $y = 0, 1$, similar to what is done in [10, 11] or [12] for the compatibility issue in parabolic problems; see [13] as well.

Defining $\delta(x)$ as a smooth cut-off function, independent of ε, such that

$$\delta(x) = \begin{cases} 0 & \text{for } 0 \le x \le \dfrac{1}{2}, \\ 1 & \text{for } x \ge \dfrac{3}{4}, \end{cases} \tag{16}$$

we set

$$\gamma_k^j(x) = \tilde{\gamma}_k^j(x)\,\delta(x), \quad k = 0, 1, \tag{17}$$

where

$$\begin{aligned} \tilde{\gamma}_k^0(x) &= \sum_{i=1}^{2n+1} \frac{(x-1)^i}{i!} \left[\partial_x^i g_{3+k}(1) - \partial_x^i u^0(1,k)\right], \\ \tilde{\gamma}_k^j(x) &= -\sum_{i=1}^{2n+1-2j} \frac{(x-1)^i}{i!} \partial_x^i u^j(1,k), \quad 1 \le j \le n. \end{aligned} \tag{18}$$

Note that the $\tilde{\gamma}_k^0$, $k = 0, 1$, depend on n, but this dependency is not made explicit to make the notations (slightly) simpler. Note also (comparing to (15)) that, at $x = 1$, for each $k = 0, 1$,

$$\begin{aligned} \tilde{\gamma}_k^j(1) &= 0, \quad 0 \le j \le n, \\ \partial_x^i \tilde{\gamma}_k^0(1) &= \partial_x^i g_{3+k}(1) - \partial_x^i u^0(1,k), \quad 1 \le i \le 2n+1, \\ \partial_x^i \tilde{\gamma}_k^j(1) &= -\partial_x^i u^j(1,k), \quad 1 \le j \le n, \ 1 \le i \le 2n+1-j. \end{aligned} \tag{19}$$

We introduce the stretched variables,

$$\overline{y} = \frac{y}{\sqrt{\varepsilon}}, \qquad \tilde{y} = \frac{(1-y)}{\sqrt{\varepsilon}}. \tag{20}$$

Then, using (13), (14), (17), and (20), we define the parabolic corrector, φ^j, $0 \le j \le n$, as

$$\varphi^j = \varphi_l^j + \varphi_u^j = \varphi_l^j(x, \overline{y}) + \varphi_u^j(x, \tilde{y}), \tag{21}$$

where φ_l^j and φ_u^j (lower and upper parts of φ^j, resp.) are defined as the solutions of the 1D heat equations below.

For $0 \le j \le n$,

$$\begin{aligned} -\partial_{\overline{y}}^2 \varphi_l^j - \partial_x \varphi_l^j &= \partial_x^2 \varphi_l^{j-1} \quad \text{for } (\overline{y}, x) \in \mathbb{R}^+ \times (0,1), \\ \varphi_l^j &= h_0^j(x) \quad \text{at } \overline{y} = 0, \\ \varphi_l^j &\longrightarrow 0 \quad \text{as } \overline{y} \longrightarrow \infty, \\ \varphi_l^j &= 0 \quad \text{at } x = 1, \\ -\partial_{\tilde{y}}^2 \varphi_u^j - \partial_x \varphi_u^j &= \partial_x^2 \varphi_u^{j-1} \quad \text{for } (\tilde{y}, x) \in \mathbb{R}^+ \times (0,1), \\ \varphi_u^j &= h_1^j(x) \quad \text{at } \tilde{y} = 0, \\ \varphi_u^j &\longrightarrow 0 \quad \text{as } \tilde{y} \longrightarrow \infty, \\ \varphi_u^j &= 0 \quad \text{at } x = 1. \end{aligned} \tag{22}$$

Here the $h_k^j(x)$, $0 \le j \le n$, $k = 0, 1$, are defined by

$$\begin{aligned} h_0^0(x) &:= g_3(x) - u^0(x,0) - \gamma_0^0(x), \\ h_0^j(x) &:= -u^j(x,0) - \gamma_0^j(x), \quad 1 \le j \le n, \\ h_1^0(x) &:= g_4(x) - u^0(x,1) - \gamma_1^0(x), \\ h_1^j(x) &:= -u^j(x,1) - \gamma_1^j(x), \quad 1 \le j \le n. \end{aligned} \tag{23}$$

Note that the smooth boundary conditions $h_k^j(x)$, $0 \le j \le n$, $k = 0, 1$, along $\overline{y}$ or $\tilde{y} = 0$ are compatible with the other 0 boundary (initial) conditions at $x = 1$ in the sense that, for each $0 \le j \le n$,

$$\partial_x^i h_k^j (1) = 0, \quad k = 0, 1, \ 0 \le i \le 2n + 1 - 2j. \tag{24}$$

From [14–16] or [1], we recall the explicit expressions of $\varphi_l^j = \varphi_l^j(x, \overline{y})$, $0 \le j \le n$,

$$\varphi_l^0 = \sqrt{\frac{2}{\pi}} \int_{\overline{y}/\sqrt{2(1-x)}}^{\infty} \exp\left(-\frac{y_1^2}{2}\right) h_0^0 \left(x + \frac{\overline{y}^2}{2y_1^2}\right) dy_1, \tag{25}$$

$$\begin{aligned}\varphi_l^j = {} & \sqrt{\frac{2}{\pi}} \int_{\overline{y}/\sqrt{2(1-x)}}^{\infty} \exp\left(-\frac{y_1^2}{2}\right) h_0^j \left(x + \frac{\overline{y}^2}{2y_1^2}\right) dy_1 \\ & + \frac{1}{2\sqrt{\pi}} \int_0^{1-x} \int_0^{\infty} \frac{1}{\sqrt{x_1}} \left\{ \exp\left(-\frac{(\overline{y} - y_1)^2}{4x_1}\right) \right. \\ & \left. - \exp\left(-\frac{(\overline{y} + y_1)^2}{4x_1}\right) \right\} \\ & \times \partial_x^2 \varphi_l^{j-1} (x + x_1, y_1)\, dy_1 dx_1. \end{aligned} \tag{26}$$

The expressions of $\varphi_u^j = \varphi_u^j(x, \tilde{y})$, $0 \le j \le n$, are identical to (25) and (26) with h_0^0, h_0^j, and $\overline{y}$, respectively, replaced by h_1^0, h_1^j and $\tilde{y}$.

3.1.2. Estimates on the Parabolic Boundary Layers. We now state some pointwise estimates for the φ_l^j and φ_u^j, $0 \le j \le n$, which are proved in the appendix.

Lemma 1. *For each $0 \le j \le n$ and $s \ge 0$, one has the following pointwise estimates:*

$$\left| y^s \partial_x^i \partial_y^m \varphi_l^j \right| \le \kappa \varepsilon^{(s-m)/2} \exp\left(-c\frac{y}{\sqrt{\varepsilon}}\right), \quad 0 \le i + m \le 2n + 2 - 2j, \tag{27}$$

$$\left| y^s \partial_x^i \partial_y^m \varphi_u^j \right| \le \kappa \varepsilon^{(s-m)/2} \exp\left(-c\frac{1-y}{\sqrt{\varepsilon}}\right), \quad 0 \le i + m \le 2n + 2 - 2j, \tag{28}$$

for a generic constant $c > 0$ independent of x, y, and ε.

From Lemma 1, it is easy to deduce the following L^p estimates.

Lemma 2. *For each $0 \le j \le n$ and $s \ge 0$, one has, for $0 \le i + m \le 2n + 2 - 2j$,*

$$\left\| y^s \partial_x^i \partial_y^m \varphi_l^j \right\|_{L^p(\Omega)} + \left\| y^s \partial_x^i \partial_y^m \varphi_u^j \right\|_{L^p(\Omega)} \kappa \varepsilon^{(s-m)/2 + 1/2p}. \tag{29}$$

Using (21) and (22), we find the equations for $\varphi^j = \varphi_l^j + \varphi_u^j$, $0 \le j \le n$:

$$\begin{aligned} -\varepsilon \partial_y^2 \varphi^j - \partial_x \varphi^j = \partial_x^2 \varphi^{j-1} & \quad \text{in } \Omega, \\ \varphi^j = h_0^j (x) + \varphi_u^j\big|_{y=0} & \quad \text{at } y = 0, \\ \varphi^j = h_1^j (x) + \varphi_l^j\big|_{y=1} & \quad \text{at } y = 1, \\ \varphi^j = 0 & \quad \text{at } x = 1. \end{aligned} \tag{30}$$

Thanks to Lemma 1, we notice that

$$\varphi_u^j\big|_{y=0} = \text{e.s.t.}, \qquad \varphi_l^j\big|_{y=1} = \text{e.s.t.}, \tag{31}$$

where e.s.t. denotes an exponentially small term with respect to the (small) parameter ε as defined in Notation 2.

3.2. Elliptic Boundary Layers (EBL). In Section 3.1, to construct the consistent parabolic boundary layer correctors φ^j, $0 \le j \le n$, we considered the modified boundary conditions (23), including γ_k^j, instead of the natural boundary conditions (14). Here, on the two sides $y = 0$ and $y = 1$ of Ω, to cancel γ_k^j which is exactly the difference of (14) and (23), we introduce elliptic boundary layer correctors ξ^j, $0 \le j \le n$, which, like γ_k^j or $\overline{\gamma}_k^j$, depend on the order n of the asymptotic expansion in (6).

Using (17), we define

$$\xi^j = \xi_l^j + \xi_u^j, \quad 0 \le j \le n, \tag{32}$$

where ξ_l^j and ξ_u^j are the solutions of the elliptic systems below.

For $0 \le j \le n$,

$$\begin{aligned} -\varepsilon\, \partial_x^2 \xi_l^j - \varepsilon \partial_y^2 \xi_l^j - \partial_x \xi_l^j = 0 & \quad \text{in } \Omega, \\ \xi_l^j = \gamma_0^j (x) & \quad \text{at } y = 0, \\ \xi_l^j = 0 & \quad \text{at } x = 0, 1, \text{ or } y = 1, \\ -\varepsilon\, \partial_x^2 \xi_u^j - \varepsilon\, \partial_y^2 \xi_u^j - \partial_x \xi_u^j = 0 & \quad \text{in } \Omega, \\ \xi_u^j = \gamma_1^j (x) & \quad \text{at } y = 1, \\ \xi_u^j = 0 & \quad \text{at } x = 0, 1, \text{ or } y = 0. \end{aligned} \tag{33}$$

The systems (33) being elliptic equations, their well-posedness are easy to verify, and hence, we omit the proof here.

The equations satisfied by the elliptic correctors $\xi^j = \xi_l^j + \xi_u^j$ read as follows.

For $0 \le j \le n$,

$$\begin{aligned} -\varepsilon \partial_x^2 \xi^j - \varepsilon \partial_y^2 \xi^j - \partial_x \xi^j = 0 & \quad \text{in } \Omega, \\ \xi^j = \gamma_k^j (x) & \quad \text{at } y = k, \ k = 0, 1, \\ \xi^j = 0 & \quad \text{at } x = 0, 1. \end{aligned} \tag{34}$$

In the error analysis below in Section 4, we will see that no extra terms related to the ξ^j appear; see (65) below. This is because the ξ^j, as appearing in (34), satisfies the same equation as for u^ε, supplemented with the (exact) boundary conditions that we need. From this observation, as it will be justified in Section 4, we notice that our main result in Theorem 9 does not require any estimate on the ξ^j.

As an extra information which might be useful elsewhere, but not in this paper, by performing the energy estimate on (34), we find that

$$\left\|\xi^j\right\|_\varepsilon \leq \kappa \varepsilon^{1/4}, \quad 0 \leq j \leq n. \tag{35}$$

In the appendix, we introduce an explicit approximate $\overline{\xi}$ of ξ. Then, thanks to the estimates on $\overline{\xi}$, one can obtain some pointwise estimates on ξ as well; see Appendix B.

At this stage, the functions u^ε and $\sum_{j=0}^n \varepsilon^j(u^j + \varphi^j + \xi^j)$ match along the sides $x = 1$ and $y = 0, 1$ (with singular derivatives of high orders for the φ^j). More precisely, from (1), (8), (30), (31), and (34), we see that

$$u^\varepsilon - \sum_{j=0}^n \varepsilon^j \left(u^j + \varphi^j + \xi^j\right) = \begin{cases} g_1 + \sum_{j=0}^n \varepsilon^j \left(u^j + \varphi^j\right) & \text{at } x = 0, \\ 0 & \text{at } x = 1, \\ \sum_{j=0}^n \varepsilon^j \varphi_u^j = \text{e.s.t.} & \text{at } y = 0, \\ \sum_{j=0}^n \varepsilon^j \varphi_l^j = \text{e.s.t.} & \text{at } y = 1. \end{cases} \tag{36}$$

We will now deal with the side $x = 0$ and construct the ordinary boundary layer θ^j and the ordinary corner layer functions (correctors) ζ^j.

3.3. Ordinary Boundary Layers (OBL). To handle the discrepancy of u^ε and $\sum_{j=0}^n \varepsilon^j u^j$ at $x = 0$, which is a part of $u^\varepsilon - \sum_{j=0}^n \varepsilon^j(u^j + \varphi^j + \xi^j)$ appearing in (36), the so-called ordinary boundary layer corrector θ^j is introduced in this section. We insert a formal expansion $u^\varepsilon \cong \sum_{j=0}^\infty \varepsilon^j \theta^j$ into the diffusive equation (1). Then, using the stretched variable $\overline{x} = x/\varepsilon$, we find the equations of θ^j:

$$-\varepsilon \partial_x^2 \theta^j - \partial_x \theta^j = \varepsilon^{-1} \partial_y^2 \theta^{j-2} \quad \text{in } \Omega, \ 0 \leq j \leq n, \tag{37}$$

where we set $\theta^{-1} = \theta^{-2} = 0$.

The main role of the θ^j is, at each order of ε^j, to cancel the error $u^\varepsilon - \sum_{j=0}^n \varepsilon^j u^j$ at $x = 0$. Hence, we supplement (37) with the boundary conditions

$$\begin{aligned} \theta^0 &= g_1(y) - u^0(0, y) \quad \text{at } x = 0, \\ \theta^j &= -u^j(0, y) \quad \text{at } x = 0, \ 1 \leq j \leq n, \\ \theta^j &\longrightarrow 0 \quad \text{as } x \longrightarrow \infty, \ 0 \leq j \leq n. \end{aligned} \tag{38}$$

The explicit expressions of θ^j, $0 \leq j \leq n$, are inductively shown to be of the form

$$\theta^j = P^k\left(\frac{x}{\varepsilon}, y\right) \exp\left(\frac{-x}{\varepsilon}\right), \quad j = 2k, 2k+1, \tag{39}$$

where $P^k(x/\varepsilon, y)$ is a polynomial in x/ε of degree k whose coefficients, independent of ε, are linear combinations of the $\partial^s u^{j-s}(0, y)/\partial y^s$, $0 \leq s \leq j$, and $P^k(0, y) = -u^j(0, y)$.

According to (39), we deduce the following lemmas.

Lemma 3. *For each $0 \leq j \leq n$, one has the pointwise estimates:*

$$\left|x^s \partial_x^i \partial_y^m \theta^j\right| \leq \kappa \varepsilon^{s-i} \exp\left(-c\frac{x}{\varepsilon}\right), \quad s, i, m \geq 0, \tag{40}$$

for a constant c independent of x, y, and ε.

Lemma 4. *For each $0 \leq j \leq n$, one has*

$$\left\|x^s \partial_x^i \partial_y^m \theta^j\right\|_{L^p(\Omega)} \leq \kappa \varepsilon^{s-i+(1/p)}, \quad s, i, m \geq 0. \tag{41}$$

Thanks to Lemma 3, we notice that the effect of θ^j near the boundary $x = 1$ is exponentially small:

$$\theta^j(1, y) = \text{e.s.t.}, \quad 0 \leq j \leq n. \tag{42}$$

From (36), (38), and (42), we infer that

$$u^\varepsilon - \sum_{j=0}^n \varepsilon^j \left(u^j + \varphi^j + \xi^j + \theta^j\right) = \begin{cases} \sum_{j=0}^n \varepsilon^j \varphi^j & \text{at } x = 0, \\ \sum_{j=0}^n \varepsilon^j \theta^j = \text{e.s.t.} & \text{at } x = 1, \\ \sum_{j=0}^n \varepsilon^j \left(\varphi_u^j + \theta^j\right) = \sum_{j=0}^n \varepsilon^j \theta^j + \text{e.s.t.} & \text{at } y = 0, \\ \sum_{j=0}^n \varepsilon^j \left(\varphi_l^j + \theta^j\right) = \sum_{j=0}^n \varepsilon^j \theta^j + \text{e.s.t.} & \text{at } y = 1. \end{cases} \tag{43}$$

3.4. Ordinary Corner Layers (OCL). To account for the value $\sum_{j=0}^n \varepsilon^j \varphi^j$ at $x = 0$, which is exactly the difference of u^ε and $\sum_{j=0}^n \varepsilon^j(u^j + \varphi^j + \xi^j + \theta^j)$ at $x = 0$ (see (43)), we introduce the ordinary corner layer correctors ζ^j, $0 \leq j \leq n$, in the form

$$\zeta^j = \zeta_l^j + \zeta_u^j, \tag{44}$$

where the ζ_l^j (or ζ_u^j) are the correctors near the corner $(0, 0)$ (or $(0, 1)$) as constructed below.

To define ζ_l^j (or ζ_u^j), we insert a formal expansion $u^\varepsilon \cong \sum_{j=0}^\infty \varepsilon^j \zeta^j$ into (1). Then, using the stretched variables $\overline{x} = x/\varepsilon$ and $\overline{y} = y/\sqrt{\varepsilon}$ near $(0, 0)$ and using $\overline{x}$ and $\widetilde{y} = (1 - y)/\sqrt{\varepsilon}$ near $(1, 1)$, we collect the terms of order ε^j, $0 \leq j \leq n$. As a result, we find the equations for ζ_l^j (or ζ_u^j):

$$-\varepsilon \partial_x^2 \zeta_k^j - \partial_x \zeta_k^j = \partial_y^2 \zeta_k^{j-1} \quad \text{in } \Omega, \ 0 \leq j \leq n, \ k = u \text{ or } l. \tag{45}$$

Here, we set $\zeta_k^{-1} = 0$ for $k = l, u$.

At each order of ε^j, $0 \le j \le n$, to cancel the error φ^j near the corner $(0, 0)$ or $(0, 1)$, we supplement (45) with the boundary conditions

$$\begin{aligned} &\zeta_k^j = -\varphi_k^j(0, y) \quad \text{at } x = 0,\ 0 \le j \le n,\ k = l, u, \\ &\zeta_k^j \longrightarrow 0 \quad \text{as } x \longrightarrow \infty,\ 0 \le j \le n,\ k = l, u. \end{aligned} \tag{46}$$

The explicit solutions ζ_l^j of (45) and (46), $0 \le j \le n$, are inductively found to be of the following form:

$$\zeta_l^j = P^j\left(\frac{x}{\varepsilon}, \frac{y}{\sqrt{\varepsilon}}\right)\exp\left(\frac{-x}{\varepsilon}\right), \tag{47}$$

where $P^j(x/\varepsilon, y/\sqrt{\varepsilon})$ is a polynomial in x/ε of degree j whose coefficients, independent of ε, are linear combinations of the quantities $\varepsilon^s \partial^{2s}\varphi_l^{j-s}(0, y/\sqrt{\varepsilon})/\partial y^{2s}$, $0 \le s \le j$. In the same manner, the explicit forms of the ζ_u^j, $0 \le j \le n$ can be found as well.

Using Lemma 1 and (47), we obtain the lemmas below.

Lemma 5. *For each $0 \le j \le n$ and $k, s \ge 0$, one has the pointwise estimates:*

$$\begin{aligned} &\left|x^k y^s \partial_x^i \partial_y^m \zeta_l^j\right| \le \kappa\varepsilon^{k-i+((s-m)/2)} \exp\left(-c\left(\frac{x}{\varepsilon} + \frac{y}{\sqrt{\varepsilon}}\right)\right), \\ &\qquad 0 \le i + m \le 2n + 2 - 2j, \\ &\left|x^k y^s \partial_x^i \partial_y^m \zeta_u^j\right| \le \kappa\varepsilon^{k-i+((s-m)/2)} \exp\left(-c\left(\frac{x}{\varepsilon} + \frac{1-y}{\sqrt{\varepsilon}}\right)\right), \\ &\qquad 0 \le i + m \le 2n + 2 - 2j, \end{aligned} \tag{48}$$

for a generic constant $c > 0$ independent of x, y, and ε.

Lemma 6. *For each $0 \le j \le n$ and $k, s \ge 0$, one has, for $0 \le i + m \le 2n + 2 - 2j$,*

$$\begin{aligned} &\left\|x^k y^s \partial_x^i \partial_y^m \zeta_l^j\right\|_{L^p(\Omega)} + \left\|x^k y^s \partial_x^i \partial_y^m \zeta_u^j\right\|_{L^p(\Omega)} \\ &\quad \le \kappa\varepsilon^{k-i+((s-m)/2)+(3/2p)}. \end{aligned} \tag{49}$$

Thanks to (46) and Lemma 5, we estimate the values of ζ^j, $0 \le j \le n$, along the sides $x = 1$, $y = 0$, and $y = 1$:

$$\zeta^j = \begin{cases} \text{e.s.t.} & \text{at } x = 1, \\ \zeta_l^j(x, 0) + \text{e.s.t.} & \text{at } y = 0, \\ \zeta_u^j(x, 0) + \text{e.s.t.} & \text{at } y = 1. \end{cases} \tag{50}$$

Now from (43), (46), and (50), we observe that

$$u^\varepsilon - \sum_{j=0}^n \varepsilon^j\left(u^j + \varphi^j + \xi^j + \theta^j + \zeta^j\right)$$

$$= \begin{cases} 0 & \text{at } x = 0, \\ \sum_{j=0}^n \varepsilon^j\left(\theta^j + \zeta^j\right) = \text{e.s.t.} & \text{at } x = 1, \\ \sum_{j=0}^n \varepsilon^j\left(\varphi_u^j + \theta^j + \zeta^j\right) = \sum_{j=0}^n \varepsilon^j\left(\theta^j + \zeta_l^j\right) + \text{e.s.t.} & \text{at } y = 0, \\ \sum_{j=0}^n \varepsilon^j\left(\varphi_l^j + \theta^j + \zeta^j\right) = \sum_{j=0}^n \varepsilon^j\left(\theta^j + \zeta_u^j\right) + \text{e.s.t.} & \text{at } y = 1. \end{cases} \tag{51}$$

3.5. Supplementary Correctors. Now, our task is to handle, at each order $0 \le j \le n$, the effects of θ^j and ζ^j along the top and bottom boundaries $y = 0, 1$. To this end, using (51), we construct the supplementary correctors η^j in the form

$$\eta^j = \eta_l^j + \eta_u^j, \quad 0 \le j \le n, \tag{52}$$

where

$$\begin{aligned} &\eta_l^j = -(1 - y)\left(\theta^j + \zeta_l^j\right) \quad (x, y = 0), \\ &\eta_u^j = -y\left(\theta^j + \zeta_u^j\right) \quad (x, y = 1). \end{aligned} \tag{53}$$

From Lemmas 3 and 5, it is easy to verify the lemmas below.

Lemma 7. *For each $0 \le j \le n$ and $s \ge 0$, one has the pointwise estimates:*

$$\left|x^s \partial_x^i \partial_y^m \eta^j\right| \le \kappa\varepsilon^{s-i} \exp\left(-c\frac{x}{\varepsilon}\right), \quad 0 \le i + m \le 2n + 2 - 2j, \tag{54}$$

for a constant c independent of x, y, and ε.

Lemma 8. *For each $0 \le j \le n$ and $s \ge 0$, one has*

$$\left\|x^s \partial_x^i \partial_y^m \eta^j\right\|_{L^p(\Omega)} \le \varepsilon^{s-i+(1/p)}, \quad 0 \le i + m \le 2n + 2 - 2j. \tag{55}$$

Using (3), (14), (17), (22), (23), (38), and (46), we find that, near the corner $(0, 0)$,

$$\begin{aligned} \left(\theta^0 + \zeta_l^0\right)(0, 0) &= g_1(0) - u^0(0, 0) - \varphi_l^0(0, 0) \\ &= g_1(0) - u^0(0, 0) - \left(g_3(0) - u^0(0, 0)\right) = 0. \end{aligned} \tag{56}$$

More generally, one can verify that

$$\left(\theta^j + \zeta_l^j\right)(0, 0) = 0, \quad \left(\theta^j + \zeta_u^j\right)(0, 1) = 0, \quad 0 \le j \le n. \tag{57}$$

Using (42), (50), and (57), we infer from (52) and (53) that, for $0 \le j \le n$,

$$\eta^j = \begin{cases} 0 & \text{at } x = 0, \\ \text{e.s.t.} & \text{at } x = 1, \\ -\left(\theta^j + \zeta_l^j\right) & \text{at } y = 0, \\ -\left(\theta^j + \zeta_u^j\right) & \text{at } y = 1. \end{cases} \quad (58)$$

Now, we finally obtain from (51) and (58) that

$$u^\varepsilon - \sum_{j=0}^{n} \varepsilon^j \left(u^j + \varphi^j + \xi^j + \theta^j + \zeta^j + \eta^j\right)$$

$$= \begin{cases} 0 & \text{at } x = 0, \\ \sum_{j=0}^{n} \varepsilon^j \left(\theta^j + \zeta^j + \eta^j\right) = \text{e.s.t.} & \text{at } x = 1, \\ \sum_{j=0}^{n} \varepsilon^j \left(\varphi_u^j + \theta^j + \zeta^j + \eta^j\right) \\ \quad = \sum_{j=0}^{n} \varepsilon^j \left(\theta^j + \zeta_l^j + \eta^j\right) + \text{e.s.t.} = \text{e.s.t.} & \text{at } y = 0, \\ \sum_{j=0}^{n} \varepsilon^j \left(\varphi_l^j + \theta^j + \zeta^j + \eta^j\right) \\ \quad = \sum_{j=0}^{n} \varepsilon^j \left(\theta^j + \zeta_u^j + \eta^j\right) + \text{e.s.t.} = \text{e.s.t.} & \text{at } y = 1. \end{cases} \quad (59)$$

The e.s.t. on the right hand side of $(59)_{2,3,4}$ are caused, respectively, by $(\theta^j + \zeta^j + \eta^j)$ at $x = 1$, $(\varphi_u^j + \zeta_u^j)$ at $y = 0$, and $(\varphi_l^j + \zeta_l^j)$ at $y = 1$.

4. Asymptotic Error Analysis: The Main Result

We define

$$\tilde{\rho} := \left(u^\varepsilon - \sum_{j=0}^{n} \varepsilon^j \left(u^j + \Theta^j\right) \right)\Bigg|_{\partial\Omega} \quad (60)$$
$$= (\text{the right-hand side of } (59)),$$

where $\Theta^j = \varphi^j + \xi^j + \theta^j + \zeta^j + \eta^j$ as defined in (11).

We construct an extension ρ of $\tilde{\rho}$ in the form

$$\rho(x, y) := r + (1 - x)(\tilde{\rho} - r)|_{x=0} + x(\tilde{\rho} - r)|_{x=1}$$
$$+ (1 - y)(\tilde{\rho} - r)|_{y=0} + y(\tilde{\rho} - r)|_{y=1}, \quad (61)$$

where

$$r(x, y) := (1 - x)(1 - y)\tilde{\rho}|_{x=y=0} + x(1 - y)\tilde{\rho}|_{x=1,y=0}$$
$$+ (1 - x)y\tilde{\rho}|_{x=0,y=1} + xy\tilde{\rho}|_{x=y=1}. \quad (62)$$

Then one can verify that $\rho|_{\partial\Omega} = \tilde{\rho}$ and

$$\rho \text{ is an e.s.t. in } \Omega. \quad (63)$$

We set

$$w_{\varepsilon n} := u^\varepsilon - \sum_{j=0}^{n} \varepsilon^j \left(u^j + \Theta^j\right) - \rho. \quad (64)$$

Then, using (1), (7), (11), (30), (34), (37), (45), (53), and (60), the equation of $w_{\varepsilon n}$ reads

$$L_\varepsilon w_{\varepsilon n} = \varepsilon^{n+1} \left(\Delta u^n + \partial_x^2 \varphi^n + \partial_y^2 \theta^n + \Xi_1^n\right)$$
$$+ \varepsilon^n \left(\partial_y^2 \theta^{n-1} + \varepsilon \partial_y^2 \zeta^n + \Xi_2^n\right) + \text{e.s.t.} \quad \text{in } \Omega, \quad (65)$$
$$w_{\varepsilon n} = 0 \quad \text{on } \partial\Omega,$$

where

$$\Xi_1 = (1 - y)\,\partial_y^2 \theta^n (x, y = 0) + y\,\partial_y^2 \theta^n \quad (x, y = 1),$$
$$\Xi_2 = (1 - y)\left(\partial_y^2 \theta^{n-1} + \varepsilon \partial_y^2 \zeta_l^n\right) \quad (x, y = 0) \quad (66)$$
$$+ y\left(\partial_y^2 \theta^{n-1} + \varepsilon \partial_y^2 \zeta_u^n\right) \quad (x, y = 1).$$

We multiply (65) by $e^x w_{\varepsilon n}$ and integrate over Ω. Then, using Lemmas 1, 3, and 5 and the Hardy inequality (see, e.g., [2]), we find that

$$\varepsilon |w_{\varepsilon n}|_{H^1}^2 + \frac{1 - \varepsilon}{2} |w_{\varepsilon n}|_{L^2}^2$$
$$\le \kappa \varepsilon^{n+1} \left| \int_\Omega \left(\Delta u^n + \partial_x^2 \varphi^n + \partial_y^2 \theta^n + \Xi_1^n\right) w_{\varepsilon n} dx\, dy \right|$$
$$+ \kappa \varepsilon^n \left| \int_\Omega x \left(\partial_y^2 \theta^{n-1} + \varepsilon \partial_y^2 \zeta^n + \Xi_2^n\right) \left(\frac{w_{\varepsilon n}}{x}\right) dx\, dy \right| \quad (67)$$
$$\le \kappa \varepsilon^{2n+2} + \frac{1}{4} |w_{\varepsilon n}|_{L^2}^2 + \frac{\varepsilon}{2} |w_{\varepsilon n}|_{H^1}^2.$$

Thanks to the elliptic regularity theory and (67), we also find that

$$\|w_{\varepsilon n}\|_{H^2(\Omega)} \le \kappa \|\Delta w_{\varepsilon n}\|_{L^2(\Omega)}$$
$$\le \kappa \varepsilon^{-1} \|\partial_x w_{\varepsilon n} + \text{lower order terms}\|_{L^2(\Omega)} \quad (68)$$
$$\le \kappa \varepsilon^{n-(1/2)}.$$

Finally, from (63), (64), (67), and (68), we obtain the convergence result on the difference of u^ε and the proposed asymptotic expansions that we summarize in Theorem 9.

Theorem 9. *For each fixed $n \ge 0$, as the diffusivity parameter ε vanishes, the difference between the solution u^ε of (1) and its asymptotic expansion (6) converges to zero in the following sense:*

$$\left\| u^\varepsilon - \sum_{j=0}^{n} \varepsilon^j \left(u^j + \Theta^j\right) \right\|_\varepsilon \le \kappa \varepsilon^{n+1}, \quad (69)$$

$$\left\| u^\varepsilon - \sum_{j=0}^{n} \varepsilon^j (u^j + \Theta^j) \right\|_{H^2(\Omega)} \le \kappa \varepsilon^{n-(1/2)}. \quad (70)$$

Remark 10. Due to the continuity of g as appearing in (3), ξ^0 can be omitted in (11) for $j = 0$ at order ε^0. Then one can verify that

$$\left\| u^{\varepsilon} - \left(u^0 + \varphi^0 + \theta^0 + \zeta^0 + \eta^0\right)\right\|_{\varepsilon} \le \kappa\varepsilon^{3/4}, \qquad (71)$$

which gives a convergence estimate that a factor of $\varepsilon^{1/4}$ is worse than what is stated in (69) for $n = 0$.

Appendix

A. Proof of Lemma 1

To prove (27), for $\overline{y} = y/\sqrt{\varepsilon}$, since $y^s \exp(-c\overline{y}) \le \kappa\varepsilon^{s/2}\exp(-c_0\overline{y})$, for all y, $0 < c_0 < c$, it suffices to prove that

$$\left|\partial_x^i \partial_{\overline{y}}^m \varphi_l^j\right| \le \kappa \exp\left(-c\frac{\overline{y}}{\sqrt{1-x}}\right), \qquad (A.1)$$

$$0 \le i + m \le 2n + 2 - 2j,\ i, m \ge 0.$$

Note that (28) would follow similarly. Since $h^j(x) = h_0^j(x) = -u^j(x,i) - \gamma_0^j(x)$ for $j \ge 1$ and $g_3(x) - u^0(x,0) - \gamma_0^0(x)$ for $j = 0$; thanks to (3), we note that $\partial_x^k h^j(1) = 0$ for $0 \le k \le 2n + 1 - 2j$.

Now, to prove (A.1), we use an induction on j.

(1) We begin with $j = 0$.

From (25), since $\partial_x^k h^0(1) = 0$ for $0 \le k \le 2n + 1$, differentiating φ_l^0 in x we find that

$$\left|\partial_x^i \varphi_l^0\right| = \sqrt{\frac{2}{\pi}}\left|\int_{\overline{y}/\sqrt{2(1-x)}}^{\infty} \exp\left(-\frac{y_1^2}{2}\right)\partial_x^i h^0\left(x + \frac{\overline{y}^2}{2y_1^2}\right)dy_1\right|$$
$$\le \kappa\int_{\overline{y}/\sqrt{2(1-x)}}^{\infty} \exp\left(-\frac{y_1^2}{2}\right)dy_1, \quad 0 \le i \le 2n+2. \qquad (A.2)$$

Using $\exp(-y_1^2/2) \le \kappa\exp(-cy_1)$, for all $y_1 \ge 0$, for some $c > 0$, (A.1) for $m = 0$ is proved. Since $-\varphi_{l\overline{y}\overline{y}}^0 - \varphi_{lx}^0 = 0$, for $m = 2(k+1)$, $k \ge 0$, from (A.2) we easily find that

$$\left|\partial_x^i \partial_{\overline{y}}^{2(k+1)} \varphi_l^0\right| = \left|\partial_x^{i+k+1}\varphi_l^0\right| \le \kappa\exp\left(-c\frac{\overline{y}}{\sqrt{1-x}}\right), \qquad (A.3)$$

$$0 \le i + k + 1 \le 2n + 2;$$

thus (A.3) holds for $0 \le i + m = i + 2(k+1) \le k + 1 + 2n + 2$, and this implies that (A.1) holds for $m = 2(k+1)$, $k \ge 0$, $j = 0$.

Since $\partial_x^{i+1}h^0(1) = 0$, $0 \le i+1 \le 2n+1$, we now notice that

$$\left|\partial_x^i \partial_{\overline{y}} \varphi_l^0\right| = \sqrt{\frac{2}{\pi}}\left|\int_{\overline{y}/\sqrt{2(1-x)}}^{\infty} \exp\left(-\frac{y_1^2}{2}\right)\partial_x^{i+1}h^0 \times\left(x + \frac{\overline{y}^2}{2y_1^2}\right)\frac{\overline{y}}{y_1^2}dy_1\right|$$
$$\le \kappa\overline{y}\int_{\overline{y}/\sqrt{2(1-x)}}^{\infty} \exp\left(-\frac{y_1^2}{2}\right)d\left(-y_1^{-1}\right), \qquad (A.4)$$
$$0 \le i + 1 \le 2n + 2.$$

Integrating by parts in the last integral, we deduce that

$$\left|\partial_x^i \partial_{\overline{y}} \varphi_l^0\right| \le \kappa\left(\sqrt{1-x}\exp\left(-\frac{\overline{y}^2}{4(1-x)}\right) + \overline{y}\int_{\overline{y}/\sqrt{2(1-x)}}^{\infty} \exp\left(-\frac{y_1^2}{2}\right)dy_1\right)$$
$$\le \kappa\left(\exp\left(-c\frac{\overline{y}}{\sqrt{1-x}}\right) + \overline{y}\exp\left(-c\frac{\overline{y}}{\sqrt{1-x}}\right)\right) \qquad (A.5)$$
$$\le (\text{since } t\ \exp(-ct) \le \kappa\exp(-c_0 t),\ \forall t \ge 0, \text{ for some } 0 < c_0 < c)$$
$$\le \kappa\exp\left(-c_0\frac{\overline{y}}{\sqrt{1-x}}\right), \quad 0 \le i + 1 \le 2n + 2.$$

Hence, for $m = 2k + 1$, $k \ge 0$, using $-\varphi_{l\overline{y}\overline{y}}^0 - \varphi_{lx}^0 = 0$ again, we note from (A.5) that

$$\left|\partial_x^i \partial_{\overline{y}}^{2k+1} \varphi_l^0\right| = \left|\partial_x^{i+k}\partial_{\overline{y}}\varphi_l^0\right| \le \kappa\exp\left(-c\frac{\overline{y}}{\sqrt{1-x}}\right), \qquad (A.6)$$

$$0 \le i + k + 1 \le 2n + 2;$$

thus (A.6) holds for $0 \le i + m = i + 2k + 1 \le k + 2n + 2$, and this proves (A.1) for $m = 2k + 1$, $k \ge 0$, $j = 0$. Hence, (A.1) is proved for $j = 0$.

(2) We now assume that (A.1) holds at any order less than or equal to $j - 1$, and we want to prove it at the order j.

At order $j \ge 1$, we perform the analysis as follows.

We first note that the homogeneous solution of φ_l^j, that is the first integral of (26) for φ_l^j, can be similarly estimated. We just replace $h_0^0(x)$ by $h_0^j(x)$ in the above analysis.

Since $\partial_x^k h^j(1) = 0$ for $0 \le k \le 2n+1-2j$, the first integral is then estimated as in (A.1) for $j = 0$. We just replace the above $2n + 2$ by $2n + 2 - 2j$. We now estimate the particular solution of φ_l^j, that is, the second integral of (26) denoted by I. For simplicity in the analysis below, we write that

$$J = \exp\left(-\frac{(\overline{y} - y_1)^2}{4x_1}\right) - \exp\left(-\frac{(\overline{y} + y_1)^2}{4x_1}\right). \qquad (A.7)$$

Using the induction assumption at order $j-1$, from (A.1) for $j-1$, we note that for $0 \le 2+i \le 2n+2-2(j-1)$, that is, $0 \le i \le 2n+2-2j$,

$$\partial_x^{2+i}\varphi_l^{j-1}(1^-, y_1) = 0, \tag{A.8}$$

$$\begin{aligned}\left|\partial_x^{2+i}\varphi_l^{j-1}(x+x_1, y_1)\right| &\le \kappa \exp\left(-c\frac{y_1}{\sqrt{1-x-x_1}}\right)\\ &\le \kappa \exp\left(-c\frac{y_1}{\sqrt{1-x}}\right).\end{aligned} \tag{A.9}$$

Thanks to (A.8), differentiating I, the second integral of (26), we find that

$$\partial_x^i I = \frac{1}{2\sqrt{\pi}}\int_0^{1-x}\int_0^{\infty}\frac{J}{\sqrt{x_1}}\partial_x^{2+i}\varphi_l^{j-1}(x+x_1, y_1)\,dy_1dx_1, \quad 0 \le i \le 2n+2-2j. \tag{A.10}$$

Since $0 \le x_1 \le 1-x \le 1$, $|J| \le 2\exp(-(\overline{y}-y_1)^2/4x_1) \le 2\exp(-(\overline{y}-y_1)^2/4(1-x))$, from (A.9) we find that for $0 \le i \le 2n+2-2j$,

$$\begin{aligned}\left|\partial_x^i I\right| &\le \kappa\int_0^{1-x}\frac{1}{\sqrt{x_1}}dx_1\\ &\quad\times\int_0^{\infty}\exp\left(-\frac{(\overline{y}-y_1)^2}{4(1-x)}\right)\exp\left(-c\frac{y_1}{\sqrt{1-x}}\right)dy_1\\ &\le \left(\text{setting } \beta = 2c\sqrt{1-x}\right)\\ &\le \kappa\int_0^{\infty}\exp\left(-\frac{(y_1-\overline{y}+\beta)^2}{4(1-x)}\right.\\ &\qquad\left.-\frac{\beta}{2(1-x)}\overline{y}+\frac{\beta^2}{4(1-x)}\right)dy_1\\ &\le \kappa\exp\left(-c\frac{\overline{y}}{\sqrt{1-x}}\right)\int_0^{\infty}\exp\left(-\frac{(y_1-\overline{y}+\beta)^2}{4(1-x)}\right)dy_1\\ &\le \kappa\exp\left(-c\frac{\overline{y}}{\sqrt{1-x}}\right).\end{aligned} \tag{A.11}$$

This proves (A.1) for $m=0$, $j \ge 1$.

For $m = 2(k+1)$, $k \ge 0$, thanks to the induction assumption on $j-1$, from (A.1) we note that for $0 \le i+s+2+2(k-s) \le 2n+2-2(j-1)$,

$$\left|\partial_x^{i+s+2}\partial_{\overline{y}}^{2(k-s)}\varphi_l^{j-1}\right| \le \kappa\exp\left(-c\frac{\overline{y}}{\sqrt{1-x}}\right), \tag{A.12}$$

from (A.11) we also note that

$$\left|\partial_x^{i+k+1} I\right| \le \kappa\exp\left(-c\frac{\overline{y}}{\sqrt{1-x}}\right), \quad \text{for } 0 \le i+k+1 \le 2n+2-2j. \tag{A.13}$$

Since $-I_{\overline{y}\overline{y}} - I_x = \varphi_{lxx}^{j-1}$, we thus find that

$$\begin{aligned}\left|\partial_x^i\partial_{\overline{y}}^{2(k+1)} I\right| &\le \left|\partial_x^{i+k+1} I\right| + \sum_{s=0}^{k}\left|\partial_x^{i+s+2}\partial_{\overline{y}}^{2(k-s)}\varphi_l^{j-1}\right|\\ &\le \kappa\exp\left(-c\frac{\overline{y}}{\sqrt{1-x}}\right),\end{aligned} \tag{A.14}$$

for $0 \le i+2(k+1) \le 2n+2-2j$.

This proves (A.1) for $m = 2(k+1)$, $k \ge 0$, $j \ge 1$.

Since $\partial_{\overline{y}}\exp(-(\overline{y}-y_1)^2/4x_1) = -\partial_{y_1}\exp(-(\overline{y}-y_1)^2/4x_1)$, differentiating (A.10) in $\overline{y}$ we observe that

$$\begin{aligned}&\left|\partial_x^i\partial_{\overline{y}} I\right|\\ &\le \kappa\left|\int_0^{1-x}\frac{1}{\sqrt{x_1}}\int_0^{\infty}\partial_{\overline{y}} J\cdot\partial_x^{2+i}\varphi_l^{j-1}(x+x_1, y_1)\,dy_1dx_1\right|\\ &= \kappa\left|\int_0^{1-x}\frac{1}{\sqrt{x_1}}\int_0^{\infty}\partial_{y_1}\widetilde{J}\cdot\partial_x^{2+i}\varphi_l^{j-1}(x+x_1, y_1)\,dy_1dx_1\right|\\ &\le (\text{integrating by parts in } y_1)\\ &\le \kappa\left|\int_0^{1-x}\frac{1}{\sqrt{x_1}}\exp\left(-\frac{\overline{y}^2}{4x_1}\right)\partial_x^{2+i}\varphi_l^{j-1}(x+x_1, 0)\,dx_1\right|\\ &\quad+\kappa\left|\int_0^{1-x}\frac{1}{\sqrt{x_1}}\int_0^{\infty}\widetilde{J}\cdot\partial_x^{2+i}\partial_{y_1}\varphi_l^{j-1}(x+x_1, y_1)\,dy_1dx_1\right|,\end{aligned} \tag{A.15}$$

where $\widetilde{J} = \exp(-(\overline{y}-y_1)^2/4x_1) + \exp(-(\overline{y}+y_1)^2/4x_1) \le 2\exp(-(\overline{y}-y_1)^2/4(1-x))$. Here, we note that $\partial_{\overline{y}}J = -\partial_{y_1}\widetilde{J}$. Thanks to the induction assumption on $j-1$, from (A.1) we thus note that for $0 \le 2+i+1 \le 2n+2-2(j-1)$, that is, $0 \le i \le 2n+1-2j$,

$$\left|\partial_x^{2+i}\partial_{y_1}\varphi_l^{j-1}(x+x_1, y_1)\right| \le \kappa\exp\left(-c\frac{y_1}{\sqrt{1-x}}\right), \tag{A.16}$$

and for $0 \le 2+i \le 2n+2-2(j-1)$, that is, $0 \le i \le 2n+2-2j$,

$$\left|\partial_x^{2+i}\partial_{y_1}\varphi_l^{j-1}(x+x_1, 0)\right| \le \kappa. \tag{A.17}$$

As in (A.11) we similarly find that for $0 \le i \le 2n+1-2j$,

$$\left|\partial_x^i\partial_{\overline{y}} I\right| \le \kappa\exp\left(-c\frac{\overline{y}}{\sqrt{1-x}}\right). \tag{A.18}$$

Hence, for $m = 2k+1$, $k \ge 0$, using the induction assumption at order $j-1$, from (A.1) we note that for $0 \le i+s+2+2(k-s)-1 \le 2n+2-2(j-1)$,

$$\left|\partial_x^{i+s+2}\partial_{\overline{y}}^{2(k-s)-1}\varphi_l^{j-1}\right| \le \kappa\exp\left(-c\frac{\overline{y}}{\sqrt{1-x}}\right), \tag{A.19}$$

from (A.18) we also note that

$$\left|\partial_x^{i+k}\partial_{\overline{y}} I\right| \le \kappa\exp\left(-c\frac{\overline{y}}{\sqrt{1-x}}\right), \quad \text{for } 0 \le i+k \le 2n+1-2j. \tag{A.20}$$

Using $-I_{\overline{yy}} - I_x = \varphi_{lxx}^{j-1}$ again we thus find that

$$\begin{aligned} &\left|\partial_x^i \partial_{\overline{y}}^{2k+1} I\right| \\ &\leq \left|\partial_x^{i+k}\partial_{\overline{y}} I\right| + \sum_{s=0}^{k-1} \left|\partial_x^{i+s+2}\partial_{\overline{y}}^{2(k-s)-1}\varphi_l^{j-1}\right| \\ &\leq \kappa \exp\left(-c\frac{\overline{y}}{\sqrt{1-x}}\right), \quad \text{for } 0 \leq i+2k+1 \leq 2n+2-2j. \end{aligned} \tag{A.21}$$

This proves (A.1) for $m = 2k+1$, $k \geq 0$, $j \geq 1$. Hence, the lemma is proved.

B. Study of the Elliptic Corner Correctors $\xi^j = \xi_l^j + \xi_u^j$

For each $0 \leq j \leq n$, using the stretched variables

$$X = \frac{1-x}{2\varepsilon}, \qquad Y = \frac{y}{2\varepsilon}, \tag{B.1}$$

one can define an approximation $\overline{\xi}_l^j$ of ξ_l^j as a solution of the system,

$$\begin{aligned} -\partial_X^2\overline{\xi}_l^j - \partial_Y^2\overline{\xi}_l^j + 2\partial_X\overline{\xi}_l^j &= 0, \\ \overline{\xi}_l^j &= 0 \quad \text{at } X = 0, \\ \overline{\xi}_l^j &= \gamma_0^j(1-2\varepsilon X) \quad \text{at } Y = 0, \\ \overline{\xi}_l^j &\longrightarrow 0 \quad \text{at } X^2+Y^2 \longrightarrow \infty,\ Y>0. \end{aligned} \tag{B.2}$$

The explicit expression of $\overline{\xi}^j$ is given as (see [1])

$$\begin{aligned} \overline{\xi}_l^j = \frac{Y}{\pi}\int_0^\infty \left[\frac{K_1(s_1)}{s_1} - \frac{K_1(s_2)}{s_2}\right] \\ \times \gamma_0^j(1-2\varepsilon s)\exp(-(s-X))\,ds, \end{aligned} \tag{B.3}$$

where K_1 is the modified Bessel function of the second kind of the first order, and

$$s_1 = \sqrt{(X-s)^2+Y^2}, \qquad s_2 = \sqrt{(X+s)^2+Y^2}. \tag{B.4}$$

Using an appropriate barrier function and the maximum principle (see [1]), one can verify that

$$\left|\overline{\xi}_l^j(X,Y)\right| \leq \kappa \exp\left(-c\left(\sqrt{X^2+Y^2}-X\right)\right), \tag{B.5}$$

for a constant c independent of ε. An approximation $\overline{\xi}_u^j$ of ξ_u^j can be constructed in the same manner.

Using (33) and (B.5), we notice that $\xi^j - (\overline{\xi}_l^j + \overline{\xi}_u^j) = \text{e.s.t.}$ at $y = 0, 1$. We deduce from (17), (18), and (B.2)$_2$ that $\gamma_k^j(1) = 0$, and hence $\xi^j - (\overline{\xi}_l^j + \overline{\xi}_u^j) = 0$ at $x = 1$. Moreover, using Lemma B.1 below, we see that $\xi^j - (\overline{\xi}_l^j + \overline{\xi}_u^j) = 0$ at $x = 0$. Then, using these observations, thanks to the maximum principle, we find that

$$\left|\xi^j - \left(\overline{\xi}_l^j + \overline{\xi}_u^j\right)\right| = \text{e.s.t.}, \quad 0 \leq j \leq n, \tag{B.6}$$

which implies that the elliptic corrector ξ^j, $0 \leq j \leq n$, satisfies the pointwise estimates similar to (B.5) as for its approximation $\overline{\xi}^j = \overline{\xi}_l^j + \overline{\xi}_u^j$.

Lemma B.1. *There exists a positive constant* κ*, independent of* ε*, such that*

$$\left|\overline{\xi}^j(X,Y)\right| \leq \kappa \min\{U(X), U(Y)\}, \quad 0 \leq j \leq n, \tag{B.7}$$

where

$$U(Z) = \begin{cases} 1 & \text{if } 0 \leq Z \leq a, \\ 1-4(Z-a)^2 & \text{if } a < Z \leq b, \\ 0 & \text{if } Z > b, \end{cases} \tag{B.8}$$

and $a = 3/8\varepsilon$, $b = a + (1/2)$.

To prove Lemma B.1, we will recall below the weak maximum principle (see, e.g., [17]).

We consider an operator L of the form,

$$Lu = \sum_{i,j=1}^m D_i\left(a^{ij}D_ju\right) + \sum_{i=1}^m \left(D_i\left(b^iu\right) + c^iD_iu\right) + du, \tag{B.9}$$

where the coefficients a^{ij}, b^i, c^i, and d are assumed to be measurable functions in a domain $\Omega_m \subset \mathbb{R}^m$, and D_i denotes $\partial/\partial x_i$, $1 \leq i \leq m$.

We assume that L is strictly elliptic in Ω_m, that is, there exists $\lambda > 0$, such that

$$\sum_{i,j=1}^m a^{ij}\xi_i\xi_j \geq \lambda|\xi|^2, \quad \forall \xi \in \mathbb{R}^m, \tag{B.10}$$

and that L has bounded coefficients, that is, for some constants Λ and $\nu \geq 0$, we have

$$\sum_{i,j=1}^m \left|a^{ij}\right|^2 \leq \Lambda^2, \qquad \lambda^{-2}\sum_{i=1}^m \left(\left|b^i\right|^2 + \left|c^i\right|^2\right) + \lambda^{-1}|d| \leq \nu^2. \tag{B.11}$$

We also assume that

$$\int_{\Omega_m}\left(dv - \sum_{i=1}^m b^iD_iv\right) \leq 0, \quad \forall v \geq 0,\ v \in \mathscr{C}_c^1(\Omega_m). \tag{B.12}$$

We say that a weakly differentiable function u satisfies $Lu = 0$, (≥ 0 or ≤ 0), in the *weak sense*, if

$$\int_{\Omega_m}\left(\sum_{i,j=1}^m a^{ij}D_juD_iv + \sum_{i=1}^m \left(b^iuD_iv - c^iD_iuv\right) - duv\right) = 0, \quad (\leq 0 \text{ or } \geq 0), \tag{B.13}$$

for all nonnegative functions $v \in \mathscr{C}_c^1(\Omega_m)$, provided that $a^{ij}D_ju + b^iu$ and $c^iD_iu + du$, $1 \le i \le m$, are locally integrable.

Now we state the following maximum principle.

Lemma B.2. *Under the conditions* (B.10), (B.11), *and* (B.12), *if* $u \in H^1(\Omega_m)$ *satisfies* $Lu \ge 0$, *or* $Lu \le 0$, *in the weak sense, then one has, respectively,*

$$\sup_{\Omega_m} u \le \sup_{\partial\Omega_m} u^+ \quad or \quad \inf_{\Omega_m} u \ge \inf_{\partial\Omega_m} u^- . \tag{B.14}$$

Thanks to Lemma B.2, we can now prove Lemma B.1.

Proof of Lemma B.1. Given $\varepsilon_0 > 0$, thanks to $(B.2)_4$, we may choose a region $\Omega_0 = \{(X,Y) \in [0,\infty)\times[0,\infty) \mid X^2+Y^2 \le R^2\}$ with R sufficiently large so that $|\overline{\xi}^j| < \varepsilon_0$ on the circle $X^2+Y^2=R^2$, $X,Y>0$. We then define the barrier function

$$\widetilde{U} = \widetilde{U}(X,Y) = CU(X) + \varepsilon_0, \quad (X,Y) \in [0,\infty)\times[0,\infty), \tag{B.15}$$

where $U(X)$ is given in (B.8). Here, C is a positive constant and it will be determined below. Note that $\widetilde{U}(X,Y) \in H^1(\Omega_0)$.

Going back to the elliptic problem (B.2) and writing the elliptic operator L as

$$L\overline{\xi}^j := \overline{\xi}^j_{XX} + \overline{\xi}^j_{YY} - 2\overline{\xi}^j_X, \tag{B.16}$$

we find that the operator L satisfies (B.10), (B.11), and (B.12). We also find that for $v \in \mathscr{C}_c^1(\Omega_0)$, $v \ge 0$,

$$\int_{\Omega_0}\left(\widetilde{U}_X v_X + \widetilde{U}_Y v_Y + 2\widetilde{U}_X v\right) = \int\int_0^R C\left(U_X v_X + 2U_X v\right)dX\,dY \ge 0. \tag{B.17}$$

Indeed, if $0 \le X \le a$ or $X > b$ (a, b as in (B.8)), since $U_X = 0$, we only have to consider the case $a < R \le b$. For this R, since $U \in H^2(\Omega_0)$, we note that $\int_0^R (U_X v_X + 2U_X v)dX = \int_0^R(-U_{XX} + 2U_X)v\,dX = \int_a^R(-U_{XX} + 2U_X)v\,dX \ge 0$ for all non-negative $v \in \mathscr{C}_c^1(\Omega_0)$.

Since $L(\pm\overline{\xi}^j) = 0$, we thus find that $L(\widetilde{U} \pm \overline{\xi}^j) \le 0$ in the weak sense.

Thanks to the cut-off function $\delta(x)$, choosing $C = \max_{x\in[1/2,1]}|\widetilde{\gamma}^j(x)|$, we find that

$$\begin{aligned}\widetilde{U}(X,0) &= CU(X) + \varepsilon_0 \ge \left|\widetilde{\gamma}^j(x)\right|\chi_{[1/2,1]}(x) \\ &= \left|\widetilde{\gamma}^j(1-2\varepsilon X)\right|\chi_{[0,1/4\varepsilon]}(X) \ge \left|P^j(X)\right| \\ &= \left|\overline{\xi}^j(X,0)\right|, \\ \widetilde{U}(0,Y) &= C + \varepsilon_0 \ge 0 = \left|\overline{\xi}^j(0,Y)\right|, \\ \widetilde{U}(X,Y) &\ge \varepsilon_0 > \left|\overline{\xi}^j(X,Y)\right| \end{aligned} \tag{B.18}$$

on $\partial\Omega_0$ with $X^2+Y^2=R^2$, $X,Y>0$.

Thus, we note that $\widetilde{U} \pm \overline{\xi}^j \ge 0$ on $\partial\Omega_0$. Using Lemma B.2 we conclude that, for $(X,Y) \in \Omega_0$,

$$\left(\widetilde{U} \pm \overline{\xi}^j\right)(X,Y) \ge \inf_{\Omega_0}\left(\widetilde{U} \pm \overline{\xi}^j\right) \ge \inf_{\partial\Omega_0}\left(\widetilde{U} \pm \overline{\xi}^j\right)^- = 0. \tag{B.19}$$

Hence, we have

$$\left|\overline{\xi}^j(X,Y)\right| \le \widetilde{U}(X,Y) = CU(X) + \varepsilon_0. \tag{B.20}$$

Letting $R \to \infty$, we deduce that $|\overline{\xi}^j(X,Y)| \le CU(X)$. Similarly, we can conclude that $|\overline{\xi}^j(X,Y)| \le CU(Y)$. This proves the lemma.

Acknowledgments

This work was supported in part by the Research Fund of Indiana University, by NSF Grants DMS 0906440, DMS 1206438, and DMS 1212141, and by Basic Science Research Program through the National Research Foundation of Korea (NRF) funded by the Ministry of Education, Science and Technology (2012001167).

References

[1] S.-D. Shih and R. B. Kellogg, "Asymptotic analysis of a singular perturbation problem," *SIAM Journal on Mathematical Analysis*, vol. 18, no. 5, pp. 1467–1511, 1987.

[2] G. H. Hardy, J. E. Littlewood, and G. Pólya, *Inequalities*, Cambridge University Press, Cambridge, UK, 1988.

[3] L. Prandtl, "Verber Flüssigkeiten bei sehr kleiner Reibung," in *Proceedings of the Verk. III Intem. Math. Kongr. Heidelberg*, pp. 484–491, Teuber, Leibzig, Germany, 1905.

[4] L. Prandtl, "Über Flüssigkeitsbewegung bei sehr kleiner Reibung," in *Verhandlung des 3 Internat. Math. Kongr. Heidelberg*, pp. 484–491, Teubner, Leipzig, Germany, 1905.

[5] L. Prandtl, *Gesammelte Abhandlungen zur angewandten Mechanik, Hydro- und Aerodynamik*, edited by W. Tollmien, H. Schlichting, and H. Görtler, Springer, Berlin, Germany, 1961.

[6] L. Prandtl, "Über Flüssigkeitsbewegung bei sehr kleiner Reibung," in *Gesammelte Abhandlungen zur angewandten Mechanik, Hydro- und Aerodynamik*, W. Tollmien, H. Schlichting, and H. Görtler, Eds., Springer, Berlin, Germany, 1961.

[7] J. B. Rauch and F. J. Massey, III, "Differentiability of solutions to hyperbolic initial-boundary value problems," *Transactions of the American Mathematical Society*, vol. 189, pp. 303–318, 1974.

[8] S. Smale, "Smooth solutions of the heat and wave equations," *Commentarii Mathematici Helvetici*, vol. 55, no. 1, pp. 1–12, 1980.

[9] R. Temam, "Behaviour at time $t = 0$ of the solutions of semilinear evolution equations," *Journal of Differential Equations*, vol. 43, no. 1, pp. 73–92, 1982.

[10] Q. Chen, Z. Qin, and R. Temam, "Numerical resolution near $t = 0$ of nonlinear evolution equations in the presence of corner singularities in space dimension 1," *Communications in Computational Physics*, vol. 9, no. 3, pp. 568–586, 2011.

[11] N. Flyer and B. Fornberg, "Accurate numerical resolution of transients in initial-boundary value problems for the heat equation," *Journal of Computational Physics*, vol. 184, no. 2, pp. 526–539, 2003.

[12] N. Flyer and B. Fornberg, "On the nature of initial-boundary value solutions for dispersive equations," *SIAM Journal on Applied Mathematics*, vol. 64, no. 2, pp. 546–564, 2004.

[13] G.-M. Gie, "Asymptotic expansion of the Stokes solutions at small viscosity: the case of non-compatible initial data," *Communications in Mathematical Sciences*, to appear.

[14] J. R. Cannon, *The One-Dimensional Heat Equation*, vol. 23 of *Encyclopedia of Mathematics and Its Applications*, Addison-Wesley Publishing Company Advanced Book Program, Reading, Mass, USA, 1984.

[15] C.-Y. Jung and R. Temam, "Numerical approximation of two-dimensional convection-diffusion equations with multiple boundary layers," *International Journal of Numerical Analysis and Modeling*, vol. 2, no. 4, pp. 367–408, 2005.

[16] G.-M. Gie, M. Hamouda, and R. Temam, "Boundary layers in smooth curvilinear domains: parabolic problems," *Discrete and Continuous Dynamical Systems Series A*, vol. 26, no. 4, pp. 1213–1240, 2010.

[17] D. Gilbarg and N. S. Trudinger, *Elliptic Partial Differential Equations of Second Order*, vol. 224 of *Fundamental Principles of Mathematical Sciences*, Springer, Berlin, Germany, 2nd edition, 1983.

8

Existence and Stability for the 3D Linearized Constant-Coefficient Incompressible Current-Vortex Sheets

Davide Catania

Dipartimento di Matematica, Facoltà di Ingegneria, Università degli Studi di Brescia, Via Valotti 9, 25133 Brescia, Italy

Correspondence should be addressed to Davide Catania; davide.catania@ing.unibs.it

Academic Editor: Helge Holden

We consider the free boundary problem for current-vortex sheets in ideal incompressible magnetohydrodynamics. The problem of current-vortex sheets arises naturally, for instance, in geophysics and astrophysics. We prove the existence of a unique solution to the constant-coefficient linearized problem and an a priori estimate with no loss of derivatives. This is a preliminary result to the study of linearized variable-coefficient current-vortex sheets, a first step to prove the existence of solutions to the nonlinear problem.

1. Introduction

1.1. The Eulerian Description. Let us consider the equations of magnetohydrodynamics (MHD) governing the motion of a perfectly conducting inviscid incompressible plasma in three-space dimension. In the case of a homogeneous plasma (i.e., the density ρ is a positive constant), the equations in a dimensionless form read

$$\begin{gathered} \partial_t u + \nabla \cdot (u \otimes u - H \otimes H) + \nabla q = 0, \\ \partial_t H - \nabla \times (u \times H) = 0, \\ \operatorname{div} u = 0, \qquad \operatorname{div} H = 0, \end{gathered} \tag{1}$$

where, using T for transposition, $u = (u_1, u_2, u_3)^T$ denotes the plasma velocity, $H = (H_1, H_2, H_3)^T$ is the magnetic field (in Alfvén velocity units), $q = p + |H|^2/2$ is the total pressure, and p is the pressure. For smooth solutions, system (1) can be written in an equivalent form as

$$\begin{gathered} \partial_t u + (u \cdot \nabla) u - (H \cdot \nabla) H + \nabla q = 0, \\ \partial_t H + (u \cdot \nabla) H - (H \cdot \nabla) u = 0, \\ \operatorname{div} u = 0, \qquad \operatorname{div} H = 0. \end{gathered} \tag{2}$$

We are interested in weak solutions, in a suitable sense, to (1) that are smooth on either side of a smooth hypersurface $\Gamma(t) = \{x_3 = f(t, x')\}$ in $[0, T] \times \Omega$, where $\Omega \subset \mathbb{R}^3$, $x' = (x_1, x_2)$, and that satisfy suitable jump conditions at each point of the front $\Gamma(t)$. For notational simplicity, we assume that the density is the same constant on either side of $\Gamma(t)$, so that we can take, with no loss of generality, $\rho \equiv 1$. In physical applications, the two densities can be very different, but such a difference intervenes only at the boundary, and it is taken into account by the jump condition for the total pressure (see below), so it does not alter the mathematical techniques applied in this paper.

Let us set $\Omega^\pm(t) = \{x_3 \gtrless f(t, x')\}$, where $\Omega = \Omega^+(t) \cup \Omega^-(t) \cup \Gamma(t)$; given any function g, we set $g^\pm = g$ in $\Omega^\pm(t)$ and denote by $[g] = g^+_{|\Gamma(t)} - g^-_{|\Gamma(t)}$ the jump across $\Gamma(t)$.

We look for current-vortex sheets solutions, that is, smooth solutions $(q^\pm, u^\pm, H^\pm)$ of (2) in $\Omega^\pm(t)$ such that $\Gamma(t)$ is a tangential discontinuity, namely, the plasma does not flow through the discontinuity front, and the magnetic field is tangent to $\Gamma(t)$, see, for example, Landau and Lifshitz [1]; thus, the boundary conditions take the form

$$\sigma_{\text{front}} = u^\pm \cdot n, \quad H^\pm \cdot n = 0, \quad [q] = 0 \quad \text{on } \Gamma(t), \tag{3}$$

where $n = n(t)$ denotes the outward unit normal on $\partial\Omega^-(t)$ and σ_{front} denotes the velocity of propagation of the interface front $\Gamma(t)$. With the given parametrization of $\Gamma(t)$, an equivalent formulation of these jump conditions is

$$\partial_t f = u^\pm \cdot N, \quad H^\pm \cdot N = 0, \quad [q] = 0 \quad \text{on } \Gamma(t), \tag{4}$$

with $N := (-\partial_1 f, -\partial_2 f, 1)^T$. Notice that the function f describing the discontinuity front is part of the unknowns of the problem, that is, this is a free boundary problem.

System (2), (4) is supplemented with initial conditions

$$u^{\pm}(0,x) = u_0^{\pm}(x), \quad H^{\pm}(0,x) = H_0^{\pm}(x), \quad x \in \Omega^{\pm}(0),$$
$$f(0,x') = f_0(x'), \quad x' \in \Gamma(0), \tag{5}$$

where $\operatorname{div} u_0^{\pm} = \operatorname{div} H_0^{\pm} = 0$ in $\Omega^{\pm}(0)$.

Current-vortex sheets have various interesting applications in geophysics and astrophysics. For instance, an accepted model in the literature for the interface region between the unperturbed flows of the interstellar plasma and the supersonic solar wind plasma is given by a current-vortex sheet separating the interstellar plasma compressed at the *bow shock* from the solar wind plasma compressed at the *termination shock*, see Ruderman and Fahr [2] and references therein. This current-vortex sheet is called the heliopause and in some sense can be considered as the outer boundary of the solar system. Similarly, the boundary separating the shocked solar wind plasma from the plasma of the magnetosphere of planets equipped with an intrinsic magnetic field (such as the Earth, Jupiter...) is a current-vortex sheet called *magnetopause*.

In order to prove the local existence of solutions to the current-vortex sheets (nonlinear) problem (2), (4), and (5), it is useful to prove existence for the linearized problem and the strong stability, that is, an a priori estimate for the solutions with no loss of derivatives (see later on). We will perform a further simplification by assuming a linearization around a constant-coefficient basic state. This must be seen as a preliminary step before considering a linearization around a general variable-coefficient basic state, in order to prove existence, uniqueness and stability for the solution to problem (2), (4), and (5) without resorting to a Nash-Moser iteration. Such result would be a rigorous confirmation of the stabilizing effect of the magnetic field on Kelvin-Helmholtz instabilities, which is well known in astrophysics.

In the last years, there has been a renewed interest for the analysis of free interface problems in fluid dynamics, especially for the Euler equations in the vacuum and the water waves problem, see Coutand and Shkoller [3, 4] and the references thereinto. This fact has produced different methodologies for obtaining a priori estimates and the proof of the existence of solutions. If the interface moves with the velocity of fluid particles, a natural approach consists in the introduction of the Lagrangian coordinates that reduces the original problem to a new one on a fixed domain. This approach has been recently employed with success in a series of papers by Coutand and Shkoller on the incompressible and compressible Euler equations in the vacuum, see [3, 4]. However, this method seems hardly applicable to problem (2), (4), and (5).

For incompressible current-vortex sheets, the precise description of the region of weak stability is known, meaning that for states outside this region the problem is ill posed. In this region, Morando et al. [5], for the constant-coefficient linearized equations, have shown an a priori estimate for solutions with loss of regularity with respect to the data. In a subset of the region of weak stability, that is, under a more restrictive stability condition (i.e., (6), introduced in the following), Trakhinin [6] has shown a similar result with no loss of regularity for the solutions to the linearized problem with respect to the source terms, but with loss of derivatives with respect to the coefficients of the problem. Moreover, in a recent paper, Coulombel et al. [7] have shown that the same more restrictive stability condition is sufficient for an a priori estimate of solutions to the nonlinear problem with no loss of regularity (here, space periodicity is assumed). However, in all the previous cases, existence of the solution is missing. Let us note that we are looking for the existence and an a priori estimate with no loss of derivatives for the linearized incompressible current-vortex sheet problem. Reference [6] already provides an estimate with no loss of regularity in the constant-coefficient case, but such a loss manifests for variable coefficients, and existence is not proved. The main novelty of our paper is the existence result; however, we hope that it is possible to extend the techniques here used in order to obtain an a priori estimate with no loss of regularity even for the variable-coefficient linearized problem. We point out that the estimate in [7] is optimal, with no loss of derivatives, but refers to the nonlinear problem and does not help in proving existence for the linearized system.

In [8], Secchi and Trakhinin have established existence and stability for the linearized compressible plasma-vacuum (free boundary) problem. In that case, the interface separates a plasma region governed by the usual MHD equations from a vacuum region with pre-Maxwell dynamics for the magnetic field, as in the case of magnetic confinement in nuclear fusion processes. This work has been employed by Morando et al. [9] to obtain the analogous result for the incompressible problem. However, the techniques used in these papers cannot be easily adapted to current-vortex sheets, since they rely on a secondary symmetrization in the vacuum part that cancels problematic boundary terms and cannot be reproduced in plasma part.

In the present paper, we follow an approach different from the one of Coutand and Shkoller. To reduce our free boundary problem to the fixed domain, we consider a change of variables inspired from Lannes [10] and performed in Secchi and Trakhinin [8]. The control of the function describing the free interface follows from a stability condition introduced by Trakhinin in [11], that is, the strong stability condition

$$\max\left(\left|H^+ \times [u]\right|, \left|H^- \times [u]\right|\right) < \left|H^+ \times H^-\right|, \tag{6}$$

which implies in particular that $H^+ \times H^-$ is not the zero vector.

We use a compressible approximation and perform some changes of variables in order to obtain a symmetric hyperbolic problem with conditions in a more suitable form, so that we can apply the results by Lax and Phillips [12] and Secchi [13, 14] for maximally nonnegative linear symmetric hyperbolic systems with characteristic boundary of constant multiplicity. Another fundamental tool is the so-called secondary symmetrization, which is a linear combination of the equations of the hyperbolic system that allows a simplification of the border term necessary to obtain the a priori estimate (see Section 4). By exploiting such techniques, we obtain the existence of a unique solution to the incompressible current-vortex sheets free boundary problem linearized about a suitable constant-coefficient state and show that the solution satisfies a priori estimates with no loss of derivatives with respect to data. As mentioned above, this is a preliminary step in order to prove existence and stability for the current-vortex sheet nonlinear problem.

1.2. The Reference Domain Ω. To avoid using local coordinate charts necessary for arbitrary geometries, and for simplicity, we will assume that the space domain Ω occupied by the fluid is given by $\Omega = \mathbb{R}^3$. We also set

$$\Omega^{\pm} := \Omega \cap \{x_3 \gtrless 0\}, \qquad \Gamma := \Omega \cap \{x_3 = 0\}. \tag{7}$$

The moving discontinuity front is given by

$$\Gamma(t) := \Omega \cap \{x_3 = f(t, x')\}. \tag{8}$$

1.3. An Equivalent Formulation in the Fixed Domain Ω. To reduce the free boundary problem (2), (4), and (5) to the fixed domains $\Omega^{\pm}$, we introduce a suitable change of variables. In all what follows, $H^s(\omega)$ denotes the Sobolev space of order s on a domain ω.

The diffeomorphism that reduces the free boundary problem (2), (4), and (5) to the fixed domains $\Omega^{\pm}$ is given in the following lemma (see [8]).

Lemma 1. *Let $m \geq 3$ be an integer. Then, for all $T > 0$ and for all $f \in \bigcap_{j=0}^{m-1} \mathscr{C}^j([0,T]; H^{m-j-0.5}(\mathbb{R}^2))$ satisfying without loss of generality $\|f\|_{\mathscr{C}([0,T];H^2(\mathbb{R}^2))} \leq 1$, there exists a function*

$$\psi \in \bigcap_{j=0}^{m-1} \mathscr{C}^j([0,T]; H^{m-j}(\mathbb{R}^3)) \tag{9}$$

such that the function

$$\Psi(t,x) := (x', x_3 + \psi(t,x)), \quad (t,x) \in [0,T] \times \Omega \tag{10}$$

defines an H^m-diffeomorphism of Ω for all $t \in [0,T]$. Moreover, there holds

$$\begin{gathered}\partial_t^j (\Psi - Id) \in \mathscr{C}([0,T]; H^{m-j}(\Omega)) \quad \text{for } j = 0, \ldots, m-1, \\ \Psi(t,x',0) = (x', f(t,x')), \qquad \partial_3 \Psi(t,x',0) = (0,0,1).\end{gathered} \tag{11}$$

We set

$$\begin{gathered}A := [D\Psi]^{-1} \text{ (inverse of the Jacobian matrix)}, \\ J := \det[D\Psi] \text{ (determinant of the Jacobian matrix)}, \\ a := JA \text{ (transpose of the cofactor matrix)},\end{gathered} \tag{12}$$

and we compute

$$A = \begin{pmatrix} 1 & 0 & 0 \\ 0 & 1 & 0 \\ -\dfrac{\partial_1 \psi}{J} & -\dfrac{\partial_2 \psi}{J} & \dfrac{1}{J} \end{pmatrix}, \quad J = 1 + \partial_3 \psi,$$
$$a = \begin{pmatrix} J & 0 & 0 \\ 0 & J & 0 \\ -\partial_1 \psi & -\partial_2 \psi & 1 \end{pmatrix}. \tag{13}$$

Now we may reduce the free boundary problem (2), (4), and (5) to a problem in the fixed domains $\Omega^{\pm}$ by the change of variables (10). Let us set

$$\begin{gathered}v^{\pm}(t,x) := u^{\pm}(t, \Psi(t,x)), \qquad B^{\pm}(t,x) := H^{\pm}(t, \Psi(t,x)), \\ Q^{\pm}(t,x) := q^{\pm}(t, \Psi(t,x)).\end{gathered} \tag{14}$$

Then, the system (2), (4), and (5) can be reformulated on the fixed reference domains $\Omega^{\pm}$ as

$$\begin{gathered}\partial_t v^{\pm} + (\tilde{v}^{\pm} \cdot \nabla) v^{\pm} - (\tilde{B}^{\pm} \cdot \nabla) B^{\pm} + A^T \nabla Q^{\pm} = 0, \\ \partial_t B^{\pm} + (\tilde{v}^{\pm} \cdot \nabla) B^{\pm} - (\tilde{B}^{\pm} \cdot \nabla) v^{\pm} = 0, \\ (A^T \nabla) \cdot v^{\pm} = 0, \quad (A^T \nabla) \cdot B^{\pm} = 0, \quad \text{in } [0,T] \times \Omega^{\pm}, \\ [Q] = 0, \quad \partial_t f = v^{\pm} \cdot N, \quad B^{\pm} \cdot N = 0, \quad \text{on } [0,T] \times \Gamma, \\ v^{\pm}_{|t=0} = v^{\pm}_0, \quad B^{\pm}_{|t=0} = B^{\pm}_0, \quad \text{on } \Omega^{\pm}, \\ f_{|t=0} = f_0, \quad \text{on } \Gamma.\end{gathered} \tag{15}$$

In (15), we have set

$$\begin{gathered}N := (-\partial_1 \psi, -\partial_2 \psi, 1)^T, \\ \tilde{v} := Av - \left(0, 0, \frac{\partial_t \psi}{J}\right)^T = \left(v_1, v_2, \frac{v \cdot N - \partial_t \psi}{J}\right)^T, \\ \tilde{B} := AB = \left(B_1, B_2, B \cdot \frac{N}{J}\right)^T.\end{gathered} \tag{16}$$

Notice that

$$J = 1, \quad N = (-\partial_1 f, -\partial_2 f, 1)^T, \quad \tilde{v}_3 = \tilde{B}_3 = 0 \quad \text{on } \Gamma. \tag{17}$$

We warn the reader that the notation A^T is used to denote the transpose of A and has nothing to do with the time interval $[0,T]$ on which the smooth solution is sought; we hope that this does not create any confusion. Vectors are written in columns.

If we set $Q_T^{\pm} = [0,T] \times \Omega^{\pm}$, $\omega_T = [0,T] \times \Gamma$, and $V^{\pm} = (Q^{\pm}, v^{\pm}, B^{\pm})^T$, problem (15) can be written in short form as

$$\begin{gathered}\mathbb{P}(V^{\pm}, \psi) = \underline{0} \quad \text{in } Q_T^{\pm}, \\ \mathbb{B}(V^{\pm}, f) = \underline{0} \quad \text{on } \omega_T, \\ (v^{\pm}, B^{\pm}, f^{\pm})|_{t=0} = (v_0^{\pm}, B_0^{\pm}, f_0^{\pm}) \quad \text{in } (\Omega^{\pm})^2 \times \Gamma,\end{gathered} \tag{18}$$

where $\underline{0}$ is the zero vector.

2. The Linearized Problem

2.1. Basic State. Let

$$\left(\widehat{V}^{\pm}, \widehat{f}\right) \tag{19}$$

be such that $\widehat{V}^{+}$ and $\widehat{V}^{-}$ are constant vectors, $\widehat{V} = (\widehat{Q}, \widehat{v}, \widehat{B})^T$, and

$$\widehat{f} \equiv 0. \tag{20}$$

We can then take $\widehat{\psi} \equiv 0$, so that $\widehat{J} = \partial_3 \widehat{\Psi}_3 = 1 + \partial_3 \widehat{\psi} = 1$. Similarly, we determine all the "hat" functions as for the corresponding functions obtained for $(V^{\pm}, f)$. Let us note, in particular, that $\widehat{A} = \widehat{a} = I_3$, $\widehat{N}^T = (0,0,1)$, $\widetilde{\widehat{v}} = \widehat{v}$, and $\widetilde{\widehat{B}} = \widehat{B}$. We assume that

$$\left[\widehat{Q}\right] = 0, \qquad \widehat{v}_3^{\pm} = \widehat{B}_3^{\pm} = 0. \tag{21}$$

In particular, $\widehat{Q}^{+} = \widehat{Q}^{-}$. Under these hypotheses, we have that the basic state $(\widehat{V}^{\pm}, \widehat{f})$ is a solution of problem (15).

Moreover, we assume that the basic state satisfies

$$\max\left(\left|\widehat{B}^{+} \times [\widehat{v}]\right|, \left|\widehat{B}^{-} \times [\widehat{v}]\right|\right) < \left|\widehat{B}^{+} \times \widehat{B}^{-}\right|. \tag{22}$$

This is the analogue of condition (6) and implies

$$\left|\widehat{B}^{+} \times \widehat{B}^{-}\right| > 0. \tag{23}$$

2.2. Linearized Problem. The linearized equations for (15) are

$$\begin{aligned} \frac{d}{ds} \mathbb{P}\left(V_s^{\pm}, \psi_s\right)\big|_{s=0} &= \mathsf{F}^{\pm}, \\ \frac{d}{ds} \mathbb{B}\left(V_s^{\pm}, f_s\right)\big|_{s=0} &= g^{\pm}, \end{aligned} \tag{24}$$

where $V_s = \widehat{V} + s\delta V$, $f_s = \widehat{f} + s\delta f$, $\psi_s = \widehat{\psi} + s\delta\psi$, and $\delta V, \delta f, \delta\psi$ are the unknowns. Here, we consider the source terms in order to make the system inhomogeneous.

If we omit δ and write explicitly the linearized equations, we get, in $Q_T^{\pm}$,

$$\begin{aligned} \partial_t v^{\pm} + \left(\widehat{v}^{\pm} \cdot \nabla\right) v^{\pm} - \left(\widehat{B}^{\pm} \cdot \nabla\right) B^{\pm} + \nabla Q^{\pm} &= \mathsf{F}_1^{\pm}, \\ \partial_t B^{\pm} + \left(\widehat{v}^{\pm} \cdot \nabla\right) B^{\pm} - \left(\widehat{B}^{\pm} \cdot \nabla\right) v^{\pm} &= \mathsf{F}_2^{\pm}, \\ \operatorname{div} v^{\pm} &= F_v^{\pm}, \\ \operatorname{div} B^{\pm} &= F_B^{\pm}. \end{aligned} \tag{25}$$

The linearized boundary conditions are obtained in a similar way (let us recall that we are omitting δ):

$$\begin{aligned} [Q] &= g_0, \\ \partial_t f - v_3^{\pm} + \widehat{v}'^{\pm} \cdot \nabla' f &= g_1^{\pm}, \\ -B_3^{\pm} + \widehat{B}'^{\pm} \cdot \nabla' f &= g_2^{\pm} \end{aligned} \tag{26}$$

on ω_T, where, as usual, if $w = (w_1, w_2, w_3)^T$ is a vector, we set $w' = (w_1, w_2)^T$, and moreover, $\nabla' = (\partial_1, \partial_2)^T$.

2.3. Reduction to Homogeneous Boundary Conditions. Let $\overline{u}^{\pm}$, $\overline{h}^{\pm}$ denote vector-valued functions which solve, respectively, the elliptic problems

$$\begin{aligned} \operatorname{div} \overline{u}^{\pm} &= F_v^{\pm} \quad \text{in } Q_T^{\pm}, \qquad \overline{u}_3^{\pm} = -g_1^{\pm} \quad \text{on } \omega_T, \\ \operatorname{div} \overline{h}^{\pm} &= F_B^{\pm} \quad \text{in } Q_T^{\pm}, \qquad \overline{h}_3^{\pm} = -g_2^{\pm} \quad \text{on } \omega_T \end{aligned} \tag{27}$$

and set $u = v - \overline{u}$, $h = B - \overline{h}$, and $q = Q - \overline{q}$, where $\overline{q}^{+} = g_0$ on ω_T^{+} and $\overline{q}^{-} = 0$ on ω_T^{-} (here, letters u and q represent different variables from the ones introduced in the original formulation (1)).

The equations in (25) can be recast in terms of (q, u, h):

$$\begin{aligned} \partial_t u^{\pm} + \left(\widehat{v}^{\pm} \cdot \nabla\right) u^{\pm} - \left(\widehat{B}^{\pm} \cdot \nabla\right) h^{\pm} + \nabla q^{\pm} &= F_1^{\pm}, \\ \partial_t h^{\pm} + \left(\widehat{v}^{\pm} \cdot \nabla\right) h^{\pm} - \left(\widehat{B}^{\pm} \cdot \nabla\right) u^{\pm} &= F_2^{\pm}, \\ \operatorname{div} u^{\pm} = 0, \qquad \operatorname{div} h^{\pm} &= 0, \end{aligned} \tag{28}$$

with new source terms $F_i = \mathsf{F}_i + \mathsf{f}_i(\widehat{v}, \widehat{B}, \nabla\overline{q}, \nabla_{t,x}\overline{u}, \nabla_{t,x}\overline{h})$, $i = 0, 1, 2$, with boundary conditions

$$\begin{aligned} [q] &= 0, \\ \partial_t f - u_3^{\pm} + \widehat{v}'^{\pm} \cdot \nabla' f &= 0, \\ -h_3^{\pm} + \widehat{B}'^{\pm} \cdot \nabla' f &= 0 \end{aligned} \tag{29}$$

and initial data

$$\left(u^{\pm}, h^{\pm}, f^{\pm}\right)\big|_{t=0} = \left(v_0^{\pm} - \overline{u}^{\pm}(0), B_0^{\pm} - \overline{h}^{\pm}(0), f_0^{\pm}\right). \tag{30}$$

3. Notations and Results

3.1. Function Spaces. We denote by $H^m = H^m(\Omega)$ the usual Sobolev space and by $\|\cdot\|_m$ its norm, while $\|\cdot\|$ is simply the norm of $L^2 = L^2(\Omega)$. We introduce a smooth odd function $\sigma = \sigma(x_3)$ defined in $\mathbb{R}$ such that $\sigma(x_3) = x_3$ for $0 \le x_3 \le 1/2$ and $\sigma(x_3) = 1$ for $x_3 \ge 1$ and set $Z_1 = \partial_{x_1}$, $Z_2 = \partial_{x_2}$ and $Z_3 = \sigma\partial_{x_3}$. For a multi-index $\alpha = (\alpha_1, \alpha_2, \alpha_3)$, $|\alpha| = \alpha_1 + \alpha_2 + \alpha_3$, we set $\partial_*^{\alpha} = Z_1^{\alpha_1} Z_2^{\alpha_2} Z_3^{\alpha_3}$,

$$\begin{aligned} \|u\|_{m,*}^2 &= \sum_{|\alpha| + 2\beta \le m} \left\|\partial_*^{\alpha} \partial_{x_3}^{\beta} u\right\|^2, \\ H_*^m &= H_*^m(\Omega) = \left\{u \in L^2 : \|u\|_{m,*} < +\infty\right\}, \end{aligned} \tag{31}$$

where $\beta \ge 0$ is an integer number (see [14–17]).

If $u(t)$, defined on $[0, T]$ with values in H_*^m, is such that $\partial_t^k u(t) \in H_*^{m-k}$ for $k = 1, 2, \ldots, m$, we set

$$\begin{aligned} |||u(t)|||_{m,*}^2 &= \sum_{k=0}^{m} \left\|\partial_t^k u(t)\right\|_{m-k,*}^2, \\ \mathscr{L}_T^p\left(H_*^m\right) &= \bigcap_{k=0}^{m} W^{k,p}\left(0, T; H_*^{m-k}\right), \end{aligned} \tag{32}$$

where $W^{k,p}(0,T;H_*^{m-k})$ is the space of functions which are in L^p together with the time derivatives up to order k on $[0,T]$, with values in H_*^{m-k}. The spaces $\mathscr{L}_T^2(H_*^m)$ and $\mathscr{L}_T^\infty(H_*^m)$ are equipped with norms, respectively,

$$[u]_{m,*,T}^2 = \int_0^T |||u(t)|||_{m,*}^2 dt, \quad |||u|||_{m,*,T} = \sup_{[0,T]} |||u(t)|||_{m,*}. \tag{33}$$

We define the same function spaces and norms for a general space domain E in the same way; we write, for instance, $||g||_{m,*;E}$ to specify the domain that we are considering.

3.2. Main Result. Now, we are able to state the main theorem. We set $U = (q,u,h)^T$ and consider problem

$$\begin{aligned} \partial_t u^\pm + (\widehat{v}^\pm \cdot \nabla) u^\pm - (\widehat{B}^\pm \cdot \nabla) h^\pm + \nabla q^\pm &= F_1^\pm, \\ \partial_t h^\pm + (\widehat{v}^\pm \cdot \nabla) h^\pm - (\widehat{B}^\pm \cdot \nabla) u^\pm &= F_2^\pm, \\ \operatorname{div} u^\pm &= 0 \end{aligned} \tag{34}$$

in $Q_T^\pm$, with boundary conditions

$$\begin{aligned} [q] &= 0, \\ \partial_t f - u_3^\pm + \widehat{v}'^\pm \cdot \nabla' f &= 0 \end{aligned} \tag{35}$$

on ω_T, and initial data

$$(u^\pm, h^\pm, f)|_{t=0} = (u_0^\pm, h_0^\pm, f_0) \quad \text{in } (\Omega^\pm)^2 \times \Gamma, \tag{36}$$

where

$$\operatorname{div} u_0^\pm = \operatorname{div} h_0^\pm = 0 \quad \text{on } \Omega^\pm. \tag{37}$$

Theorem 2. *Assume that the constant basic state $(\widehat{V}^\pm, \widehat{f})$ satisfies (20), (21), and (22), and moreover, assume that $u_0^\pm$, $h_0^\pm \in H^1(\Omega^\pm)$, $\operatorname{div} u_0^\pm = \operatorname{div} h_0^\pm = 0$ on $\Omega^\pm$, and $f_0 \in H^{1/2}(\Gamma)$. Assume that $F^\pm = (F_1^\pm, F_2^\pm) \in \mathscr{L}_T^2(H_*^1(\Omega^\pm))$, $\operatorname{div} F_2^\pm = 0$ in $Q_T^\pm$, and $(F_2^\pm)_3 = 0$ on $[0,T] \times \Gamma$, where $(F_2^\pm)_3$ denotes the third component of vector $F_2^\pm$. Finally, assume as well the compatibility conditions on $\{t=0\} \times \Gamma$:*

$$\begin{aligned} \partial_t f_0 - u_3^\pm + \widehat{v}'^\pm \cdot \nabla' f_0 &= 0, \\ -h_3^\pm + \widehat{B}'^\pm \cdot \nabla' f_0 &= 0. \end{aligned} \tag{38}$$

Then there exists a unique solution $(U^\pm, f)$ to (34), (35), and (36) ($q^\pm$ unique up to additive constants) such that $u^\pm, h^\pm \in L^\infty([0,T]; H_*^1(\Omega^\pm)) \cap W^{1,\infty}([0,T]; L^2(\Omega^\pm))$, $q^\pm \in L^\infty([0,T]; \dot{H}^1(\Omega^\pm))$ ($\dot{H}^1$ is the homogeneous Sobolev space), and $f \in \mathscr{C}^0([0,T]; H^{1/2}(\Gamma))$; in addition, for each $t \in [0,T]$, the following a priori estimate

$$\begin{aligned} &\sum_\pm \left(|||(u,h)^\pm(t)|||_{1,*;\Omega^\pm}^2 + \left\|(u_3,h_3)^\pm(t)\right\|_{H^1(\Omega^\pm)}^2 \right. \\ &\quad \left. + \left\|\nabla q^\pm(t)\right\|_{L^2(\Omega^\pm)}^2 \right) + \|f(t)\|_{H^{1/2}(\Gamma)}^2 \\ &\le Ce^{Ct} \sum_\pm \left(|||(u_0,h_0)^\pm|||_{1,*;\Omega^\pm}^2 + [F^\pm]_{1,*,t;\Omega^\pm}^2 \right. \\ &\quad \left. + \left\|F_1^\pm(0)\right\|_{L^2(\Omega^\pm)}^2 \right) + \|f_0\|_{H^{1/2}(\Gamma)}^2 \end{aligned} \tag{39}$$

is satisfied, where $C = C(\widehat{v}, \widehat{B})$ is a positive constant.

Moreover, one has $(u_3^\pm, h_3^\pm)|_\Gamma \in L^\infty([0,T]; H^{1/2}(\Gamma))$.

Remark 3. In order to consider the nonlinear problem, it might be useful to have higher regularity estimates, with no loss of derivatives, provided that data are sufficiently regular. The H^1 regularity can be obtained from the equations for $\operatorname{rot} u$ and $\operatorname{rot} h$, which allow to improve the regularity for the normal derivatives. Taking further derivatives and proceeding as in the proof of Proposition 13, one can expect to get higher regularity. The details are postponed to a future paper.

As for the regularity assumption on $u_0^\pm$, $h_0^\pm$, see Section 4.5.

Note that $\operatorname{div} h^\pm = 0$ in $Q_T^\pm$ and $-h_3^\pm + \widehat{B}'^\pm \cdot \nabla' f = 0$ on $[0,T] \times \Gamma$ follow from (34), (35), and (36) and the assumptions of Theorem 2 and, in particular, from the validity of the restrictions of such equations at time $t = 0$. More precisely, we have the following result.

Proposition 4. *Under the same assumptions of Theorem 2, the solution $U^\pm$ satisfies*

$$\operatorname{div} h^\pm = 0 \quad \text{in } Q_T^\pm, \tag{40}$$

$$-h_3^\pm + \widehat{B}'^\pm \cdot \nabla' f = 0 \quad \text{in } \omega_T. \tag{41}$$

Proof. If we take the divergence in the second equation in (34) and use $\operatorname{div} u^\pm = \operatorname{div} F_2^\pm = 0$ in $Q_T^\pm$, we get

$$(\partial_t + \widehat{v}^\pm \cdot \nabla) \operatorname{div} h^\pm = 0, \tag{42}$$

and hence $\operatorname{div} h^\pm = 0$ in $Q_T^\pm$, since at time $t = 0$, we have $\operatorname{div} h^\pm = \operatorname{div} h_0^\pm = 0$ on $\Omega^\pm$ by assumption.

On the other hand, let us take just the third component of the second equation in

$$\partial_t h_3^\pm + (\widehat{v}'^\pm \cdot \nabla') h_3^\pm - (\widehat{B}'^\pm \cdot \nabla') u_3^\pm = 0, \tag{43}$$

where we have used $(F_2^\pm)_3 = 0$ and $\widehat{v}_3^\pm = \widehat{B}_3^\pm = 0$. If we substitute $u_3^\pm$ from the second equation into (35) and observe that $(\widehat{B}'^\pm \cdot \nabla')(\widehat{v}'^\pm \cdot \nabla' f) = (\widehat{v}'^\pm \cdot \nabla')(\widehat{B}'^\pm \cdot \nabla' f)$, we deduce that

$$(\partial_t + \widehat{v}'^\pm \cdot \nabla') (h_3^\pm - \widehat{B}'^\pm \cdot \nabla' f) = 0 \tag{44}$$

in $[0,T] \times \Gamma$. Using the third equation in (38), which holds for $t = 0$, we conclude that $h_3^\pm - \widehat{B}'^\pm \cdot \nabla' f = 0$ in $[0,T] \times \Gamma$.

4. Proof

4.1. Hyperbolic Regularization. Now, let us consider the following hyperbolic compressible regularization (see Secchi [18]), where $\varepsilon > 0$:

$$\partial_t q^\pm + (\hat{v}^\pm \cdot \nabla) q^\pm - \varepsilon \hat{B}^\pm \cdot (\partial_t h^\pm + (\hat{v}^\pm \cdot \nabla) h^\pm) + \frac{1}{\varepsilon} \operatorname{div} u^\pm = 0, \tag{45}$$

$$\partial_t u^\pm + (\hat{v}^\pm \cdot \nabla) u^\pm - (\hat{B}^\pm \cdot \nabla) h^\pm + \frac{1}{\varepsilon} \nabla q^\pm = F_1^\pm, \tag{46}$$

$$\begin{aligned}&\partial_t h^\pm + (\hat{v}^\pm \cdot \nabla) h^\pm - (\hat{B}^\pm \cdot \nabla) u^\pm \\ &\quad - \varepsilon \hat{B}^\pm \left\{\partial_t q^\pm + (\hat{v} \cdot \nabla) q^\pm - \varepsilon \hat{B}^\pm \cdot (\partial_t h^\pm + (\hat{v}^\pm \cdot \nabla) h^\pm)\right\} \\ &= F_2^\pm,\end{aligned} \tag{47}$$

with boundary conditions (35).

Here, the solution $U = (q, u, h)^T$ depends on ε but, in this moment, we will not use the more precise notation $U^\varepsilon = (q^\varepsilon, u^\varepsilon, h^\varepsilon)^T$ to denote the solution to the previous system (with suitable initial data). Let us note that, at least formally, if we project the system (45), (46), and (47) on the subspace of divergence-free functions, it converges, as $\varepsilon \to 0$, to system (34) projected on the same subspace; it is sufficient to multiply (45) by ε and take the limit. We will make this statement more precise in Section 4.7.

Equations (45)–(47) in $Q_T^\pm$ form a symmetric system:

$$\begin{aligned}&\begin{pmatrix} 1 & \underline{0}^T & -\varepsilon (\hat{B}^\pm)^T \\ \underline{0} & I_3 & O_3 \\ -\varepsilon \hat{B}^\pm & O_3 & I_3 + \varepsilon^2 \hat{B}^\pm \otimes \hat{B}^\pm \end{pmatrix} \partial_t \begin{pmatrix} q^\pm \\ u^\pm \\ h^\pm \end{pmatrix} + \frac{1}{\varepsilon} \begin{pmatrix} \operatorname{div} u^\pm \\ \nabla q^\pm \\ \underline{0} \end{pmatrix} \\ &+ \begin{pmatrix} \hat{v}^\pm \cdot \nabla & \underline{0}^T & -\varepsilon (\hat{B}^\pm)^T \hat{v}^\pm \cdot \nabla \\ \underline{0} & \hat{v}^\pm \cdot \nabla & -\hat{B}^\pm \cdot \nabla \\ -\varepsilon \hat{B}^\pm \hat{v}^\pm \cdot \nabla & -\hat{B}^\pm \cdot \nabla & (I_3 + \varepsilon^2 \hat{B}^\pm \otimes \hat{B}^\pm) \hat{v}^\pm \cdot \nabla \end{pmatrix} \begin{pmatrix} q^\pm \\ u^\pm \\ h^\pm \end{pmatrix} \\ &= \begin{pmatrix} 0 \\ F_1^\pm \\ F_2^\pm \end{pmatrix},\end{aligned} \tag{48}$$

or in a compact form,

$$L^\pm U^\pm \doteq A_0^\pm \partial_t U^\pm + \sum_{j=1}^{3} \left(A_j^\pm + \frac{1}{\varepsilon} \mathscr{E}_{1,j+1} \right) \partial_j U^\pm = F^\pm, \tag{49}$$

where each matrix $A_j^\pm$ is bounded in ε and $\mathscr{E}_{h,i} = (e_{j,k}^{(h,i)})_{1 \le j,k \le 7}$, where $e_{j,k}^{(h,i)} = 1$ if $(j,k) = (h,i)$ or $(j,k) = (i,h)$, $e_{j,k}^{(h,i)} = 0$ otherwise. Notice that $A_3^\pm = O_7$ because of (21).

One can easily check that, as $\varepsilon \to 0$, this system converges to (34), at least formally. Moreover, the system is hyperbolic, since (omitting $\pm$)

$$A_0 U \cdot U = (q - \varepsilon \hat{B} \cdot h)^2 + |u|^2 + |h|^2 > 0 \tag{50}$$

for each $(q, u, h) \neq (0, \underline{0}, \underline{0})$.

4.2. Secondary Symmetrization. This new formulation is obtained by a linear combination of (45), (46), and (47) in the previous step and aims to simplify boundary terms in the following computations (see Remark 5).

More precisely, let us consider the system

$$(45) + \varepsilon \lambda \hat{B} \cdot (46) - \frac{\lambda}{\varepsilon} \operatorname{div} h = 0, \tag{51}$$

$$\varepsilon \lambda \hat{B}\,(45) + (46) - \lambda\,(54) = \underline{0}, \tag{52}$$

$$-\lambda (I_3 + \varepsilon^2 \hat{B} \otimes \hat{B})(46) + (47) = \underline{0}, \tag{53}$$

where $\lambda = \lambda^\pm$ will be chosen conveniently later on and (54) denotes the equation obtained by substituting (45) into (47), that is,

$$\partial_t h^\pm + (\hat{v}^\pm \cdot \nabla) h^\pm - (\hat{B}^\pm \cdot \nabla) u^\pm + \hat{B}^\pm \operatorname{div} u^\pm = F_2^\pm. \tag{54}$$

We get

$$\begin{aligned}&\left(d_t q - \varepsilon \hat{B} \cdot d_t h + \frac{1}{\varepsilon} \operatorname{div} u\right) + \varepsilon \lambda \hat{B} \cdot \left\{d_t u - (\hat{B} \cdot \nabla) h + \frac{1}{\varepsilon} \nabla q\right\} \\ &\quad - \frac{\lambda}{\varepsilon} \operatorname{div} h = \varepsilon \lambda \hat{B} \cdot F_1, \\ &\varepsilon \lambda \hat{B} (d_t q - \varepsilon \hat{B} \cdot d_t h) + d_t u - (\hat{B} \cdot \nabla) h + \frac{1}{\varepsilon} \nabla q \\ &\quad - \lambda \{d_t h - (\hat{B} \cdot \nabla) u\} = F_1 - \lambda F_2, \\ &- \lambda (I_3 + \varepsilon^2 \hat{B} \otimes \hat{B}) \left\{d_t u - (\hat{B} \cdot \nabla) h + \frac{1}{\varepsilon} \nabla q\right\} + d_t h - (\hat{B} \cdot \nabla) u \\ &\quad - \varepsilon \hat{B} (d_t q - \varepsilon \hat{B} \cdot d_t h) = -\lambda (I_3 + \varepsilon^2 \hat{B} \otimes \hat{B}) F_1 + F_2,\end{aligned} \tag{55}$$

where $d_t \doteq \partial_t + (\hat{v} \cdot \nabla)$. This system is symmetric; indeed, the coefficient of d_t is

$$\mathscr{A}_0 = \begin{pmatrix} 1 & \varepsilon \lambda \hat{B}^T & -\varepsilon \hat{B}^T \\ \varepsilon \lambda \hat{B} & I_3 & -\lambda (I_3 + \varepsilon^2 \hat{B} \otimes \hat{B}) \\ -\varepsilon \hat{B} & -\lambda (I_3 + \varepsilon^2 \hat{B} \otimes \hat{B}) & I_3 + \varepsilon^2 \hat{B} \otimes \hat{B} \end{pmatrix}, \tag{56}$$

the singular term is given by

$$\frac{1}{\varepsilon} \begin{pmatrix} 0 & \nabla^T & -\lambda \nabla^T \\ \nabla & O_3 & O_3 \\ -\lambda \nabla & O_3 & O_3 \end{pmatrix}, \tag{57}$$

and the remaining derivatives are included in

$$\begin{pmatrix} \lambda\widehat{B}\cdot\nabla & \underline{0}^T & -\varepsilon\lambda\widehat{B}^T\left(\widehat{B}\cdot\nabla\right) \\ \underline{0} & \lambda\widehat{B}\cdot\nabla & -\widehat{B}\cdot\nabla \\ -\varepsilon\lambda\left(\widehat{B}\otimes\widehat{B}\right)\nabla & -\widehat{B}\cdot\nabla & \lambda\left(I_3+\varepsilon^2\widehat{B}\otimes\widehat{B}\right)\left(\widehat{B}\cdot\nabla\right) \end{pmatrix}, \tag{58}$$

where I_3 and O_3 denote, respectively, the identity and null 3×3 matrix. All these matrices are symmetric, and the system can be recast as

$$\mathscr{L}^\pm U^\pm \doteq \mathscr{A}_0^\pm\partial_t U^\pm + \sum_{j=1}^{3}\left(\mathscr{A}_j^\pm + \frac{1}{\varepsilon}\mathscr{E}_{1,j+1} - \frac{\lambda}{\varepsilon}\mathscr{E}_{1,j+4}\right)\partial_j U^\pm = \mathscr{F}^\pm, \tag{59}$$

where each matrix $\mathscr{A}_j^\pm$ is bounded in ε provided $\lambda^\pm$ is bounded and $\mathscr{E}_{h,i} = (e_{j,k}^{(h,i)})_{1\le j,k\le 7}$, where $e_{j,k}^{(h,i)} = 1$ if $(j,k) = (h,i)$ or $(j,k) = (i,h)$, $e_{j,k}^{(h,i)} = 0$ otherwise. Notice that $\mathscr{A}_3^\pm = O_7$.

Remark 5. The secondary symmetrization is inspired by [11, 19], which we refer to for further details. The main idea is to consider the following variant of (49), which takes into account the divergence constraint for the magnetic field

$$D^\pm A_0^\pm\partial_t U^\pm + \sum_{j=1}^{3} D^\pm\left(A_j^\pm + \frac{1}{\varepsilon}\mathscr{E}_{1,j+1}\right)\partial_j U^\pm + \lambda^\pm E^\pm \operatorname{div} h^\pm = F^\pm, \tag{60}$$

where the matrix $D^\pm$ and the vector $E^\pm$ depend on the basic state. D is then chosen in order to obtain $\mathscr{A}_0$, whose form is useful to get the energy a priori estimate, while E is chosen to maintain the symmetry of the system. As pointed out in [19] (Remark 5), such a choice is related to the conservation of cross-helicity: $d/dt\int(v\cdot B)\,dx = 0$. On the other hand, one main feature of the secondary symmetrization is that $\lambda^\pm$ can be chosen widely arbitrarily, so that in the following we will be able to find a suitable $\lambda^\pm$ which simplifies singular boundary terms in the a priori estimate.

The following lemma guarantees that, under our assumptions and a suitable choice of $\lambda^\pm$, the matrices $\mathscr{A}_0^\pm$ are positive definite and (59) is symmetric hyperbolic.

Lemma 6. *If*

$$(\lambda^\pm)^2 < \frac{1}{1+\left|\varepsilon\widehat{B}^\pm\right|^2}, \tag{61}$$

then the symmetric matrices $\mathscr{A}_0^\pm$ are positive definite.

Proof. Omitting "±", we have

$$\begin{aligned} &\mathscr{A}_0 U\cdot U \\ &= \left(q+\varepsilon\lambda\widehat{B}\cdot u-\varepsilon\widehat{B}\cdot h\right)q + \left\{\varepsilon\lambda\widehat{B}q+u-\lambda\left(I_3+\varepsilon^2\widehat{B}\otimes\widehat{B}\right)h\right\}\cdot u \\ &\quad + \left\{-\varepsilon\widehat{B}q-\lambda\left(I_3+\varepsilon^2\widehat{B}\otimes\widehat{B}\right)u+\left(I_3+\varepsilon^2\widehat{B}\otimes\widehat{B}\right)h\right\}\cdot h \\ &= \left(q-\varepsilon\widehat{B}\cdot h\right)^2 + u\cdot u + h\cdot h + 2\varepsilon\lambda\widehat{B}\cdot u\left(q-\varepsilon\widehat{B}\cdot h\right) - 2\lambda h\cdot u \\ &\ge \left(q-\varepsilon\widehat{B}\cdot h+\varepsilon\lambda\widehat{B}\cdot u\right)^2 + |h-\lambda u|^2 + \left(1-\lambda^2-\lambda^2\left|\varepsilon\widehat{B}\right|^2\right)|u|^2. \end{aligned} \tag{62}$$

If

$$1-\lambda^2-\lambda^2\left|\varepsilon\widehat{B}\right|^2 > 0, \tag{63}$$

that is to say, if

$$\lambda^2 < \frac{1}{1+\left|\varepsilon\widehat{B}\right|^2}, \tag{64}$$

then the matrix $\mathscr{A}_0$ is positive definite.

Lemma 7. *There exists $\varepsilon_0 > 0$ such that, for each $\varepsilon \in \,]0,\varepsilon_0]$, the choice*

$$\lambda^\pm = \widehat{\lambda}^\pm \doteq \frac{[\widehat{v}]\times\widehat{B}^\mp}{\left|\widehat{B}^+\times\widehat{B}^-\right|}\operatorname{sign}\left(\widehat{B}_1^+\widehat{B}_2^- - \widehat{B}_2^+\widehat{B}_1^-\right) \tag{65}$$

satisfies (61) *and hence makes the matrices $\mathscr{A}_0^\pm$ positive definite.*

Proof. Since

$$\frac{\left|[\widehat{v}]\times\widehat{B}^\mp\right|}{\left|\widehat{B}^+\times\widehat{B}^-\right|} < 1, \tag{66}$$

which follows from (22), we deduce that there exists $\varepsilon_0 > 0$ such that, for each $\varepsilon\in[0,\varepsilon_0[$, the inequality

$$\frac{\left|[\widehat{v}]\times\widehat{B}^\mp\right|}{\left|\widehat{B}^+\times\widehat{B}^-\right|} \le \frac{1}{\sqrt{1+\left|\varepsilon\widehat{B}^\pm\right|^2}} \tag{67}$$

follows, so we conclude that (61) holds true for ε small enough.

Let us remark that $\widehat{\lambda}^\pm$ is a constant term independent of ε; moreover, since $\widehat{B}_3^\pm = 0$, this choice of $\lambda^\pm$ is the same as

$$\widehat{\lambda}^\pm = \frac{[\widehat{v}_1]\widehat{B}_2^\mp - [\widehat{v}_2]\widehat{B}_1^\mp}{\widehat{B}_1^+\widehat{B}_2^- - \widehat{B}_2^+\widehat{B}_1^-}, \tag{68}$$

which will be useful in the following steps. Moreover, this choice is equivalent to

$$\left[\widehat{v}' - \widehat{\lambda}\widehat{B}'\right] = \underline{0}. \tag{69}$$

In addition, we set

$$\Delta = \widehat{B}_1^+\widehat{B}_2^- - \widehat{B}_2^+\widehat{B}_1^-, \tag{70}$$

which is always different from zero thanks to (23).

4.3. Boundary Conditions in Algebraic Form. The boundary conditions can be recast so that the space derivatives of the front function f do not appear explicitly.

Lemma 8. *The boundary conditions* (35) *imply, on* ω_T, *the boundary conditions*

$$[q] = 0, \qquad [u_3 - \hat{\lambda} h_3] = 0 \tag{71}$$

and the front equation

$$\partial_t f = u_3^{\pm} + S_+^{\pm} h_3^- - S_-^{\pm} h_3^+, \tag{72}$$

with

$$S_+^{\pm} = \frac{1}{\Delta}\left(\hat{v}_1^{\pm}\hat{B}_2^+ - \hat{v}_2^{\pm}\hat{B}_1^+\right), \qquad S_-^{\pm} = \frac{1}{\Delta}\left(\hat{v}_1^{\pm}\hat{B}_2^- - \hat{v}_2^{\pm}\hat{B}_1^-\right),$$
$$\Delta = \hat{B}_1^+\hat{B}_2^- - \hat{B}_2^+\hat{B}_1^-. \tag{73}$$

Proof. Condition (41) can be recast as

$$\hat{B}_1^+\partial_1 f + \hat{B}_2^+\partial_2 f = h_3^+, \qquad \hat{B}_1^-\partial_1 f + \hat{B}_2^-\partial_2 f = h_3^-. \tag{74}$$

Since

$$\Delta = \det\begin{pmatrix} \hat{B}_1^+ & \hat{B}_2^+ \\ \hat{B}_1^- & \hat{B}_2^- \end{pmatrix} \neq 0 \tag{75}$$

by hypothesis, we can obtain

$$\partial_1 f = \frac{1}{\Delta}\left\{h_3^+\hat{B}_2^- - h_3^-\hat{B}_2^+\right\}, \qquad \partial_2 f = -\frac{1}{\Delta}\left\{h_3^+\hat{B}_1^- - h_3^-\hat{B}_1^+\right\}. \tag{76}$$

By substituting in the second equation of (35), we get

$$\partial_t f = u_3^{\pm} + \frac{\hat{v}_1^{\pm}\hat{B}_2^+ - \hat{v}_2^{\pm}\hat{B}_1^+}{\Delta} h_3^- - \frac{\hat{v}_1^{\pm}\hat{B}_2^- - \hat{v}_2^{\pm}\hat{B}_1^-}{\Delta} h_3^+. \tag{77}$$

Using the definition of $S_+^{\pm}$ and $S_-^{\pm}$, we deduce the conditions

$$\partial_t f = u_3^{\pm} + S_+^{\pm} h_3^- - S_-^{\pm} h_3^+, \qquad u_3^+ + S_+^+ h_3^- - S_-^+ h_3^+ = u_3^- + S_+^- h_3^- - S_-^- h_3^+. \tag{78}$$

The last condition can be recast as

$$[u_3] = -[S_+] h_3^- + [S_-] h_3^+, \tag{79}$$

where

$$[S_+] = S_+^+ - S_+^-, \qquad [S_-] = S_-^+ - S_-^-. \tag{80}$$

Recalling (68), we have

$$\hat{\lambda}^{\pm} = [S_{\mp}]. \tag{81}$$

The previous condition becomes

$$[u_3 - \hat{\lambda} h_3] = 0, \tag{82}$$

and the lemma is proved.

4.4. Equivalence of Systems. Obviously, if $(U^{\varepsilon}, f^{\varepsilon})$ is a solution to system (49) with boundary conditions (35), then it also satisfies system (59) with conditions (71) (we are assuming the same initial conditions); the converse holds true as well.

Proposition 9. *If* U^{ε} *is a solution to system* (59) *with boundary conditions* (71), *then one can find a unique* f^{ε}, *so that* $(U^{\varepsilon}, f^{\varepsilon})$ *is a solution to system* (49) *with boundary conditions* (35) *and the same initial conditions.*

Proof. First, we prove that (49) is satisfied. Since (59) holds, then we have (51), (52), and (53). If we substitute (54) = (47) + $\varepsilon\hat{B}$(45) into (52), we get (46) = $\hat{\lambda}$(47), and substituting this relation in (53), we deduce

$$\left(I_3 - \hat{\lambda}^2 I_3 - \hat{\lambda}^2\varepsilon^2\hat{B}\otimes\hat{B}\right)(47) = 0. \tag{83}$$

Since

$$\det\left(I_3 - \hat{\lambda}^2 I_3 - \hat{\lambda}^2\varepsilon^2\hat{B}\otimes\hat{B}\right) = \left(1-\hat{\lambda}^2\right)^2\left(1-\hat{\lambda}^2-\hat{\lambda}^2\varepsilon^2|\hat{B}|^2\right) \neq 0 \tag{84}$$

thanks to (61), then we get (47) = $\underline{0}$ and consequently (46) = $\hat{\lambda}$(47) = $\underline{0}$.

From (51), in order to prove (45), it is now sufficient to establish that $\operatorname{div} h^{\varepsilon} = 0$, since we already know that (46) = $\underline{0}$. With this aim, we substitute (51) into (47) and get

$$\partial_t h^{\varepsilon} + (\hat{v}\cdot\nabla) h^{\varepsilon} - \hat{\lambda}\hat{B}\operatorname{div} h^{\varepsilon} + \hat{B}\operatorname{div} u^{\varepsilon} - \left(\hat{B}\cdot\nabla\right) u^{\varepsilon} = F_2. \tag{85}$$

Taking the divergence of the above equation, exploiting the assumption $\operatorname{div} F_2 = 0$ and the fact that $\hat{v}$, $\hat{B}$, $\hat{\lambda}$ are piecewise constant, and simplifying, we deduce that

$$\left(\partial_t + \left(\hat{v} - \hat{\lambda}\hat{B}\right)\cdot\nabla\right)\operatorname{div} h^{\varepsilon} = 0. \tag{86}$$

Since $\operatorname{div} h^{\varepsilon} = 0$ in $\{t = 0\}\times\Omega^{\pm}$ by assumption on the initial data, we conclude that $\operatorname{div} h^{\varepsilon} = 0$ in $Q_T^{\pm}$, so that (45) = 0 from (51). Then (45), (46), and (47) are satisfied, which means that (49) holds.

Now, we have to prove (35). Condition $[q^{\varepsilon}] = 0$ on ω_T obviously holds, since we have the same condition in (71). We need to prove that there exists f^{ε} so that $\partial_t f^{\varepsilon} + \hat{v}'^{\pm}\cdot\nabla' f^{\varepsilon} = u_3^{\varepsilon,\pm}$ on ω_T. Given $u_3^{\varepsilon,\pm}$ and $h_3^{\varepsilon,\pm}$, let us define $f^{\pm}$ as the solutions of

$$\partial_t f^{\pm} + \hat{v}'^{\pm}\cdot\nabla' f^{\pm} = u_3^{\varepsilon,\pm} \quad \text{in } \omega_T, \qquad f^{\pm}|_{t=0} = f_0 \quad \text{on } \Gamma. \tag{87}$$

Thus, for the completion of the proof of (35), it is sufficient to show that $f^+ = f^-$ in ω_T. Proceeding as in the proof of Proposition 4, we first deduce $h_3^{\varepsilon,\pm} = (\hat{B}'^{\pm}\cdot\nabla') f^{\pm}$ in ω_T. Let us compute

$$u_3^{\varepsilon,\pm} - \hat{\lambda}^{\pm} h_3^{\varepsilon,\pm} = \partial_t f^{\pm} + \left(\hat{v}'^{\pm}\cdot\nabla'\right) f^{\pm} - \hat{\lambda}^{\pm}\left(\hat{B}'^{\pm}\cdot\nabla'\right) f^{\pm}. \tag{88}$$

Since $[u_3^\varepsilon - \widehat{\lambda} h_3^\varepsilon] = 0$ by (71), we obtain

$$\left(\partial_t+\left(\widehat{v}'^{+}-\widehat{\lambda}^{+}\widehat{B}'^{+}\right)\cdot\nabla'\right)f^{+}=\left(\partial_t+\left(\widehat{v}'^{-}-\widehat{\lambda}^{-}\widehat{B}'^{-}\right)\cdot\nabla'\right)f^{-}. \tag{89}$$

Thanks to (69), we have $\widehat{w}' \doteq \widehat{v}'^{+} - \widehat{\lambda}^{+}\widehat{B}'^{+} = \widehat{v}'^{-} - \widehat{\lambda}^{-}\widehat{B}'^{-}$, and hence,

$$\begin{aligned}\left(\partial_t + \widehat{w}'\cdot\nabla'\right)f^{+} &= \left(\partial_t + \widehat{w}'\cdot\nabla'\right)f^{-} \quad \text{in } \omega_T,\\ f^{\pm}\big|_{t=0} &= f_0 \quad \text{on } \Gamma.\end{aligned} \tag{90}$$

Thus, we can take $f^\varepsilon \doteq f^+ = f^-$ in ω_T, and (35) holds.

4.5. Existence of the Approximate Solution. Now, consider the system (59) with the boundary conditions (71) and the initial condition $U^{\varepsilon,\pm}(0) = U_0^\pm \doteq (0, u_0^\pm, h_0^\pm)^T$. This system can be written as

$$\begin{aligned}\mathscr{L}U^\varepsilon &= \mathscr{F} \quad \text{in } [0,T]\times\Omega,\\ \mathscr{M}U^\varepsilon &= \underline{0} \quad \text{in } [0,T]\times\Gamma,\\ U^\varepsilon(0) &= U_0 \quad \text{in } \Omega,\end{aligned} \tag{91}$$

where we have set $U^\varepsilon = (U^{\varepsilon,+}, U^{\varepsilon,-})^T$ and similarly for the other quantities; moreover, we set $\|U^\varepsilon\|_\Omega = \|U^{\varepsilon,+}\|_{\Omega^+} + \|U^{\varepsilon,-}\|_{\Omega^-}$ and so on. Notice that $\mathscr{M}U^\varepsilon = \underline{0}$ corresponds to the boundary conditions (71).

Let us note that, since $q_0^\pm \doteq q^{\varepsilon,\pm}|_{t=0} \equiv 0$, the condition on Γ given by $[q_0] = 0$ is automatically satisfied. Moreover, if we denote by $(\partial_t q^+, \partial_t u^+, \partial_t h^\pm)|_{t=0}$ the homonymous quantities in (91) evaluated at time $t = 0$ and calculated in terms of the initial data, from the equations and the hypotheses on $q_0^\pm, u_0^\pm, h_0^\pm$, we get the condition $(\partial_t q^\pm, \partial_t u^\pm, \partial_t h^\pm)|_{t=0} \in L^2(\Omega^\pm)$. Let us note that the hypothesis $u_0^\pm, h_0^\pm$ in $H^1(\Omega^\pm)$, rather than simply in $H_*^1(\Omega^\pm)$, is needed exactly here. We have the following result.

Proposition 10. *There exists $\varepsilon_0 > 0$ so that, for each $\varepsilon \in]0, \varepsilon_0]$, there exists a unique solution $U^\varepsilon \in \mathscr{C}^0([0,T]; H_*^1(\Omega)) \cap \mathscr{C}^1([0,T]; L^2(\Omega))$ to problem (91); moreover, one has*

$$|||U^\varepsilon(t)|||_{1,*;\Omega} \le C_\varepsilon e^{C_\varepsilon t}\left(|||U_0^\varepsilon|||_{1,*;\Omega} + [\mathscr{F}]_{1,*,t;\Omega}\right) \tag{92}$$

for each $t \in [0,T]$ and a suitable $C_\varepsilon > 0$.

Proof. The boundary matrix of system (91) is given by

$$\mathscr{A}_\nu = \begin{pmatrix}\mathscr{A}_\nu^+ & O_7\\ O_7 & \mathscr{A}_\nu^-\end{pmatrix}, \tag{93}$$

where $\mathscr{A}_\nu^\pm = \mp(\mathscr{A}_3^\pm + (1/\varepsilon)\mathscr{E}_{1,4} - (\widehat{\lambda}^\pm/\varepsilon)\mathscr{E}_{1,7})$. The matrix $\mathscr{A}_\nu$ is singular with constant rank (indeed, $\mathscr{A}_3^\pm$ is the null matrix) and, moreover, $\ker\mathscr{M}$ is maximally nonnegative for $\mathscr{A}_\nu$, that is to say,

$$\mathscr{A}_\nu U^\varepsilon\cdot U^\varepsilon \ge 0 \tag{94}$$

for each $U^\varepsilon \in \ker\mathscr{M}$ (nonnegativity) and such a property does not hold in any other vector space properly containing $\ker\mathscr{M}$ (maximality). Indeed, we have

$$\begin{aligned}\mathscr{A}_\nu\big|_\Gamma U^\varepsilon\cdot U^\varepsilon &= -\mathscr{A}_\nu^+\big|_\Gamma U^{\varepsilon,+}\cdot U^{\varepsilon,+} + \mathscr{A}_\nu^-\big|_\Gamma U^{\varepsilon,-}\cdot U^{\varepsilon,-}\\ &= -\frac{2}{\varepsilon}\left\{[q^\varepsilon]\left(u_3^{\varepsilon,+} - \widehat{\lambda}^+ h_3^{\varepsilon,+}\right) + q^{\varepsilon,-}\left[u_3^\varepsilon - \widehat{\lambda} h_3^\varepsilon\right]\right\}\\ &=0\end{aligned} \tag{95}$$

thanks to $[q^\varepsilon] = 0$ and $[u_3^\varepsilon - \widehat{\lambda} h_3^\varepsilon] = 0$ on $\ker\mathscr{M}$. Moreover, let us observe that both $\mathscr{A}_\nu^+$ and $\mathscr{A}_\nu^-$ have exactly one negative eigenvalue, so we have the same number of conditions defining $\ker\mathscr{M}$ and negative eigenvalues, that is, 2; this implies maximality.

Thus, we can apply the results in Lax and Phillips [12] (existence in L^2, see as well Friedrichs [20] for the noncharacteristic case) and Secchi [13, 14] (regularity in H_*^m), and deduce the existence of a unique global solution $U^\varepsilon \in \mathscr{C}^0([0,T]; H_*^1(\Omega)) \cap \mathscr{C}^1([0,T]; L^2(\Omega))$ to problem (91) such that (92) holds for each $t \in [0,T]$, provided $\varepsilon > 0$ is sufficiently small.

Remark 11. Let us note that, in the previous result, the smallness of ε is necessary in order to have $\mathscr{A}_0^\pm$ positive definite.

We need to have a number of boundary conditions in (71), that is, defining $\ker\mathscr{M}$, equal to the number of negative eigenvalues of $\mathscr{A}_\nu$, that is to say, 2, in order to have maximality.

Corollary 12. *There exists $\varepsilon_0 > 0$, so that, for each $\varepsilon \in]0, \varepsilon_0]$, there exists a unique solution $(U^\varepsilon, f^\varepsilon)$ to problem (49), (35), and (36) with $q^{\varepsilon,\pm}|_{t=0} \equiv 0$ such that $U^\varepsilon \in \mathscr{C}^0([0,T]; H_*^1(\Omega)) \cap \mathscr{C}^1([0,T]; L^2(\Omega))$, $(q, u_3, h_3)^{\varepsilon,\pm}|_{\omega_T} \in \mathscr{C}^0([0,T]; H^{1/2}(\Gamma))$ and $f^\varepsilon \in \mathscr{C}^0([0,T]; H^{1/2}(\Gamma))$; moreover, one has*

$$\begin{aligned}&|||U^\varepsilon(t)|||_{1,*;\Omega} + \|f^\varepsilon(t)\|_{H^{1/2}(\Gamma)}\\ &\quad\le C_\varepsilon e^{C_\varepsilon t}\left(|||U_0^\varepsilon|||_{1,*;\Omega} + [\mathscr{F}]_{1,*,t;\Omega}\right) + \|f_0\|_{H^{1/2}(\Gamma)}\end{aligned} \tag{96}$$

for each $t \in [0,T]$ and a suitable $C_\varepsilon > 0$.

Proof. Problem (91) with $f^\varepsilon|_{t=0} = f_0$ on Γ is a different way to say that (59), (71), and (36) with $q^{\varepsilon,\pm}|_{t=0} \equiv 0$ are satisfied. Using Propositions 9 and 10, we get the result for U^ε. Let us note that, since $u^{\varepsilon,\pm} \in \mathscr{C}^0([0,T]; H_*^1(\Omega))$ and the normal component $u_3^{\varepsilon,\pm}$ at the boundary is noncharacteristic, then the trace $u_3^{\varepsilon,\pm}|_{\omega_T}$ is in $\mathscr{C}^0([0,T]; H^{1/2}(\Gamma))$, as shown in Secchi [13, 14] and Shizuta [21]. In particular, $\partial_3 u_3^\varepsilon$ can be estimated using the other terms appearing in (45), so that we have

$$\|u_3^\varepsilon(t)\|_{H^1(\Omega)} \le C|||U^\varepsilon(t)|||_{1,*;\Omega}. \tag{97}$$

The same holds true for q^ε as well, for the same reason ($\partial_3 q^\varepsilon$ is obtained from (46)), and for $\partial_3 h_3^\varepsilon$, since $\operatorname{div} h^\varepsilon = 0$ is bounded (this follows from (54) proceeding similarly as in Proposition 4; note that, taking the divergence, the terms containing u^ε cancel together). Thus, from the transport equation (87), we find $f^\varepsilon \in \mathscr{C}^0([0,T]; H^{1/2}(\Gamma))$ and

$$\begin{aligned}
&\|f^\varepsilon(t)\|_{H^{1/2}(\Gamma)} \\
&\le \|f_0\|_{H^{1/2}(\Gamma)} + t\|u_3^{\varepsilon,+}\|_{L^\infty(0,t;H^{1/2}(\Gamma))} \\
&\le \|f_0\|_{H^{1/2}(\Gamma)} + Ct\|u_3^{\varepsilon,+}\|_{L^\infty(0,t;H^1(\Omega^+))} \\
&\le \|f_0\|_{H^{1/2}(\Gamma)} + C_\varepsilon e^{C_\varepsilon t}\sum_\pm \left(|||U_0^\pm|||_{1,*;\Omega^\pm} + [F^\pm]_{1,*,t;\Omega^\pm}\right).
\end{aligned} \tag{98}$$

This concludes the proof.

4.6. A Priori Energy Estimate Uniform in ε. Since estimate (96) depends on ε, we cannot use it when passing to the limit as $\varepsilon \to 0$. We can deduce the following a priori estimate for $(U^{\varepsilon,\pm}, f^\varepsilon)$, which is an L^2-estimate uniform in ε, useful to pass to the limit as $\varepsilon \to 0$. We recall that the choice $\lambda = \widehat{\lambda}$ makes $\mathscr{A}_0^\pm$ positive definite, and hence, if we fix $\varepsilon_0 > 0$ sufficiently small, there exists a suitable real number $\alpha_0 \in]0,1]$, so that $\mathscr{A}_0^\pm \ge \alpha_0 I_7$ for each $\varepsilon \in]0, \varepsilon_0]$.

Proposition 13. *There exists* $\varepsilon_0 > 0$, *such that, for each* $\varepsilon \in]0, \varepsilon_0]$, *the a priori estimate*

$$\begin{aligned}
&\sum_\pm \left(|||U^{\varepsilon,\pm}(t)|||^2_{1,*;\Omega^\pm} + \left\|(u_3,h_3)^{\varepsilon,\pm}(t)\right\|^2_{H^1(\Omega^\pm)}\right) + \|f^\varepsilon(t)\|^2_{H^{1/2}(\Gamma)} \\
&\le \frac{C}{\alpha_0}e^{Ct/\alpha_0}\sum_\pm\left(|||U_0^\pm|||^2_{1,*;\Omega^\pm} + [F^\pm]^2_{1,*,t;\Omega^\pm}\right) + \|f_0\|^2_{H^{1/2}(\Gamma)}
\end{aligned} \tag{99}$$

holds for each $t \in [0,T]$ *and for a suitable* $C > 0$ *independent of* ε, *where* U^ε *is the solution to problem* (49), (35), *and* (36) *and* f^ε *satisfies* (87) *with initial condition* $f^\varepsilon|_{t=0} = f_0$.

Proof. Let us take the scalar product of (59), which is equivalent to (49), with $U^{\varepsilon,\pm}$ and integrate over $Q_t^\pm$ (recall that $\lambda^\pm = \widehat{\lambda}^\pm$); integrating by parts the terms with spatial derivatives of $U^{\varepsilon,\pm}$, exploiting the fact that the coefficient matrices are constant, and recalling that the boundary matrix is given by $\mathscr{A}_\nu^\pm = \mp(\mathscr{A}_3^\pm + (1/\varepsilon)\mathscr{E}_{1,4} - (\widehat{\lambda}^\pm/\varepsilon)\mathscr{E}_{1,7}) = \mp((1/\varepsilon)\mathscr{E}_{1,4} - (\widehat{\lambda}^\pm/\varepsilon)\mathscr{E}_{1,7})$, since $\mathscr{A}_3^\pm = O_7$, we get

$$\begin{aligned}
&\int_{Q_t^\pm} \mathscr{A}_0^\pm\partial_t U^{\varepsilon,\pm}\cdot U^{\varepsilon,\pm}\,dx\,dt \\
&\quad\mp\frac{1}{2\varepsilon}\int_{\omega_t}\left(\mathscr{E}_{1,4} - \widehat{\lambda}^\pm\mathscr{E}_{1,7}\right)U^{\varepsilon,\pm}\cdot U^{\varepsilon,\pm}\,dx'dt \\
&= \frac{1}{2}\int_{Q_t^\pm}\left(\sum_{j=1}^3\partial_j\left(\mathscr{A}_j^\pm + \frac{1}{\varepsilon}\mathscr{E}_{1,j+1} - \frac{\widehat{\lambda}^\pm}{\varepsilon}\mathscr{E}_{1,j+4}\right)\right)U^{\varepsilon,\pm}\cdot U^{\varepsilon,\pm}\,dx\,dt \\
&\quad+\int_{Q_t^\pm}\mathscr{F}^\pm\cdot U^{\varepsilon,\pm}\,dx\,dt \\
&= \int_{Q_t^\pm}\mathscr{F}^\pm\cdot U^{\varepsilon,\pm}\,dx\,dt,
\end{aligned} \tag{100}$$

where sum over $\pm$ is assumed. Let us note that, since the basic state is piecewise constant, then $\mathscr{A}_j^\pm, \mathscr{E}_{h,k}$ and $\widehat{\lambda}^\pm$ are piecewise constant as well; thus,

$$\sum_{j=1}^3\partial_j\left(\mathscr{A}_j^\pm + \frac{1}{\varepsilon}\mathscr{E}_{1,j+1} - \frac{\widehat{\lambda}^\pm}{\varepsilon}\mathscr{E}_{1,j+4}\right) = 0 \tag{101}$$

in $Q_t^\pm$. Moreover,

$$\begin{aligned}
&\sum_\pm \mp\frac{1}{2\varepsilon}\int_{\omega_t}\left(\mathscr{E}_{1,4} - \widehat{\lambda}^\pm\mathscr{E}_{1,7}\right)U^{\varepsilon,\pm}\cdot U^{\varepsilon,\pm}\,dx'dt \\
&= -\frac{1}{\varepsilon}\int_{\omega_t}\left\{[q^\varepsilon]\left(u_3^{\varepsilon,+} - \widehat{\lambda}^+h_3^{\varepsilon,+}\right) + q^{\varepsilon,-}\left[u_3^\varepsilon - \widehat{\lambda}h_3^\varepsilon\right]\right\}dx'dt \\
&= 0,
\end{aligned} \tag{102}$$

thanks to the boundary conditions $[q^\varepsilon] = 0$ and $[u_3^\varepsilon - \widehat{\lambda}h_3^\varepsilon] = 0$, that hold in ω_t.

We deduce the equality

$$\sum_\pm\int_{Q_t^\pm}\mathscr{A}_0^\pm\partial_\tau U^{\varepsilon,\pm}\cdot U^{\varepsilon,\pm}\,dx\,d\tau = \sum_\pm\int_{Q_t^\pm}\mathscr{F}^\pm\cdot U^{\varepsilon,\pm}\,dx\,d\tau. \tag{103}$$

For the left-hand side, integrating by parts in time and exploiting the symmetry of $\mathscr{A}_0^\pm$, we have

$$\begin{aligned}
\int_{\Omega^\pm}\int_0^t\mathscr{A}_0^\pm\partial_t U^{\varepsilon,\pm}\cdot U^{\varepsilon,\pm}\,d\tau\,dx &= \frac{1}{2}\int_{\Omega^\pm}\mathscr{A}_0^\pm U^{\varepsilon,\pm}(t)\cdot U^{\varepsilon,\pm}(t)\,dx \\
&\quad-\frac{1}{2}\int_{\Omega^\pm}\mathscr{A}_0^\pm U_0^\pm\cdot U_0^\pm\,dx.
\end{aligned} \tag{104}$$

By applying the Cauchy-Schwarz inequality in the right-hand side, we obtain

$$\begin{aligned}
&\sum_\pm\int_{\Omega^\pm}\mathscr{A}_0^\pm U^{\varepsilon,\pm}(t)\cdot U^{\varepsilon,\pm}(t)\,dx \\
&\le\int_0^t\sum_\pm\|U^{\varepsilon,\pm}(\tau)\|^2_{L^2(\Omega^\pm)}d\tau + \sum_\pm\int_{\Omega^\pm}\mathscr{A}_0^\pm U_0^\pm\cdot U_0^\pm\,dx \\
&\quad+\sum_\pm\int_0^t\|\mathscr{F}^\pm(\tau)\|^2_{L^2(\Omega^\pm)}d\tau,
\end{aligned} \tag{105}$$

and hence, by exploiting $\mathscr{A}_0^\pm \ge \alpha_0 I_7$,

$$\begin{aligned}
&\sum_\pm\|U^{\varepsilon,\pm}(t)\|^2_{L^2(\Omega^\pm)} \\
&\le\frac{1}{\alpha_0}\int_0^t\sum_\pm\|U^{\varepsilon,\pm}(\tau)\|^2_{L^2(\Omega^\pm)}d\tau \\
&\quad+\frac{1}{\alpha_0}\sum_\pm\left(C\|U_0^\pm\|^2_{L^2(\Omega^\pm)} + \int_0^t\|\mathscr{F}^\pm(\tau)\|^2_{L^2(\Omega^\pm)}d\tau\right).
\end{aligned} \tag{106}$$

Through a standard application of Grönwall's lemma, we deduce that

$$\sum_{\pm}\|U^{\varepsilon,\pm}(t)\|^2_{L^2(\Omega^\pm)} \le \frac{Ce^{t/\alpha_0}}{\alpha_0}\sum_{\pm}\left(\|U_0^\pm\|^2_{L^2(\Omega^\pm)} + \int_0^t \|F^\pm(\tau)\|^2_{L^2(\Omega^\pm)}d\tau\right). \quad (107)$$

In the same way, we can apply a tangential derivative $Z = \partial_t, \partial_{x_1}, \partial_{x_2}$ to (59) and (35), take the scalar product with $ZU^{\varepsilon,\pm}$, and proceed as before to obtain

$$\sum_{\pm}\|ZU^{\varepsilon,\pm}(t)\|^2_{L^2(\Omega^\pm)} \le \frac{Ce^{t/\alpha_0}}{\alpha_0}\sum_{\pm}\left(\|ZU_0^\pm\|^2_{L^2(\Omega^\pm)} + \int_0^t \|ZF^\pm(\tau)\|^2_{L^2(\Omega^\pm)}d\tau\right). \quad (108)$$

It remains to find an estimate for $Z_3U^{\varepsilon,\pm}$, where $Z_3 = \sigma\partial_3 = \sigma\partial_{x_3}$. If we apply $\sigma\partial_3$ to (59) and recall that $\mathcal{A}_3^\pm = O_7$, we get

$$\mathcal{A}_0^\pm\partial_t(\sigma\partial_3U^{\varepsilon,\pm}) + \sum_{j=1}^3\left(\mathcal{A}_j^\pm + \frac{1}{\varepsilon}\mathcal{E}_{1,j+1} - \frac{\hat{\lambda}^\pm}{\varepsilon}\mathcal{E}_{1,j+4}\right)\partial_j(\sigma\partial_3U^{\varepsilon,\pm}) = \sigma\partial_3\mathcal{F}^\pm + \left(\frac{1}{\varepsilon}\mathcal{E}_{1,4} - \frac{\hat{\lambda}^\pm}{\varepsilon}\mathcal{E}_{1,7}\right)\sigma'\partial_3U^{\varepsilon,\pm}. \quad (109)$$

Testing by $\sigma\partial_3U^{\varepsilon,\pm}$, summing up terms with plus and minus sign, and simplifying as before, we deduce that

$$\sum_{\pm}\int_{Q_t^\pm}\mathcal{A}_0^\pm\partial_t(\sigma\partial_3U^{\varepsilon,\pm})\cdot\sigma\partial_3U^{\varepsilon,\pm}dx\,dt = \sum_{\pm}\int_{Q_t^\pm}\sigma\partial_3\mathcal{F}^\pm\cdot\sigma\partial_3U^{\varepsilon,\pm}dx\,dt + \sum_{\pm}I^\pm, \quad (110)$$

where

$$I^\pm = \frac{1}{\varepsilon}\int_{Q_t^\pm}\left(\mathcal{E}_{1,4} - \hat{\lambda}^\pm\mathcal{E}_{1,7}\right)\sigma'\partial_3U^{\varepsilon,\pm}\cdot\sigma\partial_3U^{\varepsilon,\pm}dx\,dt, \quad (111)$$

so that we get

$$\frac{\alpha_0}{2}\sum_{\pm}\|Z_3U^{\varepsilon,\pm}(t)\|^2_{L^2(\Omega^\pm)} \le \sum_{\pm}\left(C\|Z_3U_0^\pm\|^2_{L^2(\Omega^\pm)} + |I^\pm|\right) + \sum_{\pm}\frac{1}{2}\int_0^t\left(\|Z_3U^{\varepsilon,\pm}(\tau)\|^2_{L^2(\Omega^\pm)} + \|Z_3\mathcal{F}^\pm(\tau)\|^2_{L^2(\Omega^\pm)}\right)d\tau \quad (112)$$

by applying the Cauchy-Schwarz inequality. Therefore, we need to estimate $|I^\pm|$:

$$\begin{aligned}|I^\pm| &= \left|\int_{Q_t^\pm}2\sigma'\frac{\partial_3q^{\varepsilon,\pm}}{\varepsilon}\left(\sigma\partial_3u^{\varepsilon,\pm} - \hat{\lambda}^\pm\sigma\partial_3h^{\varepsilon,\pm}\right)dx\,d\tau\right| \\ &\le C\int_{Q_t^\pm}\left|\frac{\partial_3q^{\varepsilon,\pm}}{\varepsilon}\right||Z_3U^{\varepsilon,\pm}|\,dx\,d\tau \\ &\le C\int_0^t\|Z_3U^{\varepsilon,\pm}(\tau)\|^2_{L^2(\Omega^\pm)}d\tau + C\int_0^t\left\|\frac{1}{\varepsilon}\nabla q^{\varepsilon,\pm}(\tau)\right\|_{L^2(\Omega^\pm)}d\tau.\end{aligned} \quad (113)$$

From (46) and the fact that $\hat{v}_3^\pm = \hat{B}_3^\pm = 0$, we have

$$\frac{1}{\varepsilon}\nabla q^{\varepsilon,\pm} = -\partial_tu^{\varepsilon,\pm} - (\hat{v}'^\pm\cdot\nabla')u^{\varepsilon,\pm} + (\hat{B}'^\pm\cdot\nabla')h^{\varepsilon,\pm} + F_1^\pm. \quad (114)$$

Consequently, if we denote $Z_0 = \partial_t$, $Z_1 = \partial_{x_1}$, and $Z_2 = \partial_{x_2}$, while $Z_3 = \sigma\partial_{x_3}$ and use estimate (108), we obtain

$$\begin{aligned}\sum_{\pm}\left\|\frac{1}{\varepsilon}\nabla q^{\varepsilon,\pm}(\tau)\right\|^2_{L^2(\Omega^\pm)} &\le C\sum_{\pm}\left(\sum_{j=0}^2\|Z_jU^{\varepsilon,\pm}(\tau)\|^2_{L^2(\Omega^\pm)} + \|F^\pm(\tau)\|^2_{L^2(\Omega^\pm)}\right) \\ &\le C\sum_{\pm}\left(\|F^\pm(\tau)\|^2_{L^2(\Omega^\pm)} + \frac{e^{\tau/\alpha_0}}{\alpha_0}\sum_{j=0}^2\|Z_jU_0^\pm\|^2_{L^2(\Omega^\pm)} + \frac{e^{\tau/\alpha_0}}{\alpha_0}\int_0^\tau\sum_{j=0}^2\|Z_jF^\pm(s)\|^2_{L^2(\Omega^\pm)}ds\right).\end{aligned} \quad (115)$$

By combining such estimate with the previous ones for $I^\pm$ and $Z_3U^{\varepsilon,\pm}$, integrating by parts in order to simplify the double integral, and neglecting negative terms obtained in the right-hand side, we deduce

$$\sum_{\pm}\|Z_3U^{\varepsilon,\pm}(t)\|^2_{L^2(\Omega^\pm)} \le \frac{C}{\alpha_0}\int_0^t\sum_{\pm}\|Z_3U^{\varepsilon,\pm}(\tau)\|^2_{L^2(\Omega^\pm)}d\tau + \frac{C}{\alpha_0}\sum_{\pm}\beta^\pm(t), \quad (116)$$

where

$$\beta^\pm(t) = \|Z_3U_0^\pm\|^2_{L^2(\Omega^\pm)} + e^{t/\alpha_0}\sum_{j=0}^2\|Z_jU_0^\pm\|^2_{L^2(\Omega^\pm)} + \int_0^t\left(\|F^\pm(\tau)\|^2_{L^2(\Omega^\pm)} + \|Z_3F^\pm(\tau)\|^2_{L^2(\Omega^\pm)}\right)d\tau + e^{t/\alpha_0}\int_0^t\sum_{j=0}^2\|Z_jF^\pm(\tau)\|^2_{L^2(\Omega^\pm)}d\tau. \quad (117)$$

By the Grönwall's lemma, we get

$$\sum_{\pm} \|Z_3 U^{\varepsilon,\pm}(t)\|^2_{L^2(\Omega^\pm)} \le \frac{Ce^{Ct/\alpha_0}}{\alpha_0} \sum_{\pm} \beta^\pm(t). \tag{118}$$

Combining (107), (108), and (118), we obtain that

$$\begin{aligned}\sum_{\pm} \left(\|U^{\varepsilon,\pm}(t)\|^2_{H^1_*(\Omega^\pm)} + \|\partial_t U^{\varepsilon,\pm}(t)\|^2_{L^2(\Omega^\pm)} \right) &\le \frac{C}{\alpha_0} e^{Ct/\alpha_0} \sum_{\pm} \left(\|U_0^\pm\|^2_{H^1_*(\Omega^\pm)} + \|\partial_t U_0^\pm\|^2_{L^2(\Omega^\pm)} \right) \\ &+ \frac{C}{\alpha_0} e^{Ct/\alpha_0} \sum_{\pm} \int_0^t \left(\|F^\pm(\tau)\|^2_{H^1_*(\Omega^\pm)} + \|\partial_t F^\pm(\tau)\|^2_{L^2(\Omega^\pm)} \right) d\tau. \end{aligned} \tag{119}$$

To conclude, we need to estimate f^ε. From (87), proceeding as in the proof of Corollary 12 and using the previous estimate, we get

$$\|u_3^{\varepsilon,\pm}(t)\|^2_{H^1(\Omega^\pm)} + \|h_3^{\varepsilon,\pm}(t)\|^2_{H^1(\Omega^\pm)} \le C |||U^{\varepsilon,\pm}(t)|||^2_{1,*;\Omega^\pm}, \tag{120}$$

$$\|f^\varepsilon(t)\|^2_{H^{1/2}(\Gamma)} \le \|f_0\|^2_{H^{1/2}(\Gamma)} + \frac{C}{\alpha_0} e^{Ct/\alpha_0} \sum_{\pm} \left(|||U_0^\pm|||^2_{1,*;\Omega^\pm} + [F^\pm]^2_{1,*,t;\Omega^\pm} \right). \tag{121}$$

Adding up (119) and (121), we deduce the claim.

4.7. Conclusion. Let us denote by CP the compressible problem (49), (35), and (36) with $q^{\varepsilon,\pm}|_{t=0} \equiv 0$ and by IP the incompressible problem (34), (35), and (36). Corollary 12 guarantees that, for ε sufficiently small, there exists a unique solution

$$(U^\varepsilon, f^\varepsilon) \in \mathscr{C}^0([0,T]; H^1_*(\Omega)) \cap \mathscr{C}^1([0,T]; L^2(\Omega)) \times \mathscr{C}^0([0,T]; H^{1/2}(\Gamma)) \tag{122}$$

to CP, and moreover, this solution satisfies (99), which implies, in particular, that $(u^{\varepsilon,\pm}, h^{\varepsilon,\pm}, f^\varepsilon)$ is bounded in $(H^1_*(\Omega^\pm))^2 \times H^{1/2}(\Gamma)$, uniformly in ε and $t \in [0,T]$. Thus, it is possible to extract a subsequence weakly convergent, as $\varepsilon \to 0$, to a certain $(u^\pm, h^\pm, f)$. Now, we will show that it is possible to find pressure terms $q^\pm$ (unique up to additive constants), so that $(q^\pm, u^\pm, h^\pm, f)$ solves IP. Since we are considering linear systems, the weak limit can be performed easily, with the only exceptions of the singular term $(1/\varepsilon)\nabla q^{\varepsilon,\pm}$ in (46) and the boundary condition $[q] = 0$.

If we test (46) against a generic $\mathscr{C}^\infty$ divergence-free vector function $\varphi^\pm$ compactly supported in $Q_T^\pm$, we observe that

$$\frac{1}{\varepsilon} \int_{\Omega^\pm} \nabla q^{\varepsilon,\pm} \cdot \varphi = -\frac{1}{\varepsilon} \int_{\Omega^\pm} q^{\varepsilon,\pm} \operatorname{div} \varphi^\pm = 0 \tag{123}$$

and hence

$$\partial_t u^{\varepsilon,\pm} + (\hat{v}^\pm \cdot \nabla) u^{\varepsilon,\pm} - (\hat{B}^\pm \cdot \nabla) h^{\varepsilon,\pm} - F_1^\pm = 0 \tag{124}$$

in the space of solenoidal distributions defined in $Q_T^\pm$, that is to say, the left-hand side is a gradient $-\nabla Q^{\varepsilon,\pm}$ in the space of distributions. Taking the weak limit as $\varepsilon \to 0$, we obtain

$$\partial_t u^\pm + (\hat{v}^\pm \cdot \nabla) u^\pm - (\hat{B}^\pm \cdot \nabla) h^\pm - F_1^\pm = -\nabla q^\pm \tag{125}$$

in a weak sense, where $q^\pm(t) \in L^2_{\text{loc}}(\Omega^\pm)$ is unique up to additive constants, and $\nabla q^\pm$ is the weak limit of $\nabla Q^{\varepsilon,\pm}$. Note that $\nabla Q^{\varepsilon,\pm}$ converges since it is equal to a convergent quantity; moreover, its limit is still a gradient (by the fact that the subspace of gradients is closed). The weak formulation reads

$$\sum_{\pm} \int_0^T \int_{\Omega^\pm} \left(u^\pm \cdot \partial_t \varphi + (\hat{v}'^\pm \cdot \nabla') \varphi \cdot u^\pm - (\hat{B}'^\pm \cdot \nabla') \varphi \cdot h^\pm + F_1^\pm \cdot \varphi + q^\pm \operatorname{div} \varphi \right) = 0 \tag{126}$$

for each $\varphi \in \mathscr{C}_0^\infty([0,T[\times \Omega)$ (space of compactly supported functions). Now, let us consider a generic $\varphi \in \mathscr{C}_0^\infty(]0,T[\times \Omega)$; integrating by parts all terms but the ones with $F_1^\pm$, we deduce that

$$\sum_{\pm} \int_0^T \int_{\Omega^\pm} \left(\partial_t u^\pm + (\hat{v}'^\pm \cdot \nabla') u^\pm - (\hat{B}'^\pm \cdot \nabla') h^\pm + \nabla q^\pm - F_1^\pm \right) \cdot \varphi = -\int_0^T \int_\Gamma [q] \varphi_3 \tag{127}$$

(the other boundary terms are zero because of the support of φ and the fact that the unit vectors normal to the front are constant and appear derived). Because of (125), the left-hand side is zero; this means that $[q] = 0$ in ω_T.

Passing to the limit $\varepsilon \to 0$ in (99), we get

$$\begin{aligned}\sum_{\pm} \left(|||(u,h)^\pm(t)|||^2_{1,*;\Omega^\pm} + \|(u_3,h_3)^\pm(t)\|^2_{H^1(\Omega^\pm)} \right) + \|f(t)\|^2_{H^{1/2}(\Gamma)} \\ \le Ce^{Ct} \sum_{\pm} \left(|||(u_0,h_0)^\pm|||^2_{1,*;\Omega^\pm} + [F^\pm]^2_{1,*,t;\Omega^\pm} \right) + \|f_0\|^2_{H^{1/2}(\Gamma)}. \end{aligned} \tag{128}$$

Hence, $u, h \in L^\infty([0,T]; H^1_*(\Omega)) \cap W^{1,\infty}([0,T]; L^2(\Omega))$, u_3, $h_3 \in L^\infty([0,T]; H^1(\Omega))$ and consequently $(u_3, h_3)|_\Gamma \in L^\infty([0,T]; H^{1/2}(\Gamma))$, while $f \in \mathscr{C}^0([0,T]; H^{1/2}(\Gamma))$ (continuity can be achieved from (72)).

The regularity of $\nabla q^\pm$ can be obtained from the first equation in (34):

$$\|\nabla q^\pm(t)\|_{L^2(\Omega^\pm)} \le C \left(\|u^\pm(t)\|_{H^1_*(\Omega^\pm)} + \|\partial_t u^\pm(t)\|_{L^2(\Omega^\pm)} + \|h^\pm(t)\|_{H^1_*(\Omega^\pm)} + \|F_1^\pm(t)\|_{L^2(\Omega^\pm)} \right) \tag{129}$$

(recall that $\widehat{v}_3^{\pm} = \widehat{B}_3^{\pm} = 0$), where

$$\begin{aligned}\|F_1^{\pm}(t)\|_{L^2(\Omega^{\pm})} &\leq \|F_1^{\pm}(0)\|_{L^2(\Omega^{\pm})} + \int_0^t \|\partial_t F_1^{\pm}(\tau)\|_{L^2(\Omega^{\pm})} d\tau \\ &\leq \|F_1^{\pm}(0)\|_{L^2(\Omega^{\pm})} + \sqrt{t}[F_1^{\pm}]_{1,*,t;\Omega^{\pm}}\end{aligned} \tag{130}$$

by the Cauchy-Schwarz inequality. Using (128), we finally conclude (39). This ends the proof of Theorem 2.

References

[1] L. D. Landau and E. M. Lifshitz, *Course of Theoretical Physics. Vol. 8*, Pergamon Press, Oxford, UK, 1984, Electrodynamics of continuous media, Translated from the second Russian edition by J. B. Sykes, J. S. Bell and M. J. Kearsley, Second Russian edition revised by Lifshits and L. P. Pitaevskiĭ.

[2] M. S. Ruderman and H. J. Fahr, "The effect of magnetic fields on the macroscopic instability of the heliopause. II. Inclusion of solar wind magnetic fields," *Astronomy & Astrophysics*, vol. 299, pp. 258–266, 1995.

[3] D. Coutand and S. Shkoller, "Well-posedness of the free-surface incompressible Euler equations with or without surface tension," *Journal of the American Mathematical Society*, vol. 20, no. 3, pp. 829–930, 2007.

[4] D. Coutand and S. Shkoller, "Well-posedness in smooth function spaces for the moving-boundary three-dimensional compressible euler equations in physical vacuum," *Archive for Rational Mechanics and Analysis*, vol. 206, no. 2, pp. 515–616, 2012.

[5] A. Morando, Y. Trakhinin, and P. Trebeschi, "Stability of incompressible current-vortex sheets," *Journal of Mathematical Analysis and Applications*, vol. 347, no. 2, pp. 502–520, 2008.

[6] Y. Trakhinin, "On the existence of incompressible current-vortex sheets: study of a linearized free boundary value problem," *Mathematical Methods in the Applied Sciences*, vol. 28, no. 8, pp. 917–945, 2005.

[7] J.-F. Coulombel, A. Morando, P. Secchi, and P. Trebeschi, "A priori estimates for 3D incompressible current-vortex sheets," *Communications in Mathematical Physics*, vol. 311, no. 1, pp. 247–275, 2012.

[8] P. Secchi and Y. Trakhinin, "Well-posedness of the linearized plasma-vacuum interface problem." Preprint.

[9] A. Morando, Y. Trakhinin, and P. Trebeschi, "Well-posedness of the linearized plasma-vacuum interface problem in ideal incompressible mhd," *Quarterly of Applied Mathematics*. In press.

[10] D. Lannes, "Well-posedness of the water-waves equations," *Journal of the American Mathematical Society*, vol. 18, no. 3, pp. 605–654, 2005.

[11] Y. Trakhinin, "Existence of compressible current-vortex sheets: variable coefficients linear analysis," *Archive for Rational Mechanics and Analysis*, vol. 177, no. 3, pp. 331–366, 2005.

[12] P. D. Lax and R. S. Phillips, "Local boundary conditions for dissipative symmetric linear differential operators," *Communications on Pure and Applied Mathematics*, vol. 13, pp. 427–455, 1960.

[13] P. Secchi, "Linear symmetric hyperbolic systems with characteristic boundary," *Mathematical Methods in the Applied Sciences*, vol. 18, no. 11, pp. 855–870, 1995.

[14] P. Secchi, "The initial-boundary value problem for linear symmetric hyperbolic systems with characteristic boundary of constant multiplicity," *Differential and Integral Equations*, vol. 9, no. 4, pp. 671–700, 1996.

[15] A. Morando and P. Secchi, "Regularity of weakly well posed hyperbolic mixed problems with characteristic boundary," *Journal of Hyperbolic Differential Equations*, vol. 8, no. 1, pp. 37–99, 2011.

[16] A. Morando, P. Secchi, and P. Trebeschi, "Regularity of solutions to characteristic initial-boundary value problems for symmetrizable systems," *Journal of Hyperbolic Differential Equations*, vol. 6, no. 4, pp. 753–808, 2009.

[17] P. Secchi, "Some properties of anisotropic Sobolev spaces," *Archiv der Mathematik*, vol. 75, no. 3, pp. 207–216, 2000.

[18] P. Secchi, "An initial boundary value problem in ideal magneto-hydrodynamics," *Nonlinear Differential Equations and Applications*, vol. 9, no. 4, pp. 441–458, 2002.

[19] Y. Trakhinin, "The existence of current-vortex sheets in ideal compressible magnetohydrodynamics," *Archive for Rational Mechanics and Analysis*, vol. 191, no. 2, pp. 245–310, 2009.

[20] K. O. Friedrichs, "Symmetric positive linear differential equations," *Communications on Pure and Applied Mathematics*, vol. 11, pp. 333–418, 1958.

[21] Y. Shizuta, "The initial-boundary value problem for symmetric hyperbolic systems with characteristic boundary," in *Qualitative aspects and applications of nonlinear evolution equations (Trieste, 1993)*, pp. 99–108, World Scientific Publishing, River Edge, NJ, USA, 1994.

Homogenization in Sobolev Spaces with Nonstandard Growth: Brief Review of Methods and Applications

Brahim Amaziane[1] **and Leonid Pankratov**[1,2]

[1] *Laboratoire de Mathématiques et de leurs Applications, CNRS-UMR 5142, UNIV Pau & Pays Adour, Avenue de l'Université, 64000 Pau, France*
[2] *Department of Mathematics, B. Verkin Institute for Low Temperature Physics and Engineering, 47, Avenue Lenin, Kharkov 61103, Ukraine*

Correspondence should be addressed to Brahim Amaziane; brahim.amaziane@univ-pau.fr

Academic Editor: Julio Rossi

We review recent results on the homogenization in Sobolev spaces with variable exponents. In particular, we are dealing with the Γ-convergence of variational functionals with rapidly oscillating coefficients, the homogenization of the Dirichlet and Neumann variational problems in strongly perforated domains, as well as double porosity type problems. The growth functions also depend on the small parameter characterizing the scale of the microstructure. The homogenization results are obtained by the method of local energy characteristics. We also consider a parabolic double porosity type problem, which is studied by combining the variational homogenization approach and the two-scale convergence method. Results are illustrated with periodic examples, and the problem of stability in homogenization is discussed.

1. Introduction

In recent years, there has been an increasing interest in the study of the functionals with variable exponents or nonstandard $p(x)$-growth and the corresponding Sobolev spaces, see for instance [1–8] and the references therein. In particular, the conditions under which C_0^∞ functions are dense in $W^{1,p(\cdot)}$ have been found. Also, Meyers estimates, which are used in the homogenization process, have been obtained in [6]. Let us mention that such partial differential equations arise in many engineering disciplines, such as electrorheological fluids, non-Newtonian fluids with thermoconvective effects, and nonlinear Darcy flow of compressible fluids in heterogeneous porous media, see for instance [1].

This paper discusses problems of homogenization in Sobolev spaces with variable exponents. Attention is focussed on the homogenization and minimization problems for variational functionals in the framework of Sobolev spaces with nonstandard growth. The material is essentially a review with some new results.

Γ-convergence and minimization problems for functionals with periodic and locally periodic rapidly oscillating Lagrangians of p-growth with a constant p are well studied now, see for instance [9, 10] and the bibliography therein.

The works [11–15] (see also [16]) focus on the variational functionals with nonstandard growth conditions. In particular, the homogenization and Γ-convergence problems for Lagrangians with variable rapidly oscillating exponents $p(x)$ were considered in [13, 14]. It was shown that the energy minimums and the homogenized Lagrangians in the spaces $W^{1,r}$ might depend on the value of r (the so-called Lavrentiev phenomenon). For example, such a behavior can be observed for the Lagrangian $|\nabla u|^{p(x/\varepsilon)}$ with a periodic "chess board" exponent $p(y)$ and a small parameter $\varepsilon > 0$.

Another interesting example of Lagrangian with rapidly oscillating exponent was considered in [11]. Namely, for the functional

$$\mathcal{J}_\varepsilon[u] = \int |\nabla u|^{p(x/\varepsilon)} dx \qquad (1)$$

with a smooth periodic $p(y)$ such that $p(y) > 1$, it was shown that the limit functional is bounded on Sobolev-Orlicz space of functions with gradient in an L^{α} log space, where α is the fiber percolation level of $p(x)$.

Variational functionals with nonstandard growth conditions have also been considered in [9]. Chapter 21 of this book focuses on the Γ-convergence of such functionals in L^p spaces.

In this paper, we are dealing mainly with the Γ-convergence of variational functionals with periodic rapidly oscillating coefficients, the homogenization of the variational problems in strongly perforated domains (Dirichlet and Neumann problems), and nonlinear double porosity type problems, that is, with the problems where the coefficient of the differential operator asymptotically degenerates on a some specially defined subset (e.g., the set of periodically distributed inclusions) of the domain under consideration. The paper is based on the results obtained in papers [17–26].

The paper is organized as follows. In Section 2, for the sake of completeness, we recall the definition and main results on the Lebesgue and Sobolev spaces with variable exponents which will be used in the sequel. Then in auxiliary Section 3, we give some definitions which will be used in the paper. In Section 4, we study the question of the Γ-convergence and homogenization of functionals with rapidly oscillating periodic coefficients. In Section 5, we are dealing with the homogenization of the Dirichlet variational problem in strongly perforated domains. The main result of the section (see Theorem 8) is then applied to the study of nonlocal effects in the homogenization (see Theorem 9). In Section 5.3, we give a periodic example, when all the conditions of Theorems 8 and 9 are satisfied and all the coefficients of the homogenized problem are calculated explicitly. Moreover, the question of stability in homogenization is also discussed here. Theorems 8 and 9 are proved by using the so-called *method of local energy characteristics* proposed earlier by Marchenko and Khruslov for linear homogenization problems (see [27] and the references herein). This method is close to the Γ-convergence method. Briefly, it is based on the derivation of the lim-inf and lim-sup estimates for the variational functional under consideration along with the assumptions on the behavior of the local energy characteristics. In Section 6, the homogenization of the Neumann problem in strongly perforated domains is considered. In this section, the closeness of the method of local energy characteristics is shown directly. In Section 7, we are dealing with a variational problem with high contrast coefficients (nonlinear double porosity type model). The main result of the section (Theorem 23) is also obtained by the method of local energy characteristics. As an application of this result, we consider the periodic case, where we focus our attention on the question of stability in homogenization. Finally, in Section 8, we are dealing with the homogenization of a class of quasilinear parabolic equations with nonstandard growth. The main results of the section are obtained by combining the two-scale convergence method and the variational homogenization approach.

2. Sobolev Spaces with Variable Exponents

In this section, we introduce the function spaces used throughout the paper and describe their basic properties. We refer here to [1, 4, 5, 7, 8].

Let Ω be a bounded Lipschitz domain in $\mathbb{R}^n$ ($n \geq 2$). We introduce the function $p = p(x)$ and assume that this function is bounded such that

$$1 < p^- = \inf_{\Omega} p(x) \leq p(x) \leq \sup_{\Omega} p(x) = p^+ < +\infty. \tag{2}$$

We also assume that the function $p(x)$ satisfies the *log-Hölder continuity property*. Namely, for all $x, y \in \Omega$,

$$|p(x) - p(y)| \leq \omega(|x-y|) \text{ with } \overline{\lim_{\tau\to 0}}\, \omega(\tau)\ln\left(\frac{1}{\tau}\right) \leq C. \tag{3}$$

Notice that this property was introduced by Zhikov to avoid the so-called Lavrentiev phenomenon (see, e.g., [15]).

(1) By $L^{p(\cdot)}(\Omega)$ we denote the space of measurable functions f in Ω such that

$$A_{p(\cdot),\Omega}(f) = \int_{\Omega} |f(x)|^{p(x)} dx < +\infty. \tag{4}$$

The space $L^{p(\cdot)}(\Omega)$ equipped with the norm

$$\|f\|_{L^{p(\cdot)}(\Omega)} = \inf\left\{\lambda > 0 : A_{p(\cdot),\Omega}\left(\frac{f}{\lambda}\right) \leq 1\right\} \tag{5}$$

becomes a Banach space.

(2) The following inequalities hold:

$$\begin{aligned} &\min\left(\|f\|^{p^-}_{L^{p(\cdot)}(\Omega)}, \|f\|^{p^+}_{L^{p(\cdot)}(\Omega)}\right) \\ &\leq A_{p(\cdot),\Omega}(f) \\ &\leq \max\left(\|f\|^{p^-}_{L^{p(\cdot)}(\Omega)}, \|f\|^{p^+}_{L^{p(\cdot)}(\Omega)}\right), \\ &\min\left(A^{1/p^-}_{p(\cdot),\Omega}, A^{1/p^+}_{p(\cdot),\Omega}\right) \\ &\leq \|f\|_{L^{p(\cdot)}(\Omega)} \\ &\leq \max\left(A^{1/p^-}_{p(\cdot),\Omega}, A^{1/p^+}_{p(\cdot),\Omega}\right). \end{aligned} \tag{6}$$

(3) Let $f \in L^{p(\cdot)}(\Omega)$, $g \in L^{q(\cdot)}(\Omega)$ with

$$\begin{aligned} \frac{1}{p(x)} + \frac{1}{q(x)} = 1, \quad &1 < p^- \leq p(x) \leq p^+ < \infty, \\ &1 < q^- \leq q(x) \leq q^+ < +\infty. \end{aligned} \tag{7}$$

Then the Hölder's inequality holds

$$\int_{\Omega} |fg|\, dx \leq 2\|f\|_{L^{p(\cdot)}(\Omega)}\|g\|_{L^{q(\cdot)}(\Omega)}. \tag{8}$$

(4) According to (8), for every $1 \leq q = \text{const} < p^- \leq p(x) < +\infty$

$$\|f\|_{L^q(\Omega)} \leq C\,\|f\|_{L^{p(\cdot)}(\Omega)} \quad \text{with the constant } C = 2\|1\|_{p(\cdot)/(L^{p(\cdot)-q}(\Omega))}. \tag{9}$$

It is straightforward to check that for domains Ω such that $\text{meas}\,\Omega < +\infty$,

$$\|1\|_{L^{p(\cdot)}(\Omega)} \leq 2\max\left\{[\text{meas}\,\Omega]^{2/p^-}, [\text{meas}\,\Omega]^{1/2p^+}\right\}. \tag{10}$$

(5) The space $W^{1,p(\cdot)}(\Omega)$, $p(\cdot) \in [p^-, p^+] \subset]1, +\infty[$, is defined by

$$W^{1,p(\cdot)}(\Omega) = \left\{f \in L^{p(\cdot)}(\Omega) : |\nabla f| \in L^{p(\cdot)}(\Omega)\right\}. \tag{11}$$

If condition (3) is satisfied, $W_0^{1,p(\cdot)}(\Omega)$ is the closure of the set $C_0^\infty(\Omega)$ with respect to the norm of $W^{1,p(\cdot)}(\Omega)$. If the boundary of Ω is Lipschitz continuous and $p(x)$ satisfies (3), then $C_0^\infty(\Omega)$ is dense in $W_0^{1,p(\cdot)}(\Omega)$. The norm in the space $W_0^{1,p(\cdot)}$ is defined by

$$\|u\|_{W_0^{1,p(\cdot)}} = \sum_i \|D_i u\|_{L^{p(\cdot)}(\Omega)} + \|u\|_{L^{p(\cdot)}(\Omega)}. \tag{12}$$

If the boundary of Ω is Lipschitz and $p \in C^0(\Omega)$, then the norm $\|\cdot\|_{W_0^{1,p(\cdot)}(\Omega)}$ is equivalent to the norm

$$\widetilde{\|u\|}_{W_0^{1,p(x)}(\Omega)} = \sum_i \|D_i u\|_{L^{p(\cdot)}(\Omega)}. \tag{13}$$

(6) If $p \in C^0(\overline{\Omega})$, then $W^{1,p(\cdot)}(\Omega)$ is separable and reflexive.

(7) If $p, q \in C^0(\overline{\Omega})$,

$$p_*(x) = \begin{cases} \dfrac{p(x)\,n}{n - p(x)} & \text{if } p(x) < n, \\ & \qquad \text{and } 1 < q(x) \leq \sup_\Omega q(x) < \inf_\Omega p_*(x), \\ +\infty & \text{if } p(x) > n, \end{cases} \tag{14}$$

then the embedding $W_0^{1,p(\cdot)}(\Omega) \hookrightarrow L^{q(\cdot)}(\Omega)$ is continuous and compact.

(8) Friedrich's inequality is valid in the following form: if $p(x)$ satisfies conditions (2) and (3), then there exists a constant $C > 0$ such that for every $f \in W_0^{1,p(\cdot)}(\Omega)$

$$\|f\|_{L^{p(\cdot)}(\Omega)} \leq C\,\|\nabla f\|_{L^{p(\cdot)}(\Omega)}. \tag{15}$$

3. Definitions

In this auxiliary section, we introduce the necessary definitions that will be used in the paper. We start by introducing the class of the variable exponents p_ε, where ε is a small positive parameter characterizing the microscopic length scale.

Definition 1 (class $\mathfrak{L}^\varepsilon_{p_0(\cdot)}$). A sequence of functions $\{p_\varepsilon\}_{(\varepsilon>0)}$ is said to belong to the class $\mathfrak{L}^\varepsilon_{p_0(\cdot)}$ if this sequence possesses the following properties:

(i) for any $\varepsilon > 0$, p_ε is bounded in the following sense:

$$\begin{aligned} 1 < p^- \leq p_\varepsilon^- &\equiv \min_{x\in\overline{\Omega}} p_\varepsilon(x) \leq p_\varepsilon(x) \leq \max_{x\in\overline{\Omega}} p_\varepsilon(x) \\ &\equiv p_\varepsilon^+ \leq p^+ < +\infty \quad \text{in } \overline{\Omega}; \end{aligned} \tag{16}$$

(ii) for any $\varepsilon > 0$, p_ε satisfies the log-Hölder continuity property;

(iii) the function p_ε converges uniformly in Ω to a function p_0, where the limit function p_0 is assumed to satisfy the log-Hölder continuity property.

We also recall the definition of the Γ-convergence (see, e.g., [9, 10, 28] and the bibliography therein). In our case this definition takes the following form.

Definition 2 ($\Gamma_{p(\cdot)}$-convergence). The functional $I^\varepsilon : L^{p_\varepsilon(\cdot)}(\Omega) \to \mathbb{R} \cup \{\infty\}$ is said to $\Gamma_{p_0(\cdot)}$-converge to a functional $I : L^{p_0(\cdot)}(\Omega) \to \mathbb{R} \cup \{\infty\}$ if

(a) ("lim inf"-inequality) for any $u \in L^{p_0(\cdot)}(\Omega)$ and any sequence $\{u^\varepsilon\}_{\varepsilon>0} \subset L^{p_0(\cdot)}(\Omega)$ which converges to the function u strongly in the space $L^{p_0(\cdot)}(\Omega)$ we have

$$\varliminf_{\varepsilon\to0} I^\varepsilon[u^\varepsilon] \geq I[u], \tag{17}$$

(b) ("lim sup"-inequality) for any $u \in L^{p_0(\cdot)}(\Omega)$, there exists a sequence $\{w^\varepsilon\}_{\varepsilon>0} \subset L^{p_0(\cdot)}(\Omega)$ such that $\{w^\varepsilon\}_{\varepsilon>0}$ converges to the function $u(\cdot)$ strongly in the space $L^{p_0(\cdot)}(\Omega)$, and

$$\varlimsup_{\varepsilon\to0} I^\varepsilon[w^\varepsilon] \leq I[u]. \tag{18}$$

We define the strong convergence in $L^{p_0(\cdot)}(\Omega^\varepsilon)$ in the following way.

Definition 3 (strong convergence in $L^{p_0(\cdot)}(\Omega^\varepsilon)$). The sequence $\{\omega^\varepsilon\}_{\varepsilon>0} \in L^{p_0(\cdot)}(\Omega^\varepsilon)$ is said to converge strongly in the space $L^{p_0(\cdot)}(\Omega^\varepsilon)$ to a function $\omega \in L^{p_0(\cdot)}(\Omega)$ if

$$\lim_{\varepsilon\to0} \|\omega^\varepsilon - \omega\|_{L^{p_0(\cdot)}(\Omega^\varepsilon)} = 0. \tag{19}$$

Finally, we recall the definition of the two-scale convergence (see, e.g., [29]).

Definition 4. Let $Y =]0,1[^n$ be a basic cell and $\Omega_T = \Omega \times]0,T[$. A function, $\varphi \in L^2(\Omega_T; C_\#^\infty(Y))$, which is Y-periodic in y and which satisfies

$$\lim_{\varepsilon\to 0}\int_{\Omega_T}\left|\varphi\left(x,\frac{x}{\varepsilon},t\right)\right|^2 dx\,dt = \int_{\Omega_T\times Y}|\varphi(x,y,t)|^2 dy\,dx\,dt \tag{20}$$

is called an admissible test function.

Definition 5 (two-scale convergence). A sequence of functions $v^\varepsilon \in L^2(\Omega_T)$ two-scale converges to $v \in L^2(\Omega_T \times Y)$ if for any admissible test function $\varphi(x,y,t)$,

$$\lim_{\varepsilon\to 0}\int_{\Omega_T} v^\varepsilon(x,t)\,\varphi\left(x,\frac{x}{\varepsilon},t\right)dx\,dt = \int_{\Omega_T\times Y} v(x,y,t)\,\varphi(x,y,t)\,dy\,dx\,dt. \tag{21}$$

This convergence is denoted by $v^\varepsilon(x,t) \overset{2s}{\rightharpoonup} v(x,y,t)$.

The method of the local energy characteristics, generally speaking, deals with nonperiodic structures. We often make use of the following definition.

Definition 6 (distribution in an asymptotically regular way). The set $\mathcal{F}^\varepsilon$ is said to be distributed in an asymptotically regular way in Ω, if for any ball $V(y,r)$ of radius r centered at $y \in \Omega$ and $\varepsilon > 0$ small enough ($\varepsilon \le \varepsilon_0(r)$), $V(y,r) \cap \mathcal{F}^\varepsilon \neq \emptyset$ and $V(y,r) \cap (\Omega \setminus \mathcal{F}^\varepsilon) \neq \emptyset$.

4. Γ-Convergence and Homogenization of Functionals with Rapidly Oscillating Coefficients in Sobolev Spaces with Variable Exponents

Let Ω be a bounded domain in $\mathbb{R}^n$ ($n \ge 2$) with a sufficiently smooth boundary and denote that $Y =]0,1[^n$. We assume that a family of continuous functions $\{p_\varepsilon\}_{(\varepsilon>0)}$ belongs to the class, $\mathfrak{L}^\varepsilon_{p_0(\cdot)}$, and we also suppose that

(H.1) $a = a(y)$ and $b = b(y)$ are Y-periodic measurable functions such that

$$0 < a_0 \le a(y) \le a_1, \qquad 0 < b_0 \le b(y) \le b_1. \tag{22}$$

(H.2) $f \in C(\Omega)$.

For the notational convenience, we set

$$a^\varepsilon_{p_\varepsilon}(x) = \frac{1}{p_\varepsilon(x)}\,a\left(\frac{x}{\varepsilon}\right),\qquad b^\varepsilon_{p_\varepsilon}(x) = \frac{1}{p_\varepsilon(x)}\,b\left(\frac{x}{\varepsilon}\right). \tag{23}$$

In the space $L^{p_\varepsilon(\cdot)}(\Omega)$, we define the functional $J^\varepsilon : L^{p_\varepsilon(\cdot)}(\Omega) \to \mathbb{R}$:

$$J^\varepsilon[u] \overset{\text{def}}{=} \begin{cases} \displaystyle\int_\Omega \left\{a^\varepsilon_{p_\varepsilon}(x)|\nabla u|^{p_\varepsilon(x)} + b^\varepsilon_{p_\varepsilon}(x)|u|^{p_\varepsilon(x)} - f(x)\,u\right\}dx, & \text{if } u \in W^{1,p_\varepsilon(\cdot)}(\Omega),\\ +\infty, & \text{otherwise.}\end{cases} \tag{24}$$

We study the asymptotic behavior of J^ε and its minimizers as $\varepsilon \to 0$. Our analysis relies on the Γ-convergence approach in Sobolev spaces with variable exponents. The main result of the section is the following.

Theorem 7. *Let assumptions (H.1) and (H.2) be fulfilled. Then*

(i) *The functional J^ε, $\Gamma_{p_0(\cdot)}$-converges to the functional $J_{hom} : L^{p_0(\cdot)}(\Omega) \to \mathbb{R}\cup\{+\infty\}$ given by*

$$J_{hom}[u] = \begin{cases} \displaystyle\int_\Omega \left\{\mathsf{T}(x,\nabla u) + \beta_p(x)|u|^{p_0(x)} - f(x)\,u\right\}dx, & \text{if } u \in W^{1,p_0}(\Omega),\\ +\infty, & \text{otherwise,}\end{cases} \tag{25}$$

where

$$\mathsf{T}(x,\mathbf{b}) = \inf\left\{\frac{1}{p_0(x)}\int_Y a(y)\,|\nabla v(y) + \mathbf{b}|^{p_0(x)}dy : v \in W_\#^{1,p_0(x)}(Y)\right\},$$

$$\beta_p(x) = \frac{1}{p_0(x)}\int_Y b(y)\,dy. \tag{26}$$

(ii) *The minimizer u^ε of the functional J^ε converges to the minimizer u of the functional J_{hom} strongly in the space $L^{p_0(\cdot)}(\Omega)$.*

The Scheme of the Proof of Theorem 7 (See [18] for More Details) Is as Follows. We start our analysis by proving the "lim inf"-inequality. The proof of this inequality is done in two main steps. First we introduce an auxiliary functional $\widetilde{J}^\varepsilon : L^{\pi_\varepsilon(\cdot)}(\Omega) \to \mathbb{R}$:

$$\widetilde{J}^\varepsilon[u] = \begin{cases} \displaystyle\int_\Omega \left\{a^\varepsilon_{p_\varepsilon}(x)|\nabla u|^{\pi_\varepsilon(x)} + b^\varepsilon_{p_\varepsilon}(x)|u|^{p_\varepsilon(x)} - f(x)\,u\right\}dx, & \text{if } u \in W^{1,\pi_\varepsilon(\cdot)}(\Omega);\\ +\infty, & \text{otherwise,}\end{cases} \tag{27}$$

where $\pi_\varepsilon(x) = \min\{p_\varepsilon(x), p_0(x)\}$ and prove the "lim inf"-inequality for this functional. Then, at the second step, we show that the "lim inf"-inequality for the auxiliary functional $\tilde{J}^\varepsilon$ implies the "lim inf"-inequality for J^ε. Then, using a special test function and the Meyers estimate (see [6]), we obtain the "lim sup"-inequality. Finally, we prove the convergence of the minimizers. This completes the proof of Theorem 7.

5. Homogenization of the Dirichlet Problem and Related Questions

5.1. Homogenization of the Dirichlet Problem. Let Ω be a bounded domain in $\mathbb{R}^n$ ($n \geq 2$) with sufficiently smooth boundary. Let $\mathcal{F}^\varepsilon$ be an open subset in Ω. Here ε is a small parameter characterizing the scale of the microstructure. We assume that $\mathcal{F}^\varepsilon$ is distributed in an asymptotically regular way in Ω and we set

$$\Omega^\varepsilon = \Omega \setminus \overline{\mathcal{F}}^\varepsilon. \tag{28}$$

Let $p_\varepsilon = p_\varepsilon(x)$ be a continuous function defined in the domain $\overline{\Omega}$. We assume that, for any $\varepsilon > 0$, it satisfies the following conditions:

(A.1.1) the function $p_\varepsilon(x)$ is bounded in the following sense:

$$\begin{aligned} 1 < p^- \leq p_\varepsilon^- &\equiv \min_{x\in\overline{\Omega}} p_\varepsilon(x) \leq p_\varepsilon(x) \leq \max_{x\in\overline{\Omega}} p_\varepsilon(x) \\ &\equiv p_\varepsilon^+ \leq p^+ \leq n \quad \text{in } \overline{\Omega}; \end{aligned} \tag{29}$$

(A.1.2) the function p_ε satisfies the log-Hölder continuity property;

(A.1.3) the function p_ε converges uniformly in Ω to a function p_0, where the limit function p_0 is assumed to satisfy the log-Hölder continuity property;

(A.1.4) the function p_ε is such that $p_\varepsilon(x) \geq p_0(x)$ in Ω.

We consider the following variational problem:

$$J^\varepsilon[u] \stackrel{\text{def}}{=} \int_{\Omega^\varepsilon} \mathsf{F}_\varepsilon(x, u, \nabla u)\, dx \longrightarrow \min, \quad u \in W_0^{1,p_\varepsilon(\cdot)}(\Omega^\varepsilon), \tag{30}$$

where

$$\mathsf{F}_\varepsilon(x, u, \nabla u) = \frac{1}{p_\varepsilon(x)}|\nabla u|^{p_\varepsilon(x)} + \frac{1}{p_\varepsilon(x)}|u|^{p_\varepsilon(x)} - f(x)\, u, \tag{31}$$

and $f \in C^1(\Omega)$. It is known from [1–3, 6] that for each $\varepsilon > 0$, there exists a unique solution (minimizer) $u^\varepsilon \in W^{1,p_\varepsilon(\cdot)}(\Omega^\varepsilon)$ of problem (30). Let us extend u^ε in $\mathcal{F}^\varepsilon$ by zero (keeping for it the same notation). Then we obtain the family $\{u^\varepsilon\} \subset W^{1,p_\varepsilon(\cdot)}(\Omega)$. We study the asymptotic behavior of u^ε as $\varepsilon \to 0$.

Instead of the classical periodicity assumption on the microstructure of the perforated domain Ω^ε, we impose certain conditions on the local energy characteristic of the set $\mathcal{F}^\varepsilon$. To this end we introduce K_h^z an open cube centered at $z \in \Omega$ with length equal to h ($0 < \varepsilon \ll h < 1$), and we set

$$c^{\varepsilon,h}(z, b) = \inf_{v^\varepsilon} \int_{K_h^z} \left\{ \frac{1}{p_\varepsilon(x)} |\nabla v^\varepsilon|^{p_\varepsilon(x)} + h^{-p^+-\gamma} |v^\varepsilon - b|^{p_\varepsilon(x)} \right\} dx, \tag{32}$$

where $\gamma > 0$, and the infimum is taken over $v^\varepsilon \in W^{1,p_\varepsilon(\cdot)}(\Omega)$ that equal zero in $\mathcal{F}^\varepsilon$. We assume that

(C.1.1) there exists a continuous function $c(x, b)$ such that for any $x \in \Omega$, any $b \in \mathbb{R}$, and a certain $\gamma = \gamma_0 > 0$,

$$\lim_{h\to 0} \overline{\lim_{\varepsilon\to 0}}\, h^{-n} c^{\varepsilon,h}(z, b) = \lim_{h\to 0} \underline{\lim_{\varepsilon\to 0}}\, h^{-n} c^{\varepsilon,h}(z, b) = c(x, b); \tag{33}$$

(C.1.2) there exists a constant A independent of ε such that for any $x \in \Omega$,

$$\lim_{h\to 0} \overline{\lim_{\varepsilon\to 0}}\, h^{-n} c^{\varepsilon,h}(z, b) \leq A\left(1 + |b|^{p_0(x)}\right). \tag{34}$$

The first main result of Section 5 is the following.

Theorem 8. *Let conditions (A.1.1)–(A.1.4) on the function p_ε and conditions (C.1.1) and (C.1.2) on the local characteristic be satisfied. Then u^ε the solution (minimizer) of the variational problem (30) (extended by zero in $\mathcal{F}^\varepsilon$) converges weakly in $W^{1,p_0(\cdot)}(\Omega)$ to u the solution (minimizer) of*

$$J_{hom}[u] \stackrel{\text{def}}{=} \int_\Omega \{\mathsf{F}_0(x, u, \nabla u) + c(x, u) - f(x)\, u\}\, dx \longrightarrow \min, \quad u \in W_0^{1,p_0(\cdot)}(\Omega), \tag{35}$$

where

$$\mathsf{F}_0(x, w, \nabla w) \stackrel{\text{def}}{=} \frac{1}{p_0(x)} |\nabla w|^{p_0(x)} + \frac{1}{p_0(x)} |w|^{p_0(x)} - g(x)\, w. \tag{36}$$

The Scheme of the Proof of Theorem 8 (See [19] for More Details) Is as Follows. First, it follows from (6), (30), and the regularity properties of the functions f, p_ε that $\|u^\varepsilon\|_{W^{1,p_\varepsilon(\cdot)}(\Omega^\varepsilon)} \leq C$. We extend u^ε by zero to the set $\mathcal{F}^\varepsilon$ and consider $\{u^\varepsilon\}$ as a sequence in $W^{1,p_\varepsilon(\cdot)}(\Omega)$. Then $\|u^\varepsilon\|_{W^{1,p_\varepsilon(\cdot)}(\Omega)} \leq C$. Condition (A.1.4) immediately implies that

$$\|u^\varepsilon\|_{W^{1,p_0(\cdot)}(\Omega)} \leq C. \tag{37}$$

Hence, one can extract a subsequence $\{u^\varepsilon, \varepsilon = \varepsilon_k \to 0\}$ that converges weakly to a function $u \in W^{1,p_0(\cdot)}(\Omega)$. We will show that u is the solution of the variational problem (35). The proof is done in two mains steps. On the first step, we prove the "lim sup"-inequality (the upper bound for the functional J^ε). To this end, we cover the domain Ω by the cubes K_h^α of length $h > 0$ centered at the points x^α, where $\{x^\alpha\}$ be a periodic grid in Ω with a period $h' = h - h^{1+\gamma/p^+}$ ($\varepsilon \ll h \ll 1$, $0 < \gamma < p^+$). We associate with this covering a partition of unity $\{\varphi_\alpha\}$: $0 \le \varphi_\alpha(x) \le 1$; $\varphi_\alpha(x) = 0$ for $x \notin K_h^\alpha$; $\varphi_\alpha(x) = 1$ for $x \in K_h^\alpha \setminus \cup_{\beta \ne \alpha} K_h^\beta$; $\sum_\alpha \varphi_\alpha(x) = 1$ for $x \in \Omega$; $|\nabla \varphi_\alpha(x)| \le C h^{-1-\gamma/p^+}$. Then we construct the function w_h^ε using the minimizers $v_\alpha^\varepsilon = v_\alpha^\varepsilon(x)$ of (32) for $z = x^\alpha$. We show that, for any $w \in W_0^{1,p_0(\cdot)}(\Omega)$,

$$\lim_{h\to 0} \overline{\lim_{\varepsilon\to 0}} J^\varepsilon [w_h^\varepsilon] \le J_{\text{hom}} [w], \tag{38}$$

and, therefore,

$$\overline{\lim_{\varepsilon\to 0}} J^\varepsilon [u^\varepsilon] \le J_{\text{hom}} [w]. \tag{39}$$

On the second step, we prove the "lim inf"-inequality (the lower bound for the functional J^ε) using the definition of the local energy characteristic and the convexity of our functional:

$$\underline{\lim_{\varepsilon=\varepsilon_k\to 0}} J^\varepsilon [u^\varepsilon] \ge J_{\text{hom}} [u]. \tag{40}$$

Finally it follows from (39) and (40) that

$$J_{\text{hom}} [u] \le J_{\text{hom}} [w], \tag{41}$$

for any $w \in W^{1,p_0(\cdot)}(\Omega)$ such that $w = 0$ on $\partial\Omega$. This means that any weak limit of the solution of problem (30) extended to the set $\mathcal{F}^\varepsilon$ by zero is the solution of the homogenized problem (35). This completes the proof of Theorem 8. The generalization of Theorem 8 is given in [17].

5.2. Nonlocal Effects in Homogenization of $p_\varepsilon(x)$-Laplacian in Perforated Domains. Let Ω be a domain in $\mathbb{R}^n$ ($n \ge 3$) defined in the previous section. Let $\mathcal{F}^\varepsilon$ be an open connected subset in Ω like a net. We assume that the set $\mathcal{F}^\varepsilon$ satisfies the following conditions:

(F.1.1) $\mathcal{F}^\varepsilon \Subset \Omega_d$, where $\Omega_d = \{x \in \Omega : \text{dist}(x, \partial\Omega) \ge d\}$ with $d > 0$ independent of ε;

(F.1.2) $\mathcal{F}^\varepsilon$ is distributed in an asymptotically regular way in Ω_d;

(F.1.3) the subdomain $\Omega^\varepsilon \overset{\text{def}}{=} \Omega \setminus \overline{\mathcal{F}^\varepsilon}$ is a connected set.

We assume that for any $\varepsilon > 0$, the function $p_\varepsilon = p_\varepsilon(x)$ satisfies the conditions (A.1.1)–(A.1.4) from Section 5.1. In the space $W^{1,p_\varepsilon(\cdot)}(\Omega^\varepsilon)$, we define the functional $J^\varepsilon : W^{1,p_\varepsilon(\cdot)}(\Omega^\varepsilon) \to \mathbb{R}$,

$$J^\varepsilon [u] = \begin{cases} \int_{\Omega^\varepsilon} \mathsf{F}_\varepsilon (x, u, \nabla u)\, dx & \text{if } u \in W^{1,p_\varepsilon(\cdot)} (\Omega^\varepsilon), \\ +\infty & \text{otherwise}, \end{cases} \tag{42}$$

where $\mathsf{F}_\varepsilon(x, u, \nabla u)$ is defined in (31) and we consider the following variational problem:

$$\begin{gathered} J^\varepsilon [u^\varepsilon] \longrightarrow \inf, \quad u^\varepsilon \in W^{1,p_\varepsilon(\cdot)} (\Omega^\varepsilon); \\ u^\varepsilon = A^\varepsilon \text{ on } \partial\mathcal{F}^\varepsilon, \quad u^\varepsilon = 0 \quad \text{on } \partial\Omega, \end{gathered} \tag{43}$$

where A^ε is an unknown constant. It is known from [1, 2] that for each $\varepsilon > 0$, there exists a unique solution $u^\varepsilon \in W^{1,p_\varepsilon(\cdot)}(\Omega^\varepsilon)$ of the variational problem (43). We extend u^ε by the equality $u^\varepsilon = A^\varepsilon$ in $\mathcal{F}^\varepsilon$ and keep for it the same notation. Thus, we obtain the family $\{u^\varepsilon\}_{(\varepsilon>0)} \subset W^{1,p_\varepsilon(\cdot)}(\Omega)$. We study the asymptotic behavior of the family $\{u^\varepsilon\}_{(\varepsilon>0)}$ as $\varepsilon \to 0$.

Let us introduce the functional $J_{\text{hom}} : W^{1,p_0(\cdot)}(\Omega) \to \mathbb{R}$,

$$J_{\text{hom}} [u, A] = \begin{cases} \int_\Omega \{\mathsf{F}_0 (x, u, \nabla u) \\ \quad + \mathbf{1}_d (x)\, c (x, u - A)\, dx\} & \text{if } u \in W^{1,p_0(\cdot)} (\Omega), \\ +\infty & \text{otherwise}, \end{cases} \tag{44}$$

where $\mathbf{1}_d = \mathbf{1}_d(x)$ is the characteristic function of the domain Ω_d and $\mathsf{F}_0(x, u, \nabla u)$ is defined in (36).

The main result of the section is the following.

Theorem 9. *Let u^ε be a solution of (43) extended by the equality $u^\varepsilon(x) = A^\varepsilon$ in $\mathcal{F}^\varepsilon$. Assume that the conditions (F.1.1)–(F.1.3), (A.1.1)–(A.1.4), and(C.1.1)–(C.1.2) on the set $\mathcal{F}^\varepsilon$, the functions p_ε, and the local characteristic $c^{\varepsilon,h}(x, b)$ are satisfied. Then there is a subsequence $\{u^\varepsilon, \varepsilon = \varepsilon_k \to 0\}$ that converges weakly in $W^{1,p_0(\cdot)}(\Omega)$ to a function u such that the pair $\{u, A\}$ is a solution of*

$$J_{hom} [u, A] \longrightarrow \inf, \quad u \in W_0^{1,p_0(\cdot)} (\Omega) \ \textit{with}\ A = \lim_{\varepsilon\to 0} A^\varepsilon. \tag{45}$$

Remark 10. It is important to notice that the constant A in (45) remains unknown. Suppose, in addition, that the function $c(x, b)$ is differentiable with respect to the argument b. Then it is easy to see that Euler's equation for the homogenized problem (45) reads

$$\begin{gathered} -\operatorname{div} \left(|\nabla u|^{p_0(x)-2} \nabla u \right) + |u|^{p_0(x)-2} u + \mathbf{1}_d (x)\, c_u' (x, u - A) \\ = f (x) \quad \text{in } \Omega; \\ u = 0 \quad \text{on } \partial\Omega, \qquad \int_{\Omega_d} c_u' (x, u - A)\, dx = 0, \end{gathered} \tag{46}$$

where c_u' denotes the partial derivative of the function c with respect to u. This means that the homogenized problem (45) is nonlocal.

Theorem 9 is proved by arguments similar to those from the proof of Theorem 8.

5.3. A Periodic Example. Theorems 8 and 9 provide sufficient conditions for the existence of the homogenized problem (45). The goal of this section is to prove that, for appropriate examples, all the conditions of Theorems 8 and 9 are satisfied and to compute the function $c(x, u)$ and the constant A in the homogenized problem (45) explicitly.

Let Ω be a bounded Lipschitz domain in $\mathbb{R}^3$ and let Ω_d be the subdomain of Ω defined in condition (F.1.1). Let $\mathfrak{F}^\varepsilon$ be an

ε-periodic coordinate lattice in $\mathbb{R}^3$ formed by the intersecting circular cylinders of radius r_ε,

$$r_\varepsilon = e^{-1/\varepsilon^2}. \quad (47)$$

We set $\mathcal{F}^\varepsilon = \mathfrak{F}^\varepsilon \cap \Omega_d$ and $\Omega^\varepsilon = \Omega \setminus \overline{\mathcal{F}^\varepsilon}$.

Let $\{p_\varepsilon\}_{(\varepsilon>0)}$ be a class of continuous functions defined in the domain $\overline{\Omega}$. We assume that, for any $\varepsilon > 0$, p_ε satisfies the log-Hölder continuity condition and the following condition:

(B.1.2) the functions p_ε are given by

$$p_\varepsilon(x) = \begin{cases} 2 + \ell(x)\varepsilon^2 & \text{in } \mathcal{N}\left(\mathcal{F}^\varepsilon, \varepsilon^2\right), \\ 2 + \ell_\varepsilon(x) & \text{elsewhere}, \end{cases} \quad (48)$$

where $\mathcal{N}(\mathcal{O}, \delta)$ denotes the cylindrical δ-neighborhood of the set $\mathcal{O}$ and where ℓ and ℓ_ε are smooth strictly positive functions in $\overline{\Omega}$, moreover, $\max_{x\in\overline{\Omega}} \ell_\varepsilon(x) = o(1)$ as $\varepsilon \to 0$.

Consider the boundary value problem

$$\begin{gathered} -\operatorname{div}\left(|\nabla u^\varepsilon|^{p_\varepsilon(x)-2}\nabla u^\varepsilon\right) + |u^\varepsilon|^{p_\varepsilon(x)-2}u^\varepsilon = f(x) \quad \text{in } \Omega^\varepsilon, \\ u^\varepsilon = A^\varepsilon \quad \text{on } \partial\mathcal{F}^\varepsilon; \qquad u^\varepsilon = 0 \quad \text{on } \partial\Omega, \\ \int_{\partial\mathcal{F}^\varepsilon} |\nabla u^\varepsilon|^{p_\varepsilon(x)-2}\frac{\partial u^\varepsilon}{\partial\vec{\nu}} ds = 0, \end{gathered} \quad (49)$$

where A^ε is an unknown constant. It is known from [1, 2] that, for each $\varepsilon > 0$, there exists a unique solution $u^\varepsilon \in W^{1,p_\varepsilon(\cdot)}(\Omega^\varepsilon)$ of problem (49). We extend u^ε by the equality $u^\varepsilon = A^\varepsilon$ in $\mathcal{F}^\varepsilon$ and keep for it the same notation. The asymptotic behavior of the solutions of problem (49) is given now by the following theorem.

Theorem 11. *Let u^ε be a solution of (49) extended by the equality $u^\varepsilon(x) = A^\varepsilon$ in $\mathcal{F}^\varepsilon$. Assume that the condition (B.1.2) on the function p_ε is satisfied. Then the sequence $\{u^\varepsilon\}_{(\varepsilon>0)}$ converges weakly in $H^1(\Omega)$ to a function u solution of*

$$\begin{gathered} -\Delta u + u + \mu(x)(u - \mathsf{A}_\ell(u)) = g(x) \quad \text{in } \Omega, \\ u = 0 \quad \text{on } \partial\Omega, \end{gathered} \quad (50)$$

where

$$\begin{gathered} \mu(x) = 6\pi\, \mathbf{1}_d(x)\frac{e^{\ell(x)} - 1}{\ell(x)}, \\ \mathsf{A}_\ell(u) = \left(\int_{\Omega_d}\mu(x)\,dx\right)^{-1}\int_{\Omega_d}\mu(x)u(x)\,dx. \end{gathered} \quad (51)$$

Theorem 11 is proved in [23] and is based essentially on the calculation of the local energy characteristic for this periodic structure.

Remark 12. It is easy to see from the proof of Theorem 11 that if in (48) we replace $\ell(x)\varepsilon^2$ by a function $\widetilde{\ell}_\varepsilon$ with $\max_{x\in\overline{\Omega}} \widetilde{\ell}_\varepsilon = o(\varepsilon^2)$, then $\mu(x) = 6\pi\, \mathbf{1}_d(x)$ as in the linear case.

Remark 13. It is shown in [27] Paragraph 3.3 that for the integrands of growth $|\nabla u^\varepsilon|^2 \ln|\nabla u^\varepsilon|$, the 3D lattice becomes extremely thin. Moreover, for the integrands of growth $|\nabla u^\varepsilon|^{2+\delta}$, where $\delta > 0$ is a parameter independent of ε, there is no 3D lattice which admits nontrivial homogenization result, because the capacity of the lattice goes to infinity as $\varepsilon \to 0$. Theorem 11 shows the maximal possible polynomial growth of the integrand (in a small neighborhood of the lattice) which admits a nontrivial homogenization result.

6. Homogenization of the Neumann Problem

6.1. Statement of the Problem and Main Results. Let Ω be a bounded Lipschitz domain in $\mathbb{R}^n$ ($n \geq 2$). Let $\{\mathcal{F}^\varepsilon\}_{(\varepsilon>0)}$ be a family of open subsets in Ω; in the sequel ε is a small positive parameter characterizing the microscopic length scale. We assume that the set $\mathcal{F}^\varepsilon$ consists of N_ε ($N_\varepsilon \to +\infty$ as $\varepsilon \to 0$) small isolated components such that their diameters go to zero as $\varepsilon \to 0$ and $\mathcal{F}^\varepsilon$ is distributed in an asymptotically regular way in Ω. We set

$$\Omega^\varepsilon = \Omega \setminus \overline{\mathcal{F}^\varepsilon}. \quad (52)$$

We assume that a family of continuous functions $\{p_\varepsilon\}_{(\varepsilon>0)}$ belongs to the class $\mathfrak{L}^\varepsilon_{p_0(\cdot)}$. On the space $L^{p_\varepsilon(\cdot)}(\Omega^\varepsilon)$, we define the functional $J^\varepsilon : L^{p_\varepsilon(\cdot)}(\Omega^\varepsilon) \to \mathbb{R} \cup \{+\infty\}$:

$$J^\varepsilon[u] = \begin{cases} \int_{\Omega^\varepsilon} \mathsf{F}_\varepsilon(x, u, \nabla u)\,dx, & \text{if } u \in W^{1,p_\varepsilon(\cdot)}(\Omega^\varepsilon), \\ +\infty, & \text{otherwise}, \end{cases} \quad (53)$$

where $\mathsf{F}_\varepsilon(x, u, \nabla u)$ is defined in (31) and $f \in C(\overline{\Omega})$.

We study the asymptotic behavior of J^ε and their minimizers as $\varepsilon \to 0$. The classical periodicity assumption is here substituted by an abstract one covering a variety of concrete behaviors, such as the periodicity, the almost periodicity, and many more besides. For this, we assume that $\Omega^\varepsilon \subset \Omega$ is a disperse medium; that is, the following assumptions hold:

(C.2.1) the local concentration of the set Ω^ε has a positive continuous limit, that is, the indicator of Ω^ε converges weakly in $L^2(\Omega)$ to a continuous positive limit. This implies that there exists a continuous positive function $\rho = \rho(x)$ such that

$$\lim_{h\to 0}\lim_{\varepsilon\to 0} h^{-n}\operatorname{meas}\left(K_h^x \cap \Omega^\varepsilon\right) = \rho(x), \quad (54)$$

for any open cube K_h^x centered at $x \in \Omega$ with lengths equal to $h > 0$;

(C.2.2) for any $q \in [p^-, p^+]$, there exists a family of extension operators $\mathsf{P}^\varepsilon_q : W^{1,q}(\Omega^\varepsilon) \to W^{1,q}(\Omega)$ such that.

global: for any $u^\varepsilon \in W^{1,q}(\Omega^\varepsilon)$,

$$\begin{gathered} \|\mathsf{u}^\varepsilon\|_{L^q(\Omega)} \leq C\|u^\varepsilon\|_{L^q(\Omega^\varepsilon)}, \\ \|\nabla\mathsf{u}^\varepsilon\|_{L^q(\Omega)} \leq C\|\nabla u^\varepsilon\|_{L^q(\Omega^\varepsilon)}, \end{gathered} \quad (55)$$

uniformly in $\varepsilon > 0$, where $\mathrm{u}^\varepsilon = \mathrm{P}^\varepsilon u^\varepsilon$ and $\mathrm{u}^\varepsilon = u^\varepsilon$ in Ω^ε;

local: for any $h > 0$ there is $\varepsilon_0(h) > 0$ such that for all $\varepsilon < \varepsilon_0(h)$, $z \in \Omega$ and any function $u \in W^{1,q}((z+[-2h,2h]^n) \cap \Omega^\varepsilon)$, the estimates hold

$$\begin{aligned}\|\mathrm{u}^\varepsilon\|_{L^q((z+[-h,h]^n)\cap\Omega)} &\le C\|u^\varepsilon\|_{L^q((z+[-2h,2h]^n)\cap\Omega^\varepsilon)},\\ \|\nabla \mathrm{u}^\varepsilon\|_{L^q((z+[-h,h]^n)\cap\Omega)} &\le C\|\nabla u^\varepsilon\|_{L^q((z+[-2h,2h]^n)\cap\Omega^\varepsilon)}.\end{aligned} \tag{56}$$

Remark 14. Notice that in condition (C.2.2), we require the existence of extension operators only in usual Sobolev spaces $W^{1,q}$ with constant q. In this case, the extension condition is well studied in the mathematical literature (see, e.g., [27, 30–32]). For instance, it holds for a wide class of disperse media (see, for instance, [27]).

One more condition is imposed on the local characteristic of the set $\mathcal{F}^\varepsilon$ associated to the functional (53). In order to formulate this condition, we denote by K_h^z an open cube centered at $z \in \Omega$ with edge length h ($0 < \varepsilon \ll h \ll 1$) and introduce the functional

$$c^{\varepsilon,h}_{p_\varepsilon(\cdot)}(z,\mathbf{b}) = \inf_{v^\varepsilon} \int_{K_h^z\cap\Omega^\varepsilon} \left\{ \frac{1}{p_\varepsilon(x)} |\nabla v^\varepsilon|^{p_\varepsilon(x)} + h^{-p_\varepsilon(x)-\gamma} |v^\varepsilon - (x-z,\mathbf{b})|^{p_\varepsilon(x)} \right\} dx, \tag{57}$$

where $\gamma > 0$, $\mathbf{b} \in \mathbb{R}^n$, and the infimum is taken over $v^\varepsilon \in W^{1,p_\varepsilon(\cdot)}(K_h^z \cap \Omega^\varepsilon)$. Here $(\cdot,\cdot)$ stands for the scalar product in $\mathbb{R}^n$. We assume that

(C.2.3) there is a continuous, with respect to $x \in \overline{\Omega}$, function $T(x,\mathbf{b})$ and $\gamma = \gamma_0$ ($0 < \gamma_0 < p^-$) such that for any $\{p_\varepsilon\}_{(\varepsilon>0)} \subset \mathfrak{L}^\varepsilon_{p_0(\cdot)}$, any $x \in \Omega$ and any $\mathbf{b} \in \mathbb{R}^n$,

$$\lim_{h\to 0}\overline{\lim_{\varepsilon\to 0}}\, h^{-n} c^{\varepsilon,h}_{p_\varepsilon(\cdot)}(x,\mathbf{b}) = \lim_{h\to 0}\underline{\lim_{\varepsilon\to 0}}\, h^{-n} c^{\varepsilon,h}_{p_\varepsilon(\cdot)}(x,\mathbf{b}) = T(x,\mathbf{b}). \tag{58}$$

Remark 15. Condition (C.2.3) is always fulfilled for periodic and locally periodic structures.

Remark 16. It is crucial in condition (C.2.3) that the limit function $T(x,\mathbf{b})$ does not depend on the particular choice of the sequence $p_\varepsilon \to p_0$. Notice that this is always the case for periodic and locally periodic perforated media. These media will be considered in detail in the last section of the paper.

Now we are in position to formulate the first convergence result of the section.

Theorem 17. *Assume that $\{p_\varepsilon\}_{(\varepsilon>0)} \subset \mathfrak{L}^\varepsilon_{p_0(\cdot)}$, and let conditions (C.2.1)–(C.2.3) be satisfied. Then the functionals J^ε defined in (53), $\Gamma_{p_0(\cdot)}$-converge to the functional $J_{hom} : L^{p_0(\cdot)}(\Omega) \to \mathbb{R} \cup \{+\infty\}$ given by*

$$J_{hom}[u] = \begin{cases} \displaystyle\int_\Omega \left\{ T(x,\nabla u) + \frac{\rho(x)}{p_0(x)}|u|^{p_0(x)} - \rho(x) f(x) u \right\} dx, & if\ u \in W^{1,p_0(\cdot)}(\Omega), \\ +\infty, & otherwise. \end{cases} \tag{59}$$

Now let us formulate the convergence result for the minimizers of the functionals J^ε. Consider the variational problem

$$J^\varepsilon[u^\varepsilon] \longrightarrow \min, \quad u^\varepsilon \in W^{1,p_\varepsilon(\cdot)}(\Omega^\varepsilon). \tag{60}$$

According to [1–3, 6], for each $\varepsilon > 0$, problem (60) has a unique solution $u^\varepsilon \in W^{1,p_\varepsilon(\cdot)}(\Omega^\varepsilon)$.

The following convergence result holds.

Theorem 18. *Under the assumptions of Theorem 17, the solution u^ε of the variational problem (60) converges strongly in $L^{p_0(\cdot)}(\Omega^\varepsilon)$ to a solution of the problem*

$$J_{hom}[u] \longrightarrow \min, \quad u \in W^{1,p_0(\cdot)}(\Omega). \tag{61}$$

The Scheme of the Proof of Theorems 17 and 18 (See [22] for More Details) Is as Follows. The "lim inf"-inequality is proved in two steps as in Theorem 7 by introducing an auxiliary functional. The "lim sup"-inequality is proved by the arguments similar to those from the proof of Theorem 8.

6.2. A Periodic Example. Theorems 17 and 18 of Section 6.1 provide sufficient conditions for the existence of the Γ-limit functional (59) and for the convergence of minimizers of the variational problem (60) to the minimizer of the homogenized variational problem (61). It is important to show that the class of functions which satisfy the conditions of these theorems is not empty. The goal of this section is to prove that for periodic and locally periodic media all conditions of the above-mentioned theorems are satisfied and to compute the coefficients of the homogenized functional (59) in terms of solutions of auxiliary cell problems. In fact, conditions (C.2.1) and (C.2.3) are always satisfied in the periodic case if the boundary of inclusions is regular enough, and that the extension condition (C.2.2) can also be replaced with the assumption on the regularity of the inclusions geometry.

Let Ω be a bounded domain in $\mathbb{R}^n$ ($n \ge 2$) with sufficiently smooth boundary. We assume that in the periodic cell $Y =]0,1[^n$, there is an obstacle $F \Subset Y$ being an open set with a sufficiently smooth boundary ∂F. We assume that this geometry is repeated periodically in the whole $\mathbb{R}^n$. The geometric structure within the domain Ω is then obtained by intersecting the ε-multiple of this geometry with Ω. Let $\{x^{\mathbf{k},\varepsilon}\}$ be an ε-periodic grid in $\mathbb{R}^n$: $x^{\mathbf{k},\varepsilon} = \varepsilon\mathbf{k}$, $\mathbf{k} \in \mathbb{Z}^n$. Then we define $\mathscr{F}^\varepsilon$ as the union of sets $\mathscr{F}^\varepsilon_{\mathbf{k}} \subset K^{\mathbf{k}}_\varepsilon$ obtained from εF by translations with vectors $\varepsilon\mathbf{k}$, $\mathbf{k} \in \mathbb{Z}^n$, that is,

$$\mathscr{F}^\varepsilon_{\mathbf{k}} = \varepsilon\mathbf{k} + \varepsilon F, \qquad \mathscr{F}^\varepsilon = \bigcup_{\mathbf{k}} \left(\mathscr{F}^\varepsilon_{\mathbf{k}} \cap \Omega\right), \tag{62}$$

and $K^{\mathbf{k}}_\varepsilon = \varepsilon\mathbf{k} + \varepsilon Y$. Notice that the geometry of the inclusions having a nontrivial intersection with the domain boundary might be rather complicated. In particular, the extension condition (C.2.2) might be violated for these inclusions. To avoid these technical difficulties, we assume that the domain Ω is not perforated in a small neighborhood of its boundary $\partial\Omega$. We set

$$\Omega^\varepsilon \stackrel{\text{def}}{=} \Omega \setminus \overline{\mathscr{F}^\varepsilon}. \tag{63}$$

Let a family of continuous functions $\{p_\varepsilon\}_{(\varepsilon>0)}$ belongs to the class $\mathfrak{L}^\varepsilon_{p_0(\cdot)}$. On the space $L^{p_\varepsilon(\cdot)}(\Omega^\varepsilon)$, we define the functional $J^\varepsilon : L^{p_\varepsilon(\cdot)}(\Omega^\varepsilon) \to \mathbb{R} \cup \{+\infty\}$:

$$J^\varepsilon[u] = \begin{cases} \displaystyle\int_{\Omega^\varepsilon} \left\{ \frac{1}{p_\varepsilon(x)} \left(|\nabla u|^{p_\varepsilon(x)} + |u|^{p_\varepsilon(x)} \right) - f(x)\, u \right\} dx, & \text{if } u \in W^{1,p_\varepsilon(\cdot)}(\Omega^\varepsilon), \\ +\infty, & \text{otherwise}, \end{cases} \tag{64}$$

where $f \in C(\overline{\Omega})$.

We study the asymptotic behavior of the functional J^ε and its minimizer as $\varepsilon \to 0$. To formulate the main result of this section we will introduce some notation. We denote by $U^{\mathbf{b}} = U^{\mathbf{b}}(p, y)$ a minimizer of the following variational problem:

$$\int_{Y^*} \frac{1}{p} \left| \nabla_y U^{\mathbf{b}} - \mathbf{b} \right|^p dy \longrightarrow \min, \quad u \in W^{1,p}_{\text{per}}(Y^*), \tag{65}$$

where $Y^* = Y \setminus \overline{F}$, and $\mathbf{b} = (b_1, b_2, \ldots, b_n)$ is a vector in $\mathbb{R}^n$, and $p > 1$ is a parameter.

If $p \geq 2$, then the solution $U^{\mathbf{b}}$ coincides with a unique solution in $W^{1,p}_{\text{per}}(Y^*)$ of the following cell problem:

$$\begin{gathered} \operatorname{div}_y \left(\left|\nabla_y U^{\mathbf{b}}\right|^{p-2} \nabla_y U^{\mathbf{b}} \right) = 0 \quad \text{in } Y^*, \\ \left(\left|\nabla_y U^{\mathbf{b}}\right|^{p-2} \nabla_y U^{\mathbf{b}} - \mathbf{b}, \vec{\nu} \right) = 0 \quad \text{on } \partial F, \\ y \longrightarrow U^{\mathbf{b}}(y) \quad Y\text{-periodic}, \end{gathered} \tag{66}$$

here $\vec{\nu}$ is the outward normal to ∂F.

The following result holds.

Theorem 19. *The sequence of functionals $\{J^\varepsilon\}_{(\varepsilon>0)}$ defined in (64) $\Gamma_{p_0(\cdot)}$-converges to the functional $J_{hom} : L^{p_0(\cdot)}(\Omega) \to \mathbb{R} \cup \{+\infty\}$ given by*

$$J_{hom}[u] = \begin{cases} \displaystyle\int_{\Omega} \left\{ T(x, \nabla u) + \frac{\rho}{p_0(x)} |u|^{p_0(x)} - \rho f(x)\, u \right\} dx, & \text{if } u \in W^{1,p_0(\cdot)}(\Omega), \\ +\infty, & \text{otherwise}, \end{cases} \tag{67}$$

where

$$\rho = \operatorname{meas} Y^*,$$
$$T(x, \mathbf{b}) = \int_{Y^*} \frac{1}{p_0(x)} \left| \nabla_y U^{\mathbf{b}}(p_0(x), y) - \mathbf{b} \right|^{p_0(x)} dy. \tag{68}$$

Moreover, a minimizer u^ε of the functional (64) converges strongly in the space $L^{p_0(\cdot)}(\Omega^\varepsilon)$ to u the minimizer of the homogenized functional (67).

Notice that Theorem 19 (see [22]) can be proved in two different ways. One of them is to check that under the assumptions of Theorem 19 conditions (C.2.1)–(C.2.3) are satisfied and that the characteristics introduced in conditions (C.2.1) and (C.2.3) coincide with those defined in (68). On the other hand, in the periodic case, the direct Γ-convergence techniques can be applied. This allows us to obtain formula (68) by means of Γ-convergence approach used in periodic homogenization.

7. Homogenization of Quasilinear Elliptic Equations with Nonstandard Growth in High-Contrast Media

7.1. Statement of the Problem and the Main Result. Let $\Omega = \Omega^\varepsilon_f \cup \overline{\Omega^\varepsilon_m}$ be a bounded domain of $\mathbb{R}^n$ ($n \geq 2$) with Lipschitz boundary $\partial\Omega$. Here $\{\Omega^\varepsilon_m\}_{(\varepsilon>0)}$ is a family of open subsets in Ω. We assume that the set Ω^ε_m is distributed in an asymptotically regular way in Ω; moreover, for the sake of simplicity, we suppose that $\Omega^\varepsilon_m \cap \partial\Omega = \emptyset$.

Remark 20. In the framework of the method of local energy characteristics presented in the section, we do not specify the geometrical structure of the set Ω^ε_m. Generally speaking, it may consist of N_ε ($N_\varepsilon \to +\infty$ as $\varepsilon \to 0$) small isolated components such that their diameters go to zero as $\varepsilon \to 0$ or it may be defined as fibres becoming more and more dense as $\varepsilon \to 0$, such that the diameters of the fibers go to zero as $\varepsilon \to 0$.

We assume that the family of functions $\{p_\varepsilon\}_{(\varepsilon>0)}$ belongs to the class $\mathfrak{L}^\varepsilon_{p_0(\cdot)}$. Let $\sigma \in C(\overline{\Omega})$ be such that

(A.3.4) there exist two real numbers σ^- and σ^+ such that the function σ is bounded in the following sense:

$$0 < \sigma^- \equiv \min_{x\in\overline{\Omega}} \sigma(x) \le \sigma(x) \le \max_{x\in\overline{\Omega}} \sigma(x) \equiv \sigma^+ < \min_{x\in\overline{\Omega}} \frac{p_0(x)\, n}{n - p_0(x)} \quad \text{in } \overline{\Omega}, \tag{69}$$

(A.3.5) the function σ satisfies the log-Hölder continuity property.

Let us now define the variational problem under consideration. To this end, we consider the functional $J^\varepsilon : W^{1,p_\varepsilon(\cdot)}(\Omega) \to \mathbb{R} \cup \{+\infty\}$,

$$J^\varepsilon[u] \overset{\text{def}}{=} \begin{cases} \int_\Omega G_\varepsilon(x, u, \nabla u)\, dx & \text{if } u \in W^{1,p_\varepsilon(\cdot)}(\Omega), \\ +\infty & \text{otherwise,} \end{cases} \tag{70}$$

where

$$G_\varepsilon(x, u, \nabla u) \overset{\text{def}}{=} \varkappa_\varepsilon(x) |\nabla u|^{p_\varepsilon(x)} + \frac{1}{\sigma(x)} |u|^{\sigma(x)} - g^\varepsilon(x) u \quad \text{with } \varkappa_\varepsilon(x) \overset{\text{def}}{=} \frac{K_\varepsilon(x)}{p_\varepsilon(x)}. \tag{71}$$

Here the function g^ε is defined by $g^\varepsilon(x) \overset{\text{def}}{=} \mathbf{1}_f^\varepsilon(x)\, g(x)$, $g \in C(\Omega)$. We denote by $\mathbf{1}_k^\varepsilon$ the characteristic function of the set Ω_k^ε, $k = f, m$. K_ε is a measurable function in Ω such that

(K.1) there exists a real number k_0 such that $0 < k_0 \le K_\varepsilon(x) \le k_0^{-1}$ in Ω_f^ε;

(K.2) for any $\varepsilon > 0$, there exists a real number k_ε such that $\sup_{x\in\Omega_m^\varepsilon} K_\varepsilon(x) = \mathrm{k}_\varepsilon > 0$ and $\mathrm{k}_\varepsilon \to 0$ as $\varepsilon \to 0$.

We consider the following variational problem:

$$J^\varepsilon[u^\varepsilon] \longrightarrow \min, \quad u^\varepsilon \in W_0^{1,p_\varepsilon(\cdot)}(\Omega). \tag{72}$$

It is known from [1] that for each $\varepsilon > 0$, there exists a unique solution $u^\varepsilon \in W_0^{1,p_\varepsilon(\cdot)}(\Omega)$ of the variational problem (72). We aim to study the asymptotic behavior of the family $\{u^\varepsilon\}$ as $\varepsilon \to 0$, bearing in mind that the geometry of $\Omega = \Omega_f^\varepsilon \cup \overline{\Omega}_m^\varepsilon$ depends on ε. So, we have to specify this geometry. Most of the papers dealing with homogenization assume that Ω is a periodic repetition of a standard cell. This classical periodicity assumption is here substituted by an abstract one covering a variety of concrete behaviors including periodicity and almost periodicity. We thus make the following assumptions:

(C.3.1) the local concentration of the set Ω_f^ε has a positive continuous limit; that is, the indicator of Ω_f^ε converges weakly in $L^2(\Omega)$ to a continuous positive limit. This implies that there exists a continuous positive function $\rho = \rho(x)$ such that

$$\lim_{h\to 0} \lim_{\varepsilon \to 0} h^{-n} \operatorname{meas}\left(K_h^x \cap \Omega_f^\varepsilon\right) = \rho(x) \tag{73}$$

for any open cube K_h^x centered at $x \in \Omega$ with lengths equal to $h > 0$;

(C.3.2) for any $\{p_\varepsilon\}_{(\varepsilon>0)} \subset \mathfrak{L}^\varepsilon_{p_0(\cdot)}$, there is a constant $C_{p_\varepsilon} \ge 0$ such that if the function $p_\varepsilon^\star$ is defined by $p_\varepsilon^\star(x) = p_\varepsilon(x) - C_{p_\varepsilon}$ in Ω, then:

(i) the sequence $\{p_\varepsilon^\star\}_{(\varepsilon>0)}$ belongs to $\mathfrak{L}^\varepsilon_{p_0(\cdot)}$, that is, $\lim_{\varepsilon\to 0} C_{p_\varepsilon} = 0$;

(ii) there exists a family of extension operators $\mathbf{P}^\varepsilon : W^{1,p_\varepsilon^\star(\cdot)}(\Omega_f^\varepsilon) \to W^{1,p_\varepsilon^\star(\cdot)}(\Omega)$ such that, for any $v^\varepsilon \in W^{1,p_\varepsilon(\cdot)}(\Omega_f^\varepsilon)$,

$$\mathbf{P}^\varepsilon v^\varepsilon = v^\varepsilon \quad \text{in } \Omega_f^\varepsilon, \qquad \|\mathbf{P}^\varepsilon v^\varepsilon\|_{W^{1,p_\varepsilon^\star(\cdot)}(\Omega)} \le \Phi\left(\|v^\varepsilon\|_{W^{1,p_\varepsilon(\cdot)}(\Omega_f^\varepsilon)}\right), \tag{74}$$

where $\Phi = \Phi(t)$ is a strictly monotone continuous function in $\mathbb{R}^+$ such that $\Phi(0) = 0$ and $\Phi(t) \to +\infty$ as $t \to +\infty$.

We also impose several conditions on the local characteristic of the set Ω_f^ε and Ω_m^ε associated to the functional (70). Let K_h^z be an open cube centered at $z \in \Omega$ with lengths equal to h $(0 < \varepsilon \ll h \ll 1)$. We introduce the following functionals:

(i) the functional $c^{\varepsilon,h}_{p_\varepsilon(\cdot)}$ associated to the energy in Ω_f^ε is defined in $\Omega \times \mathbb{R}^n$ by

$$c^{\varepsilon,h}_{p_\varepsilon(\cdot)}(z; \mathbf{a}) \overset{\text{def}}{=} \inf_{v^\varepsilon} \int_{K_h^z \cap \Omega_f^\varepsilon} \left\{ \varkappa_\varepsilon(x) |\nabla v^\varepsilon(x)|^{p_\varepsilon(x)} + h^{-p_\varepsilon(x)-\gamma} |v^\varepsilon(x) - (x - z, \mathbf{a})|^{p_\varepsilon(x)} \right\} dx, \tag{75}$$

for $z \in \Omega$, $\mathbf{a} \in \mathbb{R}^n$, where $\gamma > 0$ and the infimum is taken over $v^\varepsilon \in W^{1,p_\varepsilon(\cdot)}(K_h^z \cap \Omega_f^\varepsilon)$.

(ii) The functional $b^{\varepsilon,h}_{p_\varepsilon(\cdot)}$ associated to the energy exchange between Ω_f^ε and Ω_m^ε is defined in $\Omega \times \mathbb{R}^n$ by

$$b^{\varepsilon,h}_{p_\varepsilon(\cdot)}(z; \beta) \overset{\text{def}}{=} \inf_{w^\varepsilon} \int_{K_h^z} \left\{ \varkappa_\varepsilon(x) |\nabla w^\varepsilon|^{p_\varepsilon(x)} + \frac{\mathbf{1}_m^\varepsilon(x)}{\sigma(x)} |w^\varepsilon|^{\sigma(x)} + h^{-p_\varepsilon(x)-\gamma} \mathbf{1}_f^\varepsilon(x) |w^\varepsilon - \beta|^{p_\varepsilon(x)} \right\} dx, \tag{76}$$

for $z \in \Omega$, $\beta \in \mathbb{R}$, the infimum being taken over $w^\varepsilon \in W^{1,p_\varepsilon(\cdot)}(K_h^z)$.

We assume that the local characteristics of Ω are such that

(C.3.3) for any $x \in \Omega$ and any $\mathbf{a} \in \mathbb{R}^n$, there is a continuous function $T(x, \mathbf{a})$ and a real number $\gamma = \gamma_0$ ($0 < \gamma_0 < p^-$) such that, for any $\{p_\varepsilon\}_{(\varepsilon>0)} \subset \mathfrak{L}^\varepsilon_{p_0(\cdot)}$,

$$\lim_{h\to 0} \overline{\lim_{\varepsilon\to 0}} h^{-n} c^{\varepsilon,h}_{p_\varepsilon(\cdot)} (x, \mathbf{a}) = \lim_{h\to 0} \underline{\lim}_{\varepsilon\to 0} h^{-n} c^{\varepsilon,h}_{p_\varepsilon(\cdot)} (x, \mathbf{a}) = T (x, \mathbf{a}) . \quad (77)$$

(C.3.4) for any $x \in \Omega$ and any $\beta \in \mathbb{R}$, there is a continuous function $\mathsf{b}(x, \beta)$ and a real number $\gamma = \gamma_1$ ($0 < \gamma_1 < p^-$) such that, for any $\{p_\varepsilon\}_{(\varepsilon>0)} \subset \mathfrak{L}^\varepsilon_{p_0(\cdot)}$,

$$\lim_{h\to 0} \overline{\lim_{\varepsilon\to 0}} h^{-n} b^{\varepsilon,h}_{p_\varepsilon(\cdot)} (x, \beta) = \lim_{h\to 0} \underline{\lim}_{\varepsilon\to 0} h^{-n} b^{\varepsilon,h}_{p_\varepsilon(\cdot)} (x, \beta) = \mathsf{b} (x, \beta) . \quad (78)$$

Remark 21. It is crucial in conditions (C.3.3) and (C.3.4) that the limit functions $T(x, \mathbf{b})$ and $\mathsf{b}(x, \beta)$ do not depend on the particular choice of the sequence $\{p_\varepsilon\}_{(\varepsilon>0)} \subset \mathfrak{L}^\varepsilon_{p_0(\cdot)}$. It is proved in [26] that these assumptions are fulfilled for periodic and locally periodic media.

Remark 22. Contrary to the standard growth setting as considered in [20, 32], the local characteristic $b^{\varepsilon,h}_{p_\varepsilon(\cdot)}(z; \beta)$ is not homogeneous with respect to the parameter β. This induces the appearance of a nonlinear function $\mathsf{b}(x, u)$ in the homogenized functional (see Theorem 23 below).

The main result of the section is the following.

Theorem 23. *Let u^ε be a solution of (72). Assume that $p^\varepsilon \in \mathfrak{L}^\varepsilon_{p_0(\cdot)}$ and conditions (A.3.4)–(A.3.5), (K.1)-(K.2), and (C.3.1)–(C.3.4) are satisfied. Then u^ε, solution of the variational problem (72), converges strongly in $L^{p_0(\cdot)}(\Omega^\varepsilon_f)$ to u, solution of the following variational problem:*

$$J_{hom} [u] \longrightarrow \min, \quad u \in W^{1,p_0(\cdot)}_0 (\Omega) , \quad (79)$$

the homogenized functional $J_{hom} : W^{1,p_0(\cdot)}_0(\Omega) \to \mathbb{R} \cup \{+\infty\}$ being defined by

$$J_{hom} [u] \overset{\text{def}}{=} \begin{cases} \int_\Omega \mathsf{G}_0 (x, u, \nabla u)\, dx & \text{if } u \in W^{1,p_0(\cdot)}_0 (\Omega) , \\ +\infty & \text{otherwise,} \end{cases} \quad (80)$$

where

$$\mathsf{G}_0 (x, u, \nabla u) \overset{\text{def}}{=} T (x, \nabla u) + \frac{\rho (x)}{\sigma (x)} |u|^{\sigma(x)} + \mathsf{b} (x, u) - g (x) \rho (x) u. \quad (81)$$

Moreover, for any smooth function ζ in Ω, we have

$$\lim_{\varepsilon\to 0} \int_{\Omega^\varepsilon_m} \left\{ \frac{1}{p_\varepsilon (x)} |u^\varepsilon|^{\sigma(x)-2} \left(u (x) u^\varepsilon - |u^\varepsilon|^2\right) + \frac{1}{\sigma (x)} |u^\varepsilon|^{\sigma(x)} \right\} \zeta (x)\, dx = \int_\Omega \mathsf{b} (x, u) \zeta (x)\, dx. \quad (82)$$

The scheme of the proof of Theorem 23 (see [26] for more details) is similar to the scheme of the proof of Theorems 17 and 18.

7.2. Periodic Examples. Theorem 23 of Section 7.1 provides sufficient conditions for the existence of the homogenized functional (80) and for the convergence of minimizers of the variational problem (72) to the minimizer of the homogenized variational problem (79) under conditions (A.3.4)-(A.3.5), (K.1)-(K.2), and (C.3.1)–(C.3.4). It is important to show that the "intersection" of these conditions is not empty. The goal of this section is to prove that for periodic media all the conditions of the above-mentioned theorem are satisfied and to compute the coefficients of the homogenized functional (80) either in an explicit form or as usually by the solution of a corresponding cell problem.

Let Ω be a bounded domain in $\mathbb{R}^n$ ($n \geq 2$) with Lipschitz boundary. We assume that in the standard periodic cell $Y \overset{\text{def}}{=}]0, 1[^n$, there is an obstacle $M \subset Y$ with Lipschitz boundary ∂M. We assume that this geometry is repeated periodically in the whole $\mathbb{R}^n$. The geometric structure within the domain Ω is then obtained by intersecting the ε-multiple of this geometry with Ω. Let $\{x^{i,\varepsilon}\}$ be an ε-periodic grid in Ω. Then we define Ω^ε_m as the union of sets $M^\varepsilon_i \subset K^i_\varepsilon$ ($i = 1, 2, \ldots, N_\varepsilon$) obtained from εM by translations of vectors $\varepsilon \sum_{j=1}^n k_j \mathbf{e}_j$, that is,

$$\Omega^\varepsilon_m = \bigcup_i^{N_\varepsilon} M^\varepsilon_i, \qquad \Omega^\varepsilon_f = \Omega \setminus \overline{\Omega^\varepsilon_m}, \quad (83)$$

where K^i_ε is the cube centered at the point $x^{i,\varepsilon}$ and of length ε, $k_j \in \mathbb{Z}$, $\{\mathbf{e}_j\}^n_{j=1}$ is the canonical basis of $\mathbb{R}^n$, and $N_\varepsilon \to +\infty$ as $\varepsilon \to 0$.

Let $p_0 = p_0(x)$ be a *log-Hölder continuous* function such that

$$2 < p^- \equiv \min_{x\in\overline{\Omega}} p_0 (x) \leq p_0 (x) \leq \max_{x\in\overline{\Omega}} p_0 (x) \equiv p^+ < +\infty \quad \text{in } \overline{\Omega}. \quad (84)$$

Let $\{p_\varepsilon\}_{(\varepsilon>0)} \subset \mathfrak{L}^\varepsilon_{p_0(\cdot)}$ be a sequence defined by

$$p_\varepsilon (x) \overset{\text{def}}{=} p_0 (x) + \mathbf{d}_\varepsilon (x) , \quad (85)$$

where the function $\mathbf{d}_\varepsilon$ is such that $\mathbf{d}_\varepsilon = o(1)$ as $\varepsilon \to 0$. The asymptotic behavior of $\mathbf{d}_\varepsilon$ will be specified below. On the space $W^{1,p_\varepsilon(\cdot)}(\Omega)$, we define the functional $J^\varepsilon : W^{1,p_\varepsilon(\cdot)}(\Omega) \to \mathbb{R} \cup \{+\infty\}$,

$$J^\varepsilon [u] \overset{\text{def}}{=} \begin{cases} \int_\Omega \left\{ \frac{K_\varepsilon (x)}{p_\varepsilon (x)} |\nabla u|^{p_\varepsilon(x)} + \frac{1}{\sigma (x)} |u|^{\sigma(x)} - g^\varepsilon (x) u \right\} dx & \text{if } u \in W^{1,p_\varepsilon(\cdot)} (\Omega) , \\ +\infty & \text{otherwise,} \end{cases} \quad (86)$$

where

$$K^{\varepsilon}(x) = \begin{cases} k_f & \text{in } \Omega_f^{\varepsilon}, \\ k_m\ \varepsilon^{p_0(x)} & \text{in } \Omega_m^{\varepsilon}, \end{cases} \tag{87}$$

the function σ satisfies conditions (A.3.4) with $\sigma^- > 2$, (A.3.5), the function g^{ε} is defined by $g^{\varepsilon}(x) \stackrel{\text{def}}{=} \mathbf{1}_f^{\varepsilon}(x)\ g(x)$, $g \in C(\Omega)$. Here k_f and k_m are strictly positive constants independent of ε.

Consider the following variational problem:

$$J^{\varepsilon}[u^{\varepsilon}] \longrightarrow \min, \quad u^{\varepsilon} \in W_0^{1,p_{\varepsilon}(\cdot)}(\Omega). \tag{88}$$

We aim to study the asymptotic behavior of u^{ε} the solution of (88).

To formulate the main result of this section, we introduce some notations. We denote by $\mathrm{u}^a = \mathrm{u}^a(x,y)$ the unique solution in $W_{\#}^{1,p_0(\cdot)}(F)$ of the following cell problem:

$$\begin{gathered} \operatorname{div}_y \left\{ k_f \left|\nabla_y \mathrm{u}^a\right|^{p_0(x)-2} \nabla_y \mathrm{u}^a \right\} = 0 \quad \text{in } F, \\ \left(k_f \left|\nabla_y \mathrm{u}^a\right|^{p_0(x)-2} \nabla_y \mathrm{u}^a - \mathbf{a}, \vec{\nu} \right) = 0 \quad \text{on } \partial M, \\ y \longrightarrow \mathrm{u}^a(y) \ \ Y\text{-periodic}, \end{gathered} \tag{89}$$

where $F = Y \setminus \overline{M}$, $\vec{\nu}$ is the outward normal vector to ∂M, and $\mathbf{a} \in \mathbb{R}^n$. We denote by $\mathrm{w}^{\beta} = \mathrm{w}^{\beta}(x,y)$ the unique solution in $W_{\#}^{1,p_0(\cdot)}(M)$ of the following cell problem:

$$\begin{gathered} -\operatorname{div}_y \left\{ k_m\, \mathbf{d}(x) \left|\nabla_y \mathrm{w}^{\beta}\right|^{p_0(x)-2} \nabla_y \mathrm{w}^{\beta} \right\} \\ + \left|\mathrm{w}^{\beta}\right|^{\sigma(x)-2} \mathrm{w}^{\beta} = 0 \quad \text{in } M, \\ \mathrm{w}^{\beta}(y) = \beta \quad \text{on } \partial M, \\ y \longrightarrow \mathrm{w}^{\beta}(y) \ \ Y\text{-periodic}. \end{gathered} \tag{90}$$

Notice that in (89) and (90) x is a parameter. Regularity results for u^a and w^{β} are thus easily deduced from [33] and [34]. We also introduce the homogenized functional $J_{\text{hom}} : W^{1,p_0(\cdot)}(\Omega) \to \mathbb{R} \cup \{+\infty\}$:

$$J_{\text{hom}}[u] \stackrel{\text{def}}{=} \begin{cases} \int_{\Omega} \left\{ T(x, \nabla u) + \dfrac{\rho^{\star}}{\sigma(x)} |u|^{\sigma(x)} + \mathrm{b}(x,u) - g(x)\rho^{\star} u \right\} dx & \text{if } u \in W_0^{1,p_0(\cdot)}(\Omega), \\ +\infty & \text{otherwise}, \end{cases} \tag{91}$$

the following results hold.

Theorem 24. *Let u^{ε} be a solution of* (88). *Assume that*

$$\lim_{\varepsilon \to 0} \varepsilon^{-\mathbf{d}_{\varepsilon}(\cdot)} = \mathbf{d}(\cdot) \tag{92}$$

uniformly in Ω. Then u^{ε} converges strongly in $L^{p_0(\cdot)}(\Omega_f^{\varepsilon})$ to u the solution of the variational problem:

$$J_{hom}[u] \longrightarrow \min, \quad u \in W_0^{1,p_0(\cdot)}(\Omega), \tag{93}$$

where

$$\rho^{\star} \stackrel{\text{def}}{=} \operatorname{meas} F, \tag{94}$$

$$T(x,\mathbf{a}) \stackrel{\text{def}}{=} \frac{1}{p_0(x)} \int_F \left|\nabla_y \mathrm{u}^a(x,y) - \mathbf{a}\right|^{p_0(x)} dy, \tag{95}$$

$$\mathrm{b}(x,\beta) \stackrel{\text{def}}{=} \int_{\mathcal{M}} \left\{ \frac{1}{p_0(x)} \left[\beta \mathrm{w}^{\beta} \left|\mathrm{w}^{\beta}\right|^{\sigma(x)-2} - \left|\mathrm{w}^{\beta}\right|^{\sigma(x)} \right] + \frac{1}{\sigma(x)} \left|\mathrm{w}^{\beta}\right|^{\sigma(x)} \right\} dy. \tag{96}$$

Theorem 25. *Let u^{ε} be a solution of* (88). *Assume that, for any $x \in \Omega$,*

$$\lim_{\varepsilon \to 0} \varepsilon^{-\mathbf{d}_{\varepsilon}(x)} = +\infty. \tag{97}$$

Then u^{ε} converges strongly in $L^{p_0(\cdot)}(\Omega_f^{\varepsilon})$ to u the solution of the variational problem (93), *where $\rho^{\star}$ and the function $T(x,\mathbf{a})$ are given in* (94), (95), *and*

$$\mathrm{b}(x,\beta) \stackrel{\text{def}}{=} \frac{\operatorname{meas} M}{\sigma(x)} |\beta|^{\sigma(x)}. \tag{98}$$

Theorem 26. *Let u^{ε} be a solution of* (88). *Assume that, for any $x \in \Omega$,*

$$\lim_{\varepsilon \to 0} \varepsilon^{-\mathbf{d}_{\varepsilon}(x)} = 0. \tag{99}$$

Then u^{ε} converges strongly in $L^{p_0(\cdot)}(\Omega_f^{\varepsilon})$ to u the solution of the variational problem (93), *where $\rho^{\star}$ and the function $T(x,\mathbf{a})$ are given in* (94), (95), *and $\mathrm{b}(x,\beta) = 0$.*

Remark 27. Notice that if $K_{\varepsilon}(x) = \mathbf{1}_f^{\varepsilon} k_f + \mathbf{1}_m^{\varepsilon} k_m\ \varepsilon^{p_{\varepsilon}(x)}$ then Theorem 24 holds true with $\mathbf{d}(x) \equiv 1$.

8. Homogenization of a Class of Quasilinear Parabolic Equations with Nonstandard Growth

8.1. Statement of the Problem and Assumptions. In this section, we describe a mesoscopic double porosity model in a periodic fractured medium. We consider a reservoir $\Omega \subset \mathbb{R}^n$ $(n \geq 2)$ to be a bounded connected domain with a periodic structure. More precisely, we will scale this periodic structure by a parameter ε which represents the ratio of the cell size to the size of the whole region Ω, and we will assume that ε is a parameter tending to zero. Let $Y \stackrel{\text{def}}{=}]0,1[^n$ represent the mesoscopic domain of the basic cell of a fractured porous medium. For the sake of simplicity and without loss of generality, we assume that Y is made up of two homogeneous porous media $M \Subset Y$ and F corresponding to parties of the mesoscopic domain occupied by the matrix block and the fracture, respectively. Thus $Y = M \cup \Gamma_{m,f} \cup F$, where $\Gamma_{m,f}$ denotes the interface between the two media and the subscripts m and f refer to the matrix and fracture, respectively. Let Ω_i^ε with $i = m$ or f denotes the open set filled with the porous medium i. Then $\Omega = \Omega_m^\varepsilon \cup \Gamma_{m,f}^\varepsilon \cup \Omega_f^\varepsilon$, where $\Gamma_{m,f}^\varepsilon = \partial\Omega_m^\varepsilon \cap \partial\Omega_f^\varepsilon$. For the sake of simplicity, we will assume that $\partial\Omega \cap \Omega_m^\varepsilon = \emptyset$.

Let us introduce the nonstandard growth function used in this section. We assume that a family of continuous functions $p_\varepsilon = p_\varepsilon(x)$, $\varepsilon > 0$, is defined in $\overline{\Omega}$ and satisfies the following conditions:

(A.4.1) functions p_ε are bounded from below such that: $p_\varepsilon(x) \geq 2$ in $\overline{\Omega}$;

(A.4.2) the function $p_\varepsilon(x)$ satisfies the log-Hölder continuity property;

(A.4.3) the function $\gamma_\varepsilon(x) \overset{\text{def}}{=} p_\varepsilon(x) - 2$ converges uniformly to zero in $\overline{\Omega}$.

Now let us introduce the permeability coefficient and the porosity of the porous medium Ω. We set

$$k^\varepsilon(x) = k_m \varepsilon^2 \mathbf{1}_m^\varepsilon(x) + k_f \mathbf{1}_f^\varepsilon(x),$$
$$\omega^\varepsilon(x) = \omega_m \mathbf{1}_m^\varepsilon(x) + \omega_f \mathbf{1}_f^\varepsilon(x), \tag{100}$$

where k_f is the permeability or the hydraulic conductivity of fissures, k_m is the permeability or the hydraulic conductivity of blocks, ω_f is the porosity of fissures, and ω_m is the porosity of blocks; $\mathbf{1}_f^\varepsilon = \mathbf{1}_f^\varepsilon(x)$ and $\mathbf{1}_m^\varepsilon = \mathbf{1}_m^\varepsilon(x)$ denote the characteristic functions of the sets Ω_f^ε and Ω_m^ε, respectively. Here $0 < k_f, k_m, \omega_f, \omega_m < +\infty$.

We consider the following initial boundary value problem for the function $u^\varepsilon : Q \mapsto \mathbb{R}$:

$$\omega^\varepsilon(x)\frac{\partial u^\varepsilon}{\partial t} - \operatorname{div}\left(k^\varepsilon(x)\nabla u^\varepsilon |\nabla u^\varepsilon|^{p_\varepsilon(x)-2}\right) = g(t,x) \quad \text{in } Q,$$
$$u^\varepsilon = 0 \quad \text{on } (0,T)\times\partial\Omega,$$
$$u^\varepsilon(0,x) = u_0(x) \quad \text{in } \Omega, \tag{101}$$

where Q denotes the cylinder $(0,T)\times\Omega$, $T>0$ is given, and g, u_0 are given functions.

For simplicity and without loss of generality, we restrict the presentation to a homogeneous Dirichlet boundary condition on $\partial\Omega$, but it is easy to see that all results also hold for other boundary conditions.

Throughout the section, C will denote a generic positive constant, independent of ε, and may take different values for different occurrences.

8.2. Preliminary Results. The goal of this section is to obtain *a priori* estimates for the solution u^ε of problem (101). We start by formulating the existence and uniqueness result for (101). It is given by the following theorem (see [35]).

Theorem 28. *Let $g \in C(0,T;L^2(\Omega))$ and $u_0 \in H^2(\Omega)$. Then, for any $\varepsilon > 0$, there exists a unique solution $u^\varepsilon = u^\varepsilon(t,x)$ of the boundary value problem (101) in the space $L^\infty(0,T;W^{1,p_\varepsilon(\cdot)}(\Omega))$. Furthermore, this solution satisfies the following a priori estimates, for a.e. $t \in (0,T)$:*

$$\|u^\varepsilon(t)\|_{L^2(\Omega)}^2 + \int_0^t dt \int_\Omega \left|\frac{\partial u^\varepsilon}{\partial t}\right|^2 dx + \int_\Omega k^\varepsilon(x)|\nabla u^\varepsilon|^{p_\varepsilon(x)dx}$$
$$\leq C, \tag{102}$$

$$\|u^\varepsilon(t+\delta t) - u^\varepsilon(t)\|_{L^2(\Omega)} \leq C(\delta t)^\kappa \quad \text{with } 0<\kappa<1, \tag{103}$$

where $\delta t > 0$ is a time step which tends to zero.

We study the asymptotic behavior of the solution u^ε of problem (101) as $\varepsilon \to 0$. For this it is convenient to introduce the following notation:

$$u^\varepsilon = \begin{cases} \rho^\varepsilon & \text{in } \Omega_f^\varepsilon, \\ \sigma^\varepsilon & \text{in } \Omega_m^\varepsilon, \end{cases} \tag{104}$$

and to rewrite (101) separately in the domains $\Omega_f^\varepsilon, \Omega_m^\varepsilon$ with appropriate interface conditions. Namely, in the domain Ω_f^ε (101) reads

$$\omega_f\frac{\partial \rho^\varepsilon}{\partial t} - \operatorname{div}\left(k_f\nabla\rho^\varepsilon|\nabla\rho^\varepsilon|^{p_\varepsilon(x)-2}\right)$$
$$= g(t,x) \quad \text{in } (0,T)\times\Omega_f^\varepsilon,$$
$$k_f\nabla\rho^\varepsilon|\nabla\rho^\varepsilon|^{p_\varepsilon(x)-2}\cdot\vec{\nu}$$
$$= k_m\varepsilon^{p_\varepsilon(x)}\nabla\sigma^\varepsilon|\nabla\sigma^\varepsilon|^{p_\varepsilon(x)-2}\cdot\vec{\nu} \quad \text{on } (0,T)\times\Gamma_{m,f}^\varepsilon, \tag{105}$$
$$\rho^\varepsilon = 0 \quad \text{on } (0,T)\times\partial\Omega,$$
$$\rho^\varepsilon(0,x) = u_0(x) \quad \text{in } \Omega_f^\varepsilon,$$

where $\vec{\nu}$ is the outward normal vector to Γ_{mf}^ε. In the domain Ω_m^ε (101) reads

$$\omega_m\frac{\partial \sigma^\varepsilon}{\partial t} - \operatorname{div}\left(k_m\varepsilon^2\nabla\sigma^\varepsilon|\nabla\sigma^\varepsilon|^{p_\varepsilon(x)-2}\right)$$
$$= g(t,x) \quad \text{in } (0,T)\times\Omega_m^\varepsilon, \tag{106}$$
$$\sigma^\varepsilon = \rho^\varepsilon \quad \text{on } (0,T)\times\Gamma_{mf}^\varepsilon,$$
$$\sigma^\varepsilon(0,x) = u_0(x) \quad \text{in } \Omega_m^\varepsilon.$$

To establish some preliminary compactness result, first we notice that the *a priori* estimate (102), conditions (A.4.1), (A.4.3) along with (110), and inequalities (106), and (107) imply the bound for a.e. $t \in (0,T)$:

$$\|u^\varepsilon(t)\|_{L^2(\Omega)} + \int_0^t dt \int_\Omega \left|\frac{\partial u^\varepsilon}{\partial t}\right|^2 dx$$
$$+ \int_{\Omega_f^\varepsilon} |\nabla u^\varepsilon|^2 dx + \varepsilon^2 \int_{\Omega_m^\varepsilon} |\nabla u^\varepsilon|^2 dx \leq C. \tag{107}$$

Therefore, from (105), (106) and (107), we have for a.e. $t \in (0,T)$:

$$\|\rho^{\varepsilon}(t)\|_{L^2(\Omega_f^{\varepsilon})} + \int_0^t dt \int_{\Omega_f^{\varepsilon}} \left|\frac{\partial \rho^{\varepsilon}}{\partial t}\right|^2 dx + \int_{\Omega_f^{\varepsilon}} |\nabla \rho^{\varepsilon}|^2 dx \le C,$$

$$\|\sigma^{\varepsilon}(t)\|_{L^2(\Omega_m^{\varepsilon})} + \int_0^t dt \int_{\Omega_m^{\varepsilon}} \left|\frac{\partial \sigma^{\varepsilon}}{\partial t}\right|^2 dx + \varepsilon^2 \int_{\Omega_m^{\varepsilon}} |\nabla \sigma^{\varepsilon}|^2 dx \le C. \tag{108}$$

Now using the bounds (108), along with the extension result [30], it is easy to prove the following compactness result.

Lemma 29. *Let $u^{\varepsilon} = \langle \rho^{\varepsilon}, \sigma^{\varepsilon}\rangle$ be the solution of problem* (101). *Then there exists a subsequence, still denoted by $\{u^{\varepsilon}\}_{\varepsilon>0}$, and functions $u_f = u_f(t,x)$, $v_f = v_f(x,y)$, $u_m = u_m(t,x,y)$ such that*

(a.1) $u_f \in H^1(0,T;L^2(\Omega)) \cap L^{\infty}(0,T;H^1(\Omega))$, $v_f \in L^2(\Omega;H^1_{\#}(F)\setminus\mathbb{R})$,

(a.2) $u_m \in H^1(0,T;L^2(\Omega\times Y)) \cap L^{\infty}(0,T;L^2(\Omega;H^1(M)))$, *with* $u_m(t,x,y) = u_f(t,x)$ *for* $y \in \partial M$,

(b) *for any* $t \in (0,T)$, $\rho^{\varepsilon} \overset{2s}{\rightharpoonup} u_f$ *and* $\sigma^{\varepsilon} \overset{2s}{\rightharpoonup} u_m$,

(c) *for any* $\varphi \in L^{\infty}(0,T;L^2(\Omega;C_{\#}(Y)))$,

$$\begin{aligned}&\int_0^T dt \int_{\Omega_f^{\varepsilon}} \frac{\partial u^{\varepsilon}}{\partial t}\varphi\left(t,x,\frac{x}{\varepsilon}\right) dx \longrightarrow \int_0^T dt \int_{\Omega\times F} \frac{\partial u_f}{\partial t}\,\varphi(t,x,y)\,dx\,dy,\\ &\int_0^T dt \int_{\Omega_m^{\varepsilon}} \frac{\partial u^{\varepsilon}}{\partial t}\varphi\left(t,x,\frac{x}{\varepsilon}\right) dx \longrightarrow \int_0^T dt \int_{\Omega\times M} \frac{\partial u_m}{\partial t}\,\varphi(t,x,y)\,dx\,dy,\end{aligned} \tag{109}$$

(d) *for a.e.* $t \in (0,T)$, $\nabla\rho^{\varepsilon} \overset{2s}{\rightharpoonup} (\nabla_x u_f + \nabla_y v_f)\mathbf{1}_f(y)$ *and* $\nabla\sigma^{\varepsilon} \overset{2s}{\rightharpoonup} \nabla_y u_m \mathbf{1}_m(y)$.

8.3. Homogenization Results. In this section, we formulate the main results of the section. We give homogenization results for the problem (101). The convergence of the homogenization process is obtained by combining the technique of two-scale convergence and the variational homogenization method (see, e.g., [9, 16, 27] and the references therein).

The idea of the proof is the following. First we will reduce our parabolic problem to an elliptic one depending on the time variable as a parameter. Then we introduce a functional corresponding to this elliptic problem and study the minimization problem for it in the limit of small ε. Then we obtain the limit functional corresponding to the homogenized problem. Regarding the variational technique, it is worth to mention one trick used in the paper. In order to obtain the lower bound for the original functional, we first replace the original exponent $p_{\varepsilon}(x)$ by a new one $p_0 = 2$ and consider the corresponding family of auxiliary functionals. Then the lower semicontinuity property of convex functionals with respect to the two-scale convergence implies the desired inequality. Finally, it is not difficult to show that the limit functional for the auxiliary family does not exceed the limiting functional for the original one.

Now we are in position to formulate the first homogenization result of the section.

Theorem 30. *Let $u^{\varepsilon} = u^{\varepsilon}(t,x)$ be the solution of the boundary value problem* (101) *and let conditions (A.4.1)–(A.4.3) be satisfied. Moreover, we assume that there exists a function $\alpha \in C(\overline{\Omega})$ such that uniformly in $x \in \overline{\Omega}$,*

$$\lim_{\varepsilon\to 0} \varepsilon^{-\gamma_{\varepsilon}(x)} = \alpha(x). \tag{110}$$

Then, for a.e. $t \in (0,T)$, u^{ε} two-scale converges to $u^* \in L^2((0,T)\times\Omega\times Y)$ such that

$$u^*(t,x,y) = \begin{cases} u_f(t,x) & \text{in } Q\times F,\\ u_m(t,x,y) & \text{in } Q\times M,\end{cases} \tag{111}$$

where the couple $(u_f,u_m) \in L^2(0,T;H^1(\Omega)) \times L^{\infty}(0,T;L^2(\Omega;H^1(M)))$ is the unique solution of the homogenized problem

$$\begin{gathered}\omega_f |F| \frac{\partial u_f}{\partial t} - \operatorname{div}_x\left(K^*\nabla_x u_f\right) = S(x,u_m) \quad \text{in } Q,\\ u_f = 0 \text{ on } (0,T)\times\partial\Omega, \qquad u_f(0,x) = u_0(x) \quad \text{in } \Omega,\\ \omega_m \frac{\partial u_m}{\partial t} - \widetilde{K}^*(x)\,\Delta_y u_m = g(t,x) \quad \text{in } Q\times M,\\ u_m(t,x,y) = u_f(t,x) \quad \text{on } Q\times\partial M,\\ u_m(0,x,y) = u_0(x) \quad \text{in } \Omega\times M,\end{gathered} \tag{112}$$

where $|F|$ is the measure of the set F and $K^* = \{k^*_{ij}\}$ is the homogenized permeability tensor defined by

$$k^*_{ij} = k_f \int_F \left(\vec{e}_i + \nabla_y w_i\right)\cdot\left(\vec{e}_j + \nabla_y w_j\right) dy \tag{113}$$

with $\{\vec{e}_1, \vec{e}_2, \ldots, \vec{e}_n\}$ the canonical basis of $\mathbb{R}^n$ and w_i being the unique solution in $H^1_{\#}(F)\setminus\mathbb{R}$ of

$$\begin{gathered}-k_f\Delta_y w_i = 0 \quad \text{in } F,\\ \left(\vec{e}_i + \nabla_y w_i\right)\cdot\vec{\nu} = 0 \quad \text{on } \partial M,\\ y \longrightarrow w_i(x,y) \quad Y\text{-periodic},\end{gathered} \tag{114}$$

where $\vec{\nu} = \vec{\nu}(y)$ is the outer normal vector at ∂M, and the coefficient $\widetilde{K}^*$ in the local problem is given by

$$\widetilde{K}^*(x) = \alpha(x)\,k_m, \tag{115}$$

and the effective source term $\mathsf{S}(x,u_m)$ *is given by*

$$\mathsf{S}(x,u_m) = |F|\, g(t,x) - \widetilde{K}^*(x) \int_{\partial M} \left(\nabla_y u_m \cdot \vec{\nu}\right) ds_y. \quad (116)$$

Remark 31. The source term which appears in the right-hand side of the first equation in (112) is well defined, since $u_m \in L^\infty(0,T;L^2(\Omega;H^1(M)))$, and it follows from the third equation of (112) that

$$-\Delta_y u_m \in L^2\left(0,T;H^{-1/2}(\partial M)\right), \quad (117)$$

which allows one to define $(\nabla_y u_m \cdot \vec{\nu})$ as an element of $L^2(0,T;H^{-1/2}(\partial M))$.

The Scheme of the Proof of Theorem 30 Is as Follows. We consider our parabolic boundary value problem (101) as an elliptic one depending on the time variable t as a parameter. Namely, we consider the following boundary value problem, for a.e. $t \in]0,T[$,

$$\begin{gathered} -\operatorname{div}\left(k^\varepsilon(x)\,\nabla u^\varepsilon |\nabla u^\varepsilon|^{p_\varepsilon(x)-2}\right) = G^\varepsilon \quad \text{in } Q, \\ u^\varepsilon = 0 \quad \text{on }]0,T[\, \times \partial\Omega, \end{gathered} \quad (118)$$

where the function G^ε,

$$G^\varepsilon = G^\varepsilon(t,x) \stackrel{\text{def}}{=} g(t,x) - \omega^\varepsilon(x) \frac{\partial u^\varepsilon}{\partial t}(t,x), \quad (119)$$

is considered as a given function. Then, for any $\Delta_t \subset [0,T]$, u^ε minimizes the functional

$$J^\varepsilon[u] \stackrel{\text{def}}{=} \int_{\Delta_t} dt \int_\Omega \left\{ \frac{k^\varepsilon(x)}{p_\varepsilon(x)} |\nabla u|^{p_\varepsilon(x)} - G^\varepsilon u \right\} dx \quad (120)$$

over $u \in L^\infty(0,T;W^{1,p_\varepsilon(\cdot)}(\Omega))$. Using the two-scale convergence arguments, we obtain the "lim sup"-inequality for the functional J^ε. The "lim inf"-inequality is obtained in two steps. First, we introduce the auxiliary functional

$$\begin{gathered} \widetilde{J}^\varepsilon[u] \stackrel{\text{def}}{=} \int_{\Delta_t} dt \int_\Omega \left\{ \frac{\widetilde{k}^\varepsilon(x)}{p_\varepsilon(x)} |\nabla u|^2 - G^\varepsilon u \right\} dx \\ \text{with } \widetilde{k}^\varepsilon(x) = \alpha(x)\, k_m \varepsilon^2\, \mathbf{1}^\varepsilon_m(x) + k_f\, \mathbf{1}^\varepsilon_f(x) \end{gathered} \quad (121)$$

and obtain the inequality

$$\varliminf_{\varepsilon \to 0} \widetilde{J}^\varepsilon[u^\varepsilon] \geq J_{\text{hom}}\left[u_f, u_m\right], \quad (122)$$

where

$$\begin{aligned} J_{\text{hom}}&\left[u_f, u_m\right] \\ &= \int_{\Delta_t} dt \int_\Omega \left\{ \frac{1}{2}\left(K^*\, \nabla_x u_f \cdot \nabla_x u_f\right) \right. \\ &\qquad \left. - |F| \left(g(t,x) - \omega_f \frac{\partial u_f}{\partial t} \right) u_f \right\} dx \\ &+ \int_{\Delta_t} dt \int_{\Omega\times M} \left\{ \alpha(x) \frac{k_m}{2} \left|\nabla_y u_m\right|^2 \right. \\ &\qquad \left. - \left(g(t,x) - \omega_m \frac{\partial u_m}{\partial t} \right) u_m \right\} dx\, dy. \end{aligned} \quad (123)$$

Then in the second step, we obtain the desired "lim inf"-inequality for the initial functional. This completes the proof of Theorem 30.

The macroscopic model corresponding to the second situation is given by the following convergence result.

Theorem 32. *Let* $u^\varepsilon = \langle p^\varepsilon, \sigma^\varepsilon \rangle$ *be the solution of the boundary value problem* (101) *and let conditions (A.4.1)–(A.4.3) be satisfied. Moreover, we assume that for any* $x \in \overline{\Omega}$,

$$\lim_{\varepsilon \to 0} \varepsilon^{-\gamma_\varepsilon(x)} = +\infty. \quad (124)$$

Then, for a.e. $t \in (0,T)$, u^ε *converges in* $L^2(\Omega)$, *as* $\varepsilon \to 0$, *to* $u^0 \in L^2(0,T;H^1(\Omega))$, *the solution of*

$$\begin{gathered} \left(\omega_f |F| + \omega_m |M|\right) \frac{\partial u^0}{\partial t} - \operatorname{div}_x \left(K^*\, \nabla_x u^0\right) = g(t,x) \quad \text{in } Q, \\ u^0 = 0 \quad \text{on } (0,T)\times\partial\Omega, \qquad u^0(0,x) = u_0(x) \quad \text{in } \Omega, \end{gathered} \quad (125)$$

where $K^* = \{k^*_{ij}\}$ *is the homogenized permeability tensor defined in* (113)-(114).

The proof of Theorem 32 is similar to that of Theorem 30.

Remark 33. Notice that the structure of the limit problem depends crucially on the rate of convergence of $(p_\varepsilon(\cdot) - 2)$ to zero. The critical rate of convergence is

$$\left(p_\varepsilon(\cdot) - 2\right) \sim \frac{1}{|\ln \varepsilon|}. \quad (126)$$

More precisely, if $\overline{\lim}_{\varepsilon\to 0} |\ln\varepsilon| (p_\varepsilon(x) - 2) < +\infty$, then the limit model is of a double porosity type. If $\lim_{\varepsilon\to 0} |\ln\varepsilon|\, (p_\varepsilon(x) - 2) = +\infty$, then in the limit we obtain a single porosity model.

The proofs of Theorems 30 and 32 are given in [21].

References

[1] S. N. Antontsev and S. I. Shmarev, "Elliptic equations with anisotropic nonlinearity and nonstandard growth conditions,"

in *Handbook of Differential Equations. Stationary Partial Differential Equations*, vol. 3, chapter 1, pp. 1–100, Elsevier, 2006.

[2] S. Antontsev and S. Shmarev, "Elliptic equations and systems with nonstandard growth conditions: existence, uniqueness and localization properties of solutions," *Nonlinear Analysis. Theory, Methods & Applications A*, vol. 65, no. 4, pp. 728–761, 2006.

[3] X.-L. Fan and Q.-H. Zhang, "Existence of solutions for $p(x)$-Laplacian Dirichlet problem," *Nonlinear Analysis. Theory, Methods & Applications A*, vol. 52, no. 8, pp. 1843–1852, 2003.

[4] H. Hudzik, "On generalized Orlicz-Sobolev space," *Functiones et Approximatio Commentarii Mathematici*, vol. 4, pp. 37–51, 1976.

[5] O. Kováčik and J. Rákosník, "On spaces $L^{p(x)}$ and $W^{k,p(x)}$," *Czechoslovak Mathematical Journal*, vol. 41, no. 4, pp. 592–618, 1991.

[6] E. Acerbi and G. Mingione, "Regularity results for a class of functionals with non-standard growth," *Archive for Rational Mechanics and Analysis*, vol. 156, no. 2, pp. 121–140, 2001.

[7] J. Musielak, *Orlicz Spaces and Modular Spaces*, vol. 1034 of *Lecture Notes in Mathematics*, Springer, Berlin, Germany, 1983.

[8] I. I. Šarapudinov, "The topology of the space $L^{p(t)}([0,1])$," *Matematicheskie Zametki*, vol. 26, pp. 613–632, 1979.

[9] A. Braides and A. Defranceschi, *Homogenization of Multiple Integrals*, vol. 12 of *Oxford Lecture Series in Mathematics and its Applications*, The Clarendon Press/Oxford University Press, 1998.

[10] G. Dal Maso, *An Introduction to Γ-Convergence*, Birkhäuser, Boston, Mass, USA, 1993.

[11] S. M. Kozlov, "Geometric aspects of averaging," *Russian Mathematical Surveys*, vol. 44, pp. 79–120, 1989.

[12] V. V. Zhikov, "Lavrentiev effect and the averaging of nonlinear variational problems," *Differential Equation*, vol. 27, pp. 32–39, 1991.

[13] V. V. Zhikov, "On passage to the limit in nonlinear variational problems," *Matematicheskiĭ Sbornik*, vol. 183, no. 8, pp. 47–84, 1992.

[14] V. Zhikov, "Lavrentiev phenomenon and homogenization for some variational problems," *Comptes Rendus de l'Académie des Sciences*, vol. 316, no. 5, pp. 435–439, 1993.

[15] V. V. Zhikov, "On Lavrentiev's phenomenon," *Russian Journal of Mathematical Physics*, vol. 3, no. 2, pp. 249–269, 1995.

[16] V. V. Jikov, S. M. Kozlov, and O. A. Oleĭnik, *Homogenization of Differential Operators and Integral Functionals*, Springer, New York, NY, USA, 1994.

[17] B. Amaziane, S. Antontsev, and L. Pankratov, "Homogenization of a class of nonlinear elliptic equations with nonstandard growth," *Comptes Rendus*, vol. 335, no. 3, pp. 138–143, 2007.

[18] B. Amaziane, S. Antontsev, L. Pankratov, and A. Piatnitski, "Γ-convergence and homogenization of functionals in Sobolev spaces with variable exponents," *Journal of Mathematical Analysis and Applications*, vol. 342, no. 2, pp. 1192–1202, 2008.

[19] B. Amaziane, S. Antontsev, L. Pankratov, and A. Piatnitski, "Homogenization of p-Laplacian in perforated domain," *Annales de l'Institut Henri Poincaré*, vol. 26, no. 6, pp. 2457–2479, 2009.

[20] B. Amaziane, L. Pankratov, and A. Piatnitski, "Homogenization of a class of quasilinear elliptic equations in high-contrast fissured media," *Proceedings of the Royal Society of Edinburgh A*, vol. 136, no. 6, pp. 1131–1155, 2006.

[21] B. Amaziane, L. Pankratov, and A. Piatnitski, "Nonlinear flow through double porosity media in variable exponent Sobolev spaces," *Nonlinear Analysis. Real World Applications*, vol. 10, no. 4, pp. 2521–2530, 2009.

[22] B. Amaziane, L. Pankratov, and A. Piatnitski, "Homogenization of variational functionals with nonstandard growth in perforated domains," *Networks and Heterogeneous Media*, vol. 5, no. 2, pp. 189–215, 2010.

[23] B. Amaziane, L. Pankratov, and V. Prytula, "Nonlocal effects in homogenization of $p_\varepsilon(x)$-Laplacian in perforated domains," *Nonlinear Analysis. Theory, Methods & Applications A*, vol. 71, no. 9, pp. 4078–4097, 2009.

[24] B. Amaziane, L. Pankratov, and V. V. Pritula, "Homogenization of $p_\varepsilon(x)$-Laplacian in perforated domains with a nonlocal transmission condition," *Comptes Rendus Mécanique*, vol. 337, pp. 173–178, 2009.

[25] C. Choquet and L. Pankratov, "Nonlinear double porosity models with non-standard growth," *Comptes Rendus Mécanique*, vol. 337, no. 9-10, pp. 659–666, 2009.

[26] C. Choquet and L. Pankratov, "Homogenization of a class of quasilinear elliptic equations with non-standard growth in high-contrast media," *Proceedings of the Royal Society of Edinburgh A*, vol. 140, no. 3, pp. 495–539, 2010.

[27] V. A. Marchenko and E. Y. Khruslov, *Homogenization of Partial Differential Equations*, Birkhäuser, Boston, Mass, USA, 2006.

[28] A. A. Kovalevskiĭ, "Conditions for Γ-convergence and homogenization of integral functionals with different domains of the definition," *Doklady Akademii Nauk Ukrainskoj SSR*, no. 4, pp. 5–8, 1991.

[29] G. Allaire, "Homogenization and two-scale convergence," *SIAM Journal on Mathematical Analysis*, vol. 23, no. 6, pp. 1482–1518, 1992.

[30] E. Acerbi, V. C. Piat, G. Dal Maso, and D. Percivale, "An extension theorem from connected sets, and homogenization in general periodic domains," *Nonlinear Analysis. Theory, Methods & Applications A*, vol. 18, no. 5, pp. 481–496, 1992.

[31] D. Cioranescu and J. S. J. Paulin, *Homogenization of Reticulated Structures*, vol. 136 of *Applied Mathematical Sciences*, Springer, New York, NY, USA, 1999.

[32] L. Pankratov and A. Piatnitski, "Nonlinear "double porosity" type model," *Comptes Rendus Mathématique*, vol. 334, no. 5, pp. 435–440, 2002.

[33] L. C. Evans, "A new proof of local $C^{1,\alpha}$ regularity for solutions of certain degenerate elliptic p.d.e," *Journal of Differential Equations*, vol. 45, no. 3, pp. 356–373, 1982.

[34] P. Tolksdorf, "Regularity for a more general class of quasilinear elliptic equations," *Journal of Differential Equations*, vol. 51, no. 1, pp. 126–150, 1984.

[35] S. Antontsev and S. Shmarev, "Parabolic equations with anisotropic nonstandard growth conditions," in *Free Boundary Problems. Theory and Applications*, vol. 154 of *International Series of Numerical Mathematics*, pp. 33–44, Birkhäuser, Basel, Switzerland, 2007.

10

Analysis of Caputo Impulsive Fractional Order Differential Equations with Applications

Lakshman Mahto,[1] Syed Abbas,[1] and Angelo Favini[2]

[1] *School of Basic Sciences, Indian Institute of Technology Mandi, Mandi, Himachal Pradesh 175001, India*
[2] *Dipartimento di Matematica, Universitá di Bologna, Pizza di Porta S. Donato 5, 40126 Bologna, Italy*

Correspondence should be addressed to Syed Abbas; sabbas.iitk@gmail.com

Academic Editor: Nikolai Leonenko

We use Sadovskii's fixed point method to investigate the existence and uniqueness of solutions of Caputo impulsive fractional differential equations of order $\alpha \in (0, 1)$ with one example of impulsive logistic model and few other examples as well. We also discuss Caputo impulsive fractional differential equations with finite delay. The results proven are new and compliment the existing one.

1. Introduction

Dynamics of many evolutionary processes from various fields such as population dynamics, control theory, physics, biology, and medicine. undergo abrupt changes at certain moments of time like earthquake, harvesting, shock, and so forth. These perturbations can be well approximated as instantaneous change of states or impulses. These processes are modeled by *impulsive differential equations*. In 1960, Milman and Myšhkis introduced impulsive differential equations in their paper [1]. Based on their work, several monographs have been published by many authors like Samoilenko and Perestyuk [2], Lakshmikantham et al. [3], Bainov and Simeonov [4, 5], Bainov and Covachev [6], and Benchohra et al. [7]. All the authors mentioned perviously have considered impulsive differential equations as ordinary differential equations coupled with impulsive effects. They considered the impulsive effects as difference equations being satisfied at impulses time. So, the solutions are piecewise continuous with discontinuities at impulses time. In the fields like biology, population dynamics, and so forth, problems with hereditary are best modeled by delay differential equations [8]. Problems associated with impulsive effects and hereditary property are modeled by impulsive delay differential equations.

The origin of fractional calculus (derivatives $(d^\alpha/dt^\alpha)f$ and integrals $I^\alpha f$ of arbitrary order $\alpha > 0$) goes back to Newton and Leibniz in the 17th century. In a letter correspondence with Leibniz, L'Hospital asked "What if the order of the derivative is $1/2$"? Leibniz replied, "Thus it follows that will be equal to $x\sqrt{dx : x}$, an apparent paradox, from which one day useful consequences will be drawn." This letter of Leibniz was in September 1695. So, 1695 is considered as the birthday of fractional calculus. Fractional order differential equations are generalizations of classical integer order differential equations and are increasingly used to model problems in fluid dynamics, finance, and other areas of application. Recent investigations have shown that sometimes physical systems can be modeled more accurately using fractional derivative formulations [9]. There are several excellent monographs available on this field [10–15]. In [11], the authors give a recent and up-to-date description of the developments of fractional differential and fractional integro-differential equations including applications. The existence and uniqueness of solutions to fractional differential equations has been considered by many authors [16–21]. Impulsive fractional differential equations represent a real framework for mathematical modeling to real world problems. Significant progress has been made in the theory of impulsive fractional differential equations [7, 22–24]. Xu et al. in their paper [25] have described an impulsive delay fishing model.

Fractional derivatives arise naturally in mathematical problems, for $\alpha > 0$ and a function $f : [0,T] \to \mathbb{R}$, recall [10, Definitions 3.1, 2.2]

(a) *the Caputo fractional derivative*

$$ {}^{C}D^{\alpha} f(t) = \frac{1}{\Gamma(1-\alpha)} \int_0^t (t-s)^{-\alpha} f'(s)\, ds, \tag{1}$$

(b) *the Riemann-Liouville fractional derivative*

$$D^{\alpha} f(t) = \frac{d^{\alpha}}{dt^{\alpha}} f(t) = \frac{1}{\Gamma(1-\alpha)} \frac{d}{dt} \int_0^t (t-s)^{-\alpha} f(s)\, ds, \tag{2}$$

provided that the right-hand sides exist pointwise on $[0,T]$ (Γ denotes the gamma function). Using the *Riemann-Liouville fractional integral* [10, Definition 2.1] $I_0^{\alpha} f(t) = (1/\Gamma(\alpha)) \int_0^t (t-s)^{\alpha-1} f(s)ds$, we have ${}^{C}D^{\alpha} f(t) = I_0^{1-\alpha}(d/dt) f(t)$ and $D^{\alpha} f(t) = (d/dt) I_0^{1-\alpha} f(t)$. $I_0^{\alpha} f$ exists, for instance, for all $\alpha > 0$, if $f \in C^0([0,T]) \cap L^1_{\text{loc}}([0,T])$; moreover, $I_0^{\alpha} f(0) = 0$. Throughout the paper, we assume that $I = [0,T]$.

One can see that both of the fractional derivatives are actually nonlocal operator because integral is a nonlocal operator. Moreover, calculating time fractional derivative of a function at some time requires all the past history, and hence fractional derivatives can be used for modeling systems with memory. Fractional differential equations can be formulated using both Caputo and Riemann-Liouville fractional derivatives. A *Riemann-Liouville initial value problem* can be stated as follows:

$$\begin{gathered} D^{\alpha} x(t) = f(t, x(t)), \quad t \in I = [0,T], \\ D^{\alpha-1} x(0) = x_0, \end{gathered} \tag{3}$$

or equivalently, $x(t) = x_0 t^{\alpha-1}/\Gamma(\alpha) + \int_0^t (t-s)^{\alpha-1} f(s, x(s)) ds$ in its integral representation [11, Theorem 3.24]. For a physical interpretation of the initial conditions in (3), see [26–28]. If derivatives of Caputo type are used instead of Riemann-Liouville type, then initial conditions for the corresponding Caputo fractional differential equations can be formulated as for classical ordinary differential equations, namely, $x(0) = x_0$.

Our main objective is to discuss existence and uniqueness of solutions of the following *impulsive fractional differential equation of Caputo type* in a Banach space X with norm $\|\cdot\|_X$:

$$\begin{gathered} {}^{C}D^{\alpha} x(t) = f(t, x(t)), \quad t \in I = [0,T],\ t \neq t_k, \\ \Delta x(t)|_{t=t_k} = I_k(x(t_k^-)), \quad k = 1, 2, \ldots, m, \\ x(0) = x_0, \end{gathered} \tag{4}$$

where $f \in C(I \times X, X)$, $I_k : X \to X$ and $x_0 \in X$. $0 = t_0 < t_1 < t_2 < \cdots < t_m < t_{m+1} = T$, $\Delta x(t)|_{t=t_k} = x(t_k^+) - x(t_K^-)$, $x(t_k^+) = \lim_{h\to 0} x(t+h)$, and $x(t_k^-) = \lim_{h\to 0} x(t-h)$. We break our function f into two components which satisfy different conditions. We observed that this kind of functions occurs in ecological modeling. We have given the example of logistic equation in the last section. Our main tool is Sadovskii's fixed point theorem.

Now, we define some important spaces and norm which will encounter frequently:

$$\begin{aligned} PC(I,X) = \Big\{ & x : [0,T] \longrightarrow X \mid x \in C([t_0,t_1], X) \\ & \cup C((t_k, t_{k+1}], X) \quad k = 1, 2, \ldots, m, \\ & x(t_k^+) \text{ and } x(t_k^-) \text{ exist, } x(t_k) = x(t_k^-) \Big\}, \end{aligned} \tag{5}$$

with sup-norm $\|\cdot\|$, defined by $\|x\| = \sup\{\|x(t)\|_X : t \in I\}$.

Definition 1. A solution of fractional differential equation (4) is a piecewise continuous function $x \in PC([0,T], X)$ which satisfied (4).

Definition 2 ([29], Definition 11.1). Kuratowskii noncompactness measure: let M be a bounded set in metric space (X,d), then Kuratowskii noncompactness measure, $\mu(M)$ is defined as $\inf\{\epsilon : M$ covered by a finite many sets such that the diameter of each set $\leq \epsilon\}$.

Definition 3 ([29], Definition 11.6). Condensing map: Let $\Phi : X \to X$ be a bounded and continuous operator on Banach space X such that $\mu(\Phi(B)) < \mu(B)$ for all bounded set $B \subset D(\Phi)$, where μ is the Kuratowskii noncompactness measure, then Φ is called condensing map.

Definition 4. Compact map: a map $f : X \to X$ is said to be compact if the image of every bounded subset of X under f is precompact (closure is compact).

Theorem 5 (see [30]). *Let B be a convex, bounded, and closed subset of a Banach space X and let $\Phi : B \to B$ be a condensing map. Then, Φ has a fixed point in B.*

Lemma 6 ([29], Example 11.7). *A map $\Phi = \Phi_1 + \Phi_2 : X \to X$ is k-contraction with $0 \leq k < 1$ if*

(a) Φ_1 is k-contraction, that is, $\|\Phi_1(x) - \Phi_1(y)\|_X \leq k\|x-y\|_X$;

(b) Φ_2 is compact,

and hence Φ is a condensing map.

The structure of the paper is as follows. In Section 2, we prove the existence (Theorem 8) and uniqueness of solutions to (4). We show in Section 3 the existence and uniqueness of solutions for a general class of impulsive functional differential equations of fractional order $\alpha \in (0,1)$. In Section 4, we give some examples in favor of our sufficient conditions.

2. Impulsive Fractional Differential Equation

Consider the initial value problem (4) on the cylinder $R = \{(t,x) \in \mathbb{R} \times X : t \in [0,T],\ x \in B(0,r)\}$ for some fixed $T > 0$, $r > 0$, and assume that there exist $p \in (0,\alpha)$, $\alpha \in (0,1)$, $M_1, M_2, L_1 \in L_{1/p}([0,T],\mathbb{R}^+)$ and functions $f_1, f_2 \in C(R \times X, X)$ such that $f = f_1 + f_2$, and the following assumptions are satisfied:

(A.1) f_1 is bounded and Lipschitz, in particular, $\|f_1(t,x)\|_X \le M_1(t)$ and $\|f_1(t,x) - f_1(t,y)\|_X \le L_1(t)\|x-y\|_X$ for all $(t,x), (t,y) \in R$,

(A.2) f_2 is compact and bounded, in particular, $\|f_2(t,x)\|_X \le M_2(t)$ for all $(t,x) \in R$,

(A.3) $I_k \in C(X,X)$ such that $\|I_k(x)\|_X \le l_1$ and $\|I_k(x) - I_k(y)\|_X \le l_2\|x-y\|_X$,

where $C(R \times X, X)$ denotes set of continuous functions from $R \times X$ to X and $L_{1/p}([0,T],\mathbb{R}^+)$ denotes space $1/p$-Lesbegue measurable functions from $[0,T]$ to $\mathbb{R}^+$ with norm $\|\cdot\|_{1/p}$.

Lemma 7 (Fečken et al. [24], Lemma 2). *The initial value problem* (4) *is equivalent to the nonlinear integral equation*

$$\begin{aligned} x(t) &= x_0 + \frac{1}{\Gamma(\alpha)}\int_0^t (t-s)^{\alpha-1} f_1(s,x(s))\,ds \\ &\quad + \frac{1}{\Gamma(\alpha)}\int_0^t (t-s)^{\alpha-1} f_2(s,x(s))\,ds, \quad t \in [0,t_1] \\ &= x_0 + I_1(x(t_1^-)) + \frac{1}{\Gamma(\alpha)}\int_0^t (t-s)^{\alpha-1} f_1(s,x(s))\,ds \\ &\quad + \frac{1}{\Gamma(\alpha)}\int_0^t (t-s)^{\alpha-1} f_2(s,x(s))\,ds, \quad t \in (t_1,t_2] \\ &= x_0 + \sum_{k=1}^{2} I_k(x(t_k^-)) + \frac{1}{\Gamma(\alpha)}\int_0^t (t-s)^{\alpha-1} f_1(s,x(s))\,ds \\ &\quad + \frac{1}{\Gamma(\alpha)}\int_0^t (t-s)^{\alpha-1} f_2(s,x(s))\,ds, \quad t \in (t_2,t_3] \\ &= x_0 + \sum_{k=1}^{m} I_k(x(t_k^-)) + \frac{1}{\Gamma(\alpha)}\int_0^t (t-s)^{\alpha-1} f_1(s,x(s))\,ds \\ &\quad + \frac{1}{\Gamma(\alpha)}\int_0^t (t-s)^{\alpha-1} f_2(s,x(s))\,ds, \quad t \in (t_m,T]. \end{aligned} \tag{6}$$

In other words, every solution of the integral equation (6) *is also solution of our original initial value problem* (4) *and conversely.*

Theorem 8 (existence of solutions). *Under the assumptions (A.1)–(A.3), the problem* (4) *has at least one solution in* $[0,T]$, *provided that*

$$\gamma_1 = ml_2 + \frac{c\|L_1\|_{1/p} T^{\alpha-p}}{\Gamma(\alpha+1)} < 1, \quad \text{where } c = \left(\frac{1-p}{\alpha-p}\right)^{1-p}. \tag{7}$$

Proof. Let B_λ be the closed bounded and convex subset of $PC([0,T],X)$, where B_λ is defined as $B_\lambda = \{x : \|x\| \le \lambda\}$, $\lambda = \max\{\lambda_0, \lambda_1, \ldots, \lambda_m\}$, and

$$\lambda_k = \|x_0\|_X + kl_1 + \frac{c\left(\|M_1\|_{1/p} + \|M_2\|_{1/p}\right)}{\Gamma(\alpha)} T^{\alpha-p}, \quad k = 0,1,2,\ldots,m. \tag{8}$$

Define a map $F : B_\lambda \to X$ such that

$$\begin{aligned} Fx(t) &= x_0 + \sum_{0<t_k<t} I_k(x(t_k^-)) \\ &\quad + \frac{1}{\Gamma(\alpha)}\int_0^t (t-s)^{\alpha-1} f_1(s,x(s))\,ds \\ &\quad + \frac{1}{\Gamma(\alpha)}\int_0^t (t-s)^{\alpha-1} f_2(s,x(s))\,ds. \end{aligned} \tag{9}$$

Let us consider that

$$\begin{aligned} F_1x(t) &= x_0 + \sum_{0<t_k<t} I_k(x(t_k^-)) \\ &\quad + \frac{1}{\Gamma(\alpha)}\int_0^t (t-s)^{\alpha-1} f_1(s,x(s))\,ds, \\ F_2x(t) &= \frac{1}{\Gamma(\alpha)}\int_0^t (t-s)^{\alpha-1} f_2(s,x(s))\,ds. \end{aligned} \tag{10}$$

Step 1 (F is self mapping). Now, we show that $F(B_r) \subset B_r$. For $t \in [0,t_1]$

$$\begin{aligned} \|Fx(t)\|_X &\le \|x_0\|_X + \frac{1}{\Gamma(\alpha)}\int_0^t (t-s)^{\alpha-1}\|f(s\cdot x(s))\|_X ds \\ &\le \|x_0\|_X + \frac{1}{\Gamma(\alpha)}\int_0^t (t-s)^{\alpha-1}\|f_1(s,x(s))\|_X ds \\ &\quad + \frac{1}{\Gamma(\alpha)}\int_0^t (t-s)^{\alpha-1}\|f_2(s,x(s))\|_X ds \\ &\le \|x_0\|_X + \frac{1}{\Gamma(\alpha)}\int_0^t (t-s)^{\alpha-1} M_1(s)\,ds \\ &\quad + \frac{1}{\Gamma(\alpha)}\int_0^t (t-s)^{\alpha-1} M_2(s)\,ds \\ &\le \|x_0\|_X + \frac{1}{\Gamma(\alpha)}\left(\int_0^t (t-s)^{(\alpha-1)/(1-p)} ds\right)^{1-p} \\ &\quad \times \left(\int_0^t M_1^{1/p}(s)\,ds\right)^p + \frac{1}{\Gamma(\alpha)} \\ &\quad \times \left(\int_0^t (t-s)^{(\alpha-1)/(1-p)} ds\right)^{1-p} \left(\int_0^t M_2^{1/p}(s)\,ds\right)^p \\ &\le \|x_0\|_X + \frac{c\left(\|M_1\|_{1/p} + \|M_2\|_{1/p}\right)}{\Gamma(\alpha)} T^{\alpha-p} = \lambda_0. \end{aligned} \tag{11}$$

For $t \in (t_1, t_2]$,

$$\begin{aligned}
\|Fx(t)\|_X &\le \|x_0\|_X + |I_1(x(t_1^-))| \\
&\quad + \frac{1}{\Gamma(\alpha)} \int_0^t (t-s)^{\alpha-1} \|f(s \cdot x(s))\|_X ds \\
&\le \|x_0\|_X + l_1 + \frac{1}{\Gamma(\alpha)} \int_0^t (t-s)^{\alpha-1} \|f_1(s, x(s))\|_X ds \\
&\quad + \frac{1}{\Gamma(\alpha)} \int_0^t (t-s)^{\alpha-1} \|f_2(s, x(s))\|_X ds \\
&\le \|x_0\|_X + l_1 + \frac{1}{\Gamma(\alpha)} \int_0^t (t-s)^{\alpha-1} M_1(s)\, ds \\
&\quad + \frac{1}{\Gamma(\alpha)} \int_0^t (t-s)^{\alpha-1} M_2(s)\, ds \\
&\le \|x_0\|_X + l_1 + \frac{1}{\Gamma(\alpha)} \\
&\quad \times \left(\int_0^t (t-s)^{(\alpha-1)/(1-p)} ds \right)^{1-p} \left(\int_0^t M_1^{1/p}(s)\, ds \right)^p \\
&\quad + \frac{1}{\Gamma(\alpha)} \left(\int_0^t (t-s)^{(\alpha-1)/(1-p)} ds \right)^{1-p} \\
&\quad \times \left(\int_0^t M_2^{1/p}(s)\, ds \right)^p \\
&\le \|x_0\|_X + l_1 + \frac{c\left(\|M_1\|_{1/p} + \|M_2\|_{1/p} \right)}{\Gamma(\alpha)} T^{\alpha-p} \\
&= \lambda_1.
\end{aligned} \tag{12}$$

For $t \in (t_2, t_3]$,

$$\begin{aligned}
\|Fx(t)\| &\le \|x_0\|_X + \sum_{k=0}^{2} |I_k(x(t_k^-))| \\
&\quad + \frac{1}{\Gamma(\alpha)} \int_0^t (t-s)^{\alpha-1} \|f(s \cdot x(s))\|_X ds \\
&\le \|x_0\|_X + 2l_1 + \frac{1}{\Gamma(\alpha)} \int_0^t (t-s)^{\alpha-1} \|f_1(s, x(s))\|_X ds \\
&\quad + \frac{1}{\Gamma(\alpha)} \int_0^t (t-s)^{\alpha-1} \, |f_2(s, x(s))|\, ds \\
&\le \|x_0\|_X + 2l_1 + \frac{1}{\Gamma(\alpha)} \int_0^t (t-s)^{\alpha-1} M_1(s)\, ds \\
&\quad + \frac{1}{\Gamma(\alpha)} \int_0^t (t-s)^{\alpha-1} M_2(s)\, ds \\
&\le \|x_0\|_X + 2l_1 + \frac{1}{\Gamma(\alpha)} \left(\int_0^t (t-s)^{(\alpha-1)/(1-p)} ds \right)^{1-p} \\
&\quad \times \left(\int_0^t M_1^{1/p}(s)\, ds \right)^p + \frac{1}{\Gamma(\alpha)} \\
&\quad \times \left(\int_0^t (t-s)^{(\alpha-1)/(1-p)} ds \right)^{1-p} \left(\int_0^t M_2^{1/p}(s)\, ds \right)^p \\
&\le \|x_0\|_X + 2l_1 + \frac{c\left(\|M_1\|_{1/p} + \|M_2\|_{1/p} \right)}{\Gamma(\alpha)} T^{\alpha-p} \\
&= \lambda_2.
\end{aligned} \tag{13}$$

For $t \in (t_m, T]$,

$$\begin{aligned}
\|Fx(t)\|_X &\le \|x_0\|_X + \sum_{k=0}^{m} |I_k(x(t_k^-))| \\
&\quad + \frac{1}{\Gamma(\alpha)} \int_0^t (t-s)^{\alpha-1} \|f(s \cdot x(s))\|_X ds \\
&\le \|x_0\|_X + ml_1 \\
&\quad + \frac{1}{\Gamma(\alpha)} \int_0^t (t-s)^{\alpha-1} \|f_1(s, x(s))\|_X ds \\
&\quad + \frac{1}{\Gamma(\alpha)} \int_0^t (t-s)^{\alpha-1} \, \|f_2(s, x(s))\|\, ds \\
&\le \|x_0\|_X + ml_1 + \frac{1}{\Gamma(\alpha)} \int_0^t (t-s)^{\alpha-1} M_1(s)\, ds \\
&\quad + \frac{1}{\Gamma(\alpha)} \int_0^t (t-s)^{\alpha-1} M_2(s)\, ds \\
&\le \|x_0\|_X + ml_1 \\
&\quad + \frac{1}{\Gamma(\alpha)} \left(\int_0^t (t-s)^{(\alpha-1)/(1-p)} ds \right)^{1-p} \\
&\quad \times \left(\int_0^t M_1^{1/p}(s)\, ds \right)^p + \frac{1}{\Gamma(\alpha)} \\
&\quad \times \left(\int_0^t (t-s)^{(\alpha-1)/(1-p)} ds \right)^{1-p} \left(\int_0^t M_2^{1/p}(s)\, ds \right)^p \\
&\le \|x_0\|_X + ml_1 + \frac{c\left(\|M_1\|_{1/p} + \|M_2\|_{1/p} \right)}{\Gamma(\alpha)} T^{\alpha-p} \\
&= \lambda_m,
\end{aligned} \tag{14}$$

and thus $F(B_\lambda) \subset B_\lambda$.

Step 2 (F_1 is continuous and γ-contraction). To prove the continuity of F_1 for $t \in [0, T]$, let us consider a sequence x_n converging to x. Taking the norm of $F_1 x_n(t) - F_1 x(t)$, we have

$$\begin{aligned}
&\|F_1 x_n(t) - F_1 x(t)\|_X \\
&\quad \le \sum_{0<t_k<t} \|I_1(x_n(t_k^-) - x(t_k^-))\|_X \\
&\quad\quad + \frac{1}{\Gamma\alpha} \int_0^t (t-s)^{\alpha-1} L(s)\, \|x_n(s) - x(s)\|_X ds \\
&\quad \le \sum_{0<t_k<t} l_2 \|(x_n(t_k^-) - x(t_k^-))\|_X
\end{aligned}$$

$$+\frac{1}{\Gamma\alpha}\left(\int_0^t (t-s)^{(\alpha-1)/(1-p)} ds\right)^{1-p}$$
$$\times\left(\int_0^t L_1^{1/p}(s)\,ds\right)^p \|x_n - x\|$$
$$\leq\left(ml_2 + \frac{c\|L_1\|_{1/p}T^{\alpha-p}}{\Gamma\alpha}\right)\|x_n - x\|. \tag{15}$$

From the pervoius analysis, we obtain

$$\|F_1x_n - F_1x\| \leq \left(ml_2 + \frac{c\|L_1\|_{1/p}T^{\alpha-p}}{\Gamma(\alpha)}\right)\|x_n - x\|. \tag{16}$$

To prove that F_1 is γ_1-contraction, let us consider that, for $x, y \in B_r$,

$$\|F_1x(t) - F_1y(t)\|_X$$
$$\leq \sum_{0<t_k<t} \|I_k(x(t_k^-) - y(t_k^-))\|_X$$
$$+\frac{1}{\Gamma(\alpha)}\int_0^t (t-s)^{\alpha-1}L_1(s)\|x(s) - y(s)\|_X ds$$
$$\leq \sum_{0<t_k<t} l_2\|(x(t_k^-) - y(t_k^-)\|_X$$
$$+\frac{1}{\Gamma(\alpha)}\left(\int_0^t (t-s)^{(\alpha-1)/(1-p)} ds\right)^{1-p} \tag{17}$$
$$\times\left(\int_0^t L_1^{1/p}(s)\,ds\right)^p \|x - y\|$$
$$\leq\left(ml_2 + \frac{c\|L_1\|_{1/p}T^{\alpha-p}}{\Gamma(\alpha)}\right)\|x - y\|.$$

Thus, for

$$\gamma_1 = \left(ml_2 + \frac{c\|L_1\|_{1/p}T^{\alpha-p}}{\Gamma(\alpha)}\right) < 1, \tag{18}$$

F_1 is γ_1-contraction.

Step 3 (F_2 is compact). For $0 \leq \tau_1 \leq \tau_2 \leq T$, we have

$$\|F_2x(\tau_2) - F_2x(\tau_1)\|_X$$
$$\leq \frac{1}{\Gamma(\alpha)}\left\|\int_0^{\tau_2}(\tau_2-s)^{\alpha-1}f_2(s,x(s))\,ds - \int_0^{\tau_1}(\tau_1-s)^{\alpha-1}f_2(s,x(s))\,ds\right\|_X$$
$$\leq \frac{1}{\Gamma(\alpha)}\left\|\int_0^{\tau_1}(\tau_2-s)^{\alpha-1}f_2(s,x(s))\,ds + \int_{\tau_1}^{\tau_2}(\tau_2-s)^{\alpha-1}f_2(s,x(s))\,ds - \int_0^{\tau_1}(\tau_1-s)^{\alpha-1}f_2(s,x(s))\,ds\right\|_X$$
$$\leq \frac{1}{\Gamma(\alpha)}\int_0^{\tau_1}\left((\tau_1-s)^{\alpha-1} - (\tau_2-s)^{\alpha-1}\right) \times\|f_2(s,x(s))\|_X ds$$
$$+\frac{1}{\Gamma(\alpha)}\int_{\tau_1}^{\tau_2}(\tau_2-s)^{\alpha-1}\|f_2(s,x(s))\|_X ds$$

($(\tau-s)^{\alpha-1}$ is a decreasing function of $\tau - s$.)

$$\leq \frac{1}{\Gamma(\alpha)}\int_0^{\tau_1}\left((\tau_1-s)^{\alpha-1} - (\tau_2-s)^{\alpha-1}\right)M_2(s)\,ds$$
$$+\frac{1}{\Gamma(\alpha)}\int_{\tau_1}^{\tau_2}(\tau_2-s)^{\alpha-1}M_2(s)\,ds$$
$$\leq \frac{1}{\Gamma(\alpha)}\left(\left(\int_0^{\tau_1}\left((\tau_1-s)^{\alpha-1} - (\tau_2-s)^{\alpha-1}\right)^{1/(1-p)}ds\right)^{1-p}\right)$$
$$\times\left(\int_0^{\tau_1}M_2^{1/p}(s)\,ds\right)^p + \frac{1}{\Gamma(\alpha)}$$
$$\times\left(\int_{\tau_1}^{\tau_2}(\tau_2-s)^{(\alpha-1)/(1-p)}ds\right)^{1-p}\left(\int_0^{\tau_1}M_2^{1/p}(s)\,ds\right)^p$$
$$\leq \frac{1}{\Gamma(\alpha)}\left(\left(\int_0^{\tau_1}\left((\tau_1-s)^{(\alpha-1)/(1-p)} - (\tau_2-s)^{(\alpha-1)/(1-p)}\right)ds\right)^{1-p}\right)$$
$$\times\left(\int_0^{\tau_1}M_2^{1/p}(s)\,ds\right)^p$$

($x^z - y^z \geq (x-y)^z$, $\forall x \geq y \geq 0$, $z > 1$.)

$$+\frac{1}{\Gamma(\alpha)}\left(\int_{\tau_1}^{\tau_2}(\tau_2-s)^{(\alpha-1)/(1-p)}ds\right)^{1-p}$$
$$\times\left(\int_0^{\tau_1}M_2^{1/p}(s)\,ds\right)^p$$
$$\leq \frac{c}{\Gamma(\alpha)}\left({\tau_1}^{(\alpha-p)/(1-p)} - {\tau_2}^{(\alpha-p)/(1-p)} + (\tau_2-\tau_1)^{(\alpha-p)/(1-p)}\right)^{1-p}$$
$$\times\|M_2\|_{1/p} + \frac{c}{\Gamma(\alpha)}(\tau_2-\tau_1)^{\alpha-p}\|M_2\|_{1/p}$$
$$\leq \frac{c}{\Gamma(\alpha)}\left((\tau_1-\tau_2)^{(\alpha-p)/(1-p)}\right)^{1-p}\|M_2\|_{1/p}$$
$$+\frac{c}{\Gamma(\alpha)}(\tau_2-\tau_1)^{\alpha-p}\|M_2\|_{1/p} \quad (\tau_2 > \tau_1.)$$
$$\leq \frac{2c\|M_2\|_{1/p}}{\Gamma(\alpha)}(\tau_2-\tau_1)^{\alpha-p}. \tag{19}$$

The right-hand side of the pervoiusly expression does not depend on x. Thus using Arzela-Ascoli theorem for equicontinuous functions (Diethelm, Theorem D.10 [10]), we conclude that $F_2(B_r)$ is relatively compact, and hence F_2 is completely continuous on $I - \{t_1, t_2, \ldots, t_m\}$. In a similar way, it can be prove the equi-continuity of F on $t = t_k^-$ and $t = t_k^+$, $k = 1, 2, \ldots, m$. And thus F_2 is compact on $[0, T]$.

Step 4 (F is condensing). As $F = F_1 + F_2$, F_1 is continuous contraction, F_2 is compact, so by using Lemma 6, F is condensing map on B_r.

And hence by using the Theorem 5, we conclude that (4) has a solution in B_r.

Theorem 9. *If f is bounded and Lipschitz, in particular, $\|f(t,x) - f(t,y)\|_X \le L_1^*(t)\|x - y\|_X$ for all $(t,x), (t,y) \in R$ and $L_1^* \in L_{1/p}([0,T], \mathbb{R}^+)$, then the problem (4) has a unique solution in B_λ, provided that*

$$\gamma_1^* = ml_2 + \frac{c\|L_1^*\|(1/p)T^{\alpha-p}}{\Gamma(\alpha+1)} < 1, \tag{20}$$

where $c = ((1-p)/(\alpha-p))^{1-p}$.

3. Impulsive Fractional Differential Equations with Finite Delay

In this section, we discuss existence and uniqueness of solutions of the following *impulsive fractional differential equations of Caputo type with finite delay* in a Banach space X with norm $|\cdot|$

$$\begin{aligned} {}^C D^\alpha x(t) &= f(t, x_t), \quad t \in I = [0,T], \ t \neq t_k, \\ \Delta x(t)|_{t=t_k} &= I_k(x(t_k^-)), \quad k = 1, 2, \ldots, m, \\ x(t) &= \phi(t), \quad t \in [-r, 0], \end{aligned} \tag{21}$$

where $f : I \times \mathscr{C} \to X$, $\mathscr{C} = C([-r,0], X)$ with norm $\|x\|_r = \sup\{\|x(t)\|_X : t \in [-r,0]\}$. $I_k \in C(X,X)$, $(k = 1, 2, \ldots, m)$, and X is a Banach space with a norm $|\cdot|_X$. For any $x : [-r,T] \to X$ and $t \in I$, $x_t \in \mathscr{C}$ and is defined by $x_t(s) = x(t+s)$, $s \in [-r,0]$. Here, our tools are Banach and Schaefer fixed point theorems.

Define a new Banach space $PC([-r,T], X)$

$$\begin{aligned} PC([-r,T], X) = \{x : [-r,T] \longrightarrow X \mid\ & x \in C((t_k, t_{k+1}], X) \\ & \cup C([-r,0], X),\ k = 0, 1, 2, \ldots, m, \\ & x(t_k^+) \text{ and } x(t_k^-) \text{ exist}, \\ & x(t_k) = x(t_k^-)\} \end{aligned} \tag{22}$$

with sup-norm $\|\cdot\|$, defined by $\|x\| = \sup\{\|x(t)\|_X : t \in [-r,T]\}$.

Definition 10. A solution of fractional differential equation (21) is a piecewise continuous function $x \in PC([-r,T], X)$ which satisfies (21).

Consider the initial value problem (21) on $I \times \mathscr{C}$ for some fixed $T > 0$ and assume that there exist $p \in (0,\alpha)$, $\alpha \in (0,1)$, $M_3, M_4, L_2 \in L_{1/p}([0,T], \mathbb{R}^+)$ such that the following assumptions are satisfied:

(A.4) $f \in C(I \times \mathscr{C}, X)$,

(A.5) f bounded, in particular, $\|f(t,\phi)\|_X \le M_3(t)$ for all $(t,\phi) \in I \times \mathscr{C}$,

(A.6) f is Lipschitz, in particular, $\|f(t,\phi) - f(t,\psi)\|_X \le L_2(t)\,\|\phi - \psi\|_r$ for all $(t,\phi), (t,\psi) \in I \times \mathscr{C}$.

Lemma 11 (Fečken et al. [24], Lemma 2). *The initial value problem (21) is equivalent to the nonlinear integral equation*

$$\begin{aligned} x(t) &= \phi(0) + \frac{1}{\Gamma(\alpha)} \int_0^t (t-s)^{\alpha-1} f(s, x_s)\, ds, \quad t \in [0, t_1] \\ &= \phi(0) + I_1(x(t_1^-)) + \frac{1}{\Gamma(\alpha)} \int_0^t (t-s)^{\alpha-1} f(s, x_s)\, ds, \quad t \in (t_1, t_2] \\ &= \phi(0) + \sum_{k=1}^{2} I_k(x(t_k^-)) + \frac{1}{\Gamma(\alpha)} \int_0^t (t-s)^{\alpha-1} f(s, x_s)\, ds, \quad t \in (t_2, t_3] \\ &= \phi(0) + \sum_{k=1}^{m} I_k(x(t_k^-)) + \frac{1}{\Gamma(\alpha)} \int_0^t (t-s)^{\alpha-1} f(s, x_s)\, ds, \quad t \in (t_m, T] \\ &= \phi(t), \quad t \in [-r, 0]. \end{aligned} \tag{23}$$

In other words, every solution of the integral equation (23) is also solution of our original initial value problem (21) and conversely.

Remark 12. Since history part/initial condition $x(t) = \phi(t)$, $t \in [-r,0]$, is known, so we will investigate the existence and uniqueness of solution in $I = [0,T]$.

Theorem 13 (existence and uniqueness of solution). *Under the assumptions (A.3)–(A.6), the problem (21) has a unique solution in $[0,T]$, provided that*

$$\gamma_2 = ml_2 + \frac{c\|L_2\|_{1/p} T^{\alpha-p}}{\Gamma(\alpha)} < 1, \tag{24}$$

where $c = ((1-p)/(\alpha-p))^{1-p}$.

Proof. In this case, we define the operator $F : PC(I,X) \to PC(I,X)$ by

$$Fx(t) = \sum_{0<t_k<t}^{m} I_k(x_n(t_k^-)) + \frac{1}{\Gamma(\alpha)} \int_0^t (t-s)^{\alpha-1} f(s, x_s)\, ds. \tag{25}$$

Step 1. To prove that F is self mapping, we need to prove that for each $x \in PC(I, X)$, $F_x \in PC(I, X)$.

The proof is similar to the proof of continuity of F_1 in Step 2 of Theorem 8, and hence we omit it.

Step 2 (F is continuous and γ_2-contraction). The proof of this step is also similar to the proof of continuous and γ_1-contraction of F_1 in Step 2 of Theorem 8.

Now by applying Banach's fixed point theorem, we get that the operator F has an unique fixed point in $PC(I, X)$, and hence the problem (21) has a unique solution in $PC([-r, T], X)$.

Our next result is based on Schaefer's fixed point theorem. In this case, we replace assumption (A.3) with the following linear growth condition:

(A.3′) I_k bounded, in particular, $\|I_k(x)\|_X \le l_1^*$,

(A.5′) f bounded, in particular, $\|f(t,\phi)\|_X \le M_4(t)(1 + \|\phi\|_r)$ for all $(t,\phi) \in I \times \mathscr{C}$.

Theorem 14. *Under the assumptions (A.3′) and (A.5′), problem (21) has at least one solution.*

Proof. We transform the problem into a fixed point problem. For this purpose, consider the operator $F : PC(I, X) \to PC(I, X)$ defined by

$$Fx(t) = \sum_{0<t_k<t}^{m} I_k(x_n(t_k^-)) + \frac{1}{\Gamma(\alpha)}\int_0^t (t-s)^{\alpha-1} f(s, x_s)\, ds. \tag{26}$$

Step 1 (F is continuous). Let $\{x^n\}$ be a sequence such that $x^n \to x$ in $PC(I, X)$. Then, for every $t \in I$, we have

$$\begin{aligned}
&\|Fx^n(t) - Fx(t)\|_X \\
&\quad\le \sum_{0<t_k<t} \|I_k(x^n(t_k^-)) - I_k(x(t_k^-))\| \\
&\qquad + \frac{1}{\Gamma\alpha}\int_0^t (t-s)^{\alpha-1}\|f(s,x_s^n) - f(s,x_s)\|_X ds \\
&\quad\le \sum_{0<t_k<t} \|I_k(x^n(t_k^-)) - I_k(x(t_k^-))\|_X \\
&\qquad + \|f(\cdot,x^n_\cdot) - f(\cdot,x_\cdot)\| \frac{1}{\Gamma\alpha}\left(\int_0^t (t-s)^{\alpha-1}\right)^{1-p} ds \\
&\quad\le \sum_{k=1}^{m} \|I_k(x^n(\cdot)) - I_k(x(\cdot))\| \\
&\qquad + \frac{T^\alpha}{\Gamma(\alpha+1)}\|f(\cdot,x^n_\cdot) - f(\cdot,x_\cdot)\|.
\end{aligned} \tag{27}$$

We can see that if $n \to \infty$, $Fx^n \to Fx$, as $x^n \to x$ and I_k and both f are continuous. Hence, F is continuous.

Step 2 (F maps bounded sets into bounded sets in $PC(I, X)$). It is enough to show that for any $\delta > 0$, there exists an $l = \|\phi(0)\|_r + ml_1^* + (c(1+\delta)(\|M_4\|_{1/p})/\Gamma(\alpha))T^{\alpha-p} > 0$ such that $x \in B_\delta = \{x \in PC(I, X) \mid \|x\| \le \delta\}$, and we have $\|Fx\| \le l$.

For $t \in [0, T]$, we have

$$\begin{aligned}
\|Fx(t)\|_X &\le \|\phi(0)\|_r + \sum_{0<t_k<t} |I_k(x(t_k^-))| \\
&\quad + \frac{1}{\Gamma(\alpha)}\int_0^t (t-s)^{\alpha-1}\|f(s,x_s)\|_X ds \\
&\le \|\phi(0)\|_r + ml_1^* \\
&\quad + \frac{1}{\Gamma(\alpha)}\int_0^t (t-s)^{\alpha-1} M_4(s)(1 + \|x_s\|_r)\, ds \\
&\le \|\phi(0)\|_r + \frac{(1+\|x\|)}{\Gamma(\alpha)}\int_0^t (t-s)^{\alpha-1} M_4(s)\, ds \\
&\le \|\phi(0)\|_r + ml_1^* + \frac{(1+\|x\|)}{\Gamma(\alpha)} \\
&\quad \times\left(\int_0^t (t-s)^{(\alpha-1)/(1-p)} ds\right)^{1-p}\left(\int_0^t M_4^{1/p}(s)\, ds\right)^p \\
&\le \|\phi(0)\|_r + ml_1^* + \frac{c(1+\delta)\left(\|M_4\|_{1/p}\right)}{\Gamma(\alpha)}T^{\alpha-p}.
\end{aligned} \tag{28}$$

Step 3 (F maps bounded sets into equicontinuous sets in $PC(I, X)$). The proof of this step is similar to the proof of compactness of F_2 in Step 3 of Theorem 8.

As a consequence of Steps 1–3 together with PC-type Arzela-Ascoli theorem (Fečken et al. [24], Theorem 2.11), the map $F : PC(I, X) \to PC(I, X)$ is completely continuous.

Step 4 (A priori bounds). Now, we prove that the set

$E(F) = \{x \in PC(I, X) \mid x = \lambda Fx$, for some $\lambda \in (0,1)\}$ is bounded.

We observe that for $t \in [0, T]$ and $x \in E(F)$, $x(t) = \lambda Fx(t)$ and

$$\begin{aligned}
\|x(t)\|_X &\le \|Fx(t)\|_X \\
&\le \|\phi(0)\|_r + \sum_{0<t_k<t} \|I_k(x(t_k^-))\|_X \\
&\quad + \frac{1}{\Gamma(\alpha)}\int_0^t (t-s)^{\alpha-1}\|f(s,x_s)\|_X ds \\
&\le \|\phi(0)\|_r + ml_1^* \\
&\quad + \frac{1}{\Gamma(\alpha)}\int_0^t (t-s)^{\alpha-1} M_4(s)(1+\|x_s\|)\, ds \\
&\le \|\phi(0)\|_r + ml_1^* + \frac{(1+\|x\|)}{\Gamma(\alpha)}
\end{aligned}$$

$$\times\left(\int_0^t (t-s)^{(\alpha-1)/(1-p)} ds\right)^{1-p}\left(\int_0^t M_4^{1/p}(s)\, ds\right)^p$$
$$\leq \|\phi(0)\|_r + ml_1^* + \frac{c(1+\|x\|)\left(\|M_4\|_{1/p}\right)}{\Gamma(\alpha)} T^{\alpha-p}, \tag{29}$$

and hence

$$\|x\| \leq \frac{\|\phi(0)\|_r + ml_1^* + \left(c\left(\|M_4\|_{1/p}\right)/\Gamma(\alpha)\right)T^{\alpha-p}}{1-\left(c\left(\|M_4\|_{1/p}\right)/\Gamma(\alpha)\right)T^{\alpha-p}}. \tag{30}$$

This shows that $E(F)$ is bounded.

As a consequence of Schaefer's fixed point theorem, the problem (21) has at least one solution in $PC([-r,T],X)$.

Theorem 15. *If f is bounded and Lipschitz, in particular, $\|f(t,\phi)-f(t,\psi)\|_X \leq L_2^*(t)\ \|\phi-\psi\|_r$, for all $(t,\phi),(t,\psi) \in I\times\mathscr{C}$ and $L_2^* \in L_{1/p}([0,T],\mathbb{R}^+)$, then problem (21) has a unique solution in $PC([-r,T],X)$, provided that*

$$\gamma_2^* = ml_2 + \frac{c\|L_2^*\|_{1/p}T^{\alpha-p}}{\Gamma(\alpha+1)} < 1, \tag{31}$$

where $c = ((1-p)/(\alpha-p))^{1-p}$.

Further, we consider the following more general Caputo fractional differential equation

$$^{C}D^{\alpha}x(t) = f(t,x(t),x_t), \quad t\in[0,T],\ t\neq t_k,$$
$$\Delta x(t)|_{t=t_k} = I_k(x(t_k^-)), \quad k=1,2,\ldots,m, \tag{32}$$
$$x(t) = \phi(t), \quad t\in[-r,0],$$

where $f: I\times X\times\mathscr{C} \to X$ and $\mathscr{C} = C([-r,0],X)$ with norm $\|x\|_r = \sup\{\|x(t)\|_X : t\in[-r,0]\}$, $I_k \in C(X,X)$, $(k = 1,2,\ldots,m)$, and X is a separable real Banach space with the norm $\|\cdot\|_X$. Here, our tools will be Banach and Schaefer fixed point theorems.

Definition 16. A solution of fractional differential equation (32) is a piecewise continuous function $x \in PC([-r,T],X)$ which satisfies (32).

Consider the initial value problem (32) on $I\times\mathscr{C}\times X$ for some fixed $T>0$ and assume that there exist $p\in(0,\alpha)$, $M_5, M_6, L_3, L_4 \in L_{1/p}([0,T],\mathbb{R}^+)$ such that the following assumptions are satisfied:

(A.7) $f \in C(I\times X\times\mathscr{C},X)$,

(A.8) f bounded, in particular, $\|f(t,x,\phi)\|_X \leq M_5(t)$ for all $(t,x,\phi)\in I\times X\times\mathscr{C}$,

(A.9) f is Lipschitz, in particular, $\|f(t,x,\phi) - f(t,y,\psi)\|_X \leq L_3(t)\|x-y\|_X + L_4(t)\|\phi-\psi\|_r$ for all $(t,x,\phi),(t,y,\psi)\in I\times X\times\mathscr{C}$,

(A.8′) $\|f(t,x,\phi)\|_X \leq M_6(t)(1+\|x\|_X+\|\phi\|_r)$.

Lemma 17 (Fečken et al. [24], Lemma 2). *The initial value problem (32) is equivalent to the following nonlinear integral equation*

$$x(t) = \phi(0) + \frac{1}{\Gamma(\alpha)}\int_0^t (t-s)^{\alpha-1} f(s,x(s),x_s)\, ds, \quad t\in[0,t_1]$$
$$= \phi(0) + I_1(x(t_1^-)) + \frac{1}{\Gamma(\alpha)}\int_0^t (t-s)^{\alpha-1} f(s,x(s),x_s)\, ds, \quad t\in(t_1,t_2]$$
$$= \phi(0) + \sum_{k=1}^{2} I_k(x(t_k^-)) + \frac{1}{\Gamma(\alpha)}\int_0^t (t-s)^{\alpha-1} f(s,x(s),x_s)\, ds, \quad t\in(t_2,t_3]$$
$$= \phi(0) + \sum_{k=1}^{m} I_k(x(t_k^-)) + \frac{1}{\Gamma(\alpha)}\int_0^t (t-s)^{\alpha-1} f(s,x(s),x_s)\, ds, \quad t\in(t_m,T]$$
$$= \phi(t), \quad t\in[-r,0]. \tag{33}$$

In other words, every solution of the integral equation (33) is also solution of our original initial value problem (32) and conversely.

Theorem 18. *Under the assumptions (A.3), (A.7), (A.8), and (A.9), the problem (32) has a unique solution, provided that*

$$\gamma_3 = ml_2 + \frac{c\left(\|L_3\|_{1/p} + \|L_4\|_{1/p}\right)T^{\alpha-p}}{\Gamma(\alpha)} < 1. \tag{34}$$

Proof. The proof is similar to the proof of Theorem 13.

Theorem 19. *Under the assumptions (A.3′), and (A.8′), the problem (32) has at least one solution in $PC([-r,T],X)$.*

Proof. The proof is similar to the proof of Theorem 14.

Theorem 20. *If f is bounded and Lipschitz, in particular, $\|f(t,x,\phi)-f(t,y,\psi)\|_X \leq L_3^*(t)\|x-y\|_X + L_4^*(t)\|\phi-\psi\|_r$, for all $(t,x,\phi),(t,y,\psi)\in I\times X\times\mathscr{C}$ and $L_3^*, L_4^* \in L_{1/p}([0,T],\mathbb{R}^+)$, then the problem (32) has a unique solution in $PC([-r,T],X)$, provided that*

$$\gamma_3^* = ml_2 + \frac{c\left(\|L_3^*\|_{1/p} + \|L_2^*\|_{1/p}\right)T^{\alpha-p}}{\Gamma(\alpha+1)} < 1, \tag{35}$$

where $c = ((1-p)/(\alpha-p))^{1-p}$.

4. Examples

Example 21 (fractional impulsive logistic equation). Consider the following class of fractional logistic equations in banach space X with norm $\|\cdot\|_X$:

$$\begin{aligned} {}^{C}D^{\alpha}x(t) &= x(t)(a(t) - b(t)x(t)), \quad t \in [0,T], \ t \neq t_k, \\ \Delta x(t)|_{t=t_k} &= I_k(x(t_k^-)), \quad k = 1,2,\dots,m, \\ x(0) &= x_0, \end{aligned} \tag{36}$$

where $a(t) \in [a_*, a^*]$ and $b(t) \in [b_*, b^*]$ with $a_*, b_* > 0$.

Lemma 22 (Fečken et al. [24], Lemma 2). *The initial value problem (36) is equivalent to the nonlinear integral equation*

$$\begin{aligned} x(t) &= x_0 + \frac{1}{\Gamma(\alpha)}\int_0^t (t-s)^{\alpha-1}(x(s)(a(s) - b(s)x(s)))\,ds, \\ &\qquad t \in [0,t_1] \\ &= x_0 + I_1(x(t_1^-)) + \frac{1}{\Gamma(\alpha)}\int_0^t (t-s)^{\alpha-1}x(s)a(s)\,ds \\ &\quad - \frac{1}{\Gamma(\alpha)}\int_0^t (t-s)^{\alpha-1}b(s)x^2(s)\,ds, \quad t \in (t_1,t_2] \\ &= x_0 + \sum_{k=1}^{2} I_k(x(t_k^-)) + \frac{1}{\Gamma(\alpha)}\int_0^t (t-s)^{\alpha-1}x(s)a(s)\,ds \\ &\quad - \frac{1}{\Gamma(\alpha)}\int_0^t (t-s)^{\alpha-1}b(s)x^2(s)\,ds, \\ &\qquad t \in (t_2,t_3] \\ &= x_0 + \sum_{k=1}^{m} I_k(x(t_k^-)) + \frac{1}{\Gamma(\alpha)}\int_0^t (t-s)^{\alpha-1}x(s)a(s)\,ds \\ &\quad - \frac{1}{\Gamma(\alpha)}\int_0^t (t-s)^{\alpha-1}b(s)x^2(s)\,ds, \quad t \in [t_m,T]. \end{aligned} \tag{37}$$

In other words, every solution of the integral equation (37) is also solution of our original initial value problem (36) and conversely.

We can easily see that for the problem (36), our functions are $f_1(t,x) = a(t)x$ and $f_2(t,x) = -b(t)x^2$. It is not difficult to deduce that

$$\begin{aligned} \|f_1(t,x)\|_X &\le a^*\|x\|_X + ml_1, \\ \|f_1(t,x) - f_1(t,y)\|_X &\le a^*\|x-y\|_X. \end{aligned} \tag{38}$$

Also $\|f_2(t,x)\|_X \le b^*\|x\|^2$. From the integral representation of problem (36), we get

$$\|x(t)\|_X \le \|x_0\|_X + ml_1 + \frac{1}{\Gamma(\alpha)}\int_0^t (t-s)^{\alpha-1}a^*\|x(s)\|_X ds. \tag{39}$$

Using Gronwall's inequality (Diethelm, Lemma 6.19 [10, 31]), we get

$$\begin{aligned} \|x(t)\|_X &\le (\|x_0\|_X + ml_1)\exp\left(\frac{1}{\Gamma(\alpha)}\int_0^t (t-s)^{\alpha-1}a^*ds\right) \\ &\le (\|x_0\|_X + ml_1)\exp\left(\frac{a^*}{\Gamma(\alpha+1)}\right). \end{aligned} \tag{40}$$

Thus x is bounded which implies that all the assumptions of Theorem 8 are satisfied, and hence there exists a solution of the problem (36).

We give some more examples which are inspired by [24].

Example 23. Consider the following Caputo impulsive delay fractional differential equations

$$\begin{aligned} {}^{C}D^{\alpha}x(t) &= \frac{e^{-\nu t}\|x_t\|}{(1+e^t)(1+\|x_t\|)}, \quad t \in [0,1], \ t \neq t_1, \ \nu > 0, \\ \Delta x(t)|_{t=t_1} &= \frac{1}{2}, \\ x(t) &= \phi(t), \quad t \in [-r,0]. \end{aligned} \tag{41}$$

Set $C_1 = C([-r,0],\mathbb{R}^+)$, $f(t,\phi) = e^{-\nu t}\phi/((1+e^t)(1+\phi))$, $(t,\phi) \in [0,1] \times C_1$.

Let $\phi_1, \phi_2 \in C_1$ and let $t \in [0,1]$. Then, we have

$$\begin{aligned} |f(t,\phi_1) - f(t,\phi_2)| &= \frac{e^{-\nu t}}{1+e^t}\left|\frac{\phi_1}{1+\phi_1} - \frac{\phi_2}{1+\phi_2}\right| \\ &= \frac{e^{-\nu t}\|\phi_1 - \phi_2\|_r}{(1+e^t)(1+\phi_1)(1+\phi_2)} \\ &\le \frac{e^{-\nu t}\|\phi_1 - \phi_2\|_r}{(1+e^t)} \le L^*(t)\|\phi_1 - \phi_2\|_r, \end{aligned} \tag{42}$$

where $L^*(t) = e^{-\nu t}/2$.

Again, for all $\phi \in C_1$ and each $t \in [0,T]$,

$$\begin{aligned} |f(t,\phi)| &= \frac{e^{-\nu t}}{1+e^t}\left|\frac{\phi}{1+\phi}\right| \le \frac{e^{-\nu t}}{(1+e^t)} \\ &< m_1(t), \end{aligned} \tag{43}$$

where $m_1(t) = e^{-\nu t}/2$.

For $t \in [0,1]$ and some $p \in (0,\alpha)$, $L^*(t) = m_1(t) = e^{-\nu t}/2 \in L_{1/p}([0,1],\mathbb{R}^+)$ with $M_1^* = \|e^{-\nu t}/10\|_{1/p}$, we can arrive at the inequality $1/4 + cM_1^*/\Gamma(\alpha) < 1$. We can see that all the assumptions of Theorem 13 are satisfied, and hence the problem (41) has a unique solution in $[0,1]$.

Example 24. Consider the following Caputo impulsive delay fractional differential equation

$$\begin{aligned} {}^{C}D^{\alpha}x(t) &= \frac{\|x_t\|}{(1+e^t)(1+\|x_t\|)}, \quad t \in [0,1], \ t \neq t_1, \ \nu > 0, \\ \Delta x(t)|_{t=t_1} &= \frac{1}{2}, \\ x(t) &= \phi(t), \quad t \in [-r, 0]. \end{aligned} \tag{44}$$

Set $C_2 = C([0,1], \mathbb{R}^+)$, $f(t,\phi) = \phi/((1+e^t)(1+\phi))$, $(t,\phi) \in [-r,0] \times C_2$.

Again, for all $\phi \in C_2$ and each $t \in [0,T]$,

$$\begin{aligned} |f(t,\phi)| &= \frac{e^{-t}}{1+e^t}\left|\frac{\phi}{1+\phi}\right| \\ &\leq \frac{e^{-t}}{(1+e^t)} < m_2(t)(1+\|\phi\|), \end{aligned} \tag{45}$$

where $m_2(t) = e^{-t}/4$.

For $t \in [0,1]$ and some $p \in (0,\alpha)$, $m_2(t) = e^{-t}/4 \in L_{1/p}([0,1], \mathbb{R}^+)$ with $M_2^* = \|e^{-t}/4\|_{1/p}$, we can arrive at the inequality $1/4 + cM/\Gamma(\alpha) < 1$. We can see that all the assumptions of Theorem 14 are satisfied, and hence the problem (44) has a solution in $[0,1]$.

Acknowledgments

The authors are thankful to the anonymous reviewer for his/her valuable comments and suggestions. S. Abbas acknowledge "Erasmus Mundus Lot13, India 4EU" for providing him fellowship to visit University of Bologna, Italy. He also acknowledges Professor Stefan Siegmund, Center for Dynamics of TU Dresden, Germany, for his fruitful discussion on this topic and for hosting his short visit.

References

[1] V. D. Milman and A. D. Myškis, "On the stability of motion in the presence of impulses," *Siberial Mathematical Journal*, vol. 1, pp. 233–237, 1960.

[2] A. M. Samoilenko and N. A. Perestyuk, *Differential Equations With Impulses*, Viska Scola, Kiev, Ukraine, 1987.

[3] V. Lakshmikantham, D. D. Baĭnov, and P. S. Simeonov, *Theory of Impulsive Differential Equations*, vol. 6 of *Series in Modern Applied Mathematics*, World Scientific, Teaneck, NJ, USA, 1989.

[4] D. D. Bainov and P. S. Simeonov, *Systems with Impulsive Effects*, Horwood, Chichister, UK, 1989.

[5] D. D. Bainov and P. S. Simeonov, *Impulsive DiffErential Equations: Periodic Solutions and Its Applications*, Longman Scientific and Technical Group, Harlow, UK, 1993.

[6] D. Bainov and V. Covachev, *Impulsive Differential Equations with a Small Parameter*, vol. 24 of *Series on Advances in Mathematics for Applied Sciences*, World Scientific, River Edge, NJ, USA, 1994.

[7] M. Benchohra, J. Henderson, and S. K. Ntonyas, *Impulsive Diffrential Equations and Inclusions*, vol. 2, Hindawi Publishing Corporation, New York, NY, USA, 2006.

[8] J. Hale, *Theory of Functional Differential Equations*, Springer, New York, NY, USA, 2nd edition, 1977.

[9] F. Mainardi, "Fractional calculus: some basic problems in continuum and statistical mechanics," in *Fractals and Fractional Calculus in Continuum Mechanics*, vol. 378 of *CISM Courses and Lectures*, pp. 291–348, Springer, Vienna, Austria, 1997.

[10] K. Diethelm, *The Analysis of Fractional Differential Equations*, vol. 2004 of *Lecture Notes in Mathematics*, Springer, Berlin, Germany, 2010.

[11] A. A. Kilbas, H. M. Srivastava, and J. J. Trujillo, *Theory and Applications of Fractional Differential Equations*, vol. 204, North-Holland Mathematics Studies, 2006.

[12] K. S. Miller and B. Ross, *An Introduction to the Fractional Calculus and Fractional Differential Equations*, A Wiley-Interscience Publication, John Wiley & Sons, New York, NY, USA, 1993.

[13] I. Podlubny, *Fractional Differential Equations*, vol. 198 of *Mathematics in Science and Engineering*, Academic Press, San Diego, Calif, USA, 1999.

[14] J. Sabatier, O. P. Agrawal, and J. A. Tenreiro Machado, *Advances in Fractional Calculus*, Springer, Dordrecht, The Netherlands, 2007.

[15] S. G. Samko, A. A. Kilbas, and O. I. Marichev, *Fractional Integrals and Derivatives*, Gordon and Breach Science, Yverdon, Switzerland, 1993.

[16] S. Abbas, "Existence of solutions to fractional order ordinary and delay differential equations and applications," *Electronic Journal of Differential Equations*, vol. 2011, no. 9, pp. 1–11, 2011.

[17] S. Abbas, M. Banerjee, and S. Momani, "Dynamical analysis of fractional-order modified logistic model," *Computers & Mathematics with Applications*, vol. 62, no. 3, pp. 1098–1104, 2011.

[18] E. Ahmed, A. M. A. El-Sayed, and H. A. A. El-Saka, "Equilibrium points, stability and numerical solutions of fractional-order predator-prey and rabies models," *Journal of Mathematical Analysis and Applications*, vol. 325, no. 1, pp. 542–553, 2007.

[19] S. B. Hadid, "Local and global existence theorems on differential equations of non-integer order," *Journal of Fractional Calculus*, vol. 7, pp. 101–105, 1995.

[20] R. W. Ibrahim and S. Momani, "On the existence and uniqueness of solutions of a class of fractional differential equations," *Journal of Mathematical Analysis and Applications*, vol. 334, no. 1, pp. 1–10, 2007.

[21] V. Lakshmikantham and A. S. Vatsala, "Basic theory of fractional differential equations," *Nonlinear Analysis: Theory, Methods & Applications*, vol. 69, no. 8, pp. 2677–2682, 2008.

[22] R. P. Agarwal, M. Benchohra, and B. A. Slimani, "Existence results for differential equations with fractional order and impulses," *Georgian Academy of Sciences. A. Razmadze Mathematical Institute. Memoirs on Differential Equations and Mathematical Physics*, vol. 44, pp. 1–21, 2008.

[23] M. Benchohra and B. A. Slimani, "Existence and uniqueness of solutions to impulsive fractional differential equations," *Electronic Journal of Differential Equations*, vol. 2009, no. 10, pp. 1–11, 2009.

[24] M. Fečkan, Y. Zhou, and J. Wang, "On the concept and existence of solution for impulsive fractional differential equations," *Communications in Nonlinear Science and Numerical Simulation*, vol. 17, no. 7, pp. 3050–3060, 2012.

[25] D. Xu, Y. Hueng, and L. Ling, "Existence of positive solutions of an Impulsive Delay Fishing model," *Bulletin of Mathematical Analysis and Applications*, vol. 3, no. 2, pp. 89–94, 2011.

[26] C. Giannantoni, "The problem of the initial conditions and their physical meaning in linear differential equations of fractional order," *Applied Mathematics and Computation*, vol. 141, no. 1, pp. 87–102, 2003.

[27] N. Heymans and I. Podlubny, "Physical interpretation of initial conditions for fractional differential equations with Riemann-Liouville fractional derivatives," *Rheologica Acta*, vol. 45, no. 5, pp. 765–771, 2006.

[28] I. Podlubny, "Geometric and physical interpretation of fractional integration and fractional differentiation," *Fractional Calculus & Applied Analysis*, vol. 5, no. 4, pp. 367–386, 2002, Dedicated to the 60th anniversary of Prof. Francesco Mainard.

[29] E. Zeidler, *Non-Linear Functional Analysis and Its Application: Fixed Point-Theorems*, vol. 1, Springer, New York, NY, USA, 1986.

[30] B. N. Sadovskiĭ, "On a fixed point principle," *Functional Analysis and Its Applications*, vol. 1, no. 2, pp. 74–76, 1967.

[31] H. Ye, J. Gao, and Y. Ding, "A generalized Gronwall inequality and its application to a fractional differential equation," *Journal of Mathematical Analysis and Applications*, vol. 328, no. 2, pp. 1075–1081, 2007.

11

Picard Type Iterative Scheme with Initial Iterates in Reverse Order for a Class of Nonlinear Three Point BVPs

Mandeep Singh and Amit K. Verma

Department of Mathematics, BITS Pilani, Pilani, Rajasthan 333031, India

Correspondence should be addressed to Amit K. Verma; amitkverma02@yahoo.co.in

Academic Editor: Yuji Liu

We consider the following class of three point boundary value problem $y''(t)+f(t,y)=0, 0<t<1, y'(0)=0, y(1)=\delta y(\eta)$, where $\delta>0, 0<\eta<1$, the source term $f(t,y)$ is Lipschitz and continuous. We use monotone iterative technique in the presence of upper and lower solutions for both well-order and reverse order cases. Under some sufficient conditions, we prove some new existence results. We use examples and figures to demonstrate that monotone iterative method can efficiently be used for computation of solutions of nonlinear BVPs.

1. Introduction

In recent years, multipoint boundary value problems have been extensively studied by many authors ([1–5] and the references there in). Multipoint BVPs have lots of applications in various branches of science and engineering; for example, Webb [6] studied a second-order nonlinear boundary value problem subject to some nonlocal boundary conditions, which models a thermostat, and Zou et al. [7] studied the design of a large size bridge with multipoint supports.

It is well known that one of the most important tools for dealing with existence results for nonlinear problems is the method of upper and lower solutions. The method of upper and lower solutions has a long history and some of its ideas can be traced back to Picard [8]. Later, it was extensively studied by Dragoni [9].

Recently, there have been numerous results in the presence of an upper solution u_0 and a lower solution v_0 with $u_0 \geq v_0$. But, in many cases, the upper and lower solutions may occur in the reversed order also, that is, $u_0 \leq v_0$. Cabada et al. [10] considered the monotone iterative method for the following BVP:

$$y'' = f(t,y), \qquad y'(a) = 0 = y'(b) \tag{1}$$

with reversed ordered upper and lower solutions. So far, there have been some results in the presence of reverse ordered upper and lower solutions [10–13]. Xian et al. [14] considered the following second-order three point BVP:

$$\begin{gathered} y''(t) + f(t,y) = 0, \quad 0<t<1, \\ y(0) = 0, \qquad y(1) - \delta y(\eta) = 0, \end{gathered} \tag{2}$$

where $f(I\times R, R)$, $I=[0,1]$, $0<\eta<1$, $0<\delta<1$. They used the fixed point index theory with non-well-ordered upper and lower solutions.

Recently, Li et al. [15] studied the existence and uniqueness of solutions of second-order three point BVP

$$\begin{gathered} y'' + f(t,y) = 0, \quad 0<t<1, \\ y'(0) = 0, \qquad y(1) = \delta y(\eta), \quad 0<\eta<1,\ \delta>0 \end{gathered} \tag{3}$$

with upper and lower solutions in the reversed order via the monotone iterative method in Banach space.

The present work proves some new existing results for three point BVPs. Our technique is based on Picard-type iterative scheme and is quite simple and efficient from computational point of view. We believe that it can be very well adapted for this type of problem. In this paper we consider the following three point BVP:

$$\begin{gathered} y''(t) + f(t,y) = 0, \quad 0<t<1, \\ y'(0) = 0, \qquad y(1) = \delta y(\eta), \end{gathered} \tag{4}$$

where $f(I \times R, R)$, $I = [0,1]$, $0 < \eta < 1$, $\delta > 0$. We have allowed $\sup(\partial f/\partial y)$ to take both negative and positive values.

The paper is divided into 4 sections. In Section 2, we construct Green's function and establish maximum and anti-maximum principle. In Section 3, we generate monotone sequences by using results of Section 2 with upper and lower solutions as initial iterates ordered in one way or the other. We prove our final result of existence. In Section 4, we show that the monotone iterative scheme is a powerful technique. For that by using iterative scheme proposed in this paper we have computed the members of sequences in both cases (well-ordered and non-well-ordered case).

2. Preliminaries

2.1. Construction of Green's Function. To investigate (4), we consider the following linear three point BVP:

$$-y''(t) - \lambda y(t) = h(t), \quad 0 < t < 1, \qquad y'(0) = 0, \quad y(1) = \delta y(\eta) + b, \tag{5}$$

where $h \in C(I)$ and b is any constant. In this section, we construct the Green's function. We divide it into two cases.

Case I ($\lambda > 0$). Let us assume that

(H_0) $0 < \lambda \le \pi^2/4$, $\sin\sqrt{\lambda} - \delta \sin\sqrt{\lambda}\eta \ge 0$, $\delta\cos\sqrt{\lambda}\eta - \cos\sqrt{\lambda} > 0$.

It is easy to see that (H_0) can be satisfied.

Lemma 1. *Green's function for the following linear three point BVP*

$$y''(t) + \lambda y(t) = 0, \quad 0 < t < 1, \qquad y'(0) = 0, \quad y(1) = \delta y(\eta), \tag{6}$$

is given by

$$G(t,s) = \frac{1}{\sqrt{\lambda}\left(\delta\cos\sqrt{\lambda}\eta - \cos\sqrt{\lambda}\right)} \times \begin{cases} \left[\sin\sqrt{\lambda}(1-s) + \delta\sin\sqrt{\lambda}(s-\eta)\right]\cos\sqrt{\lambda}t, & 0 \le t \le s \le \eta, \\ \cos\sqrt{\lambda}s\sin\sqrt{\lambda}(1-t) + \delta\cos\sqrt{\lambda}s\sin\sqrt{\lambda}(t-\eta), & s \le t,\ s \le \eta, \\ \sin\sqrt{\lambda}(1-s)\cos\sqrt{\lambda}t, & t \le s,\ \eta \le s, \\ \cos\sqrt{\lambda}s\sin\sqrt{\lambda}(1-t) + \delta\cos\sqrt{\lambda}\eta\sin\sqrt{\lambda}(t-s), & \eta \le s \le t \le 1, \end{cases} \tag{7}$$

and if H_0 holds then $G(t,s) \ge 0$.

Proof. See proof of Lemma 2.1 in [15].

Lemma 2. *When $\lambda > 0$, $y \in C^2(I)$ is a solution of boundary value problem (5) and is given by*

$$y(t) = \frac{b\cos\sqrt{\lambda}t}{\cos\sqrt{\lambda} - \delta\cos\sqrt{\lambda}\eta} - \int_0^1 G(t,s)h(s)\,ds. \tag{8}$$

Proof. See proof of Lemma 2.3 in [15].

Remark 3. Particulary $y \in C^2(I)$ is a solution of the boundary value problem (5) if and only if $y \in C(I)$ is a solution of the integral equation

$$y(t) = \frac{b\cos\sqrt{\lambda}t}{\cos\sqrt{\lambda} - \delta\cos\sqrt{\lambda}\eta} - \int_0^1 G(t,s)h(s)\,ds. \tag{9}$$

Case II ($\lambda < 0$). Assume that

(H_0') $\lambda < 0$, $\delta\cosh\sqrt{|\lambda|}\eta - \cosh\sqrt{|\lambda|} < 0$, $\sinh\sqrt{|\lambda|} - \delta\sinh\sqrt{|\lambda|}\eta \ge 0$.

It is easy to see that (H_0') can be satisfied.

Lemma 4. *Green's function for the following three point BVP*

$$y''(t) + \lambda y(t) = 0, \quad 0 < t < 1, \qquad y'(0) = 0, \quad y(1) = \delta y(\eta), \tag{10}$$

for $\lambda < 0$ is given by

$$G(t,s) = \frac{1}{D_\lambda}\begin{cases} \left[\sinh\sqrt{|\lambda|}(1-s) + \delta\sinh\sqrt{|\lambda|}(s-\eta)\right]\cosh\sqrt{|\lambda|}t, & 0 \le t \le s \le \eta, \\ \cosh\sqrt{|\lambda|}s\sinh\sqrt{|\lambda|}(1-t) + \delta\cosh\sqrt{|\lambda|}s\sinh\sqrt{|\lambda|}(t-\eta), & s \le t,\ s \le \eta, \\ \sinh\sqrt{|\lambda|}(1-s)\cosh\sqrt{|\lambda|}t, & t \le s,\ \eta \le s, \\ \cosh\sqrt{|\lambda|}s\sinh\sqrt{|\lambda|}(1-t) + \delta\cosh\sqrt{|\lambda|}\eta\sin\sqrt{|\lambda|}(t-s), & \eta \le s \le t \le 1, \end{cases} \tag{11}$$

where $D_\lambda = \sqrt{|\lambda|}(\delta\cosh\sqrt{|\lambda|}\eta - \cosh\sqrt{|\lambda|})$ and if H_0' holds then $G(t,s) \le 0$.

Proof. Proof is same as given in Lemma 1.

Lemma 5. *When $\lambda < 0$, $y \in C^2(I)$ is a solution of boundary value problem (5) and is given by*

$$y(t) = \frac{b\cosh\sqrt{|\lambda|}t}{\cosh\sqrt{|\lambda|} - \delta\cosh\sqrt{|\lambda|}\eta} - \int_0^1 G(t,s)h(s)\,ds. \tag{12}$$

Proof. Proof is same as given in Lemma 2.

2.2. Maximum and Antimaximum Principle

Proposition 6 (antimaximum principle). *Let $0 < \lambda \leq \pi^2/4$, $\sin\sqrt{\lambda} - \delta \sin\sqrt{\lambda}\eta \geq 0$, $\delta\cos\sqrt{\lambda}\eta - \cos\sqrt{\lambda} > 0$, $b \geq 0$, and $h(t) \in C[0,1]$ is such that $h(t) \geq 0$; then $y(t)$ is nonpositive on I.*

Proposition 7 (maximum principle). *Let $\lambda < 0$, $\sinh\sqrt{|\lambda|} - \delta\sinh\sqrt{|\lambda|}\eta \geq 0$, $\delta\cosh\sqrt{|\lambda|}\eta - \cosh\sqrt{|\lambda|} < 0$, $b \geq 0$ and $h(t) \in C[0,1]$ is such that $h(t) \geq 0$; then $y(t)$ is nonnegative on I.*

3. Three Point Nonlinear BVP

Based on maximum and antimaximum Principle we develop theory to solve the three point nonlinear BVP and divide it into the following two subsections.

3.1. Reverse Ordered Lower and Upper Solutions

Theorem 8. *Let there exist v_0, u_0 in $C^2[0,1]$ such that $u_0 \leq v_0$ and satisfies*

$$\begin{gathered} -u_0''(t) \geq f(t,u_0), \quad 0<t<1, \\ u_0'(0) = 0, \quad u_0(1) \geq \delta u_0(\eta), \end{gathered} \tag{13}$$

$$\begin{gathered} -v_0''(t) \leq f(t,v_0), \quad 0<t<1, \\ v_0'(0) = 0, \quad v_0(1) \leq \delta v_0(\eta). \end{gathered} \tag{14}$$

If $f : D \to R$ is continuous on $D := \{(t,y) \in [0,1]\times R^2 : u_0 \leq y \leq v_0\}$ and there exists $M \geq 0$ such that for all $(t,y), (t,w) \in D$,

$$y \leq w \Longrightarrow f(t,w) - f(t,y) \leq M(w-y), \tag{15}$$

then the boundary value problem (5) has at least one solution in the region D. Further, if $\exists$ a constant λ such that $M - \lambda \leq 0$ and (H_0) is satisfied, then the sequences $\{u_n\}$ generated by

$$\begin{gathered} -u_{n+1}''(t) - \lambda u_{n+1} = F(t,u_n), \\ u_{n+1}'(0) = 0, \quad u_{n+1}(1) = \delta u_{n+1}(\eta), \end{gathered} \tag{16}$$

where $F(t,u_n) = f(t,u_n) - \lambda u_n$, with initial iterate u_0 converge monotonically (non-decreasing) and uniformly towards a solution $u(t)$ of (5). Similarly, using v_0 as an initial iterate leads to a nonincreasing sequences $\{v_n\}$ converging to a solution $v(t)$. Any solution $z(t)$ in D must satisfy

$$u(t) \leq z(t) \leq v(t). \tag{17}$$

Proof. From (13) and (16) (for $n = 0$)

$$\begin{gathered} -(u_0-u_1)'' - \lambda(u_0-u_1) \geq 0, \\ (u_0-u_1)'(0) = 0, \quad (u_0-u_1)(1) \geq (u_0-u_1)(\eta). \end{gathered} \tag{18}$$

Since $h(t) \geq 0$ and $b \geq 0$, by using Proposition 6, we have $u_0 \leq u_1$.

In view of $\lambda \geq M$, from (16) we get

$$-u_{n+1}''(t) \geq -(M-\lambda)(u_{n+1}-u_n) + f(t,u_{n+1}) \tag{19}$$

and if $u_{n+1} \geq u_n$, then

$$-u_{n+1}''(t) \geq f(t,u_{n+1}). \tag{20}$$

Since $u_0 \leq u_1$, then from (20) (for $n = 0$) and (16) (for $n = 1$), we get

$$\begin{gathered} -(u_1-u_2)'' - \lambda(u_1-u_2) \geq 0, \\ (u_1-u_2)'(0) = 0, \quad (u_1-u_2)(1) \geq (u_1-u_2)(\eta). \end{gathered} \tag{21}$$

From Proposition 6 we have $u_1 \leq u_2$.

Now from (14) and (16) (for $n = 0$)

$$\begin{gathered} -(u_1-v_0)'' - \lambda(u_1-v_0) \geq 0, \\ (u_1-v_0)'(0) = 0, \quad (u_1-v_0)(1) \geq \delta(u_1-v_0)(\eta). \end{gathered} \tag{22}$$

Thus, $u_1 \leq v_0$ follows from Proposition 6.

Now assuming that $u_{n+1} \geq u_n$, $u_{n+1} \leq v_0$, we show that $u_{n+2} \geq u_{n+1}$ and $u_{n+2} \leq v_0$ for all n. From (16) (for $n+1$) and (20), we get

$$\begin{gathered} -(u_{n+1}-u_{n+2})'' - \lambda(u_{n+1}-u_{n+2}) \geq 0, \\ (u_{n+1}-u_{n+2})'(0) = 0, \\ (u_{n+1}-u_{n+2})(1) \geq \delta(u_{n+1}-u_{n+2})(\eta), \end{gathered} \tag{23}$$

and hence from Proposition 6 we have $u_{n+1} \leq u_{n+2}$.

From (16) (for $n+1$) and (14) we get

$$\begin{gathered} -(u_{n+2}-v_0)'' - \lambda(u_{n+2}-v_0) \geq 0, \\ (u_{n+2}-v_0)'(0) = 0, \quad (u_{n+2}-v_0)(1) \geq \delta(u_{n+2}-v_0)(\eta). \end{gathered} \tag{24}$$

Then, from Proposition 6, $u_{n+2} \leq v_0$ and hence we have

$$u_1 \leq u_2 \leq \cdots \leq u_n \leq u_{n+1} \leq \cdots \leq v_0, \tag{25}$$

and starting with v_0 it is easy to get

$$v_1 \geq v_2 \geq \cdots \geq v_n \geq v_{n+1} \geq \cdots \geq u_0. \tag{26}$$

Finally, we show that $u_n \leq v_n$ for all n. For this, by assuming $u_n \leq v_n$, we show that $u_{n+1} \leq v_{n+1}$. From (16) it is easy to get

$$\begin{gathered} -(u_{n+1}-v_{n+1})'' - \lambda(u_{n+1}-v_{n+1}) \geq 0, \\ (u_{n+1}-v_{n+1})'(0) = 0, \\ (u_{n+1}-v_{n+1})(1) \geq \delta(u_{n+1}-v_{n+1})(\eta). \end{gathered} \tag{27}$$

Hence, from Proposition 6, $u_{n+1} \leq v_{n+1}$. Thus we have

$$\begin{gathered} v_0 \geq v_1 \geq v_2 \geq \cdots \geq v_n \geq v_{n+1} \geq \cdots \geq u_{n+1} \\ \geq u_n \geq \cdots \geq u_2 \geq u_1 \geq u_0. \end{gathered} \tag{28}$$

So, the sequences u_n and v_n are monotonically nondecreasing and nonincreasing, respectively, and are bounded by u_0 and v_0. Hence by Dini's theorem they converge uniformly. Let $u(t) = \lim_{n\to\infty} u_n(t)$ and $v(t) = \lim_{n\to\infty} v_n(t)$.

Using Lemma 2, the solution u_n of (16) is given by

$$u_n = \frac{b\cos\sqrt{\lambda}t}{\cos\sqrt{\lambda} - \delta\cos\sqrt{\lambda}\eta} - \int_0^1 G(t,s)\left(f(t,u_n) - \lambda u_n\right) ds. \tag{29}$$

Then, by Lebesgue's dominated convergence theorem, taking the limit as n approaches to ∞, we get

$$u(t) = \frac{b\cos\sqrt{\lambda}t}{\cos\sqrt{\lambda} - \delta\cos\sqrt{\lambda}\eta} - \int_0^1 G(t,s)\left(f(t,u) - \lambda u\right) ds, \tag{30}$$

which is the solution of boundary value problem (5).

Any solution $z(t)$ in D can play the role of $u_0(t)$; hence, $z(t) \le v(t)$ and similarly one concludes that $z(t) \ge u(t)$.

3.2. Well-Ordered Lower and Upper Solutions

Theorem 9. *Let there exist v_0, u_0 in $C^2[0,1]$ such that $u_0 \ge v_0$ and satisfies*

$$\begin{aligned} &-u_0''(t) \ge f(t,u_0), \quad 0<t<1, \\ &u_0'(0) = 0, \qquad u_0(1) \ge \delta u_0(\eta), \\ &-v_0''(t) \le f(t,v_0), \quad 0<t<1, \\ &v_0'(0) = 0, \qquad v_0(1) \le \delta v_0(\eta). \end{aligned} \tag{31}$$

If $f : D_0 \to R$ is continuous on $D_0 := \{(t,y) \in [0,1]\times R^2 : v_0 \le y \le u_0\}$ and there exists $M \ge 0$ such that for all $(t,\tilde{y})$, $(t,\tilde{w}) \in D_0$

$$\tilde{y} \le \tilde{w} \Longrightarrow f(t,\tilde{w}) - f(t,\tilde{y}) \ge -M(\tilde{w} - \tilde{y}), \tag{32}$$

then the boundary value problem (5) has at least one solution in the region D_0. If $\exists$ a constant $\lambda < 0$, such that $\lambda + M \le 0$ and (H_0') is satisfied, then the sequences $\{u_n\}$ generated by

$$\begin{aligned} &-u_{n+1}''(t) - \lambda u_{n+1} = F(t,u_n), \qquad u_{n+1}'(0) = 0, \\ &\qquad u_{n+1}(1) = \delta u_{n+1}(\eta), \end{aligned} \tag{33}$$

where $F(t,u_n) = f(t,u_n) - \lambda u_n$, with initial iterate u_0, converge monotonically (nonincreasing) and uniformly towards a solution $\tilde{u}(t)$ of (5). Similarly, using v_0 as an initial iterate leads to a nondecreasing sequences $\{v_n\}$ converging to a solution $\tilde{v}(t)$. Any solution $\tilde{z}(t)$ in D_0 must satisfy

$$\tilde{v}(t) \le \tilde{z}(t) \le \tilde{u}(t). \tag{34}$$

Proof. Proof follows from the analysis of Theorem 8.

4. Numerical Illustration

To verify our results, we consider examples and show that there exists at least one value of $\lambda \in \mathbb{R}\setminus\{0\}$ such that iterative scheme generates monotone sequences which converge to solutions of nonlinear problem. Thus, these examples validate sufficient conditions derived in this paper.

Example 10 (reverse order). Consider the boundary value problem

$$\begin{aligned} &-y''(t) = \frac{e^y}{32} - \frac{1}{64}, \qquad 0<t<1, \\ &y'(0) = 0, \qquad y(1) = 2y\left(\frac{1}{3}\right). \end{aligned} \tag{35}$$

Here, $f(t,y) = (e^y/32) - (1/64)$, $\delta = 2$, $\eta = 1/3$. This problem has $v_0 = 1$ and $u_0 = -1$ as lower and upper solutions; that is, it is non-well-ordered case. The nonlinear term is Lipschitz in y and continuous for all value of y, and Lipschitz constants are $M = e/32$. For $0.0849463 \le \lambda \le \pi^2/4$ we can see that (H_0) will be true. To verify (H_0), in Figure 1 we plot inequalities assumed in (H_0). From Figures 2, 3, and 4, we plot members of monotone sequences u_n, v_n for different values of λ.

Example 11 (well order). Consider the boundary value problem

$$\begin{aligned} &-y''(t) = \frac{1}{32}\left[\frac{e^2}{4} - \frac{\sin t}{4} - 2(y(t))^3\right], \\ &y'(0) = 0, \qquad y(1) = \frac{1}{3}y\left(\frac{1}{2}\right). \end{aligned} \tag{36}$$

Here, $f(t,y) = (1/32)[(e^2/4) - (\sin t/4) - 2(y(t))^3]$, $\delta = 1/3$, $\eta = 1/2$. This problem has $v_0 = -1$ and $u_0 = 1$ as lower and upper solutions; that is, it is well-ordered case. The nonlinear term is Lipschitz in y and continuous for all value of y, and Lipschitz constants are $M = 3/16$. For $\lambda \le -0.1875$, we can see that (H_0') will be true. To verify (H_0'), in Figure 5 we plot inequalities assumed in (H_0'). From Figures 6, 7, and 8, we plot members of monotone sequences u_n, v_n for different values of λ.

5. Conclusion

The monotone iterative technique coupled with upper and lower solutions is a powerful tool for computation of solutions of nonlinear three point boundary value problems. It proves the existence of solutions analytically and gives us a tool so that numerical solutions can also be computed and then some real-life problems, for example, bridge design problem, thermostat problem, and so forth, can be solved. We have plotted sequences for both $\lambda > 0$ and $\lambda < 0$. The plots are quite encouraging and will motivate researchers to explore further possibilities. Employing this technique, Mathematica/Maple/MATLAB user-friendly packages can be developed (see [16]).

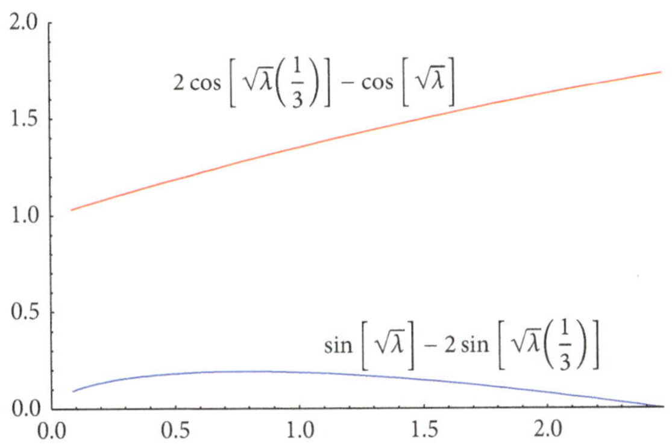

FIGURE 1: Plot of $\sin\sqrt{\lambda} - \delta\sin\sqrt{\lambda}\eta$ and $\delta\cos\sqrt{\lambda}\eta - \cos\sqrt{\lambda}$.

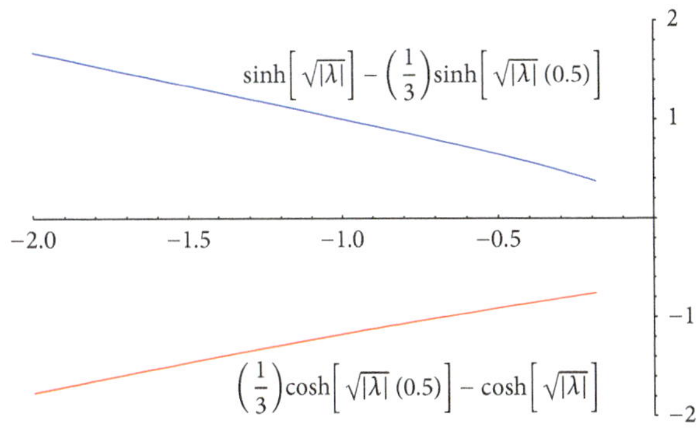

FIGURE 5: Plot of $\delta\cosh\sqrt{|\lambda|}\eta - \cosh\sqrt{|\lambda|}$ and $\sinh\sqrt{|\lambda|} - \delta\sinh\sqrt{|\lambda|}\eta$.

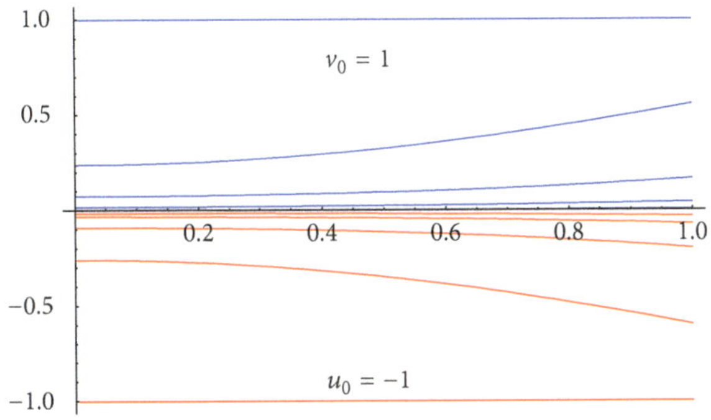

FIGURE 2: Plot of u_n (red) and v_n (blue), $n = 0, 1, 2, 3, 4$ for $\lambda = 0.9$.

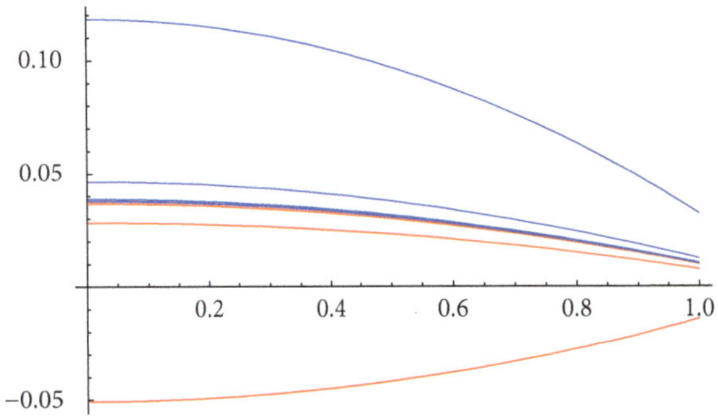

FIGURE 6: Plot of u_n (blue) and v_n (red), $n = 1, 2, 3, 4, 5$ for $\lambda = -0.2$.

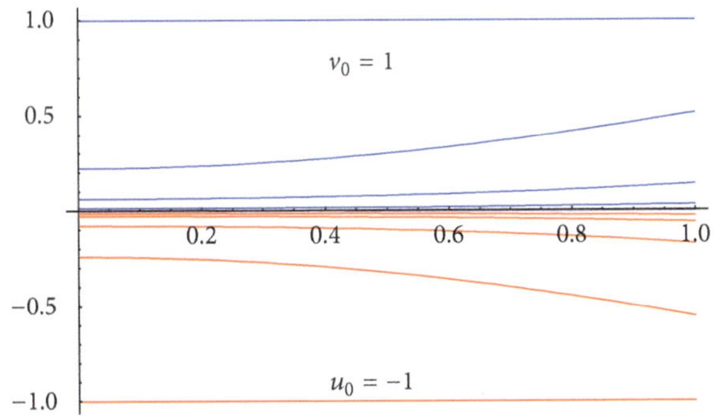

FIGURE 3: Plot of u_n (red) and v_n (blue), $n = 0, 1, 2, 3, 4$ for $\lambda = 1$.

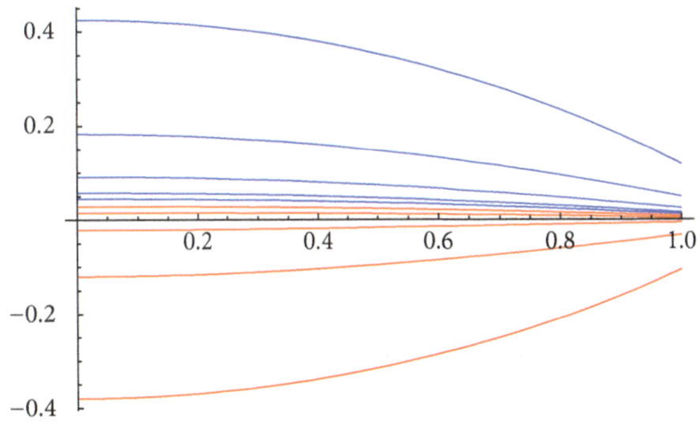

FIGURE 7: Plot of u_n (blue) and v_n (red), $n = 1, 2, 3, 4, 5$ for $\lambda = -1$.

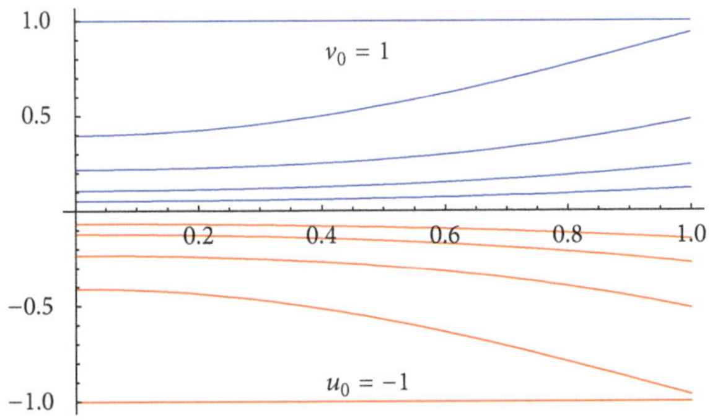

FIGURE 4: Plot of u_n (red) and v_n (blue), $n = 0, 1, 2, 3, 4$ for $\lambda = 2.3$.

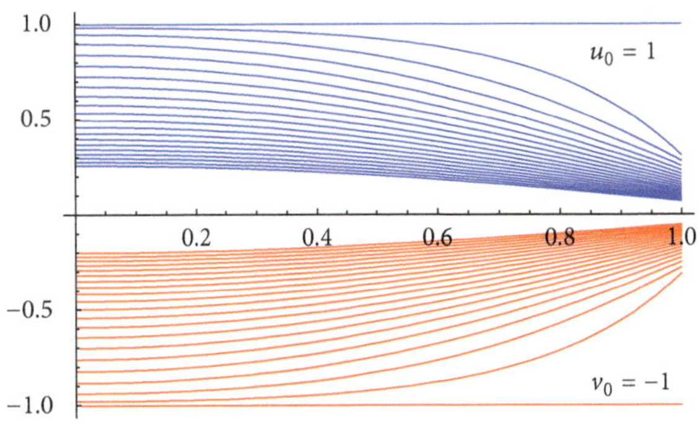

FIGURE 8: Plot of u_n (blue) and v_n (red), $n = 0, 1, 2, \ldots, 20$ for $\lambda = -20$.

Acknowledgment

This work is partially supported by Grant provided by UGC, New Delhi, India, File no. F.4-1/2006 (BSR)/7-203/2009(BSR).

References

[1] C. P. Gupta and S. I. Trofimchuk, "Existence of a solution of a three-point boundary value problem and the spectral radius of a related linear operator," *Nonlinear Analysis: Theory, Methods & Applications*, vol. 34, no. 4, pp. 489–507, 1998.

[2] Y. Liu, "Existence of three solutions to a non-homogeneous multi-point BVP of second order differential equations," *Turkish Journal of Mathematics*, vol. 35, no. 1, pp. 55–86, 2011.

[3] R. Ma and N. Castaneda, "Existence of solutions of nonlinear m-point boundary-value problems," *Journal of Mathematical Analysis and Applications*, vol. 256, no. 2, pp. 556–567, 2001.

[4] Z. Zhang and J. Wang, "The upper and lower solution method for a class of singular nonlinear second order three-point boundary value problems," *Journal of Computational and Applied Mathematics*, vol. 147, no. 1, pp. 41–52, 2002.

[5] Y. Liu, "A note on the existence of positive solutions of one-dimensional p-Laplacian boundary value problems," *Applications of Mathematics*, vol. 55, no. 3, pp. 241–264, 2010.

[6] J. R. L. Webb, "Existence of positive solutions for a thermostat model," *Nonlinear Analysis: Real World Applications*, vol. 13, no. 2, pp. 923–938, 2012.

[7] Y. Zou, Q. Hu, and R. Zhang, "On numerical studies of multi-point boundary value problem and its fold bifurcation," *Applied Mathematics and Computation*, vol. 185, no. 1, pp. 527–537, 2007.

[8] E. Picard, "Sur lapplication des metodes dapproximations succesives a letude de certains equations differentielles ordinaires," *Journal de Mathmatiques Pures et Appliques*, vol. 9, pp. 217–271, 1893.

[9] G. S. Dragoni, "II problema dei valori ai limiti studiato in grande per le equazioni differenziali del secondo ordine," *Mathematische Annalen*, vol. 105, no. 1, pp. 133–143, 1931.

[10] A. Cabada, P. Habets, and S. Lois, "Monotone method for the Neumann problem with lower and upper solutions in the reverse order," *Applied Mathematics and Computation*, vol. 117, no. 1, pp. 1–14, 2001.

[11] P. Omari and M. Trombetta, "Remarks on the lower and upper solutions method for second- and third-order periodic boundary value problems," *Applied Mathematics and Computation*, vol. 50, no. 1, pp. 1–21, 1992.

[12] A. K. Verma, "The monotone iterative method and zeros of Bessel functions for nonlinear singular derivative dependent BVP in the presence of upper and lower solutions," *Nonlinear Analysis: Theory, Methods & Applications*, vol. 74, no. 14, pp. 4709–4717, 2011.

[13] M. Cherpion, C. de Coster, and P. Habets, "A constructive monotone iterative method for second-order BVP in the presence of lower and upper solutions," *Applied Mathematics and Computation*, vol. 123, no. 1, pp. 75–91, 2001.

[14] X. Xian, D. O'Regan, and S. Jingxian, "Multiplicity results for three-point boundary value problems with a non-well-ordered upper and lower solution condition," *Mathematical and Computer Modelling*, vol. 45, no. 1-2, pp. 189–200, 2007.

[15] F. Li, M. Jia, X. Liu, C. Li, and G. Li, "Existence and uniqueness of solutions of second-order three-point boundary value problems with upper and lower solutions in the reversed order," *Nonlinear Analysis: Theory, Methods & Applications*, vol. 68, no. 8, pp. 2381–2388, 2008.

[16] A. Cabada, J. Cid, and B. Maquez-Villamarin, "Computation of Green's functions for boundary value problems with *Mathematica*," *Applied Mathematics and Computation*, vol. 219, no. 4, pp. 1919–1936, 2012.

12

Characterization for Rectifiable and Nonrectifiable Attractivity of Nonautonomous Systems of Linear Differential Equations

Yūki Naito[1] and Mervan Pašić[2]

[1] *Ehime University, Matsuyama 790-8577, Japan*
[2] *Department of Mathematics, University of Zagreb, FER, 10000 Zagreb, Croatia*

Correspondence should be addressed to Yūki Naito; ynaito@ehime-u.ac.jp

Academic Editor: Norio Yoshida

We study a new kind of asymptotic behaviour near $t = 0$ for the nonautonomous system of two linear differential equations: $\mathbf{x}'(t) = A(t)\mathbf{x}(t)$, $t \in (0, t_0]$, where the matrix-valued function $A = A(t)$ has a kind of singularity at $t = 0$. It is called rectifiable (resp., nonrectifiable) attractivity of the zero solution, which means that $\|\mathbf{x}(t)\|_2 \to 0$ as $t \to 0$ and the length of the solution curve of $\mathbf{x}$ is finite (resp., infinite) for every $\mathbf{x} \neq \mathbf{0}$. It is characterized in terms of certain asymptotic behaviour of the eigenvalues of $A(t)$ near $t = 0$. Consequently, the main results are applied to a system of two linear differential equations with polynomial coefficients which are singular at $t = 0$.

1. Introduction

We are concerned with the two-dimensional nonautonomous linear differential systems:

$$\frac{d\mathbf{x}}{dt} = A(t)\mathbf{x} \quad \text{for } t \in (0, t_0], \quad \mathbf{x}(t_0) = \mathbf{x}_0 \in \mathbb{R}^2, \quad (1)$$

where $t_0 > 0$, and $A = A(t)$, $A : (0, t_0] \to \mathcal{M}_2$ is a matrix-valued function such that $A \in C^1((0, t_0]; \mathcal{M}_2)$, where $\mathcal{M}_2$ denotes the space of all 2×2 real matrices. A solution of linear system (1) is a function $\mathbf{x} = \mathbf{x}(t)$, $\mathbf{x} : (0, t_0] \to \mathbb{R}^2$ such that $\mathbf{x} \in C^1((0, t_0], \mathbb{R}^2)$.

In appendix of the paper, we show, there exists a unique solution $\mathbf{x}$ of linear system (1). By the uniqueness of solution $\mathbf{x}$ of (1), it is clear that if $\mathbf{x}_0 = (0, 0)$, then $\mathbf{x}(t) = (0, 0)$ for all $t \in (0, t_0]$. It is called the *zero solution* of linear system (1). The zero solution of linear system (1) is said to be *attractive as* $t \to 0$ if for every $\mathbf{x}_0 \in \mathbb{R}^2$ the corresponding solution $\mathbf{x}$ of (1) satisfies $\|\mathbf{x}(t)\|_2 \to 0$ as $t \to 0$, where $\|\cdot\|_2$ denotes the standard Euclidean norm on $\mathbb{R}^2$. Since we are dealing with the attractive zero solution of (1), from $\mathbf{x} \in C^1((0, t_0], \mathbb{R}^2)$ we conclude that $\mathbf{x} \in C([0, t_0], \mathbb{R}^2) \cap C^1((0, t_0], \mathbb{R}^2)$. For a continuous function $\mathbf{x} : [0, t_0] \to \mathbb{R}^2$, let $\Gamma_{\mathbf{x}}$ denote a curve in $\mathbb{R}^2$ corresponding to $\mathbf{x}$ determined by

$$\Gamma_{\mathbf{x}} = \mathbf{x}([0, t_0]) = \{(x, y) \in \mathbb{R}^2 : x = x(t), y = y(t)\}, \quad (2)$$

where $(x(t), y(t)) \equiv \mathbf{x}(t)$. It is often said that $\Gamma_{\mathbf{x}}$ is parametrized by $\mathbf{x}$.

Let $\mathbf{x} \in C([0, t_0]; \mathbb{R}^2)$. The curve $\Gamma_{\mathbf{x}} \subseteq \mathbb{R}^2$ is said to be a *Jordan curve* if $\mathbf{x}$ is injective (one-to-one) on $[0, t_0]$. Hence, $\Gamma_{\mathbf{x}}$ is nonself-intersecting; it has two ends and it is a compact connected set in $\mathbb{R}^2$. In appendix of the paper, see Theorem A.3; we give some simple sufficient conditions on the matrix-valued function $A = A(t)$ such that the solution curve $\Gamma_{\mathbf{x}}$ of every nontrivial solution of (1) is a Jordan curve in $\mathbb{R}^2$. The assumption that $\Gamma_{\mathbf{x}}$ is a Jordan curve is important for the process of measuring the length of $\Gamma_{\mathbf{x}}$ (the rectification of $\Gamma_{\mathbf{x}}$) as follows. If $\Gamma_{\mathbf{x}}$ is a Jordan curve, then the length of $\Gamma_{\mathbf{x}}$ is denoted by length $(\Gamma_{\mathbf{x}})$ and it is defined as usual by (see [1–3])

$$\text{length}(\Gamma_{\mathbf{x}}) = \sup \left\{ \sum_{k=1}^{m} \|\mathbf{x}(t_k) - \mathbf{x}(t_{k-1})\|_2 : m \in \mathbb{N}, \right.$$
$$\left. 0 = t_m < \cdots < t_1 < t_0 \right\}, \quad (3)$$

where the supremum is taken over all finite dissections $0 = t_m < \cdots < t_1 < t_0$ of the interval $[0, t_0]$.

Definition 1. The zero solution of linear system (1) is said to be *rectifiable* (resp., *nonrectifiable*) *attractive as* $t \to 0$; if for every $\mathbf{x}_0 \in \mathbb{R}^2 \setminus \{\mathbf{0}\}$ such that the $\Gamma_{\mathbf{x}}$ of corresponding solution $\mathbf{x}$ of (1) is a Jordan curve in $\mathbb{R}^2$, one has $\|\mathbf{x}(t)\|_2 \to 0$ as $t \to 0$ and length $(\Gamma_{\mathbf{x}}) < \infty$ (resp., length $(\Gamma_{\mathbf{x}}) = \infty$).

Let $\lambda(t) = \operatorname{Re}\lambda(t) \pm i \operatorname{Im}\lambda(t)$, $t \in (0, t_0]$, be two complex conjugate eigenvalues of $A(t)$. The main purpose of the paper is to characterize attractivity, rectifiable and nonrectifiable attractivity, as $t \to 0$ of the zero solution of system (1) in the terms of regular and singular asymptotic behaviour of $A(t)$ near $t = 0$; see Theorems 2 and 3. It is illustrated on a simple two-dimensional linear differential systems with the polynomial coefficients; see Theorem 4. The key point of proofs of the main results is an asymptotic solution formula for system (1) obtained by its asymptotic integration near $t = 0$; see Section 3. The proofs of the main results are given in Sections 4, 5, and 6. Such a kind of topics has been considered for the first time in [4] but only in the case of the so-called integrable systems (systems that allow all solutions in explicit forms), where the asymptotic integration near $t = 0$ is not required. About the asymptotic integration near $t = \infty$ of differential equations and systems, we refer reader to [5–12]. On several asymptotic properties of two-dimensional differential systems near $t = \infty$, let us see, for instance, [13–18]. Recently interest for studying differential equations and systems on compact (finite time) intervals is rapidly increasing because of its application in atmosphere, fluid, and ocean dynamics and in biological science; see, for instance, [19–21] and references therein.

2. Main Results and Consequences

Let for each $t \in (0, t_0]$ the matrix elements of $A(t)$ be denoted by $(a_{i,j}(t))_{i,j\le 2}$. In the paper, we suppose the following kind of singular asymptotic behaviour of $A(t)$ near $t = 0$:

$$\lim_{\varepsilon \to 0} \int_{\varepsilon}^{t_0} \left| \frac{d}{dt} \left(t^2 a_{i,j}(t) \right) \right| dt < \infty, \quad i, j \in \{1, 2\}, \qquad \lim_{t \to 0} \left[t^2 A(t) \right] = A_0, \tag{4}$$

where A_0 is a constant matrix having two complex conjugate eigenvalues.

Some essential consequences of (4) are presented in Section 3. Since the eigenvalues of $A(t)$ are continuous on $t \in (0, t_0]$, by (4) we may assume that $A(t)$ has two complex conjugate eigenvalues $\lambda(t) = \operatorname{Re}\lambda(t) \pm i \operatorname{Im}\lambda(t)$ for $t \in (0, t_0]$.

The first two main results of the paper are the following theorems.

Theorem 2. *Let* (4) *be satisfied, and let* $\lambda(t) = \operatorname{Re}\lambda(t) \pm i \operatorname{Im}\lambda(t)$, $t \in (0, t_0]$, *be two complex conjugate eigenvalues of* $A(t)$. *Assume that* $\operatorname{Re}\lambda(t) \ge 0$ *near* $t = 0$. *Then the zero solution of linear system* (1) *is attractive as* $t \to 0$ *if and only if*

$$\lim_{t \to 0} \int_t^{t_0} \operatorname{Re}\lambda(s)\, ds = \infty. \tag{5}$$

Theorem 3. *Let* (4) *be satisfied, and let* $\lambda(t) = \operatorname{Re}\lambda(t) \pm i \operatorname{Im}\lambda(t)$, $t \in (0, t_0]$, *be two complex conjugate eigenvalues of* $A(t)$. *Assume that* $\operatorname{Re}\lambda(t) \ge 0$ *near* $t = 0$, *and that* (5) *holds. Then the zero solution of linear system* (1) *is rectifiable attractive as* $t \to 0$ *if and only if*

$$\operatorname{Im}\lambda(t)\, e^{-\int_t^{t_0} \operatorname{Re}\lambda(s) ds} \in L^1(0, t_0). \tag{6}$$

Previous theorem allows us to study the rectifiable and nonrectifiable attractivity for the following model system with polynomial singular coefficients near $t = 0$:

$$\begin{aligned} x'(t) &= e(t)\, t^{-p} x(t) + f(t)\, t^{-2} y(t), \\ y'(t) &= -g(t)\, t^{-2} x(t) + h(t)\, t^{-q} y(t), \quad t \in (0, t_0], \end{aligned} \tag{7}$$

where the parameters $p, q \in \mathbb{R}$ and coefficients $e, f, g, h \in C([0, t_0]) \cap C^1((0, t_0])$ satisfy, respectively,

$$1 \le \max\{p, q\} < 2, \tag{8}$$

$$f(0)\, g(0) > 0, \quad e(t) > 0, \quad h(t) > 0, \quad t \in [0, t_0], \tag{9}$$

$$\begin{cases} \int_0^{t_0} \left| f'(t) \right| dt < \infty, \quad \int_0^{t_0} \left| g'(t) \right| dt < \infty, \\ \int_0^{t_0} \left| \frac{d}{dt} \left(t^{2-p} e(t) \right) \right| dt < \infty, \\ \int_0^{t_0} \left| \frac{d}{dt} \left(t^{2-q} h(t) \right) \right| dt < \infty. \end{cases} \tag{10}$$

According to (8), (9), and (10), we will show that the model system (7) satisfies all assumptions of Theorem 3. Hence, we are able to derive and prove the following conditions for the rectifiable and nonrectifiable attractivity of the zero solution of model system (7).

Theorem 4. *Let* (8), (9), *and* (10) *be fulfilled.*

(i) *If* $\max\{p, q\} > 1$, *then the zero solution of* (7) *is rectifiable attractive as* $t \to 0$.

(ii) *If* $\max\{p, q\} = 1$, *then the zero solution of* (7) *is rectifiable attractive as* $t \to 0$ *provided that one of the following conditions is satisfied:*

$$p = 1, \quad q < 1, \quad e(t) > 2; \tag{11}$$

$$p < 1, \quad q = 1, \quad h(t) > 2; \tag{12}$$

$$p = 1, \quad q = 1, \quad e(t) + h(t) > 2. \tag{13}$$

(iii) *If* $\max\{p, q\} = 1$, *then the zero solution of* (7) *is nonrectifiable attractive as* $t \to 0$ *provided that one of the following conditions is satisfied:*

$$p = 1, \quad q < 1, \quad e(t) \le 2; \tag{14}$$

$$p < 1, \quad q = 1, \quad h(t) \le 2; \tag{15}$$

$$p = 1, \quad q = 1, \quad e(t) + h(t) \le 2. \tag{16}$$

As a consequence of previous theorem, we realize that the rectifiable and nonrectifiable attractivity of the zero solution of model system (7) depends on the order of growth of singular behaviour of coefficients near $t = 0$ determined by parameters p and q. Let us remark that if, besides (8), (9), and (10), it is supposed that $e(t) \equiv h(t)$, $f(t) \equiv g(t)$, and $p = q$, then system (7) is integrable, that is, all solutions of system (7) are explicitly determined, see appendix of the paper. In such a particular case of (7), Theorem 4 can be explicitly verified.

Remark 5. We suggest reader to consider the rectifiable and nonrectifiable attractivity of the zero solution of system (7) in the case of $\max\{p, q\} = 2$. It is required to assume that $4f(0)g(0) - e^2(0) > 0$ if $p = 2 > q$ as well as $4f(0)g(0) - h^2(0) > 0$ if $q = 2 > p$ and $4f(0)g(0) - (e(0) - h(0))^2 > 0$ if $p = q = 2$.

The most difficult point of this paper is to derive and prove the nonrectifiable attractivity of the zero solution of systems (1) and (7); see Sections 3 and 4. On the (non)rectifiability of the solution curve of scalar second-order differential equations, we refer reader to [22–27].

3. Some Asymptotic Properties Near $t = 0$

As a consequence of the main assumption (4), we derive in this section some important asymptotic properties of $A(t)$ as well as of its eigenvalues and solutions of system (1), which will be used in the next sections in the proof of the main results. Since the eigenvalues of $A(t)$ are continuous on $(0, t_0]$, by (4) we may assume that $A(t)$ has two complex conjugate eigenvalues $\lambda(t) = \operatorname{Re}\lambda(t) \pm i \operatorname{Im}\lambda(t)$, $t \in (0, t_0]$.

Lemma 6. *Let (4) be satisfied. Then the eigenvalues of $A(t)$ satisfy*

$$\|A(t)\| \sim |\operatorname{Im}\lambda(t)| \sim t^{-2} \quad \text{as } t \longrightarrow 0, \tag{17}$$

$$\operatorname{Im}\lambda(t) \neq 0 \quad \text{near } t = 0, \tag{18}$$

where $f(t) \sim g(t)$ means that $f(t)/g(t) \to 1$ as $t \to 0$.

Proof. Let $a_{i,j} = a_{i,j}(t)$, $a_{i,j} \in C^1((0, t_0])$, be matrix elements of $A(t)$. Also, let $a^0_{i,j}$ and $\lambda_0 = \operatorname{Re}\lambda_0 \pm i \operatorname{Im}\lambda_0$ be matrix elements and eigenvalues of A_0, respectively, where A_0 appears in (4). Rewriting (4) in terms of matrix elements and eigenvalues, we get

$$\lim_{t\to 0}\left[t^2 a_{i,j}(t)\right] = a^0_{i,j}, \quad \operatorname{Im}\lambda_0 \neq 0. \tag{19}$$

Hence, we can conclude the following:

$$\begin{aligned}
&\lim_{t\to 0}\left[t^2 |\operatorname{Im}\lambda(t)|\right] \\
&= \frac{1}{2}\lim_{t\to 0}\left[t^2\sqrt{-(a_{1,1}(t) - a_{2,2}(t))^2 - 4a_{1,2}(t)a_{2,1}(t)}\right] \\
&= \frac{1}{2}\lim_{t\to 0}\sqrt{-(t^2a_{1,1}(t) - t^2a_{2,2}(t))^2 - 4t^2a_{1,2}(t)t^2a_{2,1}(t)} \\
&= \frac{1}{2}\sqrt{-\left(a^0_{1,1} - a^0_{2,2}\right)^2 - 4a^0_{1,2}a^0_{2,1}} = |\operatorname{Im}\lambda_0| \neq 0.
\end{aligned} \tag{20}$$

It proves that $|\operatorname{Im}\lambda(t)| \sim t^{-2}$ as $t \to 0$. Hence (18) and the second claim in (17) hold. In order to complete (17), it rests to show that $\|A(t)\| \sim t^{-2}$ as $t \to 0$. In fact, from the assumption $\operatorname{Im}\lambda_0 \neq 0$, we observe that for at least one matrix element $a^0_{i,j}$ we have $a^0_{i,j} \neq 0$ and, therefore, $\|A_0\| \neq 0$. Moreover, since every matrix norm is a continuous function, from (4) we especially obtain $\lim_{t\to 0}[t^2\|A(t)\|] = \|A_0\| \neq 0$. It completes the proof of (17).

As the main consequence of (4) we have the following asymptotic formula near $t = 0$ for all solutions $\mathbf{x}$ of linear system (1):

$$\begin{aligned}
\mathbf{x}(t) = e^{-\int_t^{t_0}\operatorname{Re}\lambda(s)ds}&\left[\mathbf{h}_1(t)\cos\left(\int_t^{t_0}\operatorname{Im}\lambda(s)\,ds\right)\right. \\
&\left.+\mathbf{h}_2(t)\sin\left(\int_t^{t_0}\operatorname{Im}\lambda(s)\,ds\right)\right],
\end{aligned} \tag{21}$$

where $\mathbf{h}_1 = \mathbf{h}_1(t)$, $\mathbf{h}_2 = \mathbf{h}_2(t)$, $\mathbf{h}_i : (0, t_0] \to \mathbb{R}^2$ and there are $\mathbf{c}_1, \mathbf{c}_2 \in \mathbb{R}^2$ such that

$$\lim_{t\to 0}\mathbf{h}_i(t) = \mathbf{c}_i \neq \mathbf{0}, \quad i \in \{1, 2\}. \tag{22}$$

It is the subject of the next lemma.

Lemma 7. *Let $\lambda(t) = \operatorname{Re}\lambda(t) \pm i\operatorname{Im}\lambda(t)$ be two eigenvalues of $A(t)$. If $A(t)$ satisfies (4), then for every solution $\mathbf{x}$ of linear system (1) with $\mathbf{x}_0 \neq \{\mathbf{0}\}$ there exist two functions $\mathbf{h}_i(t)$ and two constants $\mathbf{c}_1, \mathbf{c}_2 \in \mathbb{R}^2$ such that (21) and (22) are fulfilled.*

Proof. Let $A(t)$ be a matrix satisfying all assumptions of this lemma. Let $B(\tau) = (b_{i,j}(\tau)_{i,j\le 2})$ be a new matrix depending on a new variable $\tau := 1/t$ defined by

$$B(\tau) = -\tau^{-2}A\left(\tau^{-1}\right) \quad \forall \tau \geq \tau_0 := \frac{1}{t_0}. \tag{23}$$

Then $b_{i,j} \in C^1([\tau_0, \infty))$. Let A_0 be a constant matrix determined in (4). If B_0 is a new matrix defined by $B_0 = -A_0$, then from (4), (23), and $t = 1/\tau$, we conclude

$$\begin{aligned}
&\lim_{M\to\infty}\int_{\tau_0}^{M}\left|\frac{d}{d\tau}b_{i,j}(\tau)\right|d\tau < \infty, \quad i, j \in \{1, 2\}, \\
&\lim_{\tau\to\infty}B(\tau) = B_0,
\end{aligned} \tag{24}$$

where B_0 is a constant matrix having two complex conjugate eigenvalues.

Also, the eigenvalues $\mu_1(\tau)$, $\mu_2(\tau)$ of $B(\tau)$ and $\lambda_1(t) = \operatorname{Re}\lambda(t) + i\operatorname{Im}\lambda(t)$, $\lambda_2(t) = \operatorname{Re}\lambda(t) - i\operatorname{Im}\lambda(t)$ of $A(t)$ are obviously related by the equality

$$\begin{aligned}\mu_i(\tau) &= -\tau^{-2}\lambda_i\left(\tau^{-1}\right)\\ &= -\tau^{-2}\left[\operatorname{Re}\left(\lambda\left(\tau^{-1}\right)\right) \pm i\operatorname{Im}\left(\lambda\left(\tau^{-1}\right)\right)\right], \quad i = 1,2,\end{aligned} \tag{25}$$

and hence,

$$\operatorname{Re}\left(\mu_1(\tau) - \mu_2(\tau)\right) = 0 \quad \forall \tau \geq \tau_0. \tag{26}$$

Putting $t = 1/\tau$ into system (1) and denoting by

$$\mathbf{w}(\tau) := \mathbf{x}\left(\frac{1}{\tau}\right) \quad \text{for } \tau \geq \tau_0, \tag{27}$$

we obtain the equivalent system of (1)

$$\mathbf{w}'(\tau) = B(\tau)\mathbf{w}(\tau) \quad \text{for } \tau \geq \tau_0, \tag{28}$$

where matrix $B(\tau)$ is defined in (23). For system (28), we will use the next generalized Matell's theorem appearing as Theorem 11 in Coppel's book [5, Chapter 4] and Theorem 6.5 in Kiguradze's monograph [7] (see also [12, Theorem 7] and [8]).

Theorem 8 (see [7]). *Let $\mu_1 = \mu_1(\tau)$ and $\mu_2 = \mu_2(\tau)$, $\mu_1, \mu_2 : [\tau_0, \infty) \to \mathbb{C}$ be two eigenvalues of an arbitrary 2×2 matrix $B(\tau)$, $\tau \in [\tau_0, \infty)$, with the matrix elements $b_{i,j}(\tau)$ which are absolutely continuous functions on every compact set $I \subset [\tau_0, \infty)$ and*

$$\int_{\tau_0}^{\infty}\left|\frac{d}{d\tau}b_{i,j}(\tau)\right| d\tau < \infty, \quad i, j \in \{1,2\}. \tag{29}$$

Let $c > 0$ be a constant such that

$$\begin{aligned}&\int_{\sigma_*}^{\tau_*}\operatorname{Re}\left(\mu_1(\tau) - \mu_2(\tau)\right) d\tau \leq c, \quad \tau_0 \leq \sigma_* < \tau_*, \text{ or}\\ &\int_{\tau_0}^{\infty}\operatorname{Re}\left(\mu_1(\tau) - \mu_2(\tau)\right) d\tau = \infty,\\ &\int_{\sigma_*}^{\tau_*}\operatorname{Re}\left(\mu_1(\tau) - \mu_2(\tau)\right) d\tau \geq -c, \quad \tau_0 \leq \sigma_* < \tau_*.\end{aligned} \tag{30}$$

Let $B(\tau) \to B_0$ as $\tau \to \infty$, where B_0 is a constant matrix having two different eigenvalues. Then linear differential system (28) has a fundamental system of solutions $\mathbf{w}_1(\tau) \in \mathbb{C}^2$ and $\mathbf{w}_2(\tau) \in \mathbb{C}^2$ such that

$$\begin{aligned}&\lim_{\tau\to\infty}\left[\mathbf{w}_1(\tau)e^{-\int_{\tau_0}^{\tau}\mu_1(\sigma)d\sigma}\right] = \mathbf{b}_1,\\ &\lim_{\tau\to\infty}\left[\mathbf{w}_2(\tau)e^{-\int_{\tau_0}^{\tau}\mu_2(\sigma)d\sigma}\right] = \mathbf{b}_2,\end{aligned} \tag{31}$$

where $\mathbf{b}_1, \mathbf{b}_2 \in \mathbb{C}^2$ are two eigenvectors of constant matrix B_0.

Now, with the help of (24) and (26), it is easy to check that the matrix $B(\tau)$ defined in (23) and its eigenvalues $\mu_1(\tau)$ and $\mu_2(\tau)$ defined in (25) satisfy all required assumptions of Theorem 8. Hence, by Theorem 8, we observe that linear system (1) has a fundamental system of solutions $\mathbf{x}_1(t)$ and $\mathbf{x}_2(t)$ satisfying

$$\begin{aligned}&\lim_{t\to 0}\left[\mathbf{x}_1(t)e^{\int_t^{t_0}\lambda_1(s)ds}\right] = \mathbf{b}_1,\\ &\lim_{t\to 0}\left[\mathbf{x}_2(t)e^{\int_t^{t_0}\lambda_2(s)ds}\right] = \mathbf{b}_2.\end{aligned} \tag{32}$$

Indeed, from (31) and using $\tau = 1/t$, $\tau_0 = 1/t_0$, (25), (27), and $\sigma = 1/s$, we obtain

$$\begin{aligned}\mathbf{b}_1 &= \lim_{\tau\to\infty}\left[\mathbf{w}_1(\tau)e^{-\int_{\tau_0}^{\tau}\mu_1(\sigma)d\sigma}\right]\\ &= \lim_{t\to 0}\left[\mathbf{w}_1\left(\frac{1}{t}\right)e^{-\int_{1/t_0}^{1/t}\mu_1(\sigma)d\sigma}\right]\\ &= \lim_{t\to 0}\left[\mathbf{w}_1\left(\frac{1}{t}\right)e^{\int_{1/t_0}^{1/t}\sigma^{-2}\lambda_1(\sigma^{-1})d\sigma}\right]\\ &= \lim_{t\to 0}\left[\mathbf{x}_1(t)e^{\int_t^{t_0}\lambda_1(s)ds}\right].\end{aligned} \tag{33}$$

Analogously, the second equality in (32) can be proved. Now, from (32) we particularly obtain

$$\begin{aligned}&\lim_{t\to 0}\left[\mathbf{x}_1(t)e^{\int_t^{t_0}\lambda_1(s)ds} - \mathbf{b}_1\right] = \mathbf{0},\\ &\lim_{t\to 0}\left[\mathbf{x}_2(t)e^{\int_t^{t_0}\lambda_2(s)ds} - \mathbf{b}_2\right] = \mathbf{0}.\end{aligned} \tag{34}$$

Denoting

$$\begin{aligned}&\mathbf{g}_1(t) = \mathbf{x}_1(t)e^{\int_t^{t_0}\lambda_1(s)ds} - \mathbf{b}_1,\\ &\mathbf{g}_2(t) = \mathbf{x}_2(t)e^{\int_t^{t_0}\lambda_2(s)ds} - \mathbf{b}_2,\end{aligned} \tag{35}$$

from (34), we get $\lim_{t\to 0}\mathbf{g}_i(t) = \mathbf{0}$, $i \in \{1,2\}$ and near $t = 0$,

$$\begin{aligned}&\mathbf{x}_1(t) = \left(\mathbf{b}_1 + \mathbf{g}_1(t)\right)e^{-\int_t^{t_0}\lambda_1(s)ds},\\ &\mathbf{x}_2(t) = \left(\mathbf{b}_2 + \mathbf{g}_2(t)\right)e^{-\int_t^{t_0}\lambda_2(s)ds}.\end{aligned} \tag{36}$$

Since $\lambda_1(t) = \operatorname{Re}\lambda(t) + i\operatorname{Im}\lambda(t)$, $\lambda_2(t) = \operatorname{Re}\lambda(t) - i\operatorname{Im}\lambda(t)$, $\mathbf{b}_i \in \mathbb{C}^2$ and $\mathbf{g}_i : (0, t_0] \to \mathbb{C}^2$, previous statements verify the desired asymptotic formula (21). The statement (22) holds because $\mathbf{b}_1$ and $\mathbf{b}_2$ are two eigenvectors and hence, $\mathbf{b}_i \neq \mathbf{0}$.

Now, according to the asymptotic formula (21), we are able to prove Theorem 2.

Proof of Theorem 2. Because of (4) and Lemmas 6 and 7, we can use (17) and (18) and the asymptotic solution's formula (21)-(22).

At first, we suppose that (5) holds. Taking the norm on both sides in (21) and using (22), we obtain the upper estimate

$$\|\mathbf{x}(t)\|_2 \leq ce^{-\int_t^{t_0}\operatorname{Re}\lambda(s)ds} \quad \text{near } t = 0 \tag{37}$$

with some constant $c > 0$. Now putting the limit as $t \to 0$ on both sides in (37), and using (5), we get $\|x(t)\|_2 \to 0$ as $t \to 0$. Thus, it is shown that the zero solution of system (1) is attractive as $t \to 0$ provided that (5) holds.

The proof of the reverse claim is slightly complicated. Suppose that

$$\|\mathbf{x}(t)\|_2 \longrightarrow 0 \quad \text{as } t \longrightarrow 0. \tag{38}$$

Next, from (17), and (18), we especially obtain

$$\lim_{t\to 0}\left|\int_t^{t_0} \operatorname{Im}\lambda(s)\,ds\right| = \infty. \tag{39}$$

Remarking that in (21) and (22), we use the following notations:

$$\begin{gathered}\mathbf{x}(t) = (x(t), y(t)), \quad \mathbf{h}_i(t) = (h_{i,1}(t), h_{i,2}(t)),\\ \mathbf{c}_i = (c_{i,1}, c_{i,2}), \quad i \in \{1,2\}.\end{gathered} \tag{40}$$

By (22), at least one of $c_{i,1}$ and $c_{i,2}$ is nonzero. So, for instance, $c_{1,1} \neq 0$, which together by (22) gives

$$\lim_{t\to 0}|h_{1,1}(t)| = |c_{1,1}| > 0. \tag{41}$$

Now, from (21) and (40), we derive the following inequality:

$$\begin{aligned}\|\mathbf{x}(t)\| &\geq \|\mathbf{x}(t)\|\\ &= e^{-\int_t^{t_0}\operatorname{Re}\lambda(s)ds}\left|h_{1,1}(t)\cos\left(\int_t^{t_0}\operatorname{Im}\lambda(s)\,ds\right)\right.\\ &\quad\left.+h_{1,2}(t)\sin\left(\int_t^{t_0}\operatorname{Im}\lambda(s)\,ds\right)\right|.\end{aligned} \tag{42}$$

Next, because of (18) and (39), there is a sequence $t_n \in (0, t_0]$, $t_n \to 0$ as $n \to \infty$, such that

$$\left|\int_{t_n}^{t_0}\operatorname{Im}\lambda(s)\,ds\right| = n\pi, \quad n \in \mathbb{N}. \tag{43}$$

Now, putting $t = t_n$ into (42) and using (41), (43), and $\sin(\pm n\pi) = 0$, we get the inequality

$$\|\mathbf{x}(t_n)\|_2 \geq \frac{|c_{1,1}|}{2}e^{-\int_{t_n}^{t_0}\operatorname{Re}\lambda(s)ds} \quad \forall \text{ large } n \in \mathbb{N}. \tag{44}$$

Taking the limit on both sides in previous inequality, with the help of (38), we conclude that

$$\lim_{n\to\infty} e^{-\int_{t_n}^{t_0}\operatorname{Re}\lambda(s)ds} = 0, \tag{45}$$

that is,

$$\lim_{n\to\infty}\int_{t_n}^{t_0}\operatorname{Re}\lambda(s)\,ds = \infty. \tag{46}$$

Since $\operatorname{Re}\lambda(t) \geq 0$ near $t = 0$, the function $\int_t^{t_0}\operatorname{Re}\lambda(s)ds$ is continuous (since matrix elements of $A(t)$ are continuous functions on $(0, t_0]$) and monotone. Hence, statement (46) proves (5). Thus, the statement (5) holds provided that the zero solution of system (1) is attractive.

4. Rectifiable Attractivity of the Zero Solution

Unlike the nonrectifiable attractivity which will be studied in the Section 5, the rectifiable attractivity of the zero solution is proved without any essential difficulties. It is because we can use here the following known result.

Lemma 9. *Let $\Gamma_{\mathbf{x}}$ be a Jordan curve in $\mathbb{R}^2$. Then one has length$(\Gamma_{\mathbf{x}}) < \infty$ if and only if $\mathbf{x}' \in L^1((0,t_0),\mathbb{R}^2)$.*

For the proof of previous lemma, we suggest reader to consult the appropriate scalar case shown in [28].

The following theorem derives some properties of $A(t)$ and its eigenvalues $\lambda(t)$ which ensure that the zero solution of system (1) is rectifiable attractive.

Theorem 10. *Let $\lambda(t) = \operatorname{Re}\lambda(t) \pm i\operatorname{Im}\lambda(t)$ be two complex conjugate eigenvalues of $A(t)$ for $t \in [0, t_0]$. Assume that (5) holds. Let every solution $\mathbf{x}$ of system (1) satisfies (37). If*

$$\|A(t)\|\, e^{-\int_t^{t_0}\operatorname{Re}\lambda(s)ds} \in L^1(0, t_0), \tag{47}$$

then the zero solution of (1) is rectifiable attractive as $t \to 0$.

Proof. It is clear that the attractivity of the zero solution of system (1) follows from (5), (21), and (37). Next, by taking the matrix norm in (1) and using a priori estimate (37), we obtain,

$$\begin{aligned}\|\mathbf{x}'(t)\|_2 &= \|A(t)\mathbf{x}(t)\|_2\\ &\leq \|A(t)\|\,\|\mathbf{x}(t)\|_2 \leq c\,\|A(t)\|\,e^{-\int_t^{t_0}\operatorname{Re}\lambda(s)ds}.\end{aligned} \tag{48}$$

Now, from the assumption (47) and the inequality (48), we observe that

$$\begin{aligned}&\lim_{\varepsilon\to 0}\int_\varepsilon^{t_0}\|\mathbf{x}'(t)\|_2 dt\\ &\quad\leq \lim_{\varepsilon\to 0} c\int_\varepsilon^{t_0}\|A(t)\|\,e^{-\int_t^{t_0}\operatorname{Re}\lambda(s)ds}dt < \infty.\end{aligned} \tag{49}$$

Hence, Lemma 9 proves this theorem.

If $A(t)$ admits the asymptotic behaviour given in (4), then a priori estimate (37) is fulfilled because of the asymptotic formula (21) and thus, the rectifiable attractivity of the zero solution of system (1), in such a case, holds without supposing (37), as follows.

Theorem 11. *Let (4) be satisfied, and let $\lambda(t) = \operatorname{Re}\lambda(t) \pm i\operatorname{Im}\lambda(t)$ be two complex conjugate eigenvalues of $A(t)$ for $t \in [0, t_0]$. Assume that (5) holds. If the condition (47) is fulfilled, then the zero solution of linear system (1) is rectifiable attractive as $t \to 0$.*

Proof. It is clear that all assumptions of Theorem 11 ensure that Lemma 7 can be applied. Hence, by Lemma 7, that is, by asymptotic formula (21), one can easily show that a priori estimate (37) holds for all solutions $\mathbf{x}$ of system (1). Thus, all assumptions of Theorem 10 are fulfilled. Now, Theorem 10 proves Theorem 11.

5. Nonrectifiable Attractivity of the Zero Solution

In this section we study the nonrectifiable attractivity for linear system (1). Unlike the rectifiable attractivity, the non-rectifiable attractivity is a more difficult case. It is because the required property $\mathbf{x}' \notin L^1(0,t_0)$ cannot be derived from system (1) and therefore, we are not in an opportunity to use Lemma 9. Hence, instead of Lemma 9, we state and prove the next new lemma, which plays an essential role in the proof of the main results of this section.

Lemma 12. *Let $u = u(t)$, $u \in C([0,t_0])$, and let there be a sequence $s_n \in (0,t_0]$, a number $n_0 \in \mathbb{N}$, and two functions $\varphi = \varphi(t)$, $a = a(t)$, $t \in (0,t_0]$, such that*

(H_1) sequence s_n is decreasing, $s_n \to 0$ as $n \to \infty$ and $u(s_n)u(s_{n+1}) < 0$ for all $n \geq n_0$;

(H_2) function $\varphi(t)$, $\varphi \in C^1((0,t_0])$, is strictly monotone near $t = 0$, that is, $\varphi'(t) \neq 0$ near $t = 0$ and there is a constant $c_2 > 0$ such that

$$\left|\varphi(s_{n+1}) - \varphi(s_n)\right| \leq c_2 \quad \forall n \geq n_0; \tag{50}$$

(H_3) function $a(t)$, $a \in C([0,t_0])$, is nondecreasing and there is a constant $c_1 > 0$ such that

$$\left|u(s_n)\right| \geq c_1 a(s_n) \quad \forall n \geq n_0; \tag{51}$$

(H_4) $\lim_{\varepsilon\to 0}\int_\varepsilon^{t_0} a(t)|\varphi'(t)|dt = \infty$.

Then the graph

$$\Gamma_u = \left\{(x,y) \in \mathbb{R}^2 : x = t,\ y = u(t)\ \text{for}\ t \in [0,t_0]\right\} \tag{52}$$

of function $u(t)$ is a nonrectifiable curve in $\mathbb{R}^2$.

Proof. Because of (H_1), there is a sequence $a_n \in (s_{n+1}, s_n)$ such that $u(a_n) = 0$ for all $n \geq n_0$ and hence we have

$$\begin{aligned} 0 &< \cdots < s_{n+1} < a_n < s_n < a_{n-1} \\ &< \cdots < s_{n_0+1} < a_{n_0} < s_{n_0} < t_0. \end{aligned} \tag{53}$$

Let $t_n \in (0,t_0]$ be a sequence defined by

$$t_{2n-1} := s_n, \quad t_{2n} := a_n \quad \text{for}\ n \geq 1. \tag{54}$$

Obviously for all $m \geq n_0$, the set $\{0, t_{2m-1}, \ldots, t_{2n_0-1}, t_0\}$ makes a finite partition of interval $[0,t_0]$ and since length(Γ_u) is a supremum of all sums $\sum |u(\tau_n) - u(\tau_{n-1})|$ taken over all finite partitions $\{\tau_n\}$ of $[0,t_0]$; we conclude that

$$\begin{aligned} \text{length}(\Gamma_u) &\geq \left|u(0) - u(t_{2m-1})\right| + \sum_{n=2n_0}^{2m-1}\left|u(t_n) - u(t_{n-1})\right| \\ &\quad + \left|u(t_{2n_0-1}) - u(t_0)\right| \\ &\geq \sum_{n=2n_0}^{2m-1}\left|u(t_n) - u(t_{n-1})\right| = \sum_{n=n_0}^{m}\left|u(s_n)\right|. \end{aligned} \tag{55}$$

Next, let, for instance, $\varphi(t)$ be strictly decreasing near $t = 0$ (the case when $\varphi(t)$ is strictly increasing could be analogously considered). Hence, from (H_2) and (H_3), it follows that

$$\begin{aligned} \sum_{n=n_0}^{m}\left|u(s_n)\right| &\geq c_1\sum_{n=n_0}^{m} a(s_n) \\ &\geq \frac{c_1}{c_2}\sum_{n=n_0}^{m} a(s_n)\left(\varphi(s_{n+1}) - \varphi(s_n)\right) \\ &= \frac{c_1}{c_2}\sum_{n=n_0}^{m}\int_{\varphi(s_n)}^{\varphi(s_{n+1})} a(s_n)\,d\tau \\ &= \frac{c_1}{c_2}\sum_{n=n_0}^{m}\int_{\varphi(s_n)}^{\varphi(s_{n+1})}\left[\max_{t\in[s_{n+1},s_n]} a(t)\right]d\tau \\ &\geq \frac{c_1}{c_2}\sum_{n=n_0}^{m}\int_{\varphi(s_n)}^{\varphi(s_{n+1})} a\left(\varphi^{-1}(\tau)\right)d\tau \\ &= \frac{c_1}{c_2}\sum_{n=n_0}^{m}\int_{s_{n+1}}^{s_n} a(t)\left|\varphi'(t)\right|dt. \end{aligned} \tag{56}$$

Now from (55) and previous inequality, for all $m \geq n_0$ we obtain

$$\begin{aligned} \text{length}(\Gamma_u) &\geq \frac{c_1}{c_2}\sum_{n=n_0}^{m}\int_{s_{n+1}}^{s_n} a(t)\left|\varphi'(t)\right|dt \\ &= \frac{c_1}{c_2}\int_{s_{m+1}}^{s_{n_0}} a(t)\left|\varphi'(t)\right|dt. \end{aligned} \tag{57}$$

Letting $m \to \infty$ in previous inequality, the hypothesis (H_4) proves this lemma.

Now, we are able to prove the following result.

Theorem 13. *Let $\lambda(t) = \operatorname{Re}\lambda(t) \pm i\operatorname{Im}\lambda(t)$ be two complex conjugate eigenvalues of $A(t)$ for $t \in (0,t_0]$. Assume that $\operatorname{Re}\lambda(t) \geq 0$ and $\operatorname{Im}\lambda(t) \neq 0$ near $t = 0$ and that (5) holds. Let every solution $\mathbf{x}(t) = (x(t), y(t))$ of system (1) satisfy (37) as well as the following assumption: let $u(t)$ be defined by $u(t) := x(t)$ or $u(t) := y(t)$ and let there be two constants $c_1 > 0$, $c_2 > 0$, a number $n_0 \in \mathbb{N}$, and a decreasing sequence $s_n \in (0,t_0]$, $s_n \to 0$ as $n \to \infty$, which all depend on function $u(t)$, such that for all $n \geq n_0$, one has*

$$\begin{gathered} u(s_n)u(s_{n+1}) < 0, \quad \left|u(s_n)\right| \geq c_1 e^{-\int_{s_n}^{t_0}\operatorname{Re}\lambda(t)dt}, \\ \left|\int_{s_{n+1}}^{s_n}\operatorname{Im}\lambda(t)\,dt\right| \leq c_2. \end{gathered} \tag{58}$$

If

$$\operatorname{Im}\lambda(t)\,e^{-\int_t^{t_0}\operatorname{Re}\lambda(s)ds}dt \notin L^1(0,t_0), \tag{59}$$

then the zero solution of (1) is nonrectifiable attractive as $t \to 0$.

Proof. At first, the attractivity of the zero solution of linear system (1) follows from the assumptions (5) and (37).

Next, let $\varphi(t)$ and $a(t)$ denote two new functions defined by

$$\varphi(t) = \int_t^{t_0} \operatorname{Im}\lambda(s)\,ds, \quad a(t) = e^{-\int_t^{t_0} \operatorname{Re}\lambda(s)ds}. \tag{60}$$

Let $u(t) := x(t)$ or $u(t) := y(t)$, where $\mathbf{x}(t) = (x(t), y(t))$ is a solution of system (1). With the help of notation (60) and assumption (58), it is easy to check that the function $u(t)$ satisfies all hypotheses (H_1), (H_2), and (H_3) from Lemma 12. Indeed, the existence of a sequence $s_n \in (0, t_0)$ satisfying (H_1) follows from the first inequality in (58). Also, from (18), (58), and (60) we obtain

$$\begin{aligned} &\varphi'(t) = -\operatorname{Im}\lambda(t) \neq 0 \quad \text{near } t = 0, \\ &|\varphi(s_{n+1}) - \varphi(s_n)| = \left|\int_{s_n}^{s_{n+1}} \operatorname{Im}\lambda(s)\,ds\right| \leq c_2 \quad \forall n \geq n_0, \\ &|u(s_n)| \geq c_1 e^{-\int_{s_n}^{t_0} \operatorname{Re}\lambda(s)ds} = c_1 a(s_n) \quad \forall n \geq n_0, \end{aligned} \tag{61}$$

which show that (H_2) and (H_3) from Lemma 12 are fulfilled too. Moreover, the equality

$$a(t)\left|\varphi'(t)\right| = |\operatorname{Im}\lambda(t)|\, e^{-\int_t^{t_0} \operatorname{Re}\lambda(s)ds} \tag{62}$$

together with assumption (59) ensure that the hypothesis (H_4) from Lemma 12 is also fulfilled. Now, by Lemma 12 we obtain that the graph of $u(t) := x(t)$ or $u(t) := y(t)$ is a nonrectifiable curve in $\mathbb{R}^2$. In particular, it follows that the solution curve $\Gamma_{\mathbf{x}}$ is also a nonrectifiable curve in $\mathbb{R}^2$, where $\mathbf{x}(t) = (x(t), y(t))$.

In the next theorem, the nonrectifiable attractivity is obtained without supposing the a priori estimates (37) and (58), since they immediately follow from the asymptotic behaviour of $A(t)$ near $t = 0$ given in (4).

Theorem 14. *Let* (4) *be satisfied, and let* $\lambda(t) = \operatorname{Re}\lambda(t) \pm i\operatorname{Im}\lambda(t)$ *be two complex conjugate eigenvalues of the matrix* $A(t)$ *for* $t \in (0, t_0]$. *Assume that* $\operatorname{Re}\lambda(t) \geq 0$ *near* $t = 0$ *and that* (5) *holds. If the condition* (59) *holds, then the zero solution of* (1) *is nonrectifiable attractive as* $t \to 0$.

Proof. This proof is a consequence of Theorem 13 and the asymptotic formula (21). At first, according to Lemma 7, every solution $\mathbf{x}(t) = (x(t), y(t))$ of system (1) satisfies (21). If we denote by $u(t) := x(t)$ or $u(t) := y(t)$, then (21) ensures that $u(t)$ could be written in the form

$$\begin{aligned} u(t) = &[c_1 + \varepsilon_1(t)]\, a(t)\cos\varphi(t) \\ &+ [c_2 + \varepsilon_2(t)]\, a(t)\sin\varphi(t) \quad \text{near } t = 0, \end{aligned} \tag{63}$$

where $c_1, c_2 \in \mathbb{R}$ with $c_1^2 + c_2^2 > 0$, $\varepsilon_1, \varepsilon_2 \in C(0, t_0]$, $\lim_{t\to 0}\varepsilon_1(t) = \lim_{t\to 0}\varepsilon_2(t) = 0$, and $a(t), \varphi(t)$ are defined in (60).

Since $\lambda(t)$ is a continuous complex-valued function on $(0, t_0]$, from (18), we conclude that

$$\text{either } \operatorname{Im}\lambda(t) < 0 \quad \text{or } \operatorname{Im}\lambda(t) > 0 \quad \text{near } t = 0. \tag{64}$$

Therefore, there are two real constants $c_3, c_4 \in \mathbb{R}$ and $\delta \in C(0, t_0]$ such that $c_3 \neq 0$, $\lim_{t\to 0}\delta(t) = 0$ and

$$\begin{aligned} u(t) = &[c_3 + \delta(t)]\, e^{-\int_t^{t_0} \operatorname{Re}\lambda(\sigma)d\sigma} \\ &\times \sin\left(\int_t^{t_0} |\operatorname{Im}\lambda(\sigma)|\, d\sigma + c_4\right) \quad \text{near } t = 0. \end{aligned} \tag{65}$$

Because of (59), we have $\int_t^{t_0} |\operatorname{Im}\lambda(\sigma)| d\sigma \to \infty$ as $t \to 0$ and therefore there is a decreasing sequence $s_n \in (0, t_0]$ and a number $n_0 \in \mathbb{N}$ such that

$$\int_{s_n}^{t_0} |\operatorname{Im}\lambda(\sigma)| d\sigma + c_4 = \frac{\pi}{2} + n\pi \quad \forall n \geq n_0. \tag{66}$$

Putting (66) into (65) and retaking $n_0 \in \mathbb{N}$ as sufficiently large, we get

$$\begin{aligned} &u(s_n)\, u(s_{n+1}) < 0, \\ &|u(s_n)| \geq \frac{|c_3|}{2} e^{-\int_{s_n}^{t_0} \operatorname{Re}\lambda(s)ds} \quad \forall n \geq n_0, \end{aligned} \tag{67}$$

and because of (64) and (66),

$$\begin{aligned} \left|\int_{s_{n+1}}^{s_n} \operatorname{Im}\lambda(s)\,ds\right| &= \int_{s_{n+1}}^{s_n} |\operatorname{Im}\lambda(s)|\, ds \\ &= \int_{s_{n+1}}^{t_0} |\operatorname{Im}\lambda(s)|\, ds \\ &\quad - \int_{s_n}^{t_0} |\operatorname{Im}\lambda(s)|\, ds = \pi. \end{aligned} \tag{68}$$

Now, statements (67) and (68) show that the function $u(t)$ and sequence s_n defined, respectively, in (65) and (66) satisfy the required condition (58). Therefore, the main conclusion of this theorem immediately follows from Theorem 13.

6. Proof of Theorem 4

In this section, we prove the rectifiable and nonrectifiable attractivity for the zero solution of model system (7) as a direct consequence of Theorem 3.

Let $A(t)$ be the matrix of the model system (7) given by

$$A(t) = \begin{pmatrix} e(t)t^{-p} & f(t)t^{-2} \\ -g(t)t^{-2} & h(t)t^{-q} \end{pmatrix} \quad \text{for } t \in (0, t_0], \tag{69}$$

where the real numbers p, q, and functions e, f, g, and h satisfy (8), (9), and (10), respectively. We will show that such a matrix $A(t)$ together with its eigenvalues $\lambda(t)$ satisfy the required conditions $\operatorname{Re}\lambda(t) \geq 0$ near $t = 0$, (5), and (4).

At first, because of (9), we have $\operatorname{Re}\lambda(t) = (e(t)t^{-p} + h(t)t^{-q})/2 > 0$.

Next, since $\max\{p,q\} \geq 1$, let, for instance, $\max\{p,q\} = p \geq 1$, and $t_0 \in (0,1]$; from (9) we obtain a constant $c_0 > 0$ such that

$$\begin{aligned}\int_t^{t_0} \operatorname{Re}\lambda(s)\,ds &= \frac{1}{2}\int_t^{t_0} \left(e(s)s^{-p} + h(s)s^{-q}\right)ds \\ &\geq \frac{1}{2}\int_t^{t_0} e(s)s^{-p}ds \\ &\geq c_0\int_t^{t_0} s^{-1}ds = c_0 \ln\frac{t_0}{t} \longrightarrow \infty \quad \text{as } t \longrightarrow 0.\end{aligned} \tag{70}$$

Now, it is clear that (70) proves the desired statement (5).

Finally, since $p < 2$ and $q < 2$,

$$\begin{aligned}\lim_{t\to 0}\left[t^2 A(t)\right] &= \lim_{t\to 0}\begin{pmatrix} e(t)t^{2-p} & f(t) \\ -g(t) & h(t)t^{2-q}\end{pmatrix} \\ &= \begin{pmatrix} 0 & f(0) \\ -g(0) & 0\end{pmatrix} =: A_0,\end{aligned} \tag{71}$$

where A_0 is a constant matrix. Since $f(0)g(0) > 0$, matrix A_0 has two different eigenvalues. So, the condition (4) is fulfilled too.

Thus, matrix $A(t)$ together with its eigenvalues satisfy all required conditions so that Theorem 3 can be applied to the model system (7) in all cases of p and q given in (11)–(16). Since by Theorem 3 the rectifiable and nonrectifiable attractivity of the zero solution of (1) depends on the integrability of function

$$\Phi(t) = |\operatorname{Im}\lambda(t)|\, e^{-\int_t^{t_0}\operatorname{Re}\lambda(s)ds} \quad \text{for } t \in (0,t_0], \tag{72}$$

we can conclude that the zero solutions of (7) is rectifiable attractive if $\Phi \in L^1(0,t_0)$ and nonrectifiable attractive if $\Phi \notin L^1(0,t_0)$. Let us remark that in Lemma 6, it has been proved that

$$|\operatorname{Im}\lambda(t)| \sim t^{-2} \quad \text{as } t \longrightarrow 0. \tag{73}$$

(i) Assume that $\max\{p,q\} > 1$. For instance, we assume $p > 1$. Since $e(t) > 0$ on $[0,t_0]$, we have $e(t)t^{-p+1} \to \infty$ as $t \to 0$. In particular, there exist $t_1 \in (0,t_0]$ and $\delta > 0$ such that

$$e(t)t^{-p+1} \geq 2 + \delta \quad \text{for } t \in (0,t_1]. \tag{74}$$

Since

$$e^{-\int_t^{t_0}\operatorname{Re}\lambda(s)ds} = e^{(-1/2)\int_t^{t_0}(e(s)s^{-p}+h(s)s^{-q})ds}, \tag{75}$$

we have

$$\begin{aligned}e^{-\int_t^{t_0}\operatorname{Re}\lambda(s)ds} &\leq e^{(-1/2)\int_t^{t_0}e(s)s^{-p}ds} \\ &= e^{(-1/2)\int_t^{t_0}e(s)s^{-p+1}s^{-1}ds} \\ &\leq e^{-(1+(\delta/2))\int_t^{t_0}s^{-1}ds} = ct^{1+(\delta/2)},\end{aligned} \tag{76}$$

where the constant $c = t_0^{-1-(\delta/2)}$. Now from (72), (73), and (76), we observe that

$$\Phi(t) \leq c_1 t^{-2} t^{1+(\delta/2)} = c_1 t^{-1+(\delta/2)} \in L^1(0,t_0). \tag{77}$$

Thus, it is shown that $\Phi \in L^1(0,t_0)$.

(ii) Assume that (11) is fulfilled. Since $e(t) > 2$ on $[0,t_0]$, there exists $\delta > 0$ such that $e(t) \geq 2+\delta$ on $[0,t_0]$. Since $p = 1$ and $h(t) > 0$ on $(0,t_0]$, from (75) we have

$$\begin{aligned}e^{-\int_t^{t_0}\operatorname{Re}\lambda(s)ds} &\leq e^{(-1/2)\int_t^{t_0}e(s)s^{-1}ds} \\ &\leq e^{-(1+(\delta/2))\int_t^{t_1}s^{-1}ds} = ct^{1+(\delta/2)},\end{aligned} \tag{78}$$

where the constant $c = t_0^{-1-(\delta/2)}$. Now from (72), (73), and (78), we obtain $\Phi \in L^1(0,t_0)$. When (12) is fulfilled, we obtain $\Phi \in L^1(0,t_0)$ by the similar argument.

Assume that (13) is fulfilled. Since $e(t)+h(t) > 2$ on $[0,t_0]$, there exists $\delta > 0$ such that $e(t)+h(t) \geq 2+\delta$ on $[0,t_0]$. From (75) we have

$$\begin{aligned}e^{-\int_t^{t_0}\operatorname{Re}\lambda(s)ds} &= e^{(-1/2)\int_t^{t_0}e(s)+h(s)s^{-1}ds} \\ &\leq e^{-(1+(\delta/2))\int_t^{t_0}s^{-1}ds} = ct^{1+(\delta/2)},\end{aligned} \tag{79}$$

where the constant $c = t_0^{-1-(\delta/2)}$. Thus we obtain $\Phi \in L^1(0,t_0)$ by the argument above.

(iii) Let (14) be fulfilled. Since $h \in C([0,t_0])$, there is a constant $m_0 > 0$ such that $h(t) \leq m_0$ on $[0,t_0]$, and since $q < 1$ there is a constant $c_0 > 0$ such that

$$\begin{aligned}e^{(-1/2)\int_t^{t_0}h(s)s^{-q}ds} &\geq e^{-(m_0/2)\int_t^{t_0}s^{-q}ds} \geq c_0 \\ &\text{as } t \longrightarrow 0.\end{aligned} \tag{80}$$

Hence,

$$\begin{aligned}e^{-\int_t^{t_0}\operatorname{Re}\lambda(s)ds} &= e^{(-1/2)\int_t^{t_0}(e(s)s^{-1}+h(s)s^{-q})ds} \\ &= e^{-\int_t^{t_0}s^{-1}ds}e^{(-1/2)\int_t^{t_0}h(s)s^{-q}ds} \\ &\geq c_0 e^{-\int_t^{t_0}s^{-1}ds} = c_0\frac{t}{t_0}.\end{aligned} \tag{81}$$

Now, by combining (72), (73), and (81), we observe

$$\begin{aligned}\int_t^{t_0}\Phi(s)\,ds &\geq \frac{c_1}{t_0}\int_t^{t_0}s^{-2}s\,ds = \frac{c_1}{t_0}\ln\frac{t_0}{t} \longrightarrow \infty \\ &\text{as } t \to 0.\end{aligned} \tag{82}$$

Thus, $\Phi \notin L^1(0,t_0)$. When (15) is fulfilled, we similarly obtain $\Phi \notin L^1(0,t_0)$.

Let (16) be fulfilled. Then, from (75) we have

$$\begin{aligned}e^{-\int_t^{t_0}\operatorname{Re}\lambda(s)ds} &= e^{(-1/2)\int_t^{t_0}(e(s)+h(s))s^{-1}ds} \\ &\geq e^{-\int_t^{t_0}s^{-1}ds} = \frac{t}{t_0}.\end{aligned} \tag{83}$$

By combining (72), (73), and (83), we obtain $\Phi \notin L^1(0,t_0)$. Thus, the proof is completed.

Appendix

At first, we state and prove the existence and uniqueness of solutions of linear system (1).

Theorem A.1. *There exists a unique solution* $\mathbf{x}$ *of linear system* (1).

Proof. Putting $\tau = 1/t$ into system (1) and denoting $\mathbf{w}(\tau) = \mathbf{x}(1/\tau)$, we obtain the equivalent system

$$\frac{d\mathbf{w}}{d\tau} = B(\tau)\mathbf{w} \quad \text{for } \tau \in [\tau_0, \infty), \quad \mathbf{w}(\tau_0) = \mathbf{x}_0, \tag{A.1}$$

where $\tau_0 = 1/t_0$ and the matrix $B(t)$ is defined by (23). By the standard theory for system (A.1), we obtain the existence and uniqueness of the solution to (1).

At the second, we give some simple sufficient conditions on the matrix-valued function $A = A(t)$ such that the solution curve $\Gamma_{\mathbf{x}}$ of every nontrivial solution $\mathbf{x}(t)$ of linear system (1) (as well as of (7)) is a Jordan curve.

Definition A.2. A solution $\mathbf{x}$ of linear system (1) is said to be *trivial solution* of (1) *near* $t = 0$ if there is $t_1 \in (0, t_0]$ such that $\mathbf{x}(t) = 0$ for all $t \in (0, t_1]$. Otherwise, $\mathbf{x}(t)$ is said to be *nontrivial solution* of (1) *near* $t = 0$.

Theorem A.3. *Let the matrix elements* $a_{i,j}(t)$ *of* $A(t)$ *satisfy*

$$\min\{a_{1,1}(t), a_{2,2}(t)\} > \frac{1}{2}|a_{1,2}(t) + a_{2,1}(t)| \quad t \in (0, t_0]. \tag{A.2}$$

If $\mathbf{x}$ *is a nontrivial solution of linear system* (1) *near* $t = 0$, *then* $\|\mathbf{x}(t)\|_2$ *is strictly positive and decreasing on* $(0, t_0]$. *In particular, the solution curve* $\Gamma_{\mathbf{x}}$ *is a Jordan curve.*

Remark A.4. It is clear that if $a_{1,1}(t) > 0$, $a_{2,2}(t) > 0$ and $a_{1,2}(t) = -a_{2,1}(t)$ on $(0, t_0]$, then condition (A.2) is fulfilled.

Proof of Theorem A.3. Let $\mathbf{x}(t) = (x(t), y(t))$ be a nontrivial solution of system (1) near $t = 0$. The assumption (A.2) can be rewritten as

$$\begin{aligned} a_{1,1}(t) &> \frac{1}{2}|a_{1,2}(t) + a_{2,1}(t)|, \\ a_{2,2}(t) &> \frac{1}{2}|a_{1,2}(t) + a_{2,1}(t)| \quad \text{on } (0, t_0]. \end{aligned} \tag{A.3}$$

Let $v(t)$ be a function defined by

$$v(t) = \|\mathbf{x}(t)\|^2 = x^2(t) + y^2(t) \quad \text{for } t \in (0, t_0]. \tag{A.4}$$

Taking the derivative d/dt in (A.4), it follows that

$$\frac{dv}{dt} = 2x(t)x'(t) + 2y(t)y'(t) \quad \text{for } t \in (0, t_0]. \tag{A.5}$$

Multiplying the first and second equation in system (1), respectively, by $x(t)$ and $y(t)$, we obtain

$$\begin{aligned} &x(t)x'(t) + y(t)y'(t) \\ &= a_{1,1}(t)x^2(t) + (a_{1,2}(t) + a_{2,1}(t))x(t)y(t) \\ &\quad + a_{2,2}(t)y^2(t). \end{aligned} \tag{A.6}$$

Because of inequality $ab \le (a^2 + b^2)/2$, we have

$$\begin{aligned} &|(a_{1,2}(t) + a_{2,1}(t))x(t)y(t)| \\ &\le \frac{1}{2}|a_{1,2}(t) + a_{2,1}(t)|\left(x^2(t) + y^2(t)\right), \end{aligned} \tag{A.7}$$

which together with (A.6) derives that

$$\begin{aligned} &x(t)x'(t) + y(t)y'(t) \\ &\ge a_{1,1}(t)x^2(t) - \frac{1}{2}|a_{1,2}(t) + a_{2,1}(t)|\left(x^2(t) + y^2(t)\right) \\ &\quad + a_{2,2}(t)y^2(t) = D(t), \end{aligned} \tag{A.8}$$

where $D(t)$ denotes the function

$$\begin{aligned} D(t) = &\left(a_{1,1}(t) - \frac{1}{2}|a_{1,2}(t) + a_{2,1}(t)|\right)x^2(t) \\ &+ \left(a_{2,2}(t) - \frac{1}{2}|a_{1,2}(t) + a_{2,1}(t)|\right)y^2(t). \end{aligned} \tag{A.9}$$

Putting (A.8) into (A.5), we obtain

$$\frac{dv}{dt} \ge 2D(t) \quad \text{for } t \in (0, t_0]. \tag{A.10}$$

Because of (A.3), we have $D(t) \ge 0$ on $(0, t_0]$. Hence, from (A.10), we observe $dv/dt \ge 0$ on $(0, t_0]$. Integrating this inequality over the interval $[s, t]$ for all $0 < s \le t \le t_0$, we get $v(s) \le v(t)$. If we suppose for a moment that $v(t) = 0$, then from previous inequality we get $v(s) = 0$ for all $s \in (0, t]$, which is not possible because of Definition A.2. Therefore, $v(t) > 0$ for all $t \in (0, t_0]$, which implies $\|\mathbf{x}(t)\|_2 > 0$ for all $t \in (0, t_0]$ because of (A.4). It proves the first statement of this theorem.

Next, with the help of (A.3) and $\|\mathbf{x}(t)\|_2 > 0$ for all $t \in (0, t_0]$, we deduce that $D(t) > 0$ on $(0, t_0]$, which together with (A.10) gives $v'(t) > 0$ on $(0, t_0]$. Integrating this inequality over the interval $[s, t]$ for all $0 < s < t \le t_0$, we get

$$v(s) < v(t) \quad \forall s, t \text{ such that } 0 < s < t \le t_0. \tag{A.11}$$

If, for instance, there exist s, t such that $0 < s < t \le t_0$ and $\|\mathbf{x}(s)\|_2 = \|\mathbf{x}(t)\|_2$, then involving this equality into (A.4), we obtain $v(s) = v(t)$, which is not possible because of (A.11). Therefore, $\|\mathbf{x}(t)\|_2 \ne \|\mathbf{x}(s)\|_2$. Thus, it is shown that $\Gamma_{\mathbf{x}}$ is a Jordan curve near $t = 0$. In the same way, $\|\mathbf{x}(s)\|_2 < \|\mathbf{x}(t)\|_2$ for all s, t such that $0 < s < t \le t_0$.

According to Remark A.4, previous theorem can be applied to model system (7).

Corollary A.5. *Let* $e, h \in C((0, t_0])$, $e(t) > 0$, $h(t) > 0$, *and* $f(t) = g(t)$ *near* $t = 0$. *If* $\mathbf{x}(t)$ *is a nontrivial solution of linear system (7) near* $t = 0$, *then* $\|\mathbf{x}(t)\|_2$ *is strictly positive and decreasing on* $(0, t_0]$. *In particular, the solution curve* $\Gamma_{\mathbf{x}}$ *is a Jordan curve.*

Finally, we point out that in some essential cases, the model system (7) allows the explicit form of all its solutions. Such systems are called the integrable systems; for details, see [4]. Hence, the rectifiable and nonrectifiable attractivity of the zero solution of (7) in such cases can be reproven in an explicit way. That is, the statements of Theorem 4 can be confirmed once more, but explicitly.

Proposition A.6. *Let* $h(t) > 0$ *and* $f(t) > 0$ *on* $(0, t_0]$. *The fundamental system of all solutions of linear differential system*

$$\begin{aligned} x'(t) &= h(t)t^{-p}x(t) + f(t)t^{-2}y(t), \\ y'(t) &= -f(t)t^{-2}x(t) + h(t)t^{-p}y(t), \quad t \in (0, t_0] \end{aligned} \tag{A.12}$$

is explicitly given by the formula

$$\begin{aligned} x(t) &= e^{-\int_t^{t_0} h(s)s^{-p}ds} \\ &\times \left[-c_1 \sin\left(\int_t^{t_0} f(s)s^{-2}ds\right)\right. \\ &\left.+c_2 \cos\left(\int_t^{t_0} f(s)s^{-2}ds\right)\right], \\ y(t) &= e^{-\int_t^{t_0} h(s)s^{-p}ds} \\ &\times \left[c_1 \cos\left(\int_t^{t_0} f(s)s^{-2}ds\right)\right. \\ &\left.+c_2 \sin\left(\int_t^{t_0} f(s)s^{-2}ds\right)\right]. \end{aligned} \tag{A.13}$$

Obviously for $p = q$, $e(t) = h(t) > 0$ and $g(t) = f(t) > 0$ on $(0, t_0]$, the system (A.12) is a particular case of our model system (7). Hence, according to Theorem 4, we can state the rectifiable and nonrectifiable attractivity near $x = 0$ for the system (A.12) as follows.

Corollary A.7. *Let* $h(t) > 0$ *and* $f(t) > 0$ *on* $(0, t_0]$.

(i) *If* $p > 1$, *then the zero solution of* (A.12) *is rectifiable attractive as* $t \to 0$.

(ii) *If* $p = 1$, *then the zero solution of* (A.12) *is rectifiable attractive as* $t \to 0$ *provided that* $h(t) > 1$.

(iii) *If* $p = 1$, *then the zero solution of* (A.12) *is nonrectifiable attractive as* $t \to 0$ *provided that* $h(t) \le 1$.

However, since by Proposition A.6 the system (A.12) allows the explicit form of all its solutions, this means that Corollary A.7 can be also shown by using Lemma 9, Lemma 12, and Proposition A.6.

References

[1] K. Falconer, *Fractal Geometry*, Mathematical Foundations and Applications, John Wiley & Sons, 1999.

[2] P. Mattila, *Geometry of Sets and Measures in Euclidean Spaces*, vol. 44 of *Fractals and Rectifiability*, Cambridge University Press, Cambridge, UK, 1995.

[3] C. Tricot, *Curves and Fractal Dimension*, Springer, New York, NY, USA, 1995.

[4] S. Miličić and M. Pašić, "Nonautonomous differential equations in Banach space and nonrectifiable attractivity in two-dimensional linear differential systems," *Abstract and Applied Analysis*, vol. 2013, Article ID 935089, 10 pages, 2013.

[5] W. A. Coppel, *Stability and Asymptotic Behavior of Differential Equations*, D. C. Heath and Co., Boston, Mass, USA, 1965.

[6] L. Cesari, *Asymptotic Behavior and Stability Problems in Ordinary Differential Equations*, Springer, New York, NY, USA, 3rd edition, 1971.

[7] I. T. Kiguradze, *Some Singular Boundary Value Problems for Ordinary Differential Equations*, Izdatel'stvo Tbilisskogo Universiteta, Tbilisi, Ga, USA, 1975.

[8] P. Hartman, *Ordinary Differential Equations*, Birkhäuser, Boston, Mass, USA, 2nd edition, 1982.

[9] M. Pinto, "Asymptotic integration of second-order linear differential equations," *Journal of Mathematical Analysis and Applications*, vol. 111, no. 2, pp. 388–406, 1985.

[10] J. Rovder, "Asymptotic and oscillatory behaviour of solutions of a linear differential equation," *Journal of Computational and Applied Mathematics*, vol. 41, no. 1-2, pp. 41–47, 1992.

[11] S. Castillo and M. Pinto, "Asymptotic integration of ordinary different systems," *Journal of Mathematical Analysis and Applications*, vol. 218, no. 1, pp. 1–12, 1998.

[12] I. Rachunkova and I. Rachunek, "Asymptotic formula for oscillatory solutions of some singular nonlinear differential equation," *Abstract and Applied Analysis*, vol. 2011, Article ID 981401, 9 pages, 2011.

[13] J. Jaros and T. Kusano, "Existence and precise asymptotic behavior of strongly monotone solutions of systems of nonlinear differential equations," *Difference Equations and Their Applications*, vol. 5, pp. 185–204, 2013.

[14] M. Fečkan, *Bifurcation and Chaos in Discontinuous and Continuous Systems*, Nonlinear Physical Science, Springer, 2011.

[15] O. G. Mustafa and Y. V. Rogovchenko, "Asymptotic integration of a class of nonlinear differential equations," *Applied Mathematics Letters*, vol. 19, no. 9, pp. 849–853, 2006.

[16] M. Rasmussen, *Attractivity and Bifurcation for Nonautonomous Dynamical Systems*, vol. 1907 of *Lecture Notes in Mathematics*, Springer, Berlin, Germany, 2007.

[17] L. Hatvani, "On the asymptotic stability for a two-dimensional linear nonautonomous differential system," *Nonlinear Analysis: Theory, Methods & Applications*, vol. 25, no. 9-10, pp. 991–1002, 1995.

[18] J. Sugie and M. Onitsuka, "Integral conditions on the uniform asymptotic stability for two-dimensional linear systems with time-varying coefficients," *Proceedings of the American Mathematical Society*, vol. 138, no. 7, pp. 2493–2503, 2010.

[19] F. Amato, M. Ariola, M. Carbone, and C. Cosentino, "Finite-time control of linear systems: a survey," in *Current Trends in Nonlinear Systems and Control*, pp. 195–213, Birkhäuser, Boston, Mass, USA, 2006.

[20] G. Haller, "A variational theory of hyperbolic Lagrangian coherent structures," *Physica D*, vol. 240, no. 7, pp. 574–598, 2011.

[21] K. Rateitschak and O. Wolkenhauer, "Thresholds in transient dynamics of signal transduction pathways," *Journal of Theoretical Biology*, vol. 264, no. 2, pp. 334–346, 2010.

[22] M. K. Kwong, M. Pašić, and J. S. W. Wong, "Rectifiable oscillations in second-order linear differential equations," *Journal of Differential Equations*, vol. 245, no. 8, pp. 2333–2351, 2008.

[23] M. Pašić, "Rectifiable and unrectifiable oscillations for a class of second-order linear differential equations of Euler type," *Journal of Mathematical Analysis and Applications*, vol. 335, no. 1, pp. 724–738, 2007.

[24] M. Pašić and J. S. W. Wong, "Rectifiable oscillations in second-order half-linear differential equations," *Annali di Matematica Pura ed Applicata. Series IV*, vol. 188, no. 3, pp. 517–541, 2009.

[25] M. Pašić and S. Tanaka, "Rectifiable oscillations of self-adjoint and damped linear differential equations of second-order," *Journal of Mathematical Analysis and Applications*, vol. 381, no. 1, pp. 27–42, 2011.

[26] J. S. W. Wong, "On rectifiable oscillation of Euler type second order linear differential equations," *Electronic Journal of Qualitative Theory of Differential Equations*, vol. 20, pp. 1–12, 2007.

[27] J. S. W. Wong, "On rectifiable oscillation of Emden-Fowler equations," *Memoirs on Differential Equations and Mathematical Physics*, vol. 42, pp. 127–144, 2007.

[28] L. C. Evans and R. F. Gariepy, *Measure Theory and Fine Properties of Functions*, CRC Press, New York, NY, USA, 1999.

Erratum to "Positive Solution to a Fractional Boundary Value Problem"

A. Guezane-Lakoud and R. Khaldi

Laboratory of Advanced Materials, Faculty of Sciences, University Badji Mokhtar-Annaba, P.O. Box 12, 23000 Annaba, Algeria

Correspondence should be addressed to A. Guezane-Lakoud; a_guezane@yahoo.fr

In the paper entitled "Positive solution to a fractional boundray value problems," the following problem (P1) is studied:

$$ {}^{c}D_{0^+}^{q}u(t) = f\left(t, u(t), {}^{c}D_{0^+}^{\sigma}u(t)\right), \quad 0 < t < 1, \tag{1.1}$$

$$ u(0) = u''(0) = 0, \qquad u'(1) = \alpha u''(1), \tag{1.2}$$

where $f : [0,1] \times \mathbb{R} \times \mathbb{R} \to \mathbb{R}$ is a given function, $2 < q < 3$, and $1 < \sigma < 2$. Remarking that all the calculuses in this paper are done for $0 < \sigma < 1$ and that if we take $1 < \sigma < 2$, then ${}^{c}D_{0^+}^{\sigma}Tu = (1/\Gamma(2-\sigma))\int_0^t ((Tu)''(s)/(t-s)^{\sigma-1})ds$ and the second derivative with respect to t of $G(t,s)$ is discontinuous for $s = t$, consequently we cannot apply this method to establish the existence and positivity of solution. For this reason, we correct the study of problem (P1) by taking $0 < \sigma < 1$, and then the following corrections are needed.

(1) In page 3, in Lemma 2.3, we should correct ${}^{c}D_{0^+}^{\alpha}t^{\beta-1} = (\Gamma(\beta)/\Gamma(\beta-\alpha))t^{\beta-\alpha-1}$, $\beta > n$.

(2) Equation (2.6) must be

$$ u(t) = \frac{1}{\Gamma(q-2)} \int_0^1 \frac{1}{(1-s)^{3-q}} G(t,s)\, y(s)\, ds. \tag{2.6}$$

The Green function in (2.7) is

$$ G(t,s) = \begin{cases} \dfrac{(1-s)^{3-q}(t-s)^{q-1}}{(q-1)(q-2)} + \alpha t - \dfrac{t(1-s)}{q-2}, & s < t, \\ \alpha t - \dfrac{t(1-s)}{q-2}, & t \le s. \end{cases} \tag{2.7}$$

(3) Equation (2.11) becomes

$$ u(t) = \frac{1}{\Gamma(q-2)} \times \int_0^t \left[\frac{(t-s)^{q-1}}{(q-1)(q-2)} + \frac{\alpha t}{(1-s)^{3-q}} - \frac{t(1-s)^{q-2}}{q-2}\right] \times y(s)\, ds + \frac{1}{\Gamma(q-2)} \int_t^1 \left[\frac{\alpha t}{(1-s)^{3-q}} - \frac{t(1-s)^{q-2}}{q-2}\right] y(s)\, ds. \tag{2.11}$$

Equation (2.12) must be

$$ u(t) = \frac{1}{\Gamma(q-2)} \int_0^1 \frac{1}{(1-s)^{3-q}} G(t,s)\, y(s)\, ds. \tag{2.12}$$

(4) Equation (3.1) must be

$$ Tu(t) = \frac{1}{\Gamma(q-2)} \times \int_0^1 \frac{1}{(1-s)^{3-q}} G(t,s)\, f\left(s, u(s), {}^{c}D_{0^+}^{\sigma}u(s)\right) ds. \tag{3.1}$$

In Theorem 3.2, the condition (3.5) must be

$$ C_g + C_h < \frac{1}{2}, \qquad A_g + A_h < \frac{\Gamma(2-\sigma)}{2}. \tag{3.5}$$

(5) Equation (3.12) must be

$$|Tu - Tv| < \frac{1}{2} \|u - v\|, \tag{3.12}$$

and (3.13) becomes

$$ {}^{c}D_{0^+}^{\sigma} Tu - {}^{c}D_{0^+}^{\sigma} Tv = \frac{1}{\Gamma(1-\sigma)} \int_0^t \frac{(Tu)'(s) - (Tv)'(s)}{(t-s)^{\sigma}} ds. \tag{3.13}$$

(6) Equation (3.14) is

$$G_1(t,s) = \frac{\partial G(t,s)}{\partial t} = \begin{cases} \dfrac{(1-s)^{3-q}(t-s)^{q-2}}{(q-2)} + \alpha - \dfrac{(1-s)}{q-2}, & s < t, \\ \alpha - \dfrac{(1-s)}{q-2}, & t \le s. \end{cases} \tag{3.14}$$

(7) Equation (3.15) is as follows:

$$\begin{aligned} &{}^{c}D_{0^+}^{\sigma} Tu - {}^{c}D_{0^+}^{\sigma} Tv \\ &= \frac{1}{\Gamma(q-2)\Gamma(1-\sigma)} \\ &\times \int_0^t \int_0^1 (t-s)^{-\sigma} \frac{1}{(1-r)^{3-q}} G_1(s,r) \\ &\quad \times \left(f\left(r, u(r), {}^{c}D_{0^+}^{\sigma} u(r)\right) \right. \\ &\quad \left. - f\left(r, v(r), {}^{c}D_{0^+}^{\sigma} v(r)\right)\right) dr\, ds. \end{aligned} \tag{3.15}$$

Equation (3.16) is as follows:

$$\begin{aligned} &\left|{}^{c}D_{0^+}^{\sigma} Tu - {}^{c}D_{0^+}^{\sigma} Tv\right| \\ &\le \frac{\max|u-v|}{\Gamma(q-2)\Gamma(1-\sigma)} \\ &\times \int_0^t \int_0^1 (t-s)^{-\sigma} \frac{1}{(1-r)^{3-q}} |G_1(s,r)|\, g(r)\, dr\, ds \\ &+ \frac{\max\left|{}^{c}D_{0^+}^{\sigma} u - {}^{c}D_{0^+}^{\sigma} v\right|}{\Gamma(q-2)\Gamma(1-\sigma)} \\ &\times \int_0^t \int_0^1 (t-s)^{-\sigma} \frac{1}{(1-r)^{3-q}} |G_1(s,r)|\, h(r)\, dr\, ds. \end{aligned} \tag{3.16}$$

Equation (3.17) is as follows:

$$\begin{aligned} &\int_0^1 \frac{1}{(1-r)^{3-q}} |G_1(s,r)|\, g(r)\, dr \\ &\le \Gamma(q-2)\left(2 I_{0^+}^{q-1} g(1) + |\alpha|\, I_{0^+}^{q-2} g(1)\right). \end{aligned} \tag{3.17}$$

(8) Equation (3.18) is as follows:

$$\left|{}^{c}D_{0^+}^{\sigma} Tu - {}^{c}D_{0^+}^{\sigma} Tv\right| \le \|u - v\| \frac{1}{\Gamma(2-\sigma)} \left(A_g + A_h\right). \tag{3.18}$$

Equation (3.19) becomes

$$\left|{}^{c}D_{0^+}^{\sigma} Tu - {}^{c}D_{0^+}^{\sigma} Tv\right| \le \frac{1}{2} \|u - v\|. \tag{3.19}$$

Equation (3.21) is as follows:

$$\text{a.e } (t, x, \overline{x}) \in [0,1] \times \mathbb{R}^2. \tag{3.21}$$

Equation (3.22) is as follows:

$$(\psi(r) + \phi(r) + 1)\left(\frac{C_1}{\Gamma(q-2)} + \frac{C_2}{\Gamma(2-\sigma)}\right) < r. \tag{3.22}$$

Equation (3.29) becomes

$$\left|{}^{c}D_{0^+}^{\sigma} Tu\right| \le \frac{C_2}{\Gamma(2-\sigma)} (\psi(r) + \phi(r) + 1). \tag{3.29}$$

Equation (3.30) is

$$\|Tu\| = (\psi(r) + \phi(r) + 1)\left(\frac{C_1}{\Gamma(q-2)} + \frac{C_2}{\Gamma(2-\sigma)}\right). \tag{3.30}$$

Equation (3.32) is as follows:

$$\begin{aligned} &|Tu(t_1) - Tu(t_2)| \\ &\le \frac{C}{\Gamma(q-2)} \\ &\times \int_0^{t_1} \frac{(t_2 - s)^{q-1} - (t_1 - s)^{q-1}}{(q-1)(q-2)} \\ &\qquad + (t_2 - t_1)\left(\frac{|\alpha|}{(1-s)^{3-q}} - \frac{(1-s)^{q-2}}{q-2}\right) ds \\ &+ \int_{t_1}^{t_2} \frac{(t_2 - s)^{q-1}}{(q-1)(q-2)} \\ &\qquad + (t_2 - t_1)\left(\frac{|\alpha|}{(1-s)^{3-q}} - \frac{(1-s)^{q-2}}{q-2}\right) ds \\ &+ \int_{t_2}^{1} (t_1 - t_2)\left(\frac{|\alpha|}{(1-s)^{3-q}} - \frac{(1-s)^{q-2}}{q-2}\right) ds. \end{aligned} \tag{3.32}$$

Equation (3.33) is as follows:

$$|Tu(t_1) - Tu(t_2)| \le \frac{C}{\Gamma(q-2)} \times \left[(t_2 - t_1) \times \int_0^{t_1} \frac{1}{(q-2)} + \left(\frac{|\alpha|}{(1-s)^{3-q}} + \frac{(1-s)^{q-2}}{q-2} \right) ds + \int_{t_1}^{t_2} \frac{(t_2 - s)^{q-1}}{(q-2)(q-1)} + (t_2 - t_1) \left(\frac{|\alpha|}{(1-s)^{3-q}} + \frac{(1-s)^{q-2}}{q-2} \right) ds + (t_1 - t_2) \int_{t_2}^{1} \left(\frac{|\alpha|}{(1-s)^{3-q}} + \frac{(1-s)^{q-2}}{q-2} \right) ds \right]. \tag{3.33}$$

Equation (3.34) is as follows:

$$|Tu(t_1) - Tu(t_2)| \le \frac{C(t_2 - t_1)}{\Gamma(q-2)} \frac{4 + 3|\alpha|}{(q-2)} + \frac{C}{\Gamma(q-2)} \int_{t_1}^{t_2} \frac{(t_2 - s)^{q-1}}{(q-2)(q-1)} ds. \tag{3.34}$$

Equation (3.35) is as follows:

$$\left| {}^cD_{0^+}^{\sigma} Tu(t_1) - {}^cD_{0^+}^{\sigma} Tu(t_2) \right| \le \frac{1}{\Gamma(1-\sigma)} \int_0^{t_1} \left((t_1 - s)^{-\sigma} - (t_2 - s)^{-\sigma} \right) \times \left| (Tu(s))' \right| ds + \frac{1}{\Gamma(1-\sigma)} \int_{t_1}^{t_2} (t_2 - s)^{-\sigma} \left| (Tu(s))' \right| ds. \tag{3.35}$$

Equation (3.37) is as follows:

$$\left| {}^cD_{0^+}^{\sigma} Tu(t_1) - {}^cD_{0^+}^{\sigma} Tu(t_2) \right| \le \left([\psi(r) + \phi(r) + 1] \times C_2 \left[2(t_2 - t_1)^{1-\sigma} + t_2^{1-\sigma} - t_1^{1-\sigma} \right] \right) \times (\Gamma(2-\sigma))^{-1}. \tag{3.37}$$

Equation (3.39) is as follows:

$$\left| {}^cD_{0^+}^{\sigma} u(t) \right| \le \frac{C_2}{\Gamma(2-\sigma)} (\psi(r) + \phi(r) + 1). \tag{3.39}$$

Equation (3.40) is as follows:

$$\|u\| \le (\psi(r) + \phi(r) + 1) \left(\frac{C_1}{\Gamma(q-2)} + \frac{C_2}{\Gamma(2-\sigma)} \right) < r. \tag{3.40}$$

(9) (H2) $0 < \int_0^1 (1/(1-s)^{3-q}) G(s,s) a(s) ds < \infty$.
In Lemma 4.1, we have $\alpha \ge 1/(q-2)$ and (4.1) becomes:
If $t, s \in [\tau, 1]$, $\tau > 0$, then

$$\begin{gathered} 0 < \tau G(s,s) \le G(t,s) \le \frac{2}{\tau^2} G(s,s), \\ 0 < G_1(s,s) \le G_1(t,s) \le \frac{2}{\tau} G_1(s,s). \end{gathered} \tag{4.1}$$

Equation (4.2) should be

$$\frac{\alpha t}{(1-s)^{3-q}} - \frac{t(1-s)^{q-2}}{q-2} = \frac{t}{(q-2)(1-s)^{3-q}} [(q-2)\alpha - 1 + s] \tag{4.2}$$

which is positive if $\alpha \ge 1/(q-2)$.

Equation (4.3): let $t, s \in [\tau, 1]$; it is easy to see that $G(s,s) \ne 0$, and then we have

$$\begin{aligned} \frac{G(t,s)}{G(s,s)} &= \frac{(t-s)^{q-1}(1-s)^{3-q}}{(q-1)s[(q-2)\alpha - 1 + s]} + \frac{t}{s} \\ &\le \frac{1 + (1-s)^2}{s^2} \\ &\le \frac{2}{\tau^2}, \quad 0 < \tau \le s \le t \le 1, \\ \frac{G(t,s)}{G(s,s)} &= \frac{t}{s} \le \frac{2}{\tau} \le \frac{2}{\tau^2}, \quad 0 < \tau \le t \le s \le 1. \end{aligned} \tag{4.3}$$

Finally, since $G(s,s)$ is nonnegative, we obtain $0 < \tau G(s,s) \le G(t,s) \le (2/\tau^2) G(s,s)$.

(10) In Lemma 4.3, put $\alpha \ge 1/(q-2)$ and inequality (4.5) becomes

$$\min_{t \in [\tau, 1]} \left(u(t) + {}^cD_{0^+}^{\sigma} u(t) \right) \ge \frac{\tau^3}{2} \|u\|. \tag{4.5}$$

Equation (4.7) is as follows:

$$u(t) \le \frac{2}{\tau^2 \Gamma(q-2)} \times \int_0^1 \frac{1}{(1-s)^{3-q}} G(s,s) a(s) f_1\left(u(s), {}^cD_{0^+}^{\sigma} u(s)\right) ds. \tag{4.7}$$

Equation (4.8) is as follows:

$$
\begin{aligned}
{}^{c}D_{0^+}^{\sigma}u(t) &= \frac{1}{\Gamma(q-2)\Gamma(1-\sigma)} \\
&\quad\times \int_0^t \int_0^1 (t-s)^{-\sigma} \frac{1}{(1-r)^{3-q}} G_1(s,r) \\
&\qquad\times a(r) f_1\left(u(r), {}^{c}D_{0^+}^{\sigma}u(r)\right) ds\, dr \\
&\le \frac{2}{\tau\Gamma(q-2)\Gamma(2-\sigma)} \\
&\quad\times \int_0^1 \frac{1}{(1-r)^{3-q}} G_1(r,r) a(r) \\
&\qquad\times f_1\left(u(r), {}^{c}D_{0^+}^{\sigma}u(r)\right) dr.
\end{aligned} \tag{4.8}
$$

Equation (4.9) is as follows:

$$
\begin{aligned}
\|u\| &\le \frac{2}{\tau^2\Gamma(q-2)} \\
&\quad\times \int_0^1 \frac{1}{(1-s)^{3-q}} \left[G(s,s) + \frac{G_1(s,s)}{\Gamma(2-\sigma)}\right] \\
&\qquad\times a(s) f_1\left(u(s), {}^{c}D_{0^+}^{\sigma}u(s)\right) ds.
\end{aligned} \tag{4.9}
$$

Equation (4.10) is as follows:

$$
\begin{aligned}
&\int_0^1 \frac{1}{(1-s)^{3-q}} \left[G(s,s) + \frac{G_1(s,s)}{\Gamma(2-\sigma)}\right] \\
&\qquad\times a(s) f_1\left(u(s), {}^{c}D_{0^+}^{\sigma}u(s)\right) ds \\
&\quad\ge \frac{\tau^2\Gamma(q-2)}{2} \|u\|.
\end{aligned} \tag{4.10}
$$

Equation (4.11): in view of the left hand side of (4.1), we obtain for all $t \in [\tau, 1]$

$$
\begin{aligned}
u(t) &\ge \frac{\tau}{\Gamma(q-2)} \\
&\quad\times \int_0^1 \frac{1}{(1-s)^{3-q}} G(s,s) a(s) f_1\left(u(s), {}^{c}D_{0^+}^{\sigma}u(s)\right) ds.
\end{aligned} \tag{4.11}
$$

Equation (4.12) is as follows:

$$
\begin{aligned}
{}^{c}D_{0^+}^{\sigma}u(t) &\ge \frac{1}{\Gamma(q-2)\Gamma(2-\sigma)} \\
&\quad\times \int_0^1 \frac{1}{(1-r)^{3-q}} G_1(r,r) a(r) \\
&\qquad\times f_1\left(u(r), {}^{c}D_{0^+}^{\sigma}u(r)\right) dr.
\end{aligned} \tag{4.12}
$$

Equation (4.13) is as follows:

$$
\begin{aligned}
&\min_{t\in[\tau,1]} \left(u(t) + {}^{c}D_{0^+}^{\sigma}u(t)\right) \\
&\quad\ge \frac{\tau}{\Gamma(q-2)} \int_0^1 \frac{1}{(1-s)^{3-q}} \\
&\qquad\times \left[G(s,s) + \frac{G_1(s,s)}{\Gamma(2-\sigma)}\right] \\
&\qquad\times a(s) f_1\left(u(s), {}^{c}D_{0^+}^{\sigma}u(s)\right) ds.
\end{aligned} \tag{4.13}
$$

Equation (4.14) is as follows:

$$
\min_{t\in[\tau,1]} \left(u(t) + {}^{c}D_{0^+}^{\sigma}u(t)\right) \ge \frac{\tau^3}{2} \|u\|. \tag{4.14}
$$

Equation (4.17) is as follows:

$$
K = \left\{u \in E^+, \min_{t\in[\tau,1]} \left(u(t) + {}^{c}D_{0^+}^{\sigma}u(t)\right) \ge \frac{\tau^3}{2} \|u\|\right\}. \tag{4.17}
$$

Equation (4.19) is as follows:

$$
\begin{aligned}
Tu(t) &= \frac{1}{\Gamma(q-2)} \\
&\quad\times \int_0^1 \frac{1}{(1-s)^{3-q}} \\
&\qquad\times G(t,s) a(s) f_1\left(u(s), {}^{c}D_{0^+}^{\sigma}u(s)\right) ds \\
&\le \frac{2\varepsilon\|u\|}{\tau^2\Gamma(q-2)} \int_0^1 \frac{1}{(1-s)^{3-q}} G(s,s) a(s)\, ds.
\end{aligned} \tag{4.19}
$$

Equation (4.20) is as follows:

$$
\begin{aligned}
&{}^{c}D_{0^+}^{\sigma}Tu(t) \\
&\quad= \frac{1}{\Gamma(q-2)\Gamma(1-\sigma)} \\
&\qquad\times \int_0^t \int_0^1 \frac{1}{(1-r)^{3-q}} (t-s)^{-\sigma} G_1(s,r) \\
&\qquad\times a(r) f_1\left(u(r), {}^{c}D_{0^+}^{\sigma}u(r)\right) ds\, dr \\
&\quad\le \frac{2}{\tau\Gamma(q-2)\Gamma(2-\sigma)} \\
&\qquad\times \int_0^1 \frac{1}{(1-r)^{3-q}} G_1(r,r) a(r) \\
&\qquad\times \left(|u(r)| + \left|{}^{c}D_{0^+}^{\sigma}u(r)\right|\right) dr \\
&\quad\le \frac{2\varepsilon\|u\|}{\tau^2\Gamma(q-2)\Gamma(2-\sigma)} \int_0^1 \frac{1}{(1-s)^{3-q}} G_1(s,s) a(s)\, ds.
\end{aligned} \tag{4.20}
$$

Equation (4.21) is as follows:

$$\|Tu\| \leq \frac{2\varepsilon \|u\|}{\tau^2 \Gamma(q-2)} \int_0^1 \frac{1}{(1-s)^{3-q}} \times \left[G(s,s) + \frac{G_1(s,s)}{\Gamma(2-\sigma)}\right] a(s)\,ds. \tag{4.21}$$

Equation (4.22) is as follows:

$$\varepsilon \leq \tau^2 \Gamma(q-2) \times \left(2\int_0^1 \frac{1}{(1-s)^{3-q}} \times \left[G(s,s) + \frac{G_1(s,s)}{\Gamma(2-\sigma)}\right] a(s)\,ds\right)^{-1}. \tag{4.22}$$

Let

$$R = \max\left\{2R_1, \frac{2R_2}{\tau^3}\right\}. \tag{1}$$

Equation (4.23) is as follows:

$$\min_{t\in[\tau,1]} \left(u(t) + {}^cD_{0^+}^{\sigma} u(t)\right) \geq \frac{\tau^3}{2}\|u\| = \frac{\tau^3}{2} R \geq R_2. \tag{4.23}$$

Using the left hand side of (4.1) and Lemma 4.1, we obtain (4.24):

$$Tu(t) \geq \frac{\tau^4 M \|u\|}{2\Gamma(q-2)} \int_0^1 \frac{1}{(1-s)^{3-q}} G(s,s)\, a(s)\,ds. \tag{4.24}$$

Equation (4.25) is as follows:

$${}^cD_{0^+}^{\sigma} Tu(t) \geq \frac{\tau^3 M \|u\|}{2\Gamma(q-2)\Gamma(2-\sigma)} \times \int_0^1 \frac{1}{(1-s)^{3-q}} G_1(s,s)\, a(s)\,ds. \tag{4.25}$$

Equation (4.26) is as follows:

$$Tu(t) + {}^cD_{0^+}^{\sigma} Tu(t) \geq \frac{\tau^4 M \|u\|}{2\Gamma(q-2)} \times \int_0^1 \frac{1}{(1-s)^{3-q}} \left[G(s,s) + \frac{G_1(s,s)}{\Gamma(2-\sigma)}\right] \times a(s)\,ds. \tag{4.26}$$

Equation (4.27) is as follows:

$$M \geq 2\Gamma(q-2) \times \left(\tau^4 \int_0^1 \left[\left(1/(1-s)^{3-q}\right) G(s,s) + G_1(s,s)/\Gamma(2-\sigma)\right] a(s)\,ds\right)^{-1}. \tag{4.27}$$

(11) In Example 4.6, if we choose $\sigma = 1/4 < 1$; then we get the same results with

$$C_g + C_h = 0.49821 < \frac{1}{2}, \quad A_g + A_h = 0.42552 < \frac{\Gamma(2-\sigma)}{2} = 0.459. \tag{2}$$

In Example 4.7, choose $\sigma = 1/5$, $\psi(x) = (x/10)^2 + 1$, and $\phi(\overline{x}) = \ln(1+\overline{x}^2)/9 + 1$; then we get the same results.

Remark 1. One can study the problem (P1) for $1 < \sigma < 2$ and the function f depending only on t and u instead of $f(t, u(t), {}^cD_{0^+}^{\sigma} u(t))$.

14

Fractal Oscillations of Chirp Functions and Applications to Second-Order Linear Differential Equations

Mervan Pašić[1] **and Satoshi Tanaka**[2]

[1] *Department of Applied Mathematics, Faculty of Science, University of Zagreb, 10000 Zagreb, Croatia*
[2] *Okayama University of Science, Okayama 700-0005, Japan*

Correspondence should be addressed to Satoshi Tanaka; tanaka@xmath.ous.ac.jp

Academic Editor: Norio Yoshida

We derive some simple sufficient conditions on the amplitude $a(x)$, the phase $\varphi(x)$, and the instantaneous frequency $\omega(x)$ such that the so-called chirp function $y(x) = a(x)S(\varphi(x))$ is fractal oscillatory near a point $x = x_0$, where $\varphi'(x) = \omega(x)$ and $S = S(t)$ is a periodic function on $\mathbb{R}$. It means that $y(x)$ oscillates near $x = x_0$, and its graph $\Gamma(y)$ is a fractal curve in $\mathbb{R}^2$ such that its box-counting dimension equals a prescribed real number $s \in [1,2)$ and the s-dimensional upper and lower Minkowski contents of $\Gamma(y)$ are strictly positive and finite. It numerically determines the order of concentration of oscillations of $y(x)$ near $x = x_0$. Next, we give some applications of the main results to the fractal oscillations of solutions of linear differential equations which are generated by the chirp functions taken as the fundamental system of all solutions.

1. Introduction

The brilliant heuristic approach of Tricot [1] to the fractal curves such as the graph of functions $y(x) = x^{\alpha}\sin x^{-\beta}$ and $y(x) = x^{\alpha}\cos x^{-\beta}$ gave the main motivation for studying the fractal properties near $x = 0$ of graph of oscillatory solutions of various types of differential equations: linear Euler-type equation $y'' + \lambda x^{-\sigma}y = 0$ (see [2]), general second-order linear equation $y'' + f(x)y = 0$ (see [3]) where $f(x)$ satisfies the Hartman-Wintner asymptotic condition near $x = 0$, half-linear equation $(|y'|^{p-2}y')' + f(x)|y|^{p-2}y = 0$ (see [4]), linear self-adjoint equation $(p(x)y')' + q(x)y = 0$ (see [5]), and p-Laplace differential equations in an annular domain (see [6]).

A function $y(x)$ is said to be a chirp function if it possesses the form $y(x) = a(x)S(\varphi(x))$, where $a(x)$ and $\varphi(x)$ denote, respectively, the amplitude and phase of $y(x)$, and $S(t)$ is a periodic function on $\mathbb{R}$. In all previously mentioned papers [2–5], authors are dealing with the fractal oscillations of second-order differential equations and are deriving some sufficient conditions on the coefficients of considered equations such that all their solutions $y(x)$ together with the first derivative $y'(x)$ admit asymptotic behaviour near $x = 0$. It is formally written in the form of a chirp function, that is, $y(x) = a(x)\sin(\varphi(x))$ and $y'(x) = b(x)\cos(\varphi(x))$ near $x = 0$. According to it, one can say that the asymptotic formula for solutions of considered equations satisfies the chirp-like behaviour near $x = 0$ (on the asymptotic formula for solutions near $x = \infty$, see [7, 8]). Then, in the dependence of a prescribed real number $s \in [1,2)$, authors give some asymptotic conditions on $a(x)$, $b(x)$, and $\varphi(x)$ such that all solutions $y(x)$ are fractal oscillatory near $x = 0$ with the fractal dimension s.

In this paper, independently of the asymptotic theory of differential equations, we firstly study the fractal oscillations of a chirp function; see Theorems 8 and 11. Second, taking two linearly independent chirp functions $y_1(x)$ and $y_2(x)$, we generate some new classes of fractal oscillatory linear differential equations which are not considered in [2–5] and have the general solution in the form of $y(x) = c_1y_1(x) + c_2y_2(x)$; see Theorems 16 and 17 (on some detailed description of the solution space of the second-order linear differential equations and on their constructions, we refer the reader to [9, 10]). Finally, we suggest that the reader considers the fractal oscillations near an arbitrary real point $x = x_0$ instead of $x = 0$ and studies the fractal oscillations near $x = x_0$

from the left side and from both sides; see Theorem 24. Many examples are considered to show the originality of obtained results.

The chirp functions are also appearing in the time-frequency analysis; see for instance, [11–15] as well as in several applications of the time-frequency analysis; see for instance, [16–20].

2. Statement of the Main Results

We study some local asymptotic behaviours of fractal types for the so-called chirp function as follows:

$$y(x) = a(x) S(\varphi(x)), \quad x \in (0, t_0], \tag{1}$$

where $t_0 > 0$, the *amplitude* $a = a(x)$ is a nontrivial function on $(0, t_0]$, the *phase* $\varphi = \varphi(x)$ is singular at $x = 0$ and $S = S(t)$ is a periodic function on $\mathbb{R}$ such as $\cos(t)$ and $\sin(t)$.

The chirp function (1) is sometimes written in the following form in which the so-called *instantaneous frequency* $\omega(x)$ appears instead of phase $\varphi(x)$:

$$y(x) = a(x) S\left(\int_x^{t_0} \omega(\xi)\, d\xi\right), \quad x \in (0, t_0]. \tag{2}$$

Let for some $t_0 > 0$, the functions $a \in C((0, t_0])$, $\varphi \in C^2((0, t_0])$, and $S \in C^1(\mathbb{R})$ satisfy the following basic structural conditions:

$$a(x) > 0, \quad a(x) \text{ is bounded on } (0, t_0], \tag{3}$$

$$\lim_{x\to+0} \varphi(x) = \infty, \quad \varphi(x) > 0, \quad \varphi'(x) < 0, \quad x \in (0, t_0], \tag{4}$$

$$\begin{gathered}\text{for some } \tau > 0, \quad |S(t+\tau)| = |S(t)| \quad \forall t \in \mathbb{R}, \\ \text{for some } \tau_0 \in \mathbb{R}, \quad S(\tau_0) = 0, \quad S(t) \neq 0 \\ \forall t \in (\tau_0, \tau_0 + \tau).\end{gathered} \tag{5}$$

Definition 1. A function $y \in C((0, t_0])$ is *oscillatory near* $x = 0$, if there is a decreasing sequence $x_n \in (0, t_0]$ such that $y(x_n) = 0$ for $n \in \mathbb{N}$, and $x_n \searrow 0$ as $n \to \infty$; see Figure 1.

It is easy to prove the next proposition.

Proposition 2. *Let $a \in C((0, t_0])$ satisfy $a(x) > 0$ on $(0, t_0]$, let $\varphi \in C((0, t_0])$ be strictly decreasing on $(0, t_0]$, and let $S \in C(\mathbb{R})$ satisfy* (5). *Then, the following two conditions are equivalent:*

(i) *the chirp function $y(x) = a(x)S(\varphi(x))$ is bounded and oscillatory near $x = 0$;*

(ii) *the amplitude $a(x)$ is bounded and $\lim_{x\to+0}\varphi(x) = \infty$.*

On the qualitative and oscillatory behaviours of solutions of differential equations of several types, we refer the reader to [7, 8].

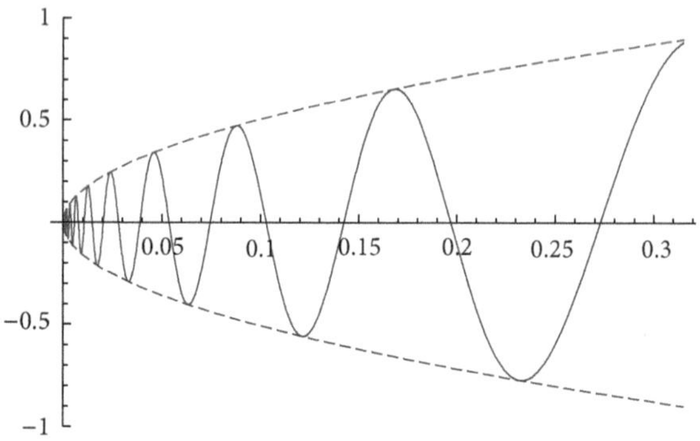

FIGURE 1: y is continuous, bounded, and oscillatory near $x = 0$.

Example 3. The following three main types of chirp functions satisfy the conditions (3), (4), and (5):

(i) the so-called (α, β)-chirps

$$y_1(x) = x^\alpha \cos\left(x^{-\beta}\right), \qquad y_2(x) = x^\alpha \sin\left(x^{-\beta}\right), \tag{6}$$

where $\alpha \geq 0$ and $\beta > 0$; the first studies on the (α, β)-chirps appeared in [1, 13, 14] from different point of views;

(ii) the so-called logarithmic chirps

$$y_1(x) = x^\gamma \cos(\rho \log x), \qquad y_2(x) = x^\gamma \sin(\rho \log x), \tag{7}$$

where $\gamma > 0$ and $\rho \in \mathbb{R}$; this type of chirps appears in definition of the Lamperti transform (see [11]) and in the fundamental system of solutions of the famous Euler equation $y'' + \lambda x^{-2} y = 0$ for $\gamma = 1/2$ and $\rho = \sqrt{\lambda - 1/4}$ (see [21]);

(iii) the chirp function of exponential type

$$\begin{gathered} y_1(x) = x e^{-1/(2x)} \cos\left(e^{1/x}\right), \\ y_2(x) = x e^{-1/(2x)} \sin\left(e^{1/x}\right). \end{gathered} \tag{8}$$

Example 4. Let $y_1(x) = a(x)\cos(\varphi(x))$ and $y_2(x) = a(x)\sin(\varphi(x))$, $x \in (0, t_0]$, where the amplitude a and phase φ satisfy (3) and (4), respectively. Then, $y(x) = c_1 y_1(x) + c_2 y_2(x)$, $c_1^2 + c_2^2 > 0$, is also a chirp function which is bounded and oscillatory near $x = 0$. Indeed, if $c_1 = 0$ or $c_2 = 0$, then $y(x) = c_2 y_2(x)$ or $y(x) = c_1 y_1(x)$; since $S(t) = \cos t$ and $S(t) = \sin t$ satisfy (5), both cases $y(x)$ are chirp functions; otherwise, we have $y(x) = Aa(x)\sin(\varphi(x) + B)$, where $A = (c_1^2 + c_2^2)^{1/2}$ and $B = \arctan(c_1/c_2)$; obviously, the amplitude $Aa(x)$ and the phase $\varphi(x) + B$ satisfy the required conditions (3) and (4), respectively.

Next, let

$$\Gamma(y) = \left\{(t_1, t_2) \in \mathbb{R}^2 : t_1 \in (0, t_0],\ t_2 = y(t_1)\right\} \tag{9}$$

denotes the graph of a function $y \in C((0, t_0])$. In this paper, we show that the graph $\Gamma(y)$ of chirp function (1) is a fractal curve in $\mathbb{R}^2$ in respect to the Minkowski-Bouligand dimension $\dim_M \Gamma(y) = s$ (box-counting dimension) and the s-dimensional upper and lower Minkowski contents $M^{*s}(\Gamma(y))$ and $M_*^s(\Gamma(y))$ defined by

$$\dim_M \Gamma(y) = \lim_{\varepsilon \to +0} \left(2 - \frac{\log |\Gamma_\varepsilon(y)|}{\log \varepsilon} \right),$$
$$M^{*s}(\Gamma(y)) = \limsup_{\varepsilon \to +0} (2\varepsilon)^{s-2} |\Gamma_\varepsilon(y)|, \tag{10}$$
$$M_*^s(\Gamma(y)) = \liminf_{\varepsilon \to +0} (2\varepsilon)^{s-2} |\Gamma_\varepsilon(y)|.$$

Here, $\Gamma_\varepsilon(y)$ denotes the ε-neighbourhood of graph $\Gamma(y)$ defined by

$$\Gamma_\varepsilon(y) = \{(t_1, t_2) \in \mathbb{R}^2 : d((t_1, t_2), \Gamma(y)) \le \varepsilon\}, \quad \varepsilon > 0, \tag{11}$$

and $d((t_1, t_2), \Gamma(y))$ denotes the distance from (t_1, t_2) to $\Gamma(y)$, and $|\Gamma_\varepsilon(y)|$ denotes the Lebesgue measure of $\Gamma_\varepsilon(y)$. On the box-counting dimension and the s-dimensional Minkowski content, we refer the reader to [22–27].

The main fractal properties considered in the paper are given in the next definitions.

Definition 5. For a given real number $s \in [1, 2)$ and a function $y \in C((0, t_0])$, which is bounded on $(0, t_0]$ and oscillatory near $x = 0$, it is said that $y(x)$ is *fractal oscillatory near $x = 0$ with the fractal dimension s*, if

$$\dim_M \Gamma(y) = s, \qquad 0 < M_*^s(\Gamma(y)) \le M^{*s}(\Gamma(y)) < \infty. \tag{12}$$

On the contrary, if there is no any $s \in [1, 2)$ such that $\Gamma(y)$ satisfies (12), then $y(x)$ *is not fractal oscillatory near $x = 0$.*

Fractal oscillations can be understood also as a refinement of rectifiable and nonrectifiable oscillations. They are recently studied in [3, 4, 21, 28–30].

Example 6. The chirp functions

$$y(x) = x \cos\left(\frac{1}{x}\right), \qquad y(x) = x \sin\left(\frac{1}{x}\right) \tag{13}$$

are not fractal oscillatory near $x = 0$. In fact, $M^{*1}(\Gamma(y)) = \infty$ and $\dim_M \Gamma(y) = 1$ (see [2]). It also implies that $M^{*s}(\Gamma(y)) = 0$ for all $s \in (1, 2)$ (see [1, 23]), and thus the statement (12) is not satisfied for any $s \in [1, 2)$. Hence, y_1 and y_2 are not fractal oscillatory near $x = 0$.

Example 7. Let $\rho > 0$. It is clear that the chirp functions

$$y(x) = \sqrt{x} \cos(\rho \log x), \qquad y(x) = \sqrt{x} \sin(\rho \log x) \tag{14}$$

are oscillatory near $x = 0$. Moreover, the length of $\Gamma(y)$ is finite (see [21]), and therefore we observe that $0 < M_*^1(\Gamma(y)) \le M^{*1}(\Gamma(y)) < \infty$ and $\dim_M \Gamma(y) = 1$ (see [22]). Thus, such chirp functions are fractal oscillatory near $x = 0$ with the fractal dimension 1.

In order to show that the chirp function (1) is fractal oscillatory near $x = 0$, we need to impose on amplitude $a(x)$ the following additional structural condition:

$$a \in C^1((0, t_0]), \quad a'(x) \ge 0 \quad \text{for } x \in (0, t_0]. \tag{15}$$

Now we are able to state the first main result of the paper.

Theorem 8. *Let the functions $a(x)$, $\varphi(x)$, and $S(t)$ satisfy the structural conditions (3), (4), (5), (15), and*

$$\varphi''(x) > 0 \quad \text{on } (0, t_0], \quad \limsup_{x \to 0} \left(\frac{1}{-\varphi'(x)} \right)' < \infty. \tag{16}$$

Let the amplitude $a(x)$ and the phase $\varphi(x)$ satisfy the following asymptotic conditions near $x = 0$:

$$\liminf_{x \to 0} \left([-\varphi'(x)]^{2-s} \int_0^x a(\xi)\, d\xi \right) > 0, \tag{17}$$

$$\limsup_{x \to 0} [x a(x)]^{(s-1)/(2-s)} \int_x^{t_0} a(\xi) [-\varphi'(\xi)]\, d\xi < \infty, \tag{18}$$

where $s \in (1, 2)$ is a given real number. Then, the chirp function (1) is fractal oscillatory near $x = 0$ with the fractal dimension s.

Remark 9. Let $s \in (1, 2)$. Assume that (15) and (17) hold. Then,

$$\limsup_{x \to 0} \left(x a(x) [-\varphi'(x)]^{2-s} \right) < \infty \tag{19}$$

implies

$$\limsup_{x \to 0} \left[\frac{1}{-\varphi'(x)} \right]^{s-1} \int_x^{t_0} a(\xi) [-\varphi'(\xi)]\, d\xi < \infty \tag{20}$$

and (18). Therefore, condition (18) is better than (19) in Theorem 8.

The proof of Remark 9 is presented in the Appendix.

Theorem 8 can be rewritten in the term of $\omega(x)$ instead of $\varphi(x)$ as follows.

Corollary 10. *Let the functions $a(x)$, $\omega(x)$, and $S(t)$ satisfy the structural conditions (3), (5), (15), and*

$$\omega \in C^1((0, t_0]), \quad \omega(x) > 0, \quad \omega'(x) < 0$$
$$\text{on } (0, t_0], \; \omega \notin L^1((0, t_0]), \tag{21}$$
$$\limsup_{x \to 0} \left(\frac{1}{\omega(x)} \right)' < \infty.$$

Let the amplitude $a(x)$ and the instantaneous frequency $\omega(x)$ satisfy the following asymptotic conditions near $x = 0$:

$$\liminf_{x \to 0} \left([\omega(x)]^{2-s} \int_0^x a(\xi)\, d\xi \right) > 0,$$
$$\limsup_{x \to 0} [x a(x)]^{(s-1)/(2-s)} \int_x^{t_0} a(\xi) \omega(\xi)\, d\xi < \infty, \tag{22}$$

where $s \in (1, 2)$ is a given real number. Then, the chirp function (2) is fractal oscillatory near $x = 0$ with the fractal dimension s.

Now, we state analogous result to Theorem 8 in the case of $s = 1$, which will be proved in Section 3.

Theorem 11. *Assume that the function $S(t)$ satisfies (5), $a \in C([0,t_0]) \cap C^1((0,t_0])$, $\varphi \in C^2((0,t_0])$ and $\lim_{x\to+0}\varphi(x) = \infty$. If $a' \in L^1((0,t_0])$, and $\varphi' a \in L^1((0,t_0])$, then the chirp function (1) is fractal oscillatory near $x = 0$ with the fractal dimension 1.*

In respect to some existing results on the fractal dimension of graph of chirp functions, previous theorems are the most simple and general. It is because in [31, 32] authors require some extra conditions on the chirp function $y(x)$ which are not easy to be satisfied in the application, for instance, the rapid convex-concave properties of $y(x)$ as in [31] and a condition on the curvature of $y(x)$ as in [32].

According to previous theorems, we can show the fractal oscillations of the so-called (α, β)-chirp as well as logarithmic chirp functions.

Example 12. We consider the (α,β)-chirp $y(x) = x^\alpha S(x^{-\beta})$, where $S(t) = \cos t$ or $S(t) = \sin t$ and $\beta > \alpha \geq 0$. It is fractal oscillatory near $x = 0$ with the fractal dimension $s = 2 - (1+\alpha)/(1+\beta)$. In fact, it is easy to see that $a(x) = x^\alpha$, $S(t)$, and $\varphi(x) = x^{-\beta}$ satisfy (3), (4), (5), (15), and (16). When $s = 2 - (1+\alpha)/(1+\beta)$, we see that

$$\left[-\varphi'(x)\right]^{2-s} \int_0^x a(\xi)\,d\xi = \frac{\beta^{(1+\alpha)/(1+\beta)}}{1+\alpha} > 0,$$

$$[x a(x)]^{(s-1)/(2-s)} \int_x^{t_0} a(\xi)\left[-\varphi'(\xi)\right] d\xi \tag{23}$$

$$= \frac{\beta}{\beta-\alpha}\left(1 - x^{\beta-\alpha} t_0^{\alpha-\beta}\right) \leq \frac{\beta}{\beta-\alpha}.$$

From Theorem 8, it follows that $y(x) = x^\alpha S(x^{-\beta})$, $\beta > \alpha \geq 0$, is fractal oscillatory near $x = 0$ with the fractal dimension $s = 2 - (1+\alpha)/(1+\beta)$.

Example 13. We consider the (α,β)-chirp $y(x) = x^\alpha S(x^{-\beta})$ again, where $S(t) = \cos t$ or $S(t) = \sin t$. Now, we assume that $0 < \beta < \alpha$. Applying Theorem 8, we easily see that it is fractal oscillatory near $x = 0$ with the fractal dimension 1.

Example 14. We consider the logarithmic chirp functions

$$y_1(x) = x^\gamma \cos(\rho \log x), \qquad y_2(x) = x^\gamma \sin(\rho \log x), \tag{24}$$

where $\gamma > 0$ and $\rho > 0$. Put $S(t) = \cos(-t)$ or $S(t) = \sin(-t)$, $a(x) = x^\gamma$, and $\varphi(x) = -\rho \log x$. Then, we easily see that $S(t)$ satisfies (5) and that $a'(x) = \gamma x^{\gamma-1} \in L^1((0,t_0])$ and $a(x)\varphi'(x) = -\rho x^{\gamma-1} \in L^1((0,t_0])$. Theorem 11 implies that $y_1(x)$ and $y_2(x)$ are fractal oscillatory near $x = 0$ with the fractal dimension 1.

Question 1. Is it possible to apply Theorem 8 on the exponential chirp given in Example 3(iii)?

Example 15. Let $y_1(x) = a(x)\cos(\varphi(x))$ and $y_2(x) = a(x)\sin(\varphi(x))$, $x \in (0,t_0]$, where the amplitude $a(x)$ and the phase $\varphi(x)$ satisfy all assumptions of Theorem 8 for an arbitrary given $s \in (1,2)$. Then, the chirp function $y(x) = c_1 y_1(x) + c_2 y_2(x)$, $c_1^2 + c_2^2 > 0$, is also fractal oscillatory near $x = 0$ with the fractal dimension s. Indeed, similarly as in Example 4, $y(x)$ can be rewritten in the form $y(x) = d_1 \sin(\varphi(x) + d_2)$, where $d_1 \neq 0$ and $d_2 \in \mathbb{R}$. It is clear that the function $S(t) = d_1 \sin(t + d_2)$ satisfies the required condition (5) and hence Theorem 8 proves that $y(x)$ is fractal oscillatory near $x = 0$ with the fractal dimension s.

Next, we pay attention to the fractal oscillations of solutions of linear differential equations generated by the system of functions as follows:

$$y(x) = c_1 y_1(x) + c_2 y_2(x), \quad c_1^2 + c_2^2 > 0,$$

where

$$y_1(x) = a(x)\cos(\varphi(x)), \; y_2(x) = a(x)\sin(\varphi(x)), \quad x \in (0,t_0]. \tag{25}$$

It is not difficult to check that (25) is the fundamental system of all solutions of the following linear differential equation:

$$y'' + \left(-2R_a - R_{\varphi'}\right) y' + \left(-R_a' + R_a^2 + R_a R_{\varphi'} + (\varphi')^2\right) y = 0, \quad x \in (0,t_0], \tag{26}$$

where $R_f = R_f(x) = f'(x)/f(x)$ for some C^1-function $f = f(x)$.

Theorem 16. *Let the functions a, $\varphi \in C^2((0,t_0])$ satisfy structural conditions (3), (4), and (15) as well as the conditions (16), (17), and (18) in respect to a given real number $s \in (1,2)$. Then, every nontrivial solution $y \in C^2((0,t_0])$ of (26) is fractal oscillatory near $x = 0$ with the fractal dimension s.*

With the help of Theorem 11, we can state analogous result to Theorem 16 in the case of $s = 1$.

Theorem 17. *Assume that $a \in C([0,t_0]) \cap C^2((0,t_0])$, $a(x) \neq 0$ for $x \in (0,t_0]$, $\varphi \in C^2((0,t_0])$, $\lim_{x\to+0}\varphi(x) = \infty$, $a' \in L^1((0,t_0])$, and $\varphi' a \in L^1((0,t_0])$. Then, every nontrivial solution $y \in C^2((0,t_0])$ of (26) is fractal oscillatory near $x = 0$ with the fractal dimension 1.*

The previous two theorems will be proved in Section 4. Assumptions on the coefficients of (26) in general are different to those considered in [2–5].

When $a'/a = -(1/2)\varphi''/\varphi'$, then (26) becomes the undamped equation

$$y'' + \left(\frac{1}{2} S_{\text{der}}(\varphi') + (\varphi')^2\right) y = 0, \quad x \in (0,t_0], \tag{27}$$

where $S_{\mathrm{der}}(f)$ denotes the Schwarzian derivative of f defined by

$$S_{\mathrm{der}}(f) = \frac{f''}{f} - \frac{3}{2}\left(\frac{f'}{f}\right)^2. \quad (28)$$

Hence, from Theorem 16, we obtain the following consequence.

Corollary 18. *Let the functions $a \in C^2((0,t_0])$ and $\varphi \in C^3((0,t_0])$ satisfy structural conditions (3), (4), and (15) as well as the conditions (16), (17), and (18) in respect to a given real number $s \in (1,2)$. Let $a'/a = -(1/2)\varphi''/\varphi'$. Then, every nontrivial solution $y \in C^2((0,t_0])$ of (27) is fractal oscillatory near $x = 0$ with the fractal dimension s.*

As a consequence of Theorem 16, and Corollary 18 we derive the following examples for linear differential equations of second order having all the solutions to be fractal oscillatory near $x = 0$.

Example 19. The so-called damped chirp equation

$$y'' + \frac{\beta - 2\alpha + 1}{x} y' + \left(\frac{\beta^2}{x^{2\beta+2}} - \frac{\alpha(\beta-\alpha)}{x^2}\right) y = 0, \quad x \in (0,t_0], \quad (29)$$

is fractal oscillatory near $x = 0$ with the fractal dimension $2 - (1+\alpha)/(1+\beta)$, where $\beta > \alpha \geq 0$. When $a(x) = x^{\alpha}$ and $\varphi(x) = x^{-\beta}$, (26) becomes (29). It is easy to see that (3), (4), (15), and (16) are satisfied. In the same as in Example 12, we see that (17) and (18) hold for $s = 2 - (1+\alpha)/(1+\beta)$. Hence, Theorem 16 proves that every nontrivial solution of (29) is fractal oscillatory near $x = 0$ with the fractal dimension $2 - (1+\alpha)/(1+\beta)$.

Now we assume that $0 < \beta < \alpha$. Then, Theorem 17 implies that (29) is fractal oscillatory near $x = 0$ with the fractal dimension 1.

Example 20. The following equation

$$y'' + \frac{1-2\gamma}{x} y' + \frac{\gamma^2 + \rho^2}{x^2} y = 0, \quad x \in (0,t_0], \quad (30)$$

is fractal oscillatory near $x = 0$ with the fractal dimension 1, where $\gamma > 0$ and $\rho > 0$. In the case where $a(x) = x^{\gamma}$ and $\varphi(x) = -\rho \log x$, (26) becomes (30). We see that $a \in C([0,t_0]) \cap C^2((0,t_0])$, $\varphi \in C^2((0,t_0])$, $\lim_{x\to+0}\varphi(x) = \infty$, $a' \in L^1((0,t_0])$, and $\varphi' a \in L^1((0,t_0])$. Therefore Theorem 17 implies that every nontrivial solution of (30) is fractal oscillatory near $x = 0$ with the fractal dimension 1.

Question 2. What can we say about the application of Theorem 16 on the case of $\varphi(x)$ given in Example 3(iii)?

At the end of this section, we suggest that the reader studies some invariant properties of fractal oscillations of the chirp function (1) in respect to the translation and reflexion. Analogously to Definitions 1 and 5, one can define the fractal oscillations near an arbitrary real point $x = x_0$ as follows.

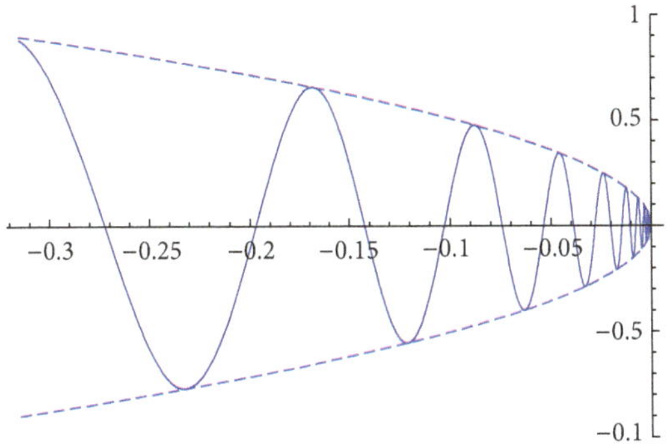

FIGURE 2: y is oscillatory near $x = 0$ from the left side.

Definition 21. Let $x_0 \in \mathbb{R}$ and $\delta > 0$. A function $y \in C((x_0, x_0+\delta])$ is *oscillatory near* $x = x_0$, if there is a decreasing sequence $b_n \in (x_0, x_0+\delta]$ such that $y(b_n) = 0$, for $n \in \mathbb{N}$, and $b_n \searrow x_0$ as $n \to \infty$. Moreover, if the graph $\Gamma(y)$ satisfies the condition (12) for some $s \in [1,2)$, then $y(x)$ is said to be *fractal oscillatory near* $x = x_0$ *with the fractal dimension s.*

Definition 22. Let $x_0 \in \mathbb{R}$ and $\delta > 0$. It is said that a function $y \in C([x_0-\delta, x_0))$ is *fractal oscillatory near* $x = 0$ *from the left side with the fractal dimension* $s \in [1,2)$, if there is an increasing sequence $a_n \in [x_0-\delta, x_0)$ such that $y(a_n) = 0$ for $n \in \mathbb{N}$, $a_n \nearrow x_0$ as $n \to \infty$, and the graph $\Gamma(y)$ satisfies the condition (12), see Figure 2.

Definition 23. Let $x_0 \in \mathbb{R}$, and let $\delta > 0$. It is said that a function $y \in C([x_0-\delta, x_0) \cup (x_0, x_0+\delta])$ is *two-sided fractal oscillatory near* $x = 0$ *with the fractal dimension* $s \in [1,2)$, if there is an increasing sequence $a_n \in [x_0-\delta, x_0)$ and a decreasing sequence $b_n \in (x_0, x_0+\delta]$ such that $y(a_n) = y(b_n) = 0$ for $n \in \mathbb{N}$, $a_n \nearrow x_0$ and $b_n \searrow x_0$ as $n \to \infty$, and the graph $\Gamma(y)$ satisfies the condition (12); see Figure 3.

With the help of Theorem 8, we state the following result.

Theorem 24. *Let $x_0 \in \mathbb{R}$, and let $\delta > 0$. Let the functions $a, \omega \in C^1((0,\delta])$ satisfy structural conditions (3), (4), (5), (15), and (16) as well as conditions (17) and (18) in respect to a given real number $s \in (1,2)$. Then, one has*

(i) *the chirp function $y(x) = a(x-x_0)S(\varphi(x-x_0))$, $x \in (x_0, x_0+\delta]$ is fractal oscillatory near $x = x_0$ with the fractal dimension s;*

(ii) *the chirp function $y(x) = a(x_0-x)S(\varphi(x_0-x))$, $x \in [x_0-\delta, x_0)$ is fractal oscillatory near $x = x_0$ from the left side with the fractal dimension s;*

(iii) *the chirp function $y(x) = a(|x-x_0|)S(\varphi(|x-x_0|))$, $x \in [x_0-\delta, x_0) \cup (x_0, x_0+\delta]$ is two-sided fractal oscillatory near $x = x_0$ with the fractal dimension s.*

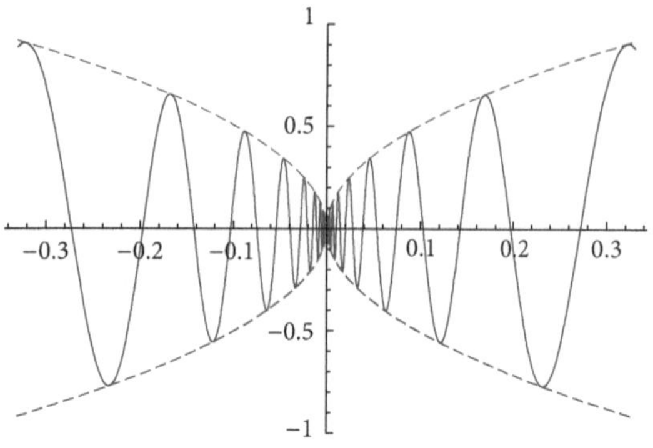

FIGURE 3: y is two-sided oscillatory near $x = 0$.

3. Proof for the Fractal Oscillations of Chirp Functions

In this section, we give the proofs of the main results dealing with the fractal oscillations of chirp functions.

By Definition 5, it follows that if for a prescribed real number $s \in (1,2)$ there are two positive constants c_1 and c_2 such that

$$c_1\varepsilon^{2-s} \le |\Gamma_\varepsilon(y)| \le c_2\varepsilon^{2-s}, \quad \varepsilon \in (0,\varepsilon_0), \qquad (31)$$

for some $\varepsilon_0 > 0$, then a function $y \in C((0,t_0])$ is fractal oscillatory near $x = 0$ with the fractal dimension s. The following lemma plays the essential role in the proof of (31).

Lemma 25. *Let $y \in C((0,t_0])$ be a bounded function on $(0,t_0]$, and let $a_n \in (0,t_0]$ be a decreasing sequence of consecutive zeros of $y(x)$ such that $a_n \to 0$. Let $\varepsilon_0 > 0$, and let $k = k(\varepsilon)$ be an index function satisfying*

$$|a_n - a_{n+1}| \le \varepsilon \quad \forall n \ge k(\varepsilon),\ \varepsilon \in (0,\varepsilon_0). \qquad (32)$$

Then

$$|\Gamma_\varepsilon(y)| \ge \sum_{n\ge k(\varepsilon)} \max_{x\in[a_{n+1},a_n]} |y(x)|(a_n - a_{n+1}) \quad \forall\varepsilon \in (0,\varepsilon_0). \qquad (33)$$

Proof. It is exactly the same as [3, pp. 2350].

Let us remark that for some $\varepsilon_0 > 0$, we say that a function $k = k(\varepsilon)$ is an index function on $(0,\varepsilon_0)$ if $k : (0,\varepsilon_0) \to \mathbb{N}$ and $\lim_{\varepsilon\to 0} k(\varepsilon) = \infty$.

Lemma 26. *Let $y \in C[a,b]$. Then,*

$$|\Gamma_\varepsilon(y)| \le 4\pi\varepsilon \text{length}(y) + 4\pi\varepsilon^2, \quad \text{for every } \varepsilon > 0. \qquad (34)$$

Proof. Let $\varepsilon > 0$. Set $x_1 = a$, and set

$$x_{i+1} = \max\{x \in [x_i,b] : d((t,y(t)),(x_i,y(x_i))) \le \varepsilon \text{ for } t \in [x_i,x]\}, \qquad (35)$$

for $i = 1,2,\ldots$. Then, there exists $n \ge 2$ such that $x_n = b$. Set $N = \max\{n \in \mathbf{N} : x_n < b\}$. We see that $N \ge 1$,

$$a = x_1 < x_2 < \cdots < x_i < x_{i+1} < \cdots < x_N < x_{N+1} = b, \quad (36)$$

and if $N \ge 2$, then

$$d((x_i,y(x_i)),(x_{i+1},y(x_{i+1}))) = \varepsilon, \quad i = 1,2,\ldots,N-1. \qquad (37)$$

We will show that

$$\Gamma_\varepsilon(y) \subset \bigcup_{i=1}^{N} B_{2\varepsilon}(x_i,y(x_i)), \qquad (38)$$

where

$$B_{2\varepsilon}(t_1,t_2) = \{(\tau_1,\tau_2) \in \mathbf{R}^2 : d((t_1,t_2),(\tau_1,\tau_2)) \le 2\varepsilon\}. \qquad (39)$$

Let $(\tau_1,\tau_2) \in \Gamma_\varepsilon(y)$. Then there exists $\xi \in [a,b]$ such that $d((\tau_1,\tau_2),(\xi,y(\xi))) \le \varepsilon$. Because of the definition of x_{i+1}, we find that $\xi \in [x_i,x_{i+1}]$ for some $i \in \{1,2,\ldots,N\}$, so that

$$d((\xi,y(\xi)),(x_i,y(x_i))) \le \varepsilon. \qquad (40)$$

Hence, it follows that

$$d((\tau_1,\tau_2),(x_i,y(x_i))) \le d((\tau_1,\tau_2),(\xi,y(\xi))) + d((\xi,y(\xi)),(x_i,y(x_i))) \le 2\varepsilon, \qquad (41)$$

which means that $(\tau_1,\tau_2) \in B_{2\varepsilon}(x_i,y(x_i))$. Therefore, we obtain (38). By (38), we conclude that

$$|\Gamma_\varepsilon(y)| \le \sum_{i=1}^{N} |B_{2\varepsilon}(x_i,y(x_i))| = 4N\pi\varepsilon^2. \qquad (42)$$

When $N = 1$, from (42), it follows that

$$|\Gamma_\varepsilon(y)| \le 4\pi\varepsilon^2 \le 4\pi\varepsilon\text{length}(y) + 4\pi\varepsilon^2. \qquad (43)$$

Now, we assume that $N \ge 2$. We observe that

$$\begin{aligned}\text{length}(y) &= \sum_{i=1}^{N}\text{length}(y|_{[x_i,x_{i+1}]}) \\ &\ge \sum_{i=1}^{N} d((x_i,y(x_i)),(x_{i+1},y(x_{i+1}))) \\ &\ge \sum_{i=1}^{N-1} d((x_i,y(x_i)),(x_{i+1},y(x_{i+1}))) \\ &= (N-1)\varepsilon,\end{aligned} \qquad (44)$$

that is,

$$N\varepsilon \le \text{length}(y) + \varepsilon. \qquad (45)$$

Combining (42) with (45), we obtain

$$|\Gamma_\varepsilon(y)| \le 4\pi\varepsilon\text{length}(y) + 4\pi\varepsilon^2. \qquad (46)$$

Lemma 27. *Let $y \in C(0,t_0]$ be bounded on $(0,t_0]$, and let $\varepsilon > 0$. Then, there exists $C > 0$ such that, for every $\varepsilon > 0$ and $t \in (0,t_0]$,*

$$|\Gamma_\varepsilon(y)| \le C\left[t \sup_{x\in(0,t]} |y(x)| + \varepsilon \text{length}\left(y|_{[t,t_0]}\right) + \varepsilon + \varepsilon^2\right]. \tag{47}$$

Proof. Let $\varepsilon > 0$, and let $t \in (0,t_0]$. From the geometry point of view, it is clear that

$$|\Gamma_\varepsilon(y|_{(0,t]})| \subset R, \tag{48}$$

where R is a rectangle in $\mathbf{R}^2$ defined by

$$R = \Big\{(t_1,t_2) : t_1 \in [-\varepsilon, t+\varepsilon], \quad t_2 \in \Big[-\varepsilon - \sup_{x\in(0,t]}|y(x)|, \varepsilon + \sup_{x\in(0,t]}|y(x)|\Big]\Big\}. \tag{49}$$

Hence, we have

$$\begin{aligned}|\Gamma_\varepsilon(y_{(0,t]})| &\le (2\varepsilon + t)\left(2\varepsilon + 2\sup_{x\in(0,t]}|y(x)|\right)\\ &= 4\varepsilon^2 + 2\left(t + 2\sup_{x\in(0,t]}|y(x)|\right)\varepsilon + 2t\sup_{x\in(0,t]}|y(x)|\\ &\le 4\varepsilon^2 + C_1\varepsilon + 2t\sup_{x\in(0,t]}|y(x)|,\end{aligned} \tag{50}$$

where $C_1 = 2(t_0 + 2\sup_{x\in(0,t_0]}|y(x)|)$. Lemma 26 implies that

$$\left|\Gamma_\varepsilon\left(y|_{[t,t_0]}\right)\right| \le 4\pi\varepsilon\text{length}\left(y|_{[t,t_0]}\right) + 4\pi\varepsilon^2. \tag{51}$$

Consequently, we see that

$$\begin{aligned}|\Gamma_\varepsilon(y)| &\le |\Gamma_\varepsilon(y|_{(0,t]})| + \left|\Gamma_\varepsilon\left(y|_{[t,t_0]}\right)\right|\\ &\le 4\varepsilon^2 + C_1\varepsilon + 2t\sup_{x\in(0,t]}|y(x)| + 4\pi\varepsilon\text{length}\left(y|_{[t,t_0]}\right) + 4\pi\varepsilon^2\\ &\le C\left[t\sup_{x\in(0,t]}|y(x)| + \varepsilon\text{length}\left(y|_{[t,t_0]}\right) + \varepsilon + \varepsilon^2\right],\end{aligned} \tag{52}$$

where $C = \max\{C_1, 4(\pi+1)\}$.

Lemma 28. *Let $y \in C^1(0,t_0]$ be bounded on $(0,t_0]$. Assume that*

$$\int_0^{t_0} |y'(x)|\,dx = \infty, \tag{53}$$

$$\limsup_{x\to 0}\left[x\sup_{\xi\in(0,x]}|y(\xi)|\right]^{(s-1)/(2-s)} \int_x^{t_0}|y'(\xi)|\,d\xi < \infty, \tag{54}$$

for some $s \in (1,2)$. Then, there exists $c_2 > 0$ such that

$$|\Gamma_\varepsilon(y)| \le c_2\varepsilon^{2-s}, \quad \varepsilon \in (0,1). \tag{55}$$

Proof. Let $\varepsilon \in (0,1)$ and let $s \in (1,2)$ be satisfy (54). By (53), there exists $x_1 \in (0,t_0)$, which depends on ε and s, such that

$$\int_{x_1}^{t_0}|y'(t)|\,dt = \varepsilon^{1-s}. \tag{56}$$

Therefore, we have

$$\begin{aligned}\text{length}\left(y|_{[x_1,t_0]}\right) &= \int_{x_1}^{t_0}\sqrt{1+|y'(x)|^2}\,dx\\ &\le \int_{x_1}^{t_0}\left(1+|y'(x)|\right)dx\\ &= t_0 - x_1 + \varepsilon^{1-s}\\ &\le t_0 + \varepsilon^{1-s}.\end{aligned} \tag{57}$$

By (54), there exists $C_1 > 0$ such that

$$\left[x\sup_{\xi\in(0,x]}|y(\xi)|\right]^{(s-1)/(2-s)}\int_x^{t_0}|y'(\xi)|\,d\xi \le C_1, \quad x \in (0,t_0], \tag{58}$$

which implies that

$$\left[x_1\sup_{\xi\in(0,x_1]}|y(\xi)|\right]^{(s-1)/(2-s)}\int_{x_1}^{t_0}|y'(\xi)|\,d\xi \le C_1, \tag{59}$$

that is,

$$x_1\sup_{\xi\in(0,x_1]}|y(\xi)| \le C_2\varepsilon^{2-s}, \tag{60}$$

where $C_2 = C_1^{(2-s)/(s-1)}$. From Lemma 27, (57), and (60), it follows that

$$\begin{aligned}|\Gamma_\varepsilon(y)| &\le C\left[C_2\varepsilon^{2-s} + \varepsilon\left(t_0+\varepsilon^{1-s}\right) + \varepsilon + \varepsilon^2\right]\\ &= C\left[(C_2+1) + (t_0+1)\varepsilon^{s-1} + \varepsilon^s\right]\varepsilon^{2-s}\\ &\le C(C_2+t_0+3)\varepsilon^{2-s}, \quad \varepsilon \in (0,1).\end{aligned} \tag{61}$$

In order to show Theorems 8 and 11, we need the following two geometric lemmas.

Lemma 29 (see [1]). *If $\Gamma \subseteq \mathbb{R}^2$ is a simple curve (i.e., its parameterization is a bijection) and* length$(\Gamma) < \infty$, *then*

$$\text{length}(\Gamma) = \lim_{\varepsilon\to 0}\frac{|\Gamma_\varepsilon|}{2\varepsilon}, \tag{62}$$

where Γ_ε denotes the ε-neighborhood of the graph Γ.

Now, we are able to prove Theorem 8.

Proof of Theorem 8. Let $s \in (1,2)$, and let $y(x)$ be a chirp function given by (1). We note here that it is enough to show that $y(x)$ satisfies (31).

At the first, let x_n be a sequence defined by $x_n = \varphi^{-1}(\tau_0 + n\tau)$ for all sufficiently large $n \in \mathbb{N}$. From (4), it follows that $\varphi^{-1}(t)$ is decreasing. Hence, x_n is decreasing as well as $x_n \to 0$ as $n \to \infty$ because of $\lim_{x\to+0}\varphi(x) = \infty$ (see (4)). We note that $y(x_n) = 0$ and $y(x) \neq 0$ on (x_{n+1}, x_n) for all sufficiently large $n \in \mathbb{N}$. Also, $-1/\varphi'(x)$ is an increasing function because of (16). The mean value theorem shows that

$$\frac{\tau}{-\varphi'(x_{n+1})} \leq x_n - x_{n+1} \leq \frac{\tau}{-\varphi'(x_n)}. \quad (63)$$

Now, let $\varepsilon_0 \in (0,1)$. Let $k(\varepsilon)$ be the smallest natural number satisfying

$$\frac{\tau}{-\varphi'(x_{k(\varepsilon)})} \leq \varepsilon \quad \forall \varepsilon \in (0, \varepsilon_0). \quad (64)$$

Such $k(\varepsilon)$ exists for every $\varepsilon \in (0,\varepsilon_0)$, since $x_n \to 0$ as $n \to \infty$ and $\lim_{x\to+0}\varphi'(x) = -\infty$ (this equality is true because $\varphi' \notin L^1(0,t_0)$ since $\lim_{x\to+0}\varphi(x) = \infty$). Moreover, since x_n is decreasing and $-1/\varphi'(x)$ is increasing, we obtain

$$-\varphi'(x_n) \geq \tau\varepsilon^{-1} \quad \forall n \geq k(\varepsilon). \quad (65)$$

Combining (63) and (65), it is easy to deduce that such defined $k(\varepsilon)$ satisfies condition (32).

By (16), there exists $L > 0$ such that $(1/(-\varphi'(x)))' \leq L$ for $x \in (0, t_0]$, which means that

$$-\frac{-\varphi''(x)}{-\varphi'(x)} \leq -L\varphi'(x), \quad x \in (0, t_0]. \quad (66)$$

Integrating (66) on $[\varphi^{-1}(t+2\tau), \varphi^{-1}(t)]$, we have

$$\log\frac{-\varphi'(\varphi^{-1}(t+2\tau))}{-\varphi'(\varphi^{-1}(t))} \leq 2L\tau, \quad t \in [\varphi(t_0), \infty), \quad (67)$$

that is,

$$-\varphi'(\varphi^{-1}(t)) \geq -e^{-2L\tau}\varphi'(\varphi^{-1}(t+\tau)), \quad t \in [\varphi(t_0), \infty). \quad (68)$$

By the definition of $k(\varepsilon)$ and (64), we see that

$$\frac{\tau}{-\varphi'(x_{k(\varepsilon)-1})} > \varepsilon. \quad (69)$$

Hence, from (68), it follows that

$$\begin{aligned} \tau\varepsilon^{-1} &> -\varphi'(x_{k(\varepsilon)-1}) \\ &= -\varphi'(\varphi^{-1}(\tau_0 + (k(\varepsilon)-1)\tau)) \\ &\geq -e^{-2L\tau}\varphi'(\varphi^{-1}(\tau_0 + (k(\varepsilon)-1)\tau + 2\tau)) \\ &= -e^{-2L\tau}\varphi'(\varphi^{-1}(\tau_0 + (k(\varepsilon)+1)\tau)) \\ &= -e^{-2L\tau}\varphi'(x_{k(\varepsilon)+1}), \end{aligned} \quad (70)$$

which implies that

$$-\varphi'(x_{k(\varepsilon)+1}) \leq M\varepsilon^{-1}, \quad \varepsilon > 0, \quad (71)$$

where $M = \tau e^{2L\tau}$.

Next, it is clear that the assumption (5) ensures a real number $M_0 > 0$ such that

$$M_0 = \max_{t\in\mathbb{R}}|S(t)| = \max_{t\in[0,\tau]}|S(t)|. \quad (72)$$

Hence, from (15), we have

$$\begin{aligned} \max_{x\in[x_{n+1},x_n]}|y(x)| &= \max_{x\in[x_{n+1},x_n]}(a(x)|S(\varphi(x))|) \\ &\geq a(x_{n+1})\max_{x\in[x_{n+1},x_n]}|S(\varphi(x))| \\ &= M_0 a(x_{n+1}). \end{aligned} \quad (73)$$

Now, from (63) and (73), it follows that

$$\begin{aligned} \max_{x\in[x_{n+1},x_n]}|y(x)|(x_n - x_{n+1}) &\geq M_0 a(x_{n+1})(x_n - x_{n+1}) \\ &\geq \tau M_0\frac{a(x_{n+1})}{-\varphi'(x_{n+1})} \\ &= \tau M_0\frac{a(\varphi^{-1}(\tau_0+(n+1)\tau))}{-\varphi'(\varphi^{-1}(\tau_0+(n+1)\tau))}. \end{aligned} \quad (74)$$

Since $\varphi^{-1}(t)$ is decreasing as well as $-1/\varphi'(x)$ is increasing and $a(x)$ is nondecreasing and positive, we conclude that the function

$$\frac{a(\varphi^{-1}(t))}{-\varphi'(\varphi^{-1}(t))} \text{ is nonincreasing.} \quad (75)$$

Hence, we observe that

$$\begin{aligned} &\sum_{n\geq k(\varepsilon)}\max_{x\in[x_{n+1},x_n]}|y(x)|(x_n - x_{n+1}) \\ &\geq \tau M_0\sum_{n\geq k(\varepsilon)}\frac{a(\varphi^{-1}(\tau_0+(n+1)\tau))}{-\varphi'(\varphi^{-1}(\tau_0+(n+1)\tau))} \\ &\geq \tau M_0\int_{k(\varepsilon)}^{\infty}\frac{a(\varphi^{-1}(\tau_0+(t+1)\tau))}{-\varphi'(\varphi^{-1}(\tau_0+(t+1)\tau))}dt \\ &= M_0\int_0^{\varphi^{-1}(\tau_0+(k(\varepsilon)+1)\tau)}a(\xi)\,d\xi \\ &= M_0\int_0^{x_{k(\varepsilon)+1}}a(\xi)\,d\xi. \end{aligned} \quad (76)$$

Next, from assumption (17), we get some $M_1 > 0$ such that

$$\int_0^x a(\xi)\,d\xi \geq M_1[-\varphi'(x)]^{-(2-s)}, \quad x \in (0, t_0]. \quad (77)$$

Now, from (76), (77), and (71), we obtain

$$\sum_{n\geq k(\varepsilon)} \max_{x\in[x_{n+1},x_n]} |y(x)|(x_n - x_{n+1}) \geq M_0 M_1 \left[-\varphi'(x_{k(\varepsilon)+1})\right]^{-(2-s)} \geq M_0 M_1 (M\varepsilon^{-1})^{-(2-s)} = M_0 M_1 M^{-(2-s)} \varepsilon^{2-s}. \tag{78}$$

Thus, using Lemma 25, we have

$$|\Gamma_\varepsilon(y)| \geq c_1 \varepsilon^{2-s}, \quad \varepsilon \in (0, \varepsilon_0), \tag{79}$$

for some $c_1 > 0$.

Next we prove that $y(x)$ satisfies (53). Assume to the contrary that

$$\int_0^{t_0} |y'(x)|\, dx < \infty. \tag{80}$$

Then, we have

$$\text{length}(y) = \int_0^{t_0} \sqrt{1 + |y'(x)|^2}\, dx \leq \int_0^{t_0} (1 + |y'(x)|)\, dx < \infty. \tag{81}$$

From Lemma 29, it follows that

$$\text{length}(y) = \lim_{\varepsilon\to+0} \frac{|\Gamma_\varepsilon(y)|}{2\varepsilon}. \tag{82}$$

On the other hand, by (79), we see that

$$\lim_{\varepsilon\to+0} \frac{|\Gamma_\varepsilon(y)|}{2\varepsilon} = \infty. \tag{83}$$

This is a contaradiction, and hence $y(x)$ satisfies (53).

Finally, we show that $y(x)$ also satisfies inequality (54). Because of (5), we obtain a constant $c > 0$ such that

$$\max_{t\in\mathbb{R}} |S(t)| = \max_{t\in[0,\tau]} |S(t)| \leq c, \quad \max_{t\in\mathbb{R}} |S'(t)| = \max_{t\in[0,\tau]} |S'(t)| \leq c. \tag{84}$$

By (15), we find that

$$\sup_{t\in(0,x]} |y(t)| = \sup_{t\in(0,x]} |a(t)|\,|S(\varphi(t))| \leq ca(x). \tag{85}$$

Hence, from (1) and (15), it follows that

$$|y'(x)| \leq c\left(a'(x) + a(x)\left[-\varphi'(x)\right]\right), \tag{86}$$

and therefore

$$\int_x^{t_0} |y'(\xi)|\, d\xi \leq ca(t_0) + c\int_x^{t_0} a(\xi)\left[-\varphi'(\xi)\right] d\xi. \tag{87}$$

Using (85) and (87), we see that

$$\begin{aligned}\left[x \sup_{\xi\in(0,x]} |y(\xi)|\right]^{(s-1)/(2-s)} \int_x^{t_0} |y'(\xi)|\, d\xi &\leq ca(t_0)\,[cxa(x)]^{(s-1)/(2-s)} + c[cxa(x)]^{(s-1)/(2-s)} \int_x^{t_0} a(\xi)\left[-\varphi'(\xi)\right] d\xi \\ &\leq ca(t_0)\left[ct_0 a(t_0)\right]^{(s-1)/(2-s)} + c^{1/(2-s)}[xa(x)]^{(s-1)/(2-s)} \int_x^{t_0} a(\xi)\left[-\varphi'(\xi)\right] d\xi.\end{aligned} \tag{88}$$

By (18) we conclude that (54) holds.

By Lemma 28, there exists $c_2 > 0$ such that

$$|\Gamma_\varepsilon(y)| \leq c_2 \varepsilon^{2-s}, \quad \varepsilon \in (0, \varepsilon_0). \tag{89}$$

Thus, we have proved that the chirp function $y(x)$ given by (1) satisfies the desired inequality (31). This completes the proof of Theorem 8.

Proof of Theorem 11. Let $y(x)$ be the chirp function (1). It is easy to see that $y(x)$ is oscillatory near $x = 0$. By (5), there exists $M > 0$ such that

$$|S(t)| \leq M, \quad |S'(t)| \leq M, \quad t \in \mathbb{R}. \tag{90}$$

Hence, we observe that

$$|y'(x)| \leq |a'(x) S(\varphi(x))| + |a(x) S(\varphi(x)) \varphi'(x)| \leq M|a'(x)| + M|a(x)\varphi'(x)|, \tag{91}$$

for $x \in (0, t_0]$. From (81), it follows that $\text{length}(y) < \infty$. Lemma 29 shows that

$$\text{length}(y) = \lim_{\varepsilon\to 0} \frac{|\Gamma_\varepsilon(y)|}{2\varepsilon}, \tag{92}$$

which implies that $\varepsilon\text{length}(\Gamma) \leq |\Gamma_\varepsilon(y)| \leq 4\varepsilon\text{length}(\Gamma)$, $\varepsilon \in (0, \varepsilon_0)$, for some $\varepsilon_0 > 0$. Therefore, $y(x)$ is fractal oscillatory near $x = 0$ with the fractal dimension 1.

4. Proof for the Fractal Oscillations of (26)

In this section, we give the proofs for the fractal oscillations of the linear second-order differential equation (26) considered as an application of the main results on the fractal oscillation of chirp functions.

Before we present the proofs of Theorems 16 and 17, we make the following observation. Since

$$y_1(x) = a(x)\cos(\varphi(x)), \qquad y_2(x) = a(x)\sin(\varphi(x)) \tag{93}$$

are solutions of (26), we see that $y(x) = c_1 y_1(x) + c_2 y_2(x)$ is a fundamental system of all solutions of (26). Assume that $c_1^2 + c_2^2 > 0$, and set $S(t) = c_1 \sin t + c_2 \cos t$. Then, $S(t)$ clearly satisfies (5).

Proof of Theorem 16. Applying Theorem 8 on $y(x) = c_1 y_1(x) + c_2 y_2(x)$, we conclude that $y(x)$ is fractal oscillatory with the fractal dimension s.

Proof of Theorem 17. Theorem 11 implies that $y(x) = c_1 y_1(x) + c_2 y_2(x)$ is fractal oscillatory near $x = 0$ with the fractal dimension 1.

Appendix

Proof of Remark 9. By (17), there exists $c_1 > 0$ such that

$$c_1 \leq \left[-\varphi'(x)\right]^{2-s} \int_0^x a(\xi)\, d\xi, \quad x \in (0, t_0]. \tag{A.1}$$

From (15), it follows that

$$c_1 \leq \left[-\varphi'(x)\right]^{2-s} a(x) \int_0^x d\xi = xa(x) \left[-\varphi'(x)\right]^{2-s}, \quad x \in (0, t_0], \tag{A.2}$$

which implies that

$$\frac{1}{-\varphi'(x)} \leq c_1^{-1/(2-s)} x^{1/(2-s)} [a(x)]^{1/(2-s)}, \quad x \in (0, t_0]. \tag{A.3}$$

By (19), there exists $c_2 > 0$ such that

$$xa(x) \left[-\varphi'(x)\right]^{2-s} \leq c_2, \quad x \in (0, t_0], \tag{A.4}$$

and hence

$$-\varphi'(x) \leq c_2^{1/(2-s)} x^{-1/(2-s)} [a(x)]^{-1/(2-s)}, \quad x \in (0, t_0]. \tag{A.5}$$

Therefore, we see that

$$\begin{aligned}
&\left[\frac{1}{-\varphi'(x)}\right]^{s-1} \int_x^{t_0} a(\xi) \left[-\varphi'(\xi)\right] d\xi \\
&\leq c_1^{-(s-1)/(2-s)} x^{(s-1)/(2-s)} [a(x)]^{(s-1)/(2-s)} c_2^{1/(2-s)} \\
&\quad \times \int_x^{t_0} \xi^{-1/(2-s)} [a(\xi)]^{-(s-1)/(2-s)} d\xi \\
&\leq c_3 x^{(s-1)/(2-s)} [a(x)]^{(s-1)/(2-s)} [a(x)]^{-(s-1)/(2-s)} \\
&\quad \times \int_x^{t_0} \xi^{-1/(2-s)} d\xi \\
&= c_3 \frac{2-s}{s-1} \left(1 - x^{(s-1)/(2-s)} t_0^{-(s-1)/(2-s)}\right) \\
&\leq c_3 \frac{2-s}{s-1}, \quad x \in (0, t_0],
\end{aligned} \tag{A.6}$$

so that (20) holds. We conclude that

$$\begin{aligned}
&[xa(x)]^{(s-1)/(2-s)} \int_x^{t_0} a(\xi) \left[-\varphi'(\xi)\right] d\xi \\
&= \left(xa(x) \left[-\varphi'(x)\right]^{2-s}\right)^{(s-1)/(2-s)} \\
&\quad \times \left(\left[\frac{1}{-\varphi'(x)}\right]^{s-1} \int_x^{t_0} a(\xi) \left[-\varphi'(\xi)\right] d\xi\right).
\end{aligned} \tag{A.7}$$

By (19) and (20), we conclude that (18) is satisfied. The proof of Remark 9 is complete.

References

[1] C. Tricot, *Curves and Fractal Dimension*, Springer, New York, NY, USA, 1995.

[2] M. Pašić, "Fractal oscillations for a class of second order linear differential equations of Euler type," *Journal of Mathematical Analysis and Applications*, vol. 341, no. 1, pp. 211–223, 2008.

[3] M. K. Kwong, M. Pašić, and J. S. W. Wong, "Rectifiable oscillations in second-order linear differential equations," *Journal of Differential Equations*, vol. 245, no. 8, pp. 2333–2351, 2008.

[4] M. Pašić and J. S. W. Wong, "Rectifiable oscillations in second-order half-linear differential equations," *Annali di Matematica Pura ed Applicata: Series 4*, vol. 188, no. 3, pp. 517–541, 2009.

[5] M. Pašić and S. Tanaka, "Fractal oscillations of self-adjoint and damped linear differential equations of second-order," *Applied Mathematics and Computation*, vol. 218, no. 5, pp. 2281–2293, 2011.

[6] Y. Naito, M. Pašić, S. Tanaka, and D. Žubrinić, "Fractal oscillations near domain boundary of radially symmetric solutions of p-Laplace equations, fractal geometry and dynamical systems in pure and applied mathematics," *Contemporary Mathematics American Mathematical Society*. In press.

[7] W. A. Coppel, *Stability and Asymptotic Behavior of Differential Equations*, D. C. Heath and Company, Boston, Mass, USA, 1965.

[8] P. Hartman, *Ordinary Differential Equations*, Birkhäuser, Boston, Mass, USA, 2nd edition, 1982.

[9] F. Neuman, "A general construction of linear differential equations with solutions of prescribed properties," *Applied Mathematics Letters*, vol. 17, no. 1, pp. 71–76, 2004.

[10] F. Neuman, "Structure of solution spaces via transformation," *Applied Mathematics Letters*, vol. 21, no. 5, pp. 529–533, 2008.

[11] P. Borgnat and P. Flandrin, "On the chirp decomposition of Weierstrass-Mandelbrot functions, and their time-frequency interpretation," *Applied and Computational Harmonic Analysis*, vol. 15, no. 2, pp. 134–146, 2003.

[12] E. J. Candès, P. R. Charlton, and H. Helgason, "Detecting highly oscillatory signals by chirplet path pursuit," *Applied and Computational Harmonic Analysis*, vol. 24, no. 1, pp. 14–40, 2008.

[13] S. Jaffard and Y. Meyer, "Wavelet methods for pointwise regularity and local oscillations of functions," *Memoirs of the American Mathematical Society*, vol. 123, no. 587, pp. 1–110, 1996.

[14] Y. Meyer and H. Xu, "Wavelet analysis and chirps," *Applied and Computational Harmonic Analysis*, vol. 4, no. 4, pp. 366–379, 1997.

[15] G. Ren, Q. Chen, P. Cerejeiras, and U. Kaehle, "Chirp transforms and chirp series," *Journal of Mathematical Analysis and Applications*, vol. 373, no. 2, pp. 356–369, 2011.

[16] M. Képesi and L. Weruaga, "Adaptive chirp-based time-frequency analysis of speech signals," *Speech Communication*, vol. 48, no. 5, pp. 474–492, 2006.

[17] L. Weruaga and M. Képesi, "The fan-chirp transform for non-stationary harmonic signals," *Signal Processing*, vol. 87, no. 6, pp. 1504–1522, 2007.

[18] E. Barlow, A. J. Mulholland, A. Nordon, and A. Gachagan, "Theoretical analysis of chirp excitation of contrast agents," *Physics Procedia*, vol. 3, no. 1, pp. 743–747, 2009.

[19] M. H. Pedersen, T. X. Misaridis, and J. A. Jensen, "Clinical evaluation of chirp-coded excitation in medical ultrasound," *Ultrasound in Medicine and Biology*, vol. 29, no. 6, pp. 895–905, 2003.

[20] T. Paavle, M. Min, and T. Parve, "Using of chirp excitation for bioimpedance estimation: theoretical aspects and modeling," in *Proceedings of the 11th International Biennial Baltic Electronics Conference (BEC'08)*, pp. 325–328, Tallinn, Estonia, October 2008.

[21] M. Pašić, "Rectifiable and unrectifiable oscillations for a class of second-order linear differential equations of Euler type," *Journal of Mathematical Analysis and Applications*, vol. 335, no. 1, pp. 724–738, 2007.

[22] K. Falconer, *Fractal Geometry: Mathematical Foundations and Applications*, John Wiley & Sons, Hoboken, NJ, USA, 1999.

[23] P. Mattila, *Geometry of Sets and Measures in Euclidean Spaces: Fractals and Rectifiability*, vol. 44 of *Cambridge Studies in Advanced Mathematics*, Cambridge University Press, Cambridge, UK, 1995.

[24] K. J. Falconer, "On the Minkowski measurability of fractals," *Proceedings of the American Mathematical Society*, vol. 123, no. 4, pp. 1115–1124, 1995.

[25] C. Q. He and M. L. Lapidus, *Generalized Minkowski Content, Spectrum of Fractal Drums, Fractal Strings and the Riemann Zeta-Function*, vol. 127 of *Memoirs of the American Mathematical Society*, American Mathematical Society, Providence, RI, USA, 1997.

[26] M. L. Lapidus and M. van Frankenhuijsen, *Fractal geometry, Complex Dimensions and Zeta Functions: Geometry and Spectra of Fractal Strings*, Springer Monographs in Mathematics, Springer, New York, NY, USA, 2006.

[27] M. Pašić, "Minkowski-Bouligand dimension of solutions of the one-dimensional p-Laplacian," *Journal of Differential Equations*, vol. 190, no. 1, pp. 268–305, 2003.

[28] J. S. W. Wong, "On rectifiable oscillation of Euler type second order linear differential equations," *Electronic Journal of Qualitative Theory of Differential Equations*, vol. 2007, no. 20, pp. 1–12, 2007.

[29] M. Pašić, "Rectifiable and unrectifiable oscillations for a generalization of the Riemann-Weber version of Euler differential equation," *Georgian Mathematical Journal*, vol. 15, no. 4, pp. 759–774, 2008.

[30] M. Pašić and S. Tanaka, "Rectifiable oscillations of self-adjoint and damped linear differential equations of second-order," *Journal of Mathematical Analysis and Applications*, vol. 381, no. 1, pp. 27–42, 2011.

[31] M. Pašić, D. Žubrinić, and V. Županović, "Oscillatory and phase dimensions of solutions of some second-order differential equations," *Bulletin des Sciences Mathématiques*, vol. 133, no. 8, pp. 859–874, 2009.

[32] L. Korkut and M. Resman, "Fractal oscillations of chirp-like functions," *Georgian Mathematical Journal*, vol. 19, no. 4, pp. 705–720, 2012.

15

An Improvement of the Differential Transformation Method and Its Application for Boundary Layer Flow of a Nanofluid

Abdelhalim Ebaid,[1] Hassan A. El-Arabawy,[1,2] and Nader Y. Abd Elazem[1]

[1] *Department of Mathematics, Faculty of Science, University of Tabuk, P.O. Box 741, Tabuk 71491, Saudi Arabia*
[2] *Department of Mathematics, Faculty of Science, Ain Shams University, Egypt*

Correspondence should be addressed to Abdelhalim Ebaid; halimgamil@yahoo.com

Academic Editor: Jürgen Geiser

The main feature of the boundary layer flow problems of nanofluids or classical fluids is the inclusion of the boundary conditions at infinity. Such boundary conditions cause difficulties for any of the series methods when applied to solve such a kind of problems. In order to solve these difficulties, the authors usually resort to either Padé approximants or the commercial numerical codes. However, an intensive work is needed to perform the calculations using Padé technique. Due to the importance of the nanofluids flow as a growing field of research and the difficulties caused by using Padé approximants to solve such problems, a suggestion is proposed in this paper to map the semi-infinite domain into a finite one by the help of a transformation. Accordingly, the differential equations governing the fluid flow are transformed into singular differential equations with classical boundary conditions which can be directly solved by using the differential transformation method. The numerical results obtained by using the proposed technique are compared with the available exact solutions, where excellent accuracy is found. The main advantage of the present technique is the complete avoidance of using Padé approximants to treat the infinity boundary conditions.

1. Introduction

Nanotechnology is an advanced technology, which deals with the synthesis of nanoparticles, processing of the nano materials and their applications. It is well known that 1 nm (nanometer) = 10^{-9} meter. Normally, if the particle sizes are in the 1–100 nm range, they are generally called nanoparticles. Nanotechnology has been widely used in industry since materials with sizes of nanometers possess unique physical and chemical properties. Nanoscale particle added fluids are called as nanofluid. The term "nanofluid" was first used by Choi [1] to describe a fluid in which nanometer-sized particles are suspended in conventional heat transfer basic fluids. Fluids such as oil, water, and ethylene glycol mixture are poor heat transfer fluids, since the thermal conductivity of these fluids plays important role on the heat transfer coefficient between the heat transfer medium and the heat transfer surface. Numerous methods have been taken to improve the thermal conductivity of these fluids by suspending nano/micro or larger-sized particle materials in liquids. An innovative technique to improve heat transfer is by using nanoscale particles in the base fluid [1]. Therefore, the effective thermal conductivity of nanofluids is expected to enhance heat transfer compared with conventional heat transfer liquids (Masuda et al. [2]). This phenomenon suggests the possibility of using nanofluids in advanced nuclear systems (Buongiorno and Hu [3]). Choi et al. [4] showed that the addition of a small amount (less than 1% by volume) of nanoparticles to conventional heat transfer liquids increased the thermal conductivity of the fluid up to approximately two times. A comprehensive survey of convective transport in nanofluids was made by Buongiorno and Hu [3] and very recently by Kakaç and Pramuanjaroenkij [5]. It may be also important to mention that a valuable book in nanofluids is published recently by Das et al. [6]. In addition, various interesting results in this regard can be found in [7–17].

Khan and Pop [18] were the first to investigate the boundary-layer flow of a nanofluid past a stretching sheet. The main feature of the boundary layer flow of nanofluids or classical fluid is the inclusion of the boundary conditions at infinity. Such conditions cause difficulties for any of the series methods when applied to solve this kind of problems. This because the infinity boundary condition cannot be applied directly to the series solution, where Padé approximants should be established before applying the boundary condition at infinity. Many authors [19–32] have been resorted to either Padé technique or some numerical commercial codes to solve the boundary value problems in unbounded domain. However, Padé technique requires a massive computational work to obtain accurate approximate solutions. Searching for a direct method to treat the boundary condition at infinity has been the main goal of many researchers for a long time to solve boundary value problems in unbounded domain. Such a direct method is proposed in this paper. The main idea is to transform the physical domain from unbounded into bounded through a transformation. Accordingly, a new system arises which is now subject to classical boundary conditions, where the boundary conditions at infinity disappeared as a result of the new transformation. The transformed system can be directly solved by the differential transformation method (DTM) [33–45] without any need to Padé approximants. In this paper, the governing system of ordinary differential equations describing the boundary-layer flow of a nanofluid past a stretching sheet is analyzed through the proposed improved version of the DTM. The main advantage of the present method is that not only it avoids the use of Padé approximants, but also gives the series solution in a straightforward manner.

2. Basic Equations

The basic equations of the steady two-dimensional boundary layer flow of a nanofluid past a stretching surface with the linear velocity $u_w(x) = ax$, where a is a constant and x is the coordinate measured along the stretching surface, as given by Kuznetsov and Nield [15] and Nield and Kuznetsov [16] and later by Khan and Pop [18], are as follows:

$$\frac{\partial u}{\partial x} + \frac{\partial v}{\partial y} = 0,$$

$$u\frac{\partial u}{\partial x} + v\frac{\partial u}{\partial y} = -\frac{1}{\rho_f}\frac{\partial p}{\partial x} + \nu\left(\frac{\partial^2 u}{\partial x^2} + \frac{\partial^2 u}{\partial y^2}\right),$$

$$u\frac{\partial v}{\partial x} + v\frac{\partial v}{\partial y} = -\frac{1}{\rho_f}\frac{\partial p}{\partial y} + \nu\left(\frac{\partial^2 v}{\partial x^2} + \frac{\partial^2 v}{\partial y^2}\right),$$

$$u\frac{\partial T}{\partial x} + v\frac{\partial T}{\partial y} = \alpha\left(\frac{\partial^2 T}{\partial x^2} + \frac{\partial^2 T}{\partial y^2}\right) + \tau\left[D_B\left(\frac{\partial C}{\partial x}\frac{\partial T}{\partial x} + \frac{\partial C}{\partial y}\frac{\partial T}{\partial y}\right) + \frac{D_T}{T_\infty}\left(\left(\frac{\partial T}{\partial x}\right)^2 + \left(\frac{\partial T}{\partial y}\right)^2\right)\right],$$

$$u\frac{\partial C}{\partial x} + v\frac{\partial C}{\partial y} = D_B\left(\frac{\partial^2 C}{\partial x^2} + \frac{\partial^2 C}{\partial y^2}\right) + \frac{D_T}{T_\infty}\left(\frac{\partial^2 T}{\partial x^2} + \frac{\partial^2 T}{\partial y^2}\right), \quad (1)$$

subject to the boundary conditions:

$$v = 0, \quad u = u_w(x) = ax, \quad T = T_w, \quad C = C_w, \quad \text{at } y = 0,$$
$$u = v = 0, \quad T = T_\infty, \quad C = C_\infty, \quad \text{as } y \longrightarrow \infty. \quad (2)$$

A complete physical description of the present problem was well presented by Khan and Pop [18] as follows. The flow takes place at $y \geq 0$, where y is the coordinate measured normal to the stretching surface. A steady uniform stress leading to equal and opposite forces is applied along the x-axis so that the sheet is stretched keeping the origin fixed. It is assumed that at the stretching surface, the temperature T and the nanoparticle fraction C take constant values T_w and C_w, respectively. The ambient values, attained as y tends to infinity, of T and C, are denoted by T_∞ and C_∞, respectively.

Here u and v are the velocity components along the axes x and y, respectively, p is the fluid pressure, ρ_f is the density of the base fluid, α is the thermal diffusivity, ν is the kinematic viscosity, a is a positive constant, D_B is the Brownian diffusion coefficient, D_T is the thermophoretic diffusion coefficient, $\tau = (\rho c)_p/(\rho c)_f$ is the ratio between the effective heat capacity of the nanoparticle material and heat capacity of the fluid with ρ being the density, c is the volumetric volume expansion coefficient, and ρ_p is the density of the particles. Khan and Pop [18] have looked for a similarity solution of (1) with the boundary conditions (2) by assuming that

$$\psi = (a\nu)^{1/2} x f(\eta), \qquad \theta(\eta) = \frac{T - T_\infty}{T_w - T_\infty},$$
$$\phi(\eta) = \frac{C - C_\infty}{C_w - C_\infty}, \qquad \eta = \left(\frac{a}{\nu}\right)^{1/2} y, \quad (3)$$

where the stream function ψ is defined in the usual way as $u = \partial\psi/\partial y$ and $v = -\partial\psi/\partial x$. Hence, a set of ordinary differential equations were obtained by [18] as

$$f''' + ff'' - (f')^2 = 0, \quad (4)$$

$$\frac{1}{\mathrm{Pr}}\theta'' + f\theta' + \mathrm{Nb}\,\phi'\theta' + \mathrm{Nt}(\theta')^2 = 0, \quad (5)$$

$$\phi'' + \mathrm{Le}\,f\phi' + \frac{\mathrm{Nt}}{\mathrm{Nb}}\theta'' = 0, \quad (6)$$

subject to the boundary conditions:

$$f(0) = 0, \quad f'(0) = 1, \quad \theta(0) = 1, \quad \phi(0) = 1,$$
$$f'(\infty) = 0, \quad \theta(\infty) = 0, \quad \phi(\infty) = 0, \quad (7)$$

where primes denote differentiation with respect to η and the four parameters are defined by

$$\mathrm{Pr} = \frac{\nu}{\alpha}, \quad \mathrm{Le} = \frac{\nu}{D_B}, \quad \mathrm{Nb} = \frac{(\rho c)_p D_B(\phi_w - \phi_\infty)}{(\rho c)_f \nu},$$
$$\mathrm{Nt} = \frac{(\rho c)_p D_T (T_w - T_\infty)}{(\rho c)_f \nu T_\infty}, \quad (8)$$

where Pr, Le, Nb, and Nt denote the Prandtl number, the Lewis number, the Brownian motion parameter, and the thermophoresis parameter, respectively. The quantities of practical interest are the Nusselt number Nu and the Sherwood number Sh which are defined as

$$\mathrm{Nu} = \frac{x q_w}{k\left(T_w - T_\infty\right)}, \qquad \mathrm{Sh} = \frac{x q_m}{D_B\left(C_w - C_\infty\right)}, \tag{9}$$

where q_w and q_m are the wall heat and mass fluxes, respectively. According to Kuznetsov and Nield [15], $\mathrm{Re}_x^{-1/2}\mathrm{Nu}$ and $\mathrm{Re}_x^{-1/2}\mathrm{Sh}$ are known as the reduced Nusselt number Nur and reduced Sherwood number Shr, respectively,

$$\mathrm{Nur} = \mathrm{Re}_x^{-1/2}\mathrm{Nu} = -\theta'(0), \qquad \mathrm{Shr} = \mathrm{Re}_x^{-1/2}\mathrm{Sh} = -\phi'(0), \tag{10}$$

where $\mathrm{Re}_x = x u_w(x)/\nu$ is the local Reynolds number based on the stretching velocity $u_w(x)$. It should be noted that the exact solution of (4) with the boundary conditions given in (7) was first obtained by Crane [46] and given as

$$f(\eta) = 1 - e^{-\eta}. \tag{11}$$

Substituting $f(\eta)$ into (5)-(6), the given system reduces to a system of two coupled differential equations as

$$\theta'' + \Pr\left(1 - e^{-\eta} + \mathrm{Nb}\phi'\right)\theta' + \mathrm{Nt}\left(\theta'\right)^2 = 0, \qquad \phi'' + \mathrm{Le}\left(1 - e^{-\eta}\right)\phi' + \frac{\mathrm{Nt}}{\mathrm{Nb}}\theta'' = 0, \tag{12}$$

subject to the boundary conditions:

$$\theta(0) = 1, \quad \theta(\infty) = 0, \qquad \phi(0) = 1, \quad \phi(\infty) = 0. \tag{13}$$

3. Transformed Equations

In order to solve boundary value problems in unbounded domain by using the DTM, authors are usually resort to Padé approximant due to the boundary condition at infinity, and an approximate solution is only available in this case. As a well-known fact, Padé approximant requires a huge amount of computational work to find out the approximate solution. In this regard, we think that if it is possible to transform the unbounded domain into a bounded one then the BVPs may be easily solved without any need to Padé approximant. A first step in this direction is to transform the unbounded domain of the independent variable $\eta \in [0,\infty)$ into a bounded one $t \in [0,1)$. Such a transformation is found as $t = 1 - e^{-\eta}$, accordingly the governing equations should be changed to be in terms of the new variable t. The effectiveness of this procedure shall be discussed in the next subsection to show the possibility of obtaining very accurate numerical solutions. In view of the mentioned transformation, the system (1)–(3) with the boundary conditions (7) is transformed into a new system in bounded domain given by

$$\begin{aligned}&(1-t)^2\theta'' - (1-t)(1-\Pr t)\theta' + \Pr(1-t)^2\left[\mathrm{Nb}\theta'\phi' + \mathrm{Nt}\left(\theta'\right)^2\right] = 0,\\ &(1-t)^2\phi'' - (1-t)(1-\mathrm{Le}t)\phi' + \frac{\mathrm{Nt}}{\mathrm{Nb}}\left[(1-t)^2\theta'' - (1-t)\theta'\right] = 0,\end{aligned} \tag{14}$$

with the boundary conditions:

$$\theta(0) = 1, \quad \theta(1) = 0, \tag{15}$$

$$\phi(0) = 1, \quad \phi(1) = 0, \tag{16}$$

where primes denote differentiation with respect to t.

4. Analysis and Results

In this section, the application of the DTM is discussed without resorting to Padé approximants. Applying the DTM to the previously mentioned system yielded the following recurrence scheme:

$$\begin{aligned}&\sum_{m=0}^{k}(k-m+1)(k-m+2)\left[\delta(m)-\delta(m-1)\right]\Theta(k-m+2)\\ &-\sum_{m=0}^{k}(k-m+1)\times\left[\delta(m)-\Pr\delta(m-1)\right]\Theta(k-m+1)\\ &+\mathrm{Nb}\Pr\sum_{r=0}^{k}\sum_{m=0}^{r}(m+1)(r-m+1)\times\left[\delta(k-r)-\delta(k-r-1)\right]\Theta(m+1)\times\Phi(r-m+1)\\ &+\mathrm{Nt}\Pr\sum_{r=0}^{k}\sum_{m=0}^{r}(m+1)(r-m+1)\left[\delta(k-r)-\delta(k-r-1)\right]\times\Theta(m+1)\Theta(r-m+1) = 0,\end{aligned} \tag{17a}$$

$$\begin{aligned}&\sum_{m=0}^{k}(k-m+1)(k-m+2)\left[\delta(m)-\delta(m-1)\right]\Phi(k-m+2)\\ &-\sum_{m=0}^{k}(k-m+1)\times\left[\delta(m)-\mathrm{Le}\,\delta(m-1)\right]\Phi(k-m+1)\\ &+\frac{\mathrm{Nt}}{\mathrm{Nb}}\left(\sum_{m=0}^{k}(k-m+1)(k-m+2)\left[\delta(m)-\delta(m-1)\right]\times\Theta(k-m+2)-(k+1)\Theta(k+1)\right) = 0,\end{aligned} \tag{17b}$$

with the transformed boundary conditions

$$\Theta(0) = 1, \qquad \sum_{k=0}^{N} \Theta(k) = 0, \tag{18}$$

$$\Phi(0) = 1, \qquad \sum_{k=0}^{N} \Phi(k) = 0. \tag{19}$$

Equations (17a), (17b), (18), and (19) are used to obtain very accurate approximate numerical solutions, where two different cases are derived and discussed in the next two subsections.

4.1. Case 1: At Nt=0 and Nb=0. At Nt $= 0$ and Nb $= 0$, the boundary value problem for ϕ becomes ill-posed and consequently the boundary value problem for θ becomes

$$(1-t)^2\theta'' - (1-t)(1-\mathrm{Pr}\,t)\theta' = 0, \tag{20}$$

subject to the boundary conditions in (15). Accordingly, a simple recurrence scheme is obtained form (17a) and (17b), where Nt $=$ Nb $= 0$ are inserted, hence

$$\begin{aligned}\sum_{m=0}^{k}(k-m+1) &\\ \times ((k-m+2)&[\delta(m)-\delta(m-1)]\,\Theta(k-m+2) \\ &-[\delta(m)-\mathrm{Pr}\delta(m-1)]\,\Theta(k-m+1)).\end{aligned} \tag{21}$$

Using the recurrence scheme (21) with the transformed initial conditions (18) for $k = 0, 1, 2, \ldots, 8$, a system of algebraic equations is obtained in $\Theta(1), \Theta(2), \ldots$, and $\Theta(10)$. Solving this system, the 10-term approximate solution at any Prandtl number is given in terms of the original similarity variable η as

$$\begin{aligned}\Theta_{10}(\eta) &\\ = 1 + \Delta\Big[&-3628800\left(1-e^{-\eta}\right) - 1814400\left(1-e^{-\eta}\right)^2 \\ &+ 604800\,(\mathrm{Pr}-2)\left(1-e^{-\eta}\right)^3 \\ &+ 151200\,(5\mathrm{Pr}-6)\left(1-e^{-\eta}\right)^4 \\ &- 30240\left(3\mathrm{Pr}^2 - 26\mathrm{Pr} + 24\right)\left(1-e^{-\eta}\right)^5 \\ &- 5040\left(35\mathrm{Pr}^2 - 154\mathrm{Pr} + 120\right)\left(1-e^{-\eta}\right)^6 \\ &+ 720\left(15\mathrm{Pr}^3 - 340\mathrm{Pr}^2 + 1044\mathrm{Pr} - 720\right) \\ &\times\left(1-e^{-\eta}\right)^7 \\ &+ 90\left(315\mathrm{Pr}^3 - 3304\mathrm{Pr}^2 + 8028\mathrm{Pr} - 5040\right) \\ &\times\left(1-e^{-\eta}\right)^8 \\ &- 10\left(105\mathrm{Pr}^4 - 4900\mathrm{Pr}^3 + 33740\mathrm{Pr}^2\right. \\ &\qquad\left. - 69264\mathrm{Pr} + 40320\right)\left(1-e^{-\eta}\right)^9 \\ &- \left(3465\mathrm{Pr}^4 - 70532\mathrm{Pr}^3 + 367884\mathrm{Pr}^2\right. \\ &\qquad\left. - 663696\mathrm{Pr} + 362880\right)\left(1-e^{-\eta}\right)^{10}\Big],\end{aligned} \tag{22}$$

where

$$\Delta = \frac{1}{4515{.}^{4} - 158682{.}^{3} + 1514564{.}^{2} - 5753736{.} + 10628640}. \tag{23}$$

Here, it may be useful to mention that an exact solution for the current case is obtained very recently by the first author in [47] and given by

$$\theta(\eta) = \frac{\Gamma\left(\mathrm{Pr}, 0, \mathrm{Pr}\,e^{-\eta}\right)}{\Gamma(\mathrm{Pr}, 0, \mathrm{Pr})}, \tag{24}$$

where $\Gamma(a, z_0, z_1) = \int_{z_0}^{z_1} \mu^{a-1} e^{-\mu} d\mu$ is the generalized Gamma function which can be expressed in terms of the incomplete Gamma function as $\Gamma(a, z_0, z_1) = \Gamma(a, z_0) - \Gamma(a, z_1)$, where $\Gamma(a, z) = \int_{z}^{\infty} \mu^{a-1} e^{-\mu} d\mu$. There is no doubt that the availability of the exact solution gives the opportunity to validate the accuracy of the suggested approach. For the purpose of illustration, the approximate solutions are compared in Figures 1, 2, and 3 with the exact one given by (24) at different values of Prandtl number. These primary results reveal that the 10-term approximate solution is sufficient to obtain numerical solutions of high accuracy for certain range of Prandtl number, mainly Pr $= 1, 2, 3$. However, at Pr $= 4$ the 10-term approximate solution is not accurate as can be seen in Figure 4. This observation leads to the conclusion that with increasing Pr more terms in the approximate series solution are in fact needed. For example, at Pr $= 5$ the 20-term approximate solution is found sufficient in Figure 5, while 30-term approximate solution is found identical to the exact one at Pr $= 10$ in Figure 6. However, the main advantage of the suggested approach is the avoidance of Padé-approximants which has been used a long time to treat the boundary condition at infinity.

4.2. Case 2: At Nt=0, Pr=Le=1, and Nb≠0. Substituting Nt $= 0$, Pr $= 1$, and Le $= 1$ into (17a) and (17b) yields

$$\begin{aligned}&\sum_{m=0}^{k}(k-m+1)(k-m+2)[\delta(m)-\delta(m-1)]\,\Theta(k-m+2) \\ &\quad - \sum_{m=0}^{k}(k-m+1)\times[\delta(m)-\mathrm{Pr}\,\delta(m-1)]\,\Theta(k-m+1) \\ &\quad + \mathrm{Nb}\,\mathrm{Pr}\sum_{r=0}^{k}\sum_{m=0}^{r}(m+1)(r-m+1) \\ &\qquad\quad \times[\delta(k-r)-\delta(k-r-1)]\,\Theta(m+1) \\ &\qquad\quad \times\Phi(r-m+1) = 0,\end{aligned}$$

$$\sum_{m=0}^{k}(k-m+1)(k-m+2)[\delta(m)-\delta(m-1)]\Phi(k-m+2)$$
$$-\sum_{m=0}^{k}(k-m+1)\times[\delta(m)-\mathrm{Le}\,\delta(m-1)]$$
$$\times\Phi(k-m+1)=0. \tag{25}$$

Using the recurrence scheme (25) with the transformed initial conditions (18) and (19) for $k = 0, 1, 2, \ldots, 4$, we get a system of algebraic equations in $\Theta(1), \Theta(2), \ldots, \Theta(6)$ and $\Phi(1), \Phi(2), \ldots, \Phi(6)$. The solution of the required system leads to the following 6-term approximate solution for the θ equation:

$$\Theta_6(\eta)=1+\Theta(1)(1-e^{-\eta})+\Theta(2)(1-e^{-\eta})^2$$
$$+\Theta(3)(1-e^{-\eta})^3+\Theta(4)(1-e^{-\eta})^4 \tag{26}$$
$$+\Theta(5)(1-e^{-\eta})^5+\Theta(6)(1-e^{-\eta})^6,$$

where $\Theta(1), \Theta(2), \ldots,$ and $\Theta(6)$ are expressed in terms of Nb but ignored here for lengthy results. However, the 6-term series solution for the ϕ equation is given explicitly as

$$\Phi_6(\eta)$$
$$=1-\frac{1}{1237}$$
$$\times\left[720(1-e^{-\eta})+360(1-e^{-\eta})^2+120(1-e^{-\eta})^3\right.$$
$$\left.+30(1-e^{-\eta})^4+6(1-e^{-\eta})^5+(1-e^{-\eta})^6\right]. \tag{27}$$

The exact solutions are obtained in [47] as

$$\phi(\eta)=\frac{1-e^{-e^{-\eta}}}{1-e^{-1}},$$
$$\theta(\eta)=\frac{1-e^{-\alpha\mathrm{Nb}(1-e^{-e^{-\eta}})}}{1-e^{-\mathrm{Nb}}}. \tag{28}$$

The obtained truncated series solution $\Theta_6(\eta)$ is compared with the exact one in Figures 7–9 at several values of Nb. As observed from Figures 7 and 8, the approximate solution is coincided with the exact one at certain values, Nb = 0.1 and Nb = 0.3. However, it approaches the exact curve at Nb = 0.5, where more terms are needed in this case. In addition, the approximate solution $\Phi_6(\eta)$ is found identical to the exact curve as shown in Figure 10.

5. Conclusions

A system of ordinary differential equations describing the boundary layer flow of a nanofluid past a stretching sheet is investigated in this paper via a new approach. The suggested approach is based on transforming the boundary conditions

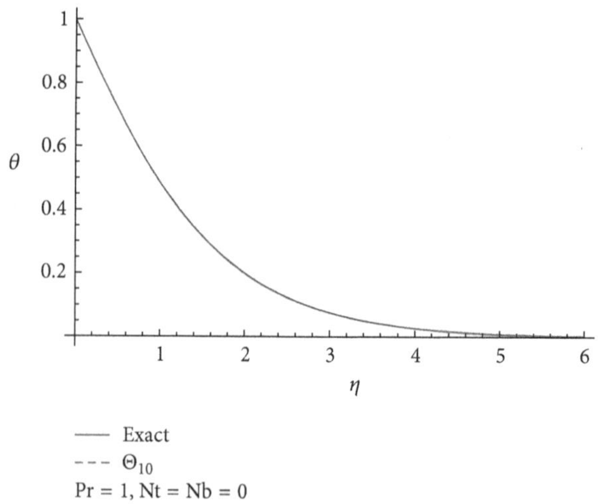

FIGURE 1

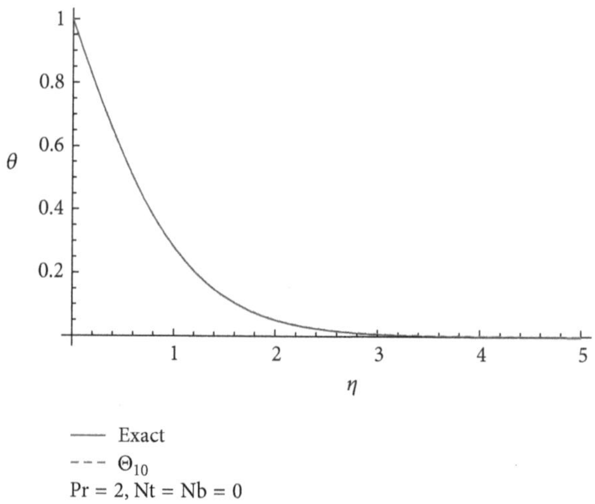

FIGURE 2

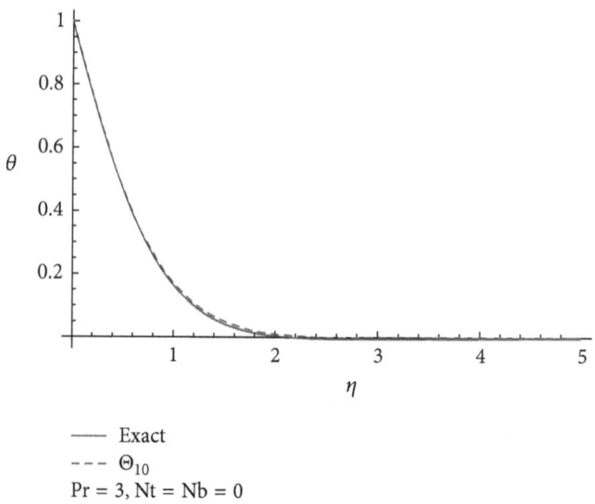

FIGURE 3

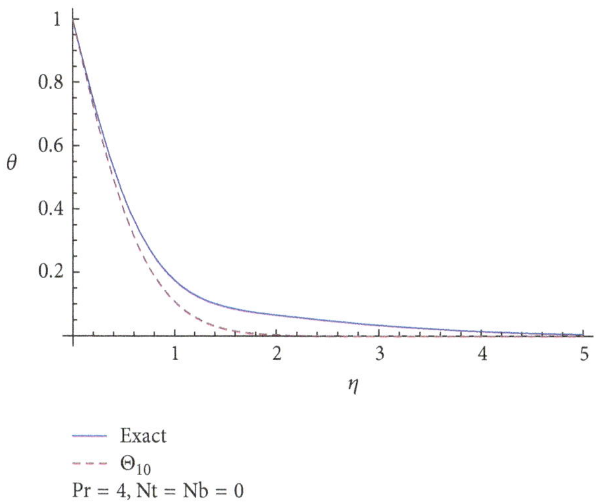

FIGURE 4

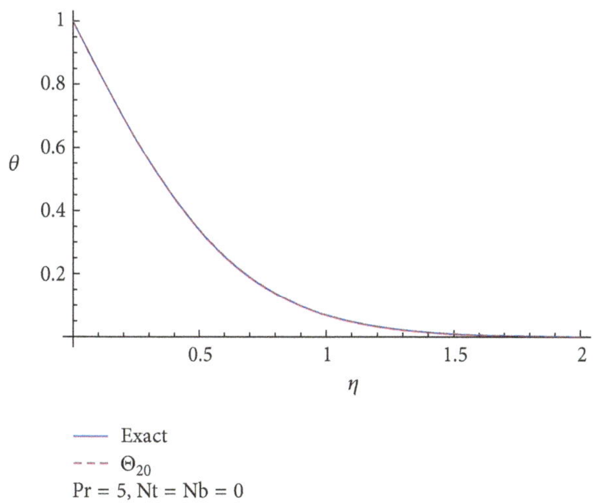

FIGURE 5

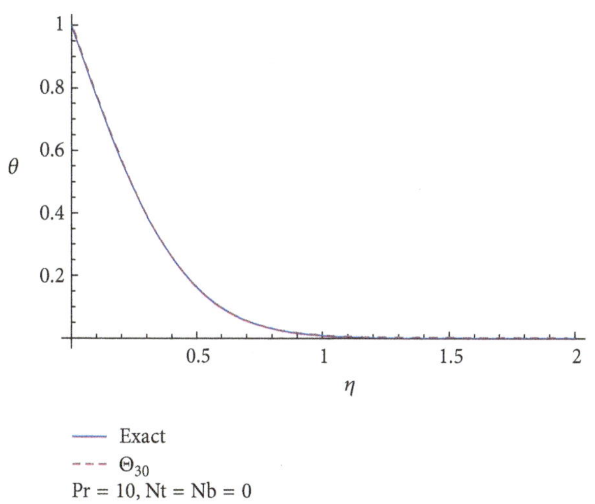

FIGURE 6

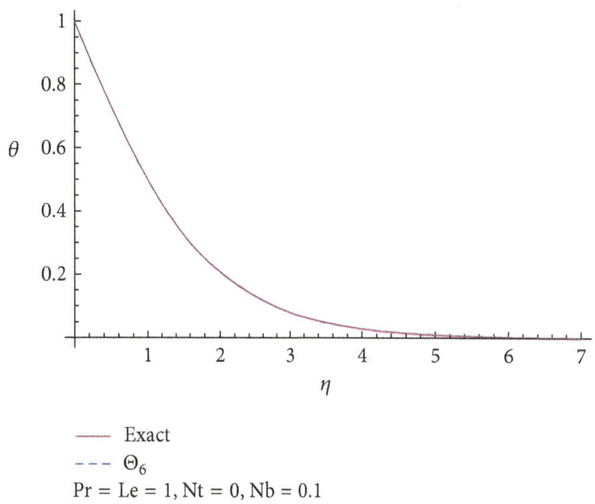

FIGURE 7

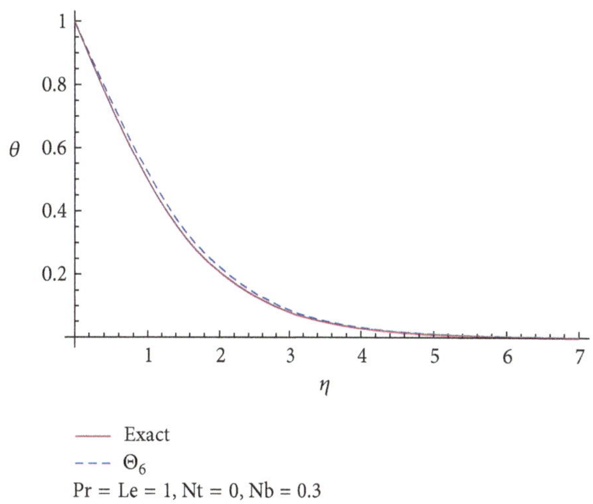

FIGURE 8

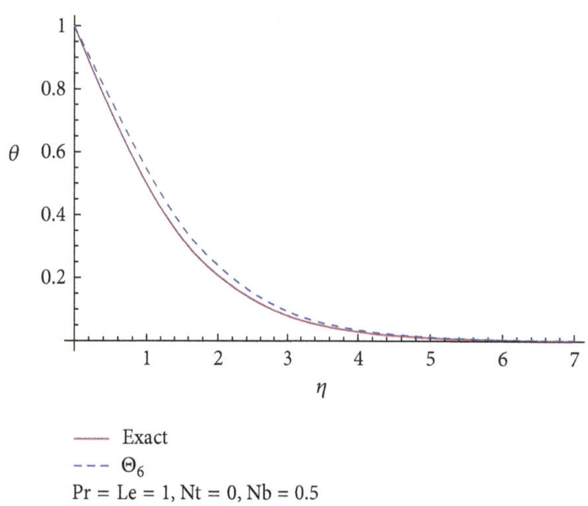

FIGURE 9

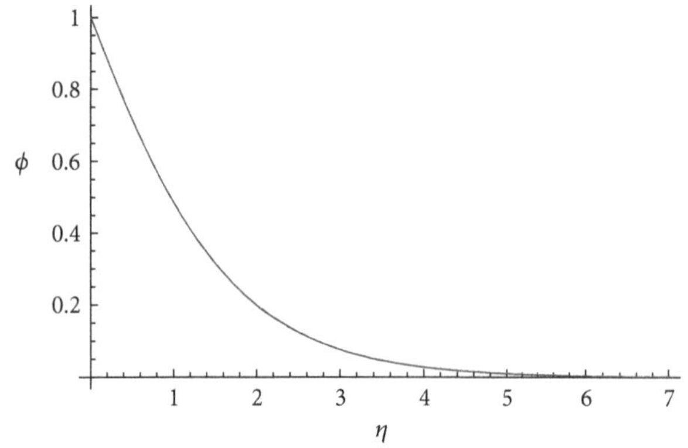

FIGURE 10

at infinity into classical conditions prior to the application of the differential transformation method. A transformation is successfully used to map the unbounded physical domain into a bounded one. In addition, the current results are validated through various comparisons with the available exact solutions. In comparison with Padé technique, the new method of solution is found not only straightforward but also effective in obtaining accurate numerical solutions, where Padé approximant was completely avoided.

Acknowledgments

This paper was funded by the Deanship of Scientific Research (DSR), University of Tabuk, Tabuk, under Grant no. (S-84-1432). The authors, therefore, acknowledge with thanks DSR technical and financial support.

References

[1] S. U. S. Choi, "Enhancing thermal conductivity of fluids with nanoparticles," in *The Proceedings of the ASME International Mechanical Engineering Congress and Exposition*, ASME, FED 231/MD 66, pp. 99–105, San Francisco, Calif, USA, 1995.

[2] H. Masuda, A. Ebata, K. Teramae, and N. Hishinuma, "Alteration of thermal conductivity and viscosity of liquid by dispersing ultra-fine particles (Dispersion of g-Al_2O_3, SiO_2, and TiO_2 ultra-fine particles)," *Netsu Bussei*, vol. 7, no. 4, pp. 227–233, 1993.

[3] J. Buongiorno and W. Hu, "Nanofluid coolants for advanced nuclear power plants," in *Proceedings of ICAPP '05*, Paper no. 5705, Seoul, Republic of Korea, May 2005.

[4] S. U. S. Choi, Z. G. Zhang, W. Yu, F. E. Lockwood, and E. A. Grulke, "Anomalous thermal conductivity enhancement in nanotube suspensions," *Applied Physics Letters*, vol. 79, no. 14, pp. 2252–2254, 2001.

[5] S. Kakaç and A. Pramuanjaroenkij, "Review of convective heat transfer enhancement with nanofluids," *International Journal of Heat and Mass Transfer*, vol. 52, no. 13-14, pp. 3187–3196, 2009.

[6] S. K. Das, S. U. S. Choi, W. Yu, and T. Pradeep, *Nanofluids: Science and Technology*, Wiley-Interscience, Hoboken, NJ, USA, 2007.

[7] J. Buongiorno, "Convective transport in nanofluids," *Journal of Heat Transfer*, vol. 128, no. 3, pp. 240–250, 2006.

[8] D. A. Nield and A. V. Kuznetsov, "The Cheng-Minkowycz problem for natural convective boundary-layer flow in a porous medium saturated by a nanofluid," *International Journal of Heat and Mass Transfer*, vol. 52, no. 25-26, pp. 5792–5795, 2009.

[9] K. Khanafer, K. Vafai, and M. Lightstone, "Buoyancy-driven heat transfer enhancement in a two-dimensional enclosure utilizing nanofluids," *International Journal of Heat and Mass Transfer*, vol. 46, no. 19, pp. 3639–3653, 2003.

[10] W. Daungthongsuk and S. Wongwises, "A critical review of convective heat transfer of nanofluids," *Renewable and Sustainable Energy Reviews*, vol. 11, no. 5, pp. 797–817, 2007.

[11] R. K. Tiwari and M. K. Das, "Heat transfer augmentation in a two-sided lid-driven differentially heated square cavity utilizing nanofluids," *International Journal of Heat and Mass Transfer*, vol. 50, no. 9-10, pp. 2002–2018, 2007.

[12] L. Wang and X. Wei, "Heat conduction in nanofluids," *Chaos, Solitons and Fractals*, vol. 39, no. 5, pp. 2211–2215, 2009.

[13] H. F. Oztop and E. Abu-Nada, "Numerical study of natural convection in partially heated rectangular enclosures filled with nanofluids," *International Journal of Heat and Fluid Flow*, vol. 29, no. 5, pp. 1326–1336, 2008.

[14] A. V. Kuznetsov and D. A. Nield, "Natural convective boundary-layer flow of a nanofluid past a vertical plate," *International Journal of Thermal Sciences*, vol. 49, no. 2, pp. 243–247, 2010.

[15] A. V. Kuznetsov and D. A. Nield, "Effect of local thermal non-equilibrium on the onset of convection in a porous medium layer saturated by a nanofluid," *Transport in Porous Media*, vol. 83, no. 2, pp. 425–436, 2010.

[16] E. H. Aly and A. Ebaid, "New exact solutions for boundary-layer flow of a nanofluid past a stretching sheet," *Journal of Computational and Theoretical Nanoscience*. In press.

[17] P. Cheng and W. J. Minkowycz, "Free convection about a vertical flat plate embedded in a porous medium with application to heat transfer from a dike," *Journal of Geophysical Research*, vol. 82, no. 14, pp. 2040–2044, 1977.

[18] W. A. Khan and I. Pop, "Boundary-layer flow of a nanofluid past a stretching sheet," *International Journal of Heat and Mass Transfer*, vol. 53, no. 11-12, pp. 2477–2483, 2010.

[19] J. P. Boyd, "Padé-approximant algorithm for solving nonlinear ordinary differential equation boundary value problems on an unbounded domain," *Computers in Physics*, vol. 11, no. 3, pp. 299–303, 1997.

[20] A. M. Wazwaz, "The modified decomposition method and Padé approximants for solving the Thomas-Fermi equation," *Applied Mathematics and Computation*, vol. 105, no. 1, pp. 11–19, 1999.

[21] A. M. Wazwaz, "The modified decomposition method and Padé approximants for a boundary layer equation in unbounded domain," *Applied Mathematics and Computation*, vol. 177, no. 2, pp. 737–744, 2006.

[22] A. M. Wazwaz, "Padé approximants and Adomian decomposition method for solving the Flierl-Petviashivili equation and its variants," *Applied Mathematics and Computation*, vol. 182, no. 2, pp. 1812–1818, 2006.

[23] A. M. Wazwaz, "The variational iteration method for solving two forms of Blasius equation on a half-infinite domain," *Applied Mathematics and Computation*, vol. 188, no. 1, pp. 485–491, 2007.

[24] S. A. Kechil and I. Hashim, "Non-perturbative solution of free-convective boundary-layer equation by Adomian decomposition method," *Physics Letters A*, vol. 363, no. 1-2, pp. 110–114, 2007.

[25] S. A. Kechil and I. Hashim, "Series solution of flow over nonlinearly stretching sheet with chemical reaction and magnetic field," *Physics Letters A*, vol. 372, no. 13, pp. 2258–2263, 2008.

[26] E. Alizadeh, K. Sedighi, M. Farhadi, and H. R. Ebrahimi-Kebria, "Analytical approximate solution of the cooling problem by Adomian decomposition method," *Communications in Nonlinear Science and Numerical Simulation*, vol. 14, no. 2, pp. 462–472, 2009.

[27] T. Hayat, Q. Hussain, and T. Javed, "The modified decomposition method and Padé approximants for the MHD flow over a non-linear stretching sheet," *Nonlinear Analysis: Real World Applications*, vol. 10, no. 2, pp. 966–973, 2009.

[28] S. Abbasbandy and T. Hayat, "Solution of the MHD Falkner-Skan flow by Hankel-Padé method," *Physics Letters A*, vol. 373, no. 7, pp. 731–734, 2009.

[29] M. M. Rashidi and E. Erfani, "A novel analytical solution of the thermal boundary-layer over a flat plate with a convective surface boundary condition using DTM-Padé," in *Proceedings of the International Conference on Signal Processing Systems (ICSPS '09)*, pp. 905–909, Singapore, May 2009.

[30] M. M. Rashidi and S. A. M. Pour, "A novel analytical solution of steady flow over a rotating disk in porous medium with heat transfer by DTM-Padé," *African Journal of Mathematics and Computer Science Research*, vol. 3, no. 6, pp. 93–100, 2010.

[31] M. M. Rashidi and S. A. M. Pour, "Explicit solution of axisymmetric stagnation flow towards a shrinking sheet by DTM-Padé," *Applied Mathematical Sciences*, vol. 4, no. 53–56, pp. 2617–2632, 2010.

[32] M. M. Rashidi and M. Keimanesh, "Using differential transform method and Padé approximant for solving mhd flow in a laminar liquid film from a horizontal stretching surface," *Mathematical Problems in Engineering*, vol. 2010, Article ID 491319, 14 pages, 2010.

[33] J. K. Zhou, *Differential Transformation and Its Applications for Electrical circuIts*, Huazhong University Press, Wuhan, China, 1986.

[34] S. H. Ho and C. K. Chen, "Analysis of general elastically end restrained non-uniform beams using differential transform," *Applied Mathematical Modelling*, vol. 22, no. 4-5, pp. 219–234, 1998.

[35] C. K. Chen and S. H. Ho, "Transverse vibration of a rotating twisted Timoshenko beams under axial loading using differential transform," *International Journal of Mechanical Sciences*, vol. 41, no. 11, pp. 1339–1356, 1999.

[36] C. K. Chen and S. H. Ho, "Solving partial differential equations by two-dimensional differential transform method," *Applied Mathematics and Computation*, vol. 106, no. 2-3, pp. 171–179, 1999.

[37] M. J. Jang, C. L. Chen, and Y. C. Liy, "On solving the initial-value problems using the differential transformation method," *Applied Mathematics and Computation*, vol. 115, no. 2-3, pp. 145–160, 2000.

[38] M. J. Jang, C. L. Chen, and Y. C. Liu, "Two-dimensional differential transform for partial differential equations," *Applied Mathematics and Computation*, vol. 121, no. 2-3, pp. 261–270, 2001.

[39] M. Köksal and S. Herdem, "Analysis of nonlinear circuits by using differential Taylor transform," *Computers and Electrical Engineering*, vol. 28, no. 6, pp. 513–525, 2002.

[40] F. Ayaz, "Solutions of the system of differential equations by differential transform method," *Applied Mathematics and Computation*, vol. 147, no. 2, pp. 547–567, 2004.

[41] A. Arikoglu and I. Ozkol, "Solution of boundary value problems for integro-differential equations by using differential transform method," *Applied Mathematics and Computation*, vol. 168, no. 2, pp. 1145–1158, 2005.

[42] S. H. Chang and I. L. Chang, "A new algorithm for calculating one-dimensional differential transform of nonlinear functions," *Applied Mathematics and Computation*, vol. 195, no. 2, pp. 799–808, 2008.

[43] A. S. V. R. Kanth and K. Aruna, "Two-dimensional differential transform method for solving linear and non-linear Schrödinger equations," *Chaos, Solitons and Fractals*, vol. 41, no. 5, pp. 2277–2281, 2009.

[44] A. S. V. R. Kanth and K. Aruna, "Differential transform method for solving the linear and nonlinear Klein-Gordon equation," *Computer Physics Communications*, vol. 180, no. 5, pp. 708–711, 2009.

[45] A. Ebaid, "Approximate periodic solutions for the non-linear relativistic harmonic oscillator via differential transformation method," *Communications in Nonlinear Science and Numerical Simulation*, vol. 15, no. 7, pp. 1921–1927, 2010.

[46] L. J. Crane, "Flow past a stretching plate," *Zeitschrift für Angewandte Mathematik und Physik*, vol. 21, no. 4, pp. 645–647, 1970.

[47] A. Ebaid and M. D. Aljoufi, "New theoretical and numerical results for the boundary-layer flow of a nanofluid past a stretching sheet," *Advanced Studies in Theoretical Physics*. In press.

16

Dynamics of a Gross-Pitaevskii Equation with Phenomenological Damping

Renato Colucci, Gerardo R. Chacón, and Andrés Vargas

Departamento de Matemáticas, Pontificia Universidad Javeriana, Carrera 7 No. 40-62, Bogotá, Colombia

Correspondence should be addressed to Andrés Vargas; a.v4rg4s@gmail.com

Academic Editor: Norio Yoshida

We study the dynamical behavior of solutions of an n-dimensional nonlinear Schrödinger equation with potential and linear derivative terms under the presence of phenomenological damping. This equation is a general version of the dissipative Gross-Pitaevskii equation including terms with first-order derivatives in the spatial coordinates which allow for rotational contributions. We obtain conditions for the existence of a global attractor and find bounds for its dimension.

1. Introduction

Nonlinear Schrödinger equations have enjoyed a considerable amount of attention during several decades due to their frequent appearance in the modeling of interesting physical phenomena in many different areas (e.g., optics, fluid mechanics, condensed matter, etc.). Additionally, the rigorous mathematical treatment of these equations has produced a great deal of new insights and techniques accompanied by a voluminous bibliography. We refer the reader to excellent sources as [1–4] for a detailed account of these aspects.

Here, for complex coefficients α and β, our purpose is to study the long-term dynamics of a general nonlinear Schrödinger evolution equation with potential of the form

$$iu_t - \alpha\Delta u + \beta|u|^2 u + f(x)u = v(x)\cdot\nabla u, \qquad (1)$$

where $u(x,t)$ and $f(x)$ are complex-valued functions and $v(x)$ is a complex vector-valued map, all defined on a bounded open domain $U \subset \mathbb{R}^n$ (for $n \geqslant 1$) with regular boundary ∂U. We prove the existence of a global attractor for (1), under Dirichlet and U-periodic boundary conditions, assuming $\mathrm{Im}(\alpha) > 0$ and some additional requirements on $f(x)$ and $v(x)$ for which existence and uniqueness of solutions are guaranteed. Our choice of signs for the terms in this equation was made such that the hypotheses needed for our analysis make, if necessary, the imaginary parts of constant and functional coefficients positive.

Particular instances of (1) have been extensively considered in the literature, specially in two and three dimensions. In fact, when the coefficients α, and β and the function $f(x)$ are real and assuming $v(x) = 0$, this is the celebrated time-dependent *Gross-Pitaevskii equation* (GP) employed to model nonlinear behavior in several physical systems. Of special relevance is the mean-field description of Bose-Einstein condensation achieved in dilute atomic gases, a state of matter which exhibits peculiar phenomena characteristic of the superfluid nature of the system.

If for the previous situation in three dimensions we choose $v(x) = i\Omega \wedge x \in i\mathbb{R}^3$, where $\Omega \in \mathbb{R}^3$ is an angular velocity vector, we recover the GP equation with angular momentum rotational term. Indeed, the right-hand side of (1) becomes

$$v\cdot\nabla u = i(\Omega\wedge x)\cdot\nabla u = i\Omega\cdot(x\wedge\nabla)u = -\Omega\cdot Lu, \qquad (2)$$

where $L := -i(x\wedge\nabla)$ is precisely the quantum mechanical angular momentum operator. This equation has been used to model physical experiments where the condensate is set into rotation with an angular frequency $\|\Omega\|$, making it known as the *Gross-Pitaevskii equation* for rotating Bose gases. In this situation, the observation of quantized vortices has been a topic of considerable experimental and theoretical studies; the reader is referred to [5] and references therein for details on vortices in Bose-Einstein condensates and the recent review [6] for a treatment that includes numerical aspects (see also [7] for the simulation of rotating condensates). Existence and uniqueness results for solutions of GP equation with rotation in $\mathbb{R}^n$ for dimensions $n = 2$ and $n = 3$ and under different assumptions can be found in [8–10]. Note that rotating and nonnecessarily nonlinear Schrödinger equations appear also in several other contexts related to wave propagation (see references in [11]).

The general case of fully complex coefficients α, β and complex-valued functions $f(x)$, $v(x)$ in (1) can be thought of as a mixture between the aforementioned GP equation and a gradient flow which accounts for dissipation in the system proportional to the rate of change of the wave function $u(x,t)$. This is indeed the case if (1) has the particular form

$$iu_t = (1 + i\gamma)\left[\alpha\Delta u - \beta|u|^2 u - f(x)u + v(x)\cdot\nabla u\right], \quad (3)$$

for real constants α, β, γ, a real-valued function $f(x)$, and an imaginary vector field $v(x)$ as in the rotational term explained earlier. In the physical literature phenomenological damping has been introduced to study Bose-Einstein condensates (both in the rotating and nonrotating cases) affected by complicated sources of dissipation related to their interaction with the thermal cloud around the condensate. For example, [12] studies the case of trapped atomic condensates and [13] analyzes dark matter wave solitons in the same context, while [14–16] study the formation of vortex lattices in rotating condensates, including numerical simulations. A theoretical explanation of the phenomenological damping terms can be found in [17] (see also [18, 19]).

From the mathematical side, equations similar to (3), but with $v = 0$, have been used to study the behavior of Ginzburg-Landau vortices in [20] and, in the case of low dissipation, the asymptotics for perturbation of constant solutions in [21]. Long-time behavior, establishing the existence of attractors under several assumptions, has also been the topic of several studies; see, for example, [22] for Ginzburg-Landau equations on bounded domains, [23] for weakly damped nonlinear Schrödinger equations in $\mathbb{R}^n$ with $n \leqslant 3$, and for dissipative nonlinearities see [24] and references therein. To our knowledge attractors for the dissipative Gross-Pitaevskii equation with rotation have not been explicitly considered.

Since the arguments needed for our analysis work in the fully complex coefficients situation, we adopt this broader setting and, in addition, do not necessarily assume that the right-hand side of (1) has the particular form of an angular momentum contribution. This situation corresponds to the case where the ratio between imaginary and real parts of α, β, $f(x)$, and $v(x)$ in (3) is not a fixed constant γ. In fact, this can include cases where there are sources of dissipation which are proportional to some of the individual terms in the equation, like the nonlinearity or the gradient terms. Such contributions are important, for example, to study exploding and collapsing Bose-Einstein condensates (see [25] and references therein for examples of pure nonlinear dissipation in that context, as well as [26] for simulation of collapse and explosion).

We point out that numerical studies in [27] show that cubic or quintic damping terms prevent blowup of solutions to focusing nonlinear Schrödinger equations without "rotational" term, while just linear damping is not enough below a certain threshold. This agrees with Theorem 8 on existence of attractors (that covers such situation if one assumes an arbitrarily small amount of linear damping) and provides some justification for the hypotheses under which this result is proved.

Finally, to summarize the content of the following sections we mention that, under appropriate assumptions introduced later, Theorem 4 establishes the existence and uniqueness of solutions to the dissipative GP equation on bounded open domains, Theorem 8 shows the existence of a compact global attractor, and Theorem 11 provides bounds for its Hausdorff and fractal dimensions.

1.1. Precise Setting. To ease notation, for any complex vector $v \in \mathbb{C}^n$ with $n \geqslant 1$, we define $v_1 := \mathrm{Re}(v) \in \mathbb{R}^n$ and $v_2 := \mathrm{Im}(v) \in \mathbb{R}^n$. Let $U \subset \mathbb{R}^n$; if $u(x,t) = u_1(x,t) + iu_2(x,t) : U \times \mathbb{R}^+ \to \mathbb{C}$ is a complex-valued function, the gradient ∇u and Laplacian Δu (both with respect to spatial coordinates) are given by $\nabla u := \nabla u_1 + i\nabla u_2 \in \mathbb{C}^n$ and $\Delta u := \Delta u_1 + i\Delta u_2 \in \mathbb{C}$. In addition, we extend bilinearly the Euclidean inner product "$\cdot$" in $\mathbb{R}^n$ to a symmetric inner product on $\mathbb{C}^n$ denoted for simplicity by the same dot; that is, for vectors $v, w \in \mathbb{C}^n$, we have

$$v\cdot w := (v_1\cdot w_1 - v_2\cdot w_2) + i(v_1\cdot w_2 + v_2\cdot w_1). \quad (4)$$

With this product, the standard Hermitian norm of $v \in \mathbb{C}^n$ satisfies $\|v\|^2 = \|v_1\|^2 + \|v_2\|^2 = v\cdot\overline{v}$, where $\overline{v} := v_1 - iv_2$.

Given coefficients $\alpha, \beta \in \mathbb{C}$ and a constant $p \geqslant 2$, we consider the following dissipative (fully complex) GP equation with nonlinearity of order $(p+1)$ on a bounded open domain $U \subset \mathbb{R}^n$, for arbitrary $n \geqslant 1$:

$$\begin{gathered} iu_t - \alpha\Delta u + \beta|u|^p u + f(x)u = v(x)\cdot\nabla u, \\ u(x,0) = u_0(x), \end{gathered} \quad (5)$$

with unknown $u(x,t) : U \times \mathbb{R}^+ \to \mathbb{C}$, a scalar function $f(x) : U \to \mathbb{C}$, and a vector-valued map $v(x) : U \to \mathbb{C}^n$. Our hypotheses on the constant coefficients are

$$\alpha_2 > 0, \quad \text{(H1)}$$

$$\beta_2 \geqslant p|\beta_1| \geqslant 0 \quad \text{(H2)}$$

and on the functional coefficients and initial data are

$$f(x) \in L^\infty(U), \quad \text{(H3)}$$

$$v(x) \in L^\infty(U, \mathbb{C}^n), \quad \text{(H4)}$$

$$u_0(x) \in L^2(U). \quad \text{(H5)}$$

Additionally, we assume that

$$\mu_2 := \operatorname*{ess\,inf}_{x\in U} f_2(x) > 0 \quad \text{(H6)}$$

and require that

$$4\alpha_2\mu_2 > \|v\|_\infty^2. \quad \text{(H7)}$$

Remark 1. Notice that hypothesis (H2) allows for $\beta = 0$ in which case there is no nonlinearity term in (5). Independently, hypotheses (H1) and (H6) already imply (H7) when $v(x) = 0$ and this includes the standard GP equation (i.e., no rotational term) but with dissipation.

Remark 2. As a consequence of hypothesis (H6), (5) has a nonzero linear damping term of the form $if_2(x)u$. Obviously, such term is always present in the case of GP with phenomenological damping (3) and is not the only source of dissipation.

For (5) we will consider Dirichlet boundary conditions on the domain $U \subset \mathbb{R}^n$, which is assumed to have regular boundary ∂U. Explicitly, we suppose that for all $t \geqslant 0$,

$$u(x,t)|_{\partial U} = 0, \tag{D1}$$

$$\Delta u(x,t)|_{\partial U} = 0. \tag{D2}$$

Alternatively we also consider U-periodic boundary conditions when the domain has the form $U = (0,L)^n \subset \mathbb{R}^n$. This means that for any pair of points $a, b \in \overline{U} = [0,L]^n$ which are equal except at an arbitrary fixed component $j \in \{1,\ldots,n\}$ for which $a_j = 0$ and $b_j = L$, it holds that

$$\partial^k u(a,t) = \partial^k u(b,t), \tag{P}$$

for all $t \geqslant 0$ and admissible integers $k \geqslant 0$. In this expression, $\partial^0 u := u$ and $\partial^k u$ denotes any k-order partial (weak) derivative admissible for u.

Remark 3. In the case of U-periodic boundary conditions, it is actually assumed that points of the form $a, b \in [0,L]^n$ described earlier are identified, so in fact any function defined on U must take the same values on those points but, in contrast to Dirichlet conditions, the precise value does not have to be fixed.

Existence and uniqueness of solutions to (5) with Dirichlet (D1)-(D2) or U-periodic (P) boundary conditions hold from classical results.

Theorem 4. *Given hypotheses (H1)-(H7), under Dirichlet or U-periodic boundary conditions, there exists a unique solution u of (5) such that*

$$u \in \mathscr{C}_b\left(\mathbb{R}^+, H^1(U)\right), \qquad u_t \in \mathscr{C}_b\left(\mathbb{R}^+, H^{-1}(U)\right), \tag{6}$$

where H^{-1} is the dual space of H^1 and $\mathscr{C}_b(\mathbb{R}^+, H^1(U))$ is the set of continuous bounded functions from $\mathbb{R}^+$ into $H^1(U)$.

The proof of the previous theorem follows by application of the standard Faedo-Galerkin method with small modifications to the proof of [28, Theorems IV.5.1 and IV.6.1]. The necessary estimates are similar to the ones given later to show the existence of an absorbing set in $L^2(U)$. It is important to notice that the estimates use only the L^2-norm of u_0, which exists by (H5).

2. Existence of a Global Attractor

In this section we prove the existence of a compact global attractor $\mathscr{A} \subset L^2(U)$. That is, a compact subset of $L^2(U)$ such that if $S(t)$ denotes the semigroup generated by solutions of (5), then

(i) $S(t)\mathscr{A} = \mathscr{A}$ for all $t \geqslant 0$;

(ii) for every bounded set $B \subset L^2(U)$, it holds that the L^2-distance $d(S(t)B, \mathscr{A}) \to 0$ as $t \to \infty$.

In order to prove the existence of the attractor, we will show first the existence of an absorbing set in $L^2(U)$, that is, a set $\mathscr{B}$ such that any bounded set of $L^2(U)$ enters into $\mathscr{B}$ after a certain time.

Theorem 5. *Assume that hypotheses (H1)-(H7) hold for (5) under Dirichlet or U-periodic boundary conditions. Then there exist positive real constants τ_0 and τ_1 such that*

(i) *the semigroup $S(t)$ possesses an absorbing ball in $L^2(U)$ as*

$$\|S(t)u_0\|_{L^2} \leqslant r_0 \quad if\ t \geqslant \tau_0, \tag{7}$$

where $r_0 := 1$ and $\tau_0 := (1/K)\log\|u_0\|_{L^2}$, with $K := \mu_2 - (1/4\alpha_2)\|v\|_\infty^2$;

(ii) *the semigroup $S(t)$ possesses an absorbing ball in $H_0^1(U)$ as*

$$\|S(t)u_0\|_{H_0^1} \leqslant r_1 \quad if\ t \geqslant \tau_1, \tag{8}$$

where $r_1 := \left((1/\alpha_2)\, e^{(1/2)\|v\|_\infty^2}(\|v\|_\infty^2 + \alpha_2^2\|f\|_\infty^2 + 1)\right)^{1/2}$ and $\tau_1 := \tau_0 + \alpha_2$.

From the previous theorem we have that the set

$$\mathscr{B} := \left\{u \in L^2(U) : \|u\|_{L^2} \leqslant r_0,\ \|\nabla u\|_{L^2} \leqslant r_1\right\} \subset H^1(U) \tag{9}$$

is absorbing for all the bounded sets of $L^2(U)$ provided that $t \geqslant \tau_1$.

Proof. We divide the proof into two steps. Before that, we recall the following inequality for the convenience of the reader (see [28, Lemma III.1.1] for a proof).

Lemma 6 (uniform Gronwall's lemma [29]). *Let f, g, and h be locally integrable functions on (t_0,∞) such that f' is locally integrable on (t_0,∞) and $f' \leqslant fg + h$, for all $t \geqslant t_0$. Assume additionally that there are real positive constants a_1, a_2, a_3 such that for every $t \geqslant t_0$,*

$$\int_t^{t+\varrho} f(s)\,ds \leqslant a_1, \qquad \int_t^{t+\varrho} g(s)\,ds \leqslant a_2,$$
$$\int_t^{t+\varrho} h(s)\,ds \leqslant a_3. \tag{10}$$

Then it holds that

$$f(t+\varrho) \leqslant \left(\frac{a_1}{\varrho} + a_3\right)e^{a_2}, \quad \forall t \geqslant t_0. \tag{11}$$

2.1. Absorbing Ball in $L^2(U)$. Multiplying (5) by $\overline{u}$, integrating over $U \subset \mathbb{R}^n$ using integration by parts with any set of the proposed boundary conditions, and taking the imaginary part,

$$\begin{aligned}\frac{1}{2}\frac{d}{dt}\|u\|_{L^2}^2 + \alpha_2\|\nabla u\|_{L^2}^2 + \beta_2\|u\|_{L^{p+2}}^{p+2} + \int_U f_2(x)|u|^2dx \\ = \operatorname{Im}\left(\int_U (v\cdot\nabla u)\overline{u}dx\right) \\ \leqslant \|v\cdot\nabla u\|_{L^2}\|u\|_{L^2} \\ \leqslant \|\nabla u\|_{L^2}\|v\|_\infty\|u\|_{L^2} \\ \leqslant \alpha_2\|\nabla u\|_{L^2}^2 + \frac{1}{4\alpha_2}\|v\|_\infty^2\|u\|_{L^2}^2.\end{aligned} \tag{12}$$

After eliminating the third term to the left-hand side of the previous calculation, we get

$$\frac{d}{dt}\|u\|_{L^2}^2 \leqslant -2K\|u\|_{L^2}^2, \tag{13}$$

where

$$K = K(\alpha_2, f_2, v) := \mu_2 - \frac{1}{4\alpha_2}\|v\|_\infty^2. \tag{14}$$

Then, applying Gronwall's lemma and since (H7) implies that $K > 0$, we obtain

$$\|u\|_{L^2}^2 \leqslant \|u_0\|_{L^2}^2 e^{-2Kt}. \tag{15}$$

Setting $\tau_0 := (1/K)\log\|u_0\|_{L^2}$, we have that

$$\|u(t)\|_{L^2} \leqslant 1 =: r_0, \quad \text{if } t \geqslant \tau_0. \tag{16}$$

This proves the existence of an absorbing ball in $L^2(U)$.

2.2. Absorbing Ball in $H_0^1(U)$. Now we multiply (5) by $\Delta\overline{u}$, integrate as before over $U \subset \mathbb{R}^n$ using boundary conditions (after integration by parts) to simplify, and take the imaginary part to get

$$\begin{aligned}\frac{1}{2}\frac{d}{dt}\|\nabla u\|_{L^2}^2 + \alpha_2\|\Delta u\|_{L^2}^2 + \operatorname{Im}\left(\int_U (v\cdot\nabla u)\Delta\overline{u}dx\right) \\ = \operatorname{Im}\left(\int_U f(x)u\Delta\overline{u}dx\right) + \operatorname{Im}\left(\beta\int_U |u|^p u\Delta\overline{u}dx\right).\end{aligned} \tag{17}$$

Integrating by parts the last term of the previous equality and using that for any $z, w \in \mathbb{C}$ it holds that $\operatorname{Im}(zw) = z_1w_2 + z_2w_1$, we get

$$\begin{aligned}\operatorname{Im}\left(\beta\int_U |u|^p u\Delta\overline{u}dx\right) \\ = -\beta_2\int_U |u|^p\|\nabla u\|^2dx \\ - p\operatorname{Im}\left(\beta\int_U |u|^{p-2}(u\nabla\overline{u})\cdot\operatorname{Re}(u\nabla\overline{u})dx\right) \\ = -\beta_2\int_U |u|^p\|\nabla u\|^2dx \\ - p\beta_1\int_U |u|^{p-2}\operatorname{Im}(u\nabla\overline{u})\cdot\operatorname{Re}(u\nabla\overline{u})dx \\ - p\beta_2\int_U |u|^{p-2}\|\operatorname{Re}(u\nabla\overline{u})\|^2dx \\ \leqslant -\beta_2\int_U |u|^p\|\nabla u\|^2dx \\ + p|\beta_1|\int_U |u|^{p-2}\|u\nabla\overline{u}\|^2dx \\ \leqslant -(\beta_2 - p|\beta_1|)\int_U |u|^p\|\nabla u\|^2dx,\end{aligned} \tag{18}$$

where the inequality appears after eliminating the last term since $\beta_2 \geqslant 0$. Replacing this in the original equation, we can write

$$\begin{aligned}\frac{1}{2}\frac{d}{dt}\|\nabla u\|_{L^2}^2 + \alpha_2\|\Delta u\|_{L^2}^2 + (\beta_2 - p|\beta_1|)\int_U |u|^p\|\nabla u\|^2dx \\ \leqslant \operatorname{Im}\left(\int_U f(x)u\Delta\overline{u}dx\right) - \operatorname{Im}\left(\int_U (v\cdot\nabla u)\ \Delta\overline{u}dx\right) \\ \leqslant \|f\|_\infty\|u\|_{L^2}\|\Delta u\|_{L^2} + \|v\|_\infty\|\nabla u\|_{L^2}\|\Delta u\|_{L^2} \\ \leqslant \frac{1}{2\alpha_2}\|f\|_\infty^2\|u\|_{L^2}^2 + \frac{\alpha_2}{2}\|\Delta u\|_{L^2}^2 \\ + \frac{1}{2\alpha_2}\|v\|_\infty^2\|\nabla u\|_{L^2}^2 + \frac{\alpha_2}{2}\|\Delta u\|_{L^2}^2.\end{aligned} \tag{19}$$

Simplifying, using $\beta_2 \geqslant p|\beta_1|$ and working with $t \geqslant \tau_0$ so that $\|u\|_{L^2} \leqslant r_0$, we arrive at

$$\frac{d}{dt}\|\nabla u\|_{L^2}^2 \leqslant \frac{1}{\alpha_2}\left(r_0^2\|f\|_\infty^2 + \|v\|_\infty^2\|\nabla u\|_{L^2}^2\right). \tag{20}$$

To apply the uniform Gronwall lemma, it is necessary to have an upper bound on $\int_t^{t+\varrho}\|\nabla u\|_{L^2}^2ds$. From (12), we have

$$\begin{aligned}\frac{1}{2}\frac{d}{dt}\|u\|_{L^2}^2 + \alpha_2\|\nabla u\|_{L^2}^2 + \beta_2\|u\|_{L^{p+2}}^{p+2} + \int_U f_2(x)|u|^2dx \\ \leqslant \|\nabla u\|_{L^2}\|v\|_\infty\|u\|_{L^2} \\ \leqslant \frac{\alpha_2}{2}\|\nabla u\|_{L^2}^2 + \frac{1}{2\alpha_2}\|v\|_\infty^2\|u\|_{L^2}^2.\end{aligned} \tag{21}$$

Now, eliminating the last two positive terms at the left-hand side, using $t \geqslant \tau_0$ again, and integrating, we find that

$$\int_t^{t+\varrho} \|\nabla u\|_{L^2}^2 ds \leqslant r_0^2 \left(\frac{\varrho}{\alpha_2^2} \|v\|_\infty^2 + \frac{1}{\alpha_2} \right) =: a_1. \quad (22)$$

There are also obvious bounds given by

$$\int_t^{t+\varrho} \frac{1}{\alpha_2} \|v\|_\infty^2 ds \leqslant \frac{\varrho}{\alpha_2} \|v\|_\infty^2 =: a_2,$$
$$\int_t^{t+\varrho} \frac{1}{\alpha_2} r_0^2 \|f\|_\infty^2 ds \leqslant \frac{\varrho}{\alpha_2} r_0^2 \|f\|_\infty^2 =: a_3. \quad (23)$$

With this, Lemma 6 immediately implies

$$\|\nabla u(t+\varrho)\|_{L^2}^2 \leqslant r_0^2 \left(\frac{1}{\alpha_2^2} \|v\|_\infty^2 + \frac{1}{\alpha_2 \varrho} + \frac{\varrho}{\alpha_2} \|f\|_\infty^2 \right) e^{(\varrho/\alpha_2)\|v\|_\infty^2}. \quad (24)$$

Choosing $\varrho = \alpha_2$ and assuming $t \geqslant \tau_1 := \tau_0 + \alpha_2$, the previous inequality becomes

$$\|\nabla u(t)\|_{L^2} \leqslant \frac{r_0}{\alpha_2} e^{(1/2)\|v\|_\infty^2} \left(1 + \|v\|_\infty^2 + \alpha_2^2 \|f\|_\infty^2\right)^{1/2} =: r_1. \quad (25)$$

This proves the existence of an absorbing ball in $H_0^1(U)$.

To show the existence of a compact global attractor in $L^2(U)$, we will make use of the following result (see [28, Theorem I.1.1]) with $H = L^2(U)$.

Theorem 7. *Suppose that $S(t)$ is a continuous semigroup acting on a metric space $\mathscr{H}$ such that there exists a bounded absorbing set $\mathscr{B}$ and for every bounded set $B \subset \mathscr{H}$ there exists $t_0 > 0$ such that $\bigcup_{t \geqslant t_0} S(t)B$ is relatively compact in $\mathscr{H}$. Then the ω-limit set*

$$\omega(\mathscr{B}) := \bigcap_{s \geqslant 0} \overline{\bigcup_{t \geqslant s} S(t)\mathscr{B}} \quad (26)$$

is a compact global attractor for the semigroup $S(t)$ in $\mathscr{H}$.

Theorem 8. *Under hypotheses (H1)–(H7) for (5), supplemented with Dirichlet or U-periodic boundary conditions, there exists a compact global attractor $\mathscr{A} \subset L^2(U)$.*

Proof. We will make use of Theorem 7. Since we have already shown the existence of the inertial set, we only need to prove that for any bounded set $B \in L^2(U)$, there exists $t_0 > 0$ such that the set $\bigcup_{t \geqslant t_0} S(t)B$ is relatively compact in $L^2(U)$. As a consequence of Theorem 5, we have proved the existence of a ball

$$\mathscr{B} := \left\{ u \in L^2(U) : \|u\|_{L^2} \leqslant r_0,\ \|\nabla u\|_{L^2} \leqslant r_1 \right\} \subset H^1(U), \quad (27)$$

such that for any bounded set $B \subset L^2(U)$, it holds that $S(t)B \subset \mathscr{B}$ for $t \geqslant \tau_1$. Since $\mathscr{B}$ is bounded in $H^1(U)$ and the embedding of $H^1(U)$ in $L^2(U)$ is compact, we conclude that the set

$$\bigcup_{t \geqslant \tau_1} S(t)B \quad (28)$$

is relatively compact in $L^2(U)$. This finishes the proof.

Remark 9. We point out here that when no dissipation is assumed in (5), which requires in particular $\alpha_2 = \beta_2 = 0$, $f_2(x) = 0$, and $v_1(x) = 0$, a very different situation may occur. For example, for the nondissipative Gross-Pitaevskii equation with rotation, it has been proved in [10, Theorem 2.2] that there are conditions on $f_1(x)$ and $v_2(x)$ such that (5) has initial data $u_0(x)$ for which solutions present finite time blowup in dimensions $n = 2$ or $n = 3$; that is, there is a finite time $\tau_* > 0$ such that

$$\lim_{t \to \tau_*} \|\nabla u(t)\|_{L^2} = \infty. \quad (29)$$

This can happen, for example, with an axial-symmetric parabolic potential for $n = 2, 3$ in the case of a focusing ($\beta_1 > 0$) and energy subcritical nonlinearity of appropriate order (more precisely $4/n \leqslant p < 4/(n-2)$ with p even). In our situation, the hypotheses employed (in particular the kind of dissipation) eliminate the possibility of such phenomena.

3. Dimension of the Attractor

In this section we provide an estimate for the dimension of the attractor by the method of evolution of infinitesimal m-dimensional volume elements in the phase space $L^2(U)$ under the flow induced by the action of the semigroup $S(t)$ of Theorem 5 (see [28, Chapter V] for details). For this estimate we will explicitly assume that the bounded open domain $U \subset \mathbb{R}^n$ is at least of class $\mathscr{C}^1$; that is, U has a boundary ∂U with a structure of an $(n-1)$-dimensional manifold of class $\mathscr{C}^1$.

Given an orthonormal basis $\varphi_1(t), \ldots, \varphi_m(t)$ for a time-dependent m-dimensional subspace of $L^2(U)$, it is possible (cf. [28, Appendix, Corollary 4.1]) to find constants $c_1, c_2 \geqslant 0$, depending on n, k and the shape of $U \subset \mathbb{R}^n$ but independent of the size $|U|$, the value m, and of the chosen orthonormal basis, such that

$$\sum_{i=1}^{m} \|\nabla \varphi_i\|_{L^2}^2 \geqslant \frac{1}{|U|^{2/n}} \left(c_1 m^{1+2/n} - c_2 m \right). \quad (30)$$

Moreover, if the functions $\varphi_i \in H_0^1(U)$, it is possible to choose $c_2 = 0$. This holds for Dirichlet boundary conditions.

In order to study the evolution of infinitesimal volume elements, we linearize (5) around a solution $u(t) = S(t)u_0$, with $u_0 \in \mathscr{A}$ as follows:

$$i\phi_t = F'(u)\phi := \alpha\Delta\phi - \beta(p+1)|u|^p\phi - f(x)\phi + v \cdot \nabla\phi,$$
$$\phi(0) = \xi. \quad (31)$$

Then, noting that $\mathrm{Re}(\phi_t) = \mathrm{Im}(F'(u)\phi)$, the evolution of the m-dimensional volume elements is given by (cf. [30, Section 13.2])

$$V_m(t) = V_m(0)\exp\left[\mathrm{Im}\left(\int_0^t \mathrm{Tr}\left(F'(u)P_m(s)\right)ds\right)\right], \quad (32)$$

where P_m is the orthogonal projection of $L^2(U)$ over the subspace spanned by the first m eigenvectors of the Laplacian. The asymptotic growth rate of $V_m(t)$ is given by

$$\lim_{t\to\infty}\frac{1}{t}\int_0^t \mathrm{Im}\left[\mathrm{Tr}\left(F'(u)P_m(s)\right)\right]ds. \quad (33)$$

Since we need the maximal growth rate, we compute the supremum over all $u_0 \in \mathscr{A}$ and over all the m-dimensional projectors as follows:

$$\sup_{u_0\in\mathscr{A}}\sup_{P_m}\lim_{t\to+\infty}\sup\frac{1}{t}\int_0^t \mathrm{Im}\left[\mathrm{Tr}\left(F'(u)P_m(s)\right)\right]ds =: \mathrm{Tr}_m(\mathscr{A}). \quad (34)$$

In order to have exponential decay, we look for the smallest m such that the number $\mathrm{Tr}_m(\mathscr{A})$ is negative. In detail, let $\varphi_1(t),\ldots,\varphi_m(t)$ be an orthonormal basis for $P_mL^2(U)$; then

$$\mathrm{Im}\left[\mathrm{Tr}\left(F'(u)P_m\right)\right] = \sum_{i=1}^m \mathrm{Im}\left[\left\langle F'(u)\varphi_i,\varphi_i\right\rangle_{L^2(U)}\right], \quad (35)$$

with

$$\begin{aligned}\mathrm{Im}&\left[\left\langle F'(u)\varphi_i,\varphi_i\right\rangle_{L^2(U)}\right]\\ &= -\alpha_2\|\nabla\varphi_i\|_{L^2}^2 - \beta_2(p+1)\int_U |u|^p|\varphi_i|^2dx\\ &\quad - \int_U f_2(x)|\varphi_i|^2dx + \mathrm{Im}\left(\int_U (v\cdot\nabla\varphi_i)\overline{\varphi}_i dx\right).\end{aligned} \quad (36)$$

Now, we estimate each of the last three terms separately. For the first, since $\beta_2 \geqslant 0$, we can simply write

$$-\beta_2(p+1)\int_U |u|^p|\varphi_i|^2dx \leqslant 0, \quad (37)$$

for the second, we have

$$-\int_U f_2(x)|\varphi_i|^2dx \leqslant -\mu_2\|\varphi_i\|_{L^2}^2 = -\mu_2 < 0, \quad (38)$$

and for the last term,

$$\begin{aligned}\mathrm{Im}\int_U (v\cdot\nabla\varphi_i)\overline{\varphi}_i dx &\leqslant \int_U |v\cdot\nabla\varphi_i||\varphi_i|dx\\ &\leqslant \int_U \|v\|\,\|\nabla\varphi_i\|\,|\varphi_i|dx\\ &\leqslant \|v\|_\infty\int_U \|\nabla\varphi_i\|\,|\varphi_i|dx\\ &\leqslant \|v\|_\infty\|\nabla\varphi_i\|_{L^2}\|\varphi_i\|_{L^2}\\ &\leqslant \frac{\alpha_2}{2}\|\nabla\varphi_i\|_{L^2}^2 + \frac{\|v\|_\infty^2}{2\alpha_2}.\end{aligned} \quad (39)$$

Replacing these bounds in (35), we get for the imaginary part of the trace

$$\begin{aligned}\mathrm{Im}&\left[\mathrm{Tr}\left(F'(u)P_m\right)\right]\\ &\leqslant -\frac{\alpha_2}{2}\sum_{i=1}^m\|\nabla\varphi_i\|_{L^2}^2 - m\mu_2 + m\frac{\|v\|_\infty^2}{2\alpha_2}\\ &\leqslant -\frac{\alpha_2c_1}{2|U|^{2/n}}m^{1+2/n} + m\left(\frac{\alpha_2c_2}{2|U|^{2/n}} + \frac{\|v\|_\infty^2}{2\alpha_2} - \mu_2\right)\\ &\leqslant -\kappa_1m^{1+2/n} + \kappa_2,\end{aligned} \quad (40)$$

where we have used (30) and a Young inequality to obtain the required constants κ_i given by

$$\begin{aligned}\kappa_1 &:= \left(\frac{\alpha_2c_1}{n+2}\right)|U|^{-2/n},\\ \kappa_2 &:= \frac{2|U|^{-1}}{(n+2)}\left(\frac{\alpha_2c_1}{2}\right)^{n/2}\left(\frac{\alpha_2c_2}{2|U|^{2/n}} + \frac{\|v\|_\infty^2}{2\alpha_2} - \mu_2\right)^{1+n/2}.\end{aligned} \quad (41)$$

Remark 10. Notice, however, that in (40) the use of a Young inequality requires

$$\mu_2 < \frac{\alpha_2c_2}{2|U|^{2/n}} + \frac{\|v\|_\infty^2}{2\alpha_2}, \quad (42)$$

which produces a well-defined positive real constant κ_2. In particular, for Dirichlet boundary conditions (where c_2 can be taken to be zero), in addition to hypothesis (H7), this requirement reads $\|v\|_\infty > 2\alpha_2\mu_2$. It is also possible to replace μ_2 by zero in the expression for κ_2 and the bound is still fine but not as sharp. See Remark 12 if (42) does not hold.

This provides an estimate for the dimension of the attractor $\mathscr{A}$. In fact, similarly to what is done in [28, Section VI.7] using Proposition V.2.1 and Theorem V.3.3 there (see also [30, Theorem 2.4]), we have the following theorem.

Theorem 11. *Under the assumptions of Theorem 8 and if* (42) *holds, let $m \in \mathbb{N}$ be such that*

$$m - 1 < \left(\frac{2\kappa_2}{\kappa_1}\right)^{n/(n+2)} \leqslant m. \quad (43)$$

Then the infinitesimal m-dimensional volume elements in phase space $L^2(U)$ decay exponentially as $t \to \infty$. Moreover, the global attractor $\mathscr{A}$ has finite Hausdorff dimension $\dim_H(\mathscr{A}) \leqslant m$ and finite fractal dimension $\dim_F(\mathscr{A}) \leqslant 2m$.

Explicitly, we have that

$$m \sim 2^{n/(n+2)} \left[\left(\frac{\alpha_2 c_1}{4|U|} \right)^{(n-2)/(n+2)} \left(\frac{\alpha_2 c_2}{2|U|^{2/n}} + \frac{\|v\|_\infty^2}{2\alpha_2} - \mu_2 \right) \right]^{n/2}. \quad (44)$$

Remark 12. Actually, a bound on the Hausdorff dimension is already obtained by finding the minimum value of $m \in \mathbb{N}$ in (40), such that $\mathrm{Im}[\mathrm{Tr}(F'(u)P_m)] < 0$ (see aforementioned references). Therefore, $\dim_H(\mathscr{A})$ is strictly smaller than (44) without the $2^{n/(n+2)}$ factor. This also shows that if (42) does not hold, the Hausdorff dimension is strictly less than 1.

As a final conclusion, we stress the fact that the vector coefficient $v(x)$ of the "rotational" term does indeed increase the dimension of the attractor in a factor roughly proportional to $\|v\|_\infty^n$, where n is the dimension of the ambient space. This should be compared with the role played by this term in proving finite time blowup for the nondissipative focusing case in [10], alluded in Remark 9.

Acknowledgments

The authors are indebted to the referee for indicating relevant references concerning numerical studies of Bose-Einstein condensates. This work was partially supported by Research Project "*Dynamics of a nonlinear Schrödinger equation*", ID-PROY 004850 of the Faculty of Sciences of Pontificia Universidad Javeriana, Bogotá, Colombia.

References

[1] J. Bourgain, *Global Solutions of Nonlinear Schrödinger Equations*, vol. 46 of *American Mathematical Society Colloquium Publications*, American Mathematical Society, Providence, RI, USA, 1999.

[2] R. Carles, *Semi-Classical Analysis for Nonlinear Schrödinger Equations*, World Scientific, Hackensack, NJ, USA, 2008.

[3] T. Tao, *Nonlinear Dispersive Equations*, vol. 106 of *CBMS Regional Conference Series in Mathematics*, Published for the Conference Board of the Mathematical Sciences, Washington, DC, USA, 2006.

[4] F. Linares and G. Ponce, *Introduction to Nonlinear Dispersive Equations*, Universitext, Springer, New York, NY, USA, 2009.

[5] A. Aftalion, *Vortices in Bose-Einstein Condensates*, vol. 67 of *Progress in Nonlinear Differential Equations and their Applications*, Birkhäuser, Boston, Mass, USA, 2006.

[6] W. Bao and Y. Cai, "Mathematical theory and numerical methods for Bose-Einstein condensation," *Kinetic and Related Models*, vol. 6, no. 1, pp. 1–135, 2013.

[7] W. Bao, Q. Du, and Y. Zhang, "Dynamics of rotating Bose-Einstein condensates and its efficient and accurate numerical computation," *SIAM Journal on Applied Mathematics*, vol. 66, no. 3, pp. 758–786, 2006.

[8] C. Hao, L. Hsiao, and H.-L. Li, "Global well posedness for the Gross-Pitaevskii equation with an angular momentum rotational term," *Mathematical Methods in the Applied Sciences*, vol. 31, no. 6, pp. 655–664, 2008.

[9] C. Hao, L. Hsiao, and H.-L. Li, "Global well posedness for the Gross-Pitaevskii equation with an angular momentum rotational term in three dimensions," *Journal of Mathematical Physics*, vol. 48, no. 10, Article ID 102105, 11 pages, 2007.

[10] P. Antonelli, D. Marahrens, and C. Sparber, "On the Cauchy problem for nonlinear Schrödinger equations with rotation," *Discrete and Continuous Dynamical Systems A*, vol. 32, no. 3, pp. 703–715, 2012.

[11] H. Liu, "Critical thresholds in the semiclassical limit of 2-D rotational Schrödinger equations," *Zeitschrift für Angewandte Mathematik und Physik*, vol. 57, no. 1, pp. 42–58, 2006.

[12] S. Choi, S. A. Morgan, and K. Burnett, "Phenomenological damping in trapped atomic Bose-Einstein condensates," *Physical Review A*, vol. 57, no. 5, pp. 4057–4060, 1998.

[13] P. G. Kevrekidis and D. J. Frantzeskakis, "Multiple dark solitons in Bose-Einstein condensates at finite temperatures," *Discrete and Continuous Dynamical Systems S*, vol. 4, no. 5, pp. 1199–1212, 2011.

[14] M. Tsubota, K. Kasamatsu, and M. Ueda, "Vortex lattice formation in a rotating Bose-Einstein condensate," *Physical Review A*, vol. 65, no. 2, Article ID 023603, 2002.

[15] K. Kasamatsu, M. MacHida, N. Sasa, and M. Tsubota, "Three-dimensional dynamics of vortex-lattice formation in Bose-Einstein condensates," *Physical Review A*, vol. 71, no. 6, Article ID 063616, 2005.

[16] L. H. Wen and X. B. Luo, "Formation and structure of vortex lattices in a rotating double-well Bose-Einstein condensate," *Laser Physics Letters*, vol. 9, no. 8, pp. 618–624, 2012.

[17] C. W. Gardiner, J. R. Anglin, and T. I. A. Fudge, "The stochastic Gross-Pitaevskii equation," *Journal of Physics B*, vol. 35, no. 6, pp. 1555–1582, 2002.

[18] C. W. Gardiner and M. J. Davis, "The stochastic Gross-Pitaevskii equation: II," *Journal of Physics B*, vol. 36, no. 23, pp. 4731–4753, 2003.

[19] A. S. Bradley and C. W. Gardiner, "The stochastic Gross-Pitaevskii equation III," http://arxiv.org/abs/cond-mat/0602162.

[20] M. Kurzke, C. Melcher, R. Moser, and D. Spirn, "Dynamics for Ginzburg-Landau vortices under a mixed flow," *Indiana University Mathematics Journal*, vol. 58, no. 6, pp. 2597–2622, 2009.

[21] E. Miot, "Damped wave dynamics for a complex Ginzburg-Landau equation with low dissipation," http://128.84.158.119/abs/1003.5375v1.

[22] J.-M. Ghidaglia and B. Héron, "Dimension of the attractors associated to the Ginzburg-Landau partial differential equation," *Physica D*, vol. 28, no. 3, pp. 282–304, 1987.

[23] P. Laurençot, "Long-time behaviour for weakly damped driven nonlinear Schrödinger equations in $\mathbb{R}^N$, $N \leq 3$," *NoDEA. Nonlinear Differential Equations and Applications*, vol. 2, no. 3, pp. 357–369, 1995.

[24] N. Kita and A. Shimomura, "Large time behavior of solutions to Schrödinger equations with a dissipative nonlinearity for arbitrarily large initial data," *Journal of the Mathematical Society of Japan*, vol. 61, no. 1, pp. 39–64, 2009.

[25] P. Antonelli and C. Sparber, "Global well-posedness for cubic NLS with nonlinear damping," *Communications in Partial Differential Equations*, vol. 35, no. 12, pp. 2310–2328, 2010.

[26] W. Bao, D. Jaksch, and P. A. Markowich, "Three-dimensional simulation of jet formation in collapsing condensates," *Journal of Physics B*, vol. 37, no. 2, pp. 329–343, 2004.

[27] W. Bao and D. Jaksch, "An explicit unconditionally stable numerical method for solving damped nonlinear Schrödinger

equations with a focusing nonlinearity," *SIAM Journal on Numerical Analysis*, vol. 41, no. 4, pp. 1406–1426, 2003.

[28] R. Temam, *Infinite-Dimensional Dynamical Systems in Mechanics and Physics*, vol. 68 of *Applied Mathematical Sciences*, Springer, New York, NY, USA, 2nd edition, 1997.

[29] C. Foiaş and G. Prodi, "Sur le comportement global des solutions non-stationnaires des équations de Navier-Stokes en dimension 2," *Rendiconti del Seminario Matematico della Università di Padova*, vol. 39, pp. 1–34, 1967.

[30] J. C. Robinson, *Infinite-Dimensional Dynamical Systems. An Introduction to Dissipative Parabolic PDEs and the Theory of Global Attractors*, Cambridge Texts in Applied Mathematics, Cambridge University Press, Cambridge, Mass, USA, 2001.

Existence of Positive Periodic Solutions for Periodic Neutral Lotka-Volterra System with Distributed Delays and Impulses

Zhenguo Luo[1,2] **and Liping Luo**[2]

[1] *Department of Mathematics, National University of Defense Technology, Changsha 410073, China*
[2] *Department of Mathematics, Hengyang Normal University, Hengyang, Hunan 421008, China*

Correspondence should be addressed to Zhenguo Luo; robert186@163.com

Academic Editor: Norio Yoshida

By using a fixed-point theorem of strict-set-contraction, we investigate the existence of positive periodic solutions for a class of the following impulsive neutral Lotka-Volterra system with distributed delays: $x_i'(t) = x_i(t)[r_i(t) - \sum_{j=1}^{n} a_{ij}(t)x_j(t) - \sum_{j=1}^{n} b_{ij}(t)\int_{-\tau_{ij}}^{0} f_{ij}(\xi)x_j(t+\xi)d\xi - \sum_{j=1}^{n} c_{ij}(t)\int_{-\sigma_{ij}}^{0} g_{ij}(\xi)x_j'(t+\xi)d\xi]$, $\Delta x_i(t_k) = -I_{ik}(x_i(t_k))$, $i = 1, 2, \ldots, n$, $k = 1, 2, \ldots$. Some verifiable criteria are established easily.

1. Introduction

It is well known that natural environments are physically highly variable, and in response, birth rates, death rates, and other vital rates of populations vary greatly in time. Theoretical evidence suggests that many population and community patterns represent intricate interactions between biology and variation in the physical environment (see [1–4]). Thus, the focus in theoretical models of population and community dynamics must be not only on how populations depend on their own population densities or the population densities of other organisms, but also on how populations change in response to the physical environment. It is reasonable to study the models of population with periodic coefficients. In addition to the theoretical and practical significance, the Lotka-Volterra model is one of the famous models for dynamics of population; therefore it has been studied extensively [5–9]. In view of the above effects, by applying a fixed-point theorem of strict-set-contraction, Li [10] established criteria to guarantee the existence of positive periodic solutions of the following neutral Lotka-Volterra system with distributed delays:

$$x_i'(t) = x_i(t)\left[a_i(t) - \sum_{j=1}^{n} b_{ij}(t)\int_{-T_{ij}}^{0} K_{ij}(\theta)\, x_j(t+\theta)\, d\theta - \sum_{j=1}^{n} c_{ij}(t)\int_{-\widehat{T}_{ij}}^{0} \widehat{K}_{ij}(\theta)\, x_j'(t+\theta)\, d\theta\right], \quad i = 1, 2, \ldots, n, \tag{1}$$

where $a_i, b_{ij}, c_{ij} \in C(R, R^+)$ $(i, j = 1, 2, \ldots, n)$ are ω-periodic functions and $T_{ij}, \widehat{T}_{ij} \in (0, +\infty)$ $(i, j = 1, 2, \ldots, n)$ and $K_{ij}, \widehat{K}_{ij} \in (R, R^+)$ satisfying $\int_{-T_{ij}}^{0} K_{ij}(\theta)d\theta = 1$, $\int_{-\widehat{T}_{ij}}^{0} \widehat{K}_{ij}(\theta)d\theta$, $i, j = 1, 2, \ldots, n$. On the other side, birth of many species is an annual birth pulse or harvesting. To have a more accurate description of many mathematical ecology systems, we need to consider the use of impulsive differential equations [11–13]. Some qualitative properties such as oscillation, periodicity, asymptotic behavior, and stability properties have been investigated extensively by many authors over the past few years [14–18]. However, to our knowledge, there are few published papers discussing the existence of periodic solutions for neutral Lotka-Volterra system with distributed delays and impulses. In this paper, we are concerned with the following neutral Lotka-Volterra system with distributed delays and impulses:

$$x_i'(t) = x_i(t)\left[r_i(t) - \sum_{j=1}^{n} a_{ij}(t)\, x_j(t) - \sum_{j=1}^{n} b_{ij}(t)\int_{-\tau_{ij}}^{0} f_{ij}(\xi)\, x_j(t+\xi)\, d\xi - \sum_{j=1}^{n} c_{ij}(t)\int_{-\sigma_{ij}}^{0} g_{ij}(\xi)\, x_j'(t+\xi)\, d\xi\right], \quad i = 1,2,\ldots,n,\ t \neq t_k,$$
$$\Delta x_i(t_k) = -I_{ik}(x_i(t_k)), \quad i = 1,2,\ldots,n,\ k = 1,2,\ldots, \tag{2}$$

where $r_i, a_{ij}, b_{ij}, c_{ij} \in C(R, R^+)$ $(i, j = 1,2,\ldots,n)$ are ω-periodic functions and $\tau_{ij}, \sigma_{ij} \in (0,+\infty)$ $(i, j = 1,2,\ldots,n)$ and $f_{ij}, g_{ij} \in C(R, R^+)$ satisfying $\int_{-\tau_{ij}}^{0} f_{ij}(\xi)d\xi = 1$, $\int_{-\sigma_{ij}}^{0} g_{ij}(\xi)d\xi = 1$, $i, j = 1,2,\ldots,n$. Moreover, $\Delta x(t_k) = x(t_k^+) - x(t_k)$ (here $x(t_k^+)$ represents the right limit of $x(t)$ at the point t_k), $I_{ik} \in C(R^+, R^+)$; that is, x changes decreasingly suddenly at times t_k; $\omega > 0$ is a constant, $R = (-\infty,+\infty)$, $R^+ = [0,+\infty)$. We assume that there exists an integer $q > 0$ such that $t_{k+q} = t_k + \omega$, $I_{i(k+q)} = I_{ik}$, where $0 < t_1 < t_2 < \cdots < t_q < \omega$.

The main purpose of this paper is by using a fixed-point theorem of strict-set-contraction [19, 20] to establish new criteria to guarantee the existence of positive periodic solutions of the system (2).

For convenience, we introduce the notation

$$f^M = \max_{t\in[0,\omega]}\{|f(t)|\}, \qquad f^L = \min_{t\in[0,\omega]}\{|f(t)|\},$$
$$\gamma_i = \limsup_{u\to 0} \sum_{t\le t_k\le t+\omega} \frac{I_{ik}(u)}{u}, \quad i = 1,2,\ldots,n,$$
$$\delta_i = e^{-\int_0^\omega r_i(t)dt}, \quad i = 1,2,\ldots,n, \tag{3}$$
$$\alpha_{ij} = \int_0^\omega \left[\delta_j a_{ij}(t) - b_{ij}(t) - c_{ij}(t)\right] dt,$$
$$\beta_{ij} = \int_0^\omega \left[a_{ij}(t) + b_{ij}(t) + c_{ij}(t)\right] dt, \quad i, j = 1,2,\ldots,n,$$

where f is a continuous ω-periodic function.

Throughout this paper, we assume that

(A_1) $\delta_i = e^{-\int_0^\omega r_i(t)dt} < 1$, $i = 1,2,\ldots,n$;

(A_2) $\delta_j a_{ij}(t) - b_{ij}(t) - c_{ij}(t) \ge 0$, $i, j = 1,2,\ldots,n$;

(A_3) $(1 + r_i^L)(\delta_i^2/(1-\delta_i))\alpha_{ij} \ge \max_{t\in[0,\omega]}\{a_{ij}(t) + b_{ij}(t) + c_{ij}(t)\}$, $i, j = 1,2,\ldots,n$;

(A_4) $(r_i^M - 1)(1/\delta_i(1-\delta_i))\beta_{ij} \le \min_{t\in[0,\omega]}\{\delta_j a_{ij}(t) - b_{ij}(t) - c_{ij}(t)\}$, $i, j = 1,2,\ldots,n$;

(A_5) $\max_{1\le i\le n}\{\sum_{j=1}^n c_{ij}^M\} \le \min_{1\le i\le n}\{(\delta_i^2/(1-\delta_i))\min_{1\le j\le n}\{\alpha_{ij}\}\}$.

The paper is organized as follows. In Section 2, we give some definitions and lemmas to prove the main results of this paper. In Section 3, by using a fixed-point theorem of strict-set-contraction, we established some criteria to guarantee the existence of at least one positive periodic solution of system (2). Finally, in Section 4, we give an example to show the validity of our result.

2. Preliminaries

In order to obtain the existence of a periodic solution of system (2), we first introduce some definitions and lemmas.

Definition 1 (see [13]). A function $x_i : R \to (0,+\infty)$ is said to be a positive solution of (2), if the following conditions are satisfied:

(a) $x_i(t)$ is absolutely continuous on each (t_k, t_{k+1});

(b) for each $k \in Z_+$, $x_i(t_k^+)$ and $x_i(t_k^-)$ exist and $x_i(t_k^-) = x_i(t_k)$;

(c) $x_i(t)$ satisfies the first equation of (2) for almost everywhere in R and $x_i(t_k)$ satisfies the second equation of (2) at impulsive point t_k, $k \in Z_+$.

Definition 2 (see [21]). Let X be a real Banach space and E a closed, nonempty subset of X. E is a cone provided

(i) $\alpha x + \beta y \in E$ for all $x, y \in E$ and all $\alpha, \beta \ge 0$,

(ii) $x, -x \in E$ imply $x = 0$.

Definition 3 (see [21]). Let A be a bounded subset in X. Define

$$\alpha_X(A) = \inf\left\{\delta > 0 : \text{there is a finite number of subsets } A_i \subset A \text{ such that } A = \bigcup_i A_i \text{ and } \operatorname{diam}(A_i) \le \delta\right\}, \tag{4}$$

where $\operatorname{diam}(A_i)$ denotes the diameter of the set A_i, obviously, $0 \le \alpha_X(A) < \infty$. So $\alpha_X(A)$ is called the Kuratowski measure of noncompactness of X.

Definition 4 (see [21]). Let X, Y be two Banach spaces and $D \subset X$, a continuous and bounded map $T : D \to Y$, is called k-set contractive if, for any bounded set $S \subset D$, we have

$$\alpha_Y(T(S)) \le k\alpha_X(S); \tag{5}$$

ϕ is called strict-set-contractive if it is k-set-contractive for some $0 \le k < 1$.

Definition 5 (see [22]). The set $F \in PC_\omega$ is said to be quasi-equicontinuous in $[0,\omega]$ if for any $\epsilon > 0$ there exists $\delta > 0$ such that if $x \in F$, $k \in N^+$, $t_1, t_2 \in (t_{k-1}, t_k) \cap [0,\omega]$, and $|t_1 - t_2| < \delta$, then $|x(t_1) - x(t_2)| < \epsilon$.

Lemma 6 (see [22]). *The set $F \subset PC_\omega$ is relatively compact if and only if*

(1) *F is bounded; that is, $\|x\| \le M$, for each $x \in F$, and some $M > 0$;*

(2) *F is quasi-equicontinuous in $[0, \omega]$.*

Lemma 7. *$x_i(t)$ is an ω-periodic solution of (2) is equivalent to $x_i(t)$ is an ω-periodic solution of the following equation:*

$$x_i(t) = \int_t^{t+\omega} G_i(t,s)\, x_i(s) \times \left[\sum_{j=1}^n a_{ij}(s)\, x_j(s) + \sum_{j=1}^n b_{ij}(s) \int_{-\tau_{ij}}^0 f_{ij}(\xi)\, x_j(s+\xi)\, d\xi + \sum_{j=1}^n c_{ij}(s) \int_{-\sigma_{ij}}^0 g_{ij}(\xi)\, x_j'(s+\xi)\, d\xi \right] ds + \sum_{t \le t_k < t+\omega} G_i(t,t_k)\, I_{ik}(x_i(t_k)), \tag{6}$$

where

$$G_i(t,s) = \frac{e^{-\int_t^s r_i(\theta)d\theta}}{1 - e^{-\int_0^\omega r_i(\theta)d\theta}}, \quad s \in [t, t+\omega],\ i = 1, 2, \ldots, n. \tag{7}$$

Proof. Assume that $x_i(t) \in X$, $i = 1, 2, \ldots, n$, is a periodic solution of (2). Then, we have

$$\frac{d}{dt}\left[x_i(t)\, e^{-\int_0^t r_i(\theta)d\theta}\right] = e^{-\int_0^t r_i(\theta)d\theta} x_i(t) \times \left[-\sum_{j=1}^n a_{ij}(t)\, x_j(t) - \sum_{j=1}^n b_{ij}(t) \int_{-\tau_{ij}}^0 f_{ij}(\xi)\, x_j(t+\xi)\, d\xi - \sum_{j=1}^n c_{ij}(t) \int_{-\sigma_{ij}}^0 g_{ij}(\xi)\, x_j'(t+\xi)\, d\xi \right], \quad t \ne t_k,\ i = 1, 2, \ldots, n. \tag{8}$$

Integrating the previous equation with $[t, t+\omega]$, we can have

$$x_i(s)\, e^{-\int_0^s r_i(\theta)d\theta}\Big|_t^{t_{m_1}+n\omega} + x_i(s)\, e^{-\int_0^s r_i(\theta)d\theta}\Big|_{t_{m_1}+n\omega}^{t_{m_2}+n\omega} + \cdots + x_i(s)\, e^{-\int_0^s r_i(\theta)d\theta}\Big|_{t_{m_q}+n\omega}^{t+\omega} = \int_t^{t+\omega} e^{-\int_0^s r_i(\theta)d\theta} x_i(s) \times \left[-\sum_{j=1}^n a_{ij}(s)\, x_j(s) - \sum_{j=1}^n b_{ij}(s) \int_{-\tau_{ij}}^0 f_{ij}(\xi)\, x_j(s+\xi)\, d\xi - \sum_{j=1}^n c_{ij}(s) \int_{-\sigma_{ij}}^0 g_{ij}(\xi)\, x_j'(s+\xi)\, d\xi \right] ds, \tag{9}$$

where $t_{m_k} + n\omega \in (t, t+\omega)$, $m_k \in \{1, 2, \ldots, q\}$, $k = 1, 2, \ldots, q$, $n \in Z_+$. Therefore,

$$x_i(t)\, e^{-\int_0^t r_i(\theta)d\theta}\left[1 - e^{-\int_t^{t+\omega} r_i(\theta)d\theta}\right] + \sum_{t \le t_k < t+\omega} \Delta x_i\left(t_{m_k}\right) e^{-\int_0^{t_{m_k}+n\omega} r_i(\theta)d\theta} = \int_t^{t+\omega} e^{-\int_0^s r_i(\theta)d\theta} x_i(s) \times \left[\sum_{j=1}^n a_{ij}(t)\, x_j(t) + \sum_{j=1}^n b_{ij}(s) \int_{-\tau_{ij}}^0 f_{ij}(\xi)\, x_j(s+\xi)\, d\xi + \sum_{j=1}^n c_{ij}(s) \int_{-\sigma_{ij}}^0 g_{ij}(\xi)\, x_j'(s+\xi)\, d\xi \right] ds, \tag{10}$$

which can be transformed into

$$x_i(t) = \int_t^{t+\omega} G_i(t,s)\, x_i(s) \times \left[\sum_{j=1}^n a_{ij}(s)\, x_j(s) + \sum_{j=1}^n b_{ij}(s) \int_{-\tau_{ij}}^0 f_{ij}(\xi)\, x_j(s+\xi)\, d\xi + \sum_{j=1}^n c_{ij}(s) \int_{-\sigma_{ij}}^0 g_{ij}(\xi)\, x_j'(s+\xi)\, d\xi \right] ds + \sum_{t \le t_k < t+\omega} G_i(t,t_k)\, I_{ik}(x_i(t_k)), \quad i = 1, 2, \ldots, n. \tag{11}$$

Thus, x_i is a periodic solution of (6). If $x_i(t) \in E$, $i = 1, 2, \ldots, n$, is a periodic solution of (6), for any $t = t_k$, from (6) we have

$$\begin{aligned}
x'(t) &= \frac{d}{dt}\left\{\int_t^{t+\omega} G_i(t,s)\,x_i(s) \times\left[\sum_{j=1}^n a_{ij}(s)\,x_j(s) + \sum_{j=1}^n b_{ij}(s)\int_{-\tau_{ij}}^0 f_{ij}(\xi)\,x_j(s+\xi)\,d\xi + \sum_{j=1}^n c_{ij}(s)\int_{-\sigma_{ij}}^0 g_{ij}(\xi)\,x_j'(s+\xi)\,d\xi\right]ds\right\} \\
&= \Bigg[G_i(t,t+\omega)\,x_i(t+\omega) \times\left(\sum_{j=1}^n a_{ij}(t+\omega)\,x_j(t+\omega) + \sum_{j=1}^n b_{ij}(t+\omega)\int_{-\tau_{ij}}^0 f_{ij}(\xi)\,x_j(t+\omega+\xi)\,d\xi + \sum_{j=1}^n c_{ij}(t+\omega)\int_{-\sigma_{ij}}^0 g_{ij}(\xi)\,x_j'(t+\omega+\xi)\,d\xi\right) \\
&\quad - G_i(t,t)\,x_i(t) \times\left(\sum_{j=1}^n a_{ij}(t)\,x_j(t) + \sum_{j=1}^n b_{ij}(t) \times\int_{-\tau_{ij}}^0 f_{ij}(\xi)\,x_j(t+\xi)\,d\xi + \sum_{j=1}^n c_{ij}(t)\int_{-\sigma_{ij}}^0 g_{ij}(\xi)\,x_j'(t+\xi)\,d\xi\right)\Bigg] + r_i(t)\,x_i(t) \\
&= x_i(t)\left[r_i(t) - \sum_{j=1}^n a_{ij}(t)\,x_j(t) - \sum_{j=1}^n b_{ij}(t)\int_{-\tau_{ij}}^0 f_{ij}(\xi)\,x_j(t-\xi)\,d\xi - \sum_{j=1}^n c_{ij}(t)\int_{-\sigma_{ij}}^0 g_{ij}(\xi)\,x_j'(t+\xi)\,d\xi\right].
\end{aligned} \tag{12}$$

For any $t = t_j$, $j \in Z_+$, we have from (6) that

$$\begin{aligned}
&x_i(t_j^+) - x_i(t_j) \\
&= \int_{t_j}^{t_j+\omega}\left[G_i(t_j^+,s) - G_i(t_j,s)\right]x_i(s) \times\left[\sum_{j=1}^n a_{ij}(s)\,x_j(s) + \sum_{j=1}^n b_{ij}(s)\int_{-\tau_{ij}}^0 f_{ij}(\xi)\,x_j(s+\xi)\,d\xi + \sum_{j=1}^n c_{ij}(s)\int_{-\sigma j}^0 g_{ij}(\xi)\,x_j'(s+\xi)\,d\xi\right]ds \\
&\quad + \sum_{t_j^+\le t_k<t_j+\omega} G_i(t_j^+,t_k)\,I_{ik}(x_i(t_k)) - \sum_{t_j\le t_k<t_j+\omega} G_i(t_j,t_k)\,I_{ik}(x_i(t_k)) \\
&= -I_{ik}(x(t_k)).
\end{aligned} \tag{13}$$

Hence $x_i(t)$ is a positive ω-periodic solution of (2). Thus we complete the proof of Lemma 7.

Lemma 8 (see [20, 23]). *Let E be a cone of the real Banach space X and $E_{r,R} = \{x \in E : r \le \|x\| \le R\}$ with $0 < r < R$. Assume that $A : E_{r,R} \to E$ is strict-set-contractive such that one of the following two conditions is satisfied:*

(a) *$Ax \nleq x$, for all $x \in E$, $\|x\| = r$ and $Ax \ngeq x$, for all $x \in E$, $\|x\| = R$;*

(b) *$Ax \ngeq x$, for all $x \in E$, $\|x\| = r$ and $Ax \nleq x$, for all $x \in E$, $\|x\| = R$.*

Then A has at least one fixed point in $E_{r,R}$.
In order to apply Lemma 7 to system (1), *one sets*

$$\begin{aligned}
PC(R) = \{&x = (x_1, x_2, \ldots, x_n)^T : R \longrightarrow R \mid x \in C((t_k, t_{k+1}), R), \\
&\exists x(t_k^-) = x(t_k), x(t_k^+), k \in N\}, \\
PC^1(R) = \{&x = (x_1, x_2, \cdots, x_n)^T : R \longrightarrow R \mid x \in C^1((t_k, t_{k+1}), R), \\
&\exists x'(t_k^-) = x'(t_k), x(t_k^+), k \in N, t \in R\}.
\end{aligned} \tag{14}$$

Define

$$X = \{x = (x_1, x_2, \ldots, x_n)^T : x \in PC(R) \mid x(t+\omega) = x(t)\} \tag{15}$$

with the norm defined by $\|x\|_0 = \sum_{i=1}^{n} |x_i|_0$, *where* $|x_i|_0 = \sup_{t\in[0,\omega]}\{|x_i(t)|\}$, $i = 1, 2, \ldots, n$, *and*

$$Y = \left\{x = (x_1, x_2, \ldots, x_n)^T : x \in PC^1(R) \mid x(t+\omega) = x(t), t \in R\right\} \tag{16}$$

with the norm defined by $\|x\|_1 = \sum_{i=1}^{n} |x_i|_1$, *where* $|x_i|_1 = \max\{|x_i|_0, |x_i'|_0\}$. *Then X and Y are both Banach spaces. Define the cone E in Y by*

$$E = \left\{x = (x_1, x_2, \ldots, x_n)^T : x \in PC^1(R) \mid x_i(t) = \delta_j |x_i(t)|_1, t \in [0,\omega], i, j = 1, 2 \ldots, n\right\}. \tag{17}$$

Let the map T be defined by

$$(Tx)(t) = ((T_1x)(t), (T_2x)(t), \ldots, (T_nx)(t))^T, \tag{18}$$

where $x \in E, t \in R$,

$$\begin{aligned}(T_ix)(t) &= \int_t^{t+\omega} G_i(t,s)\, x_i(s) \times \left[\sum_{j=1}^{n} a_{ij}(s)\, x_j(s) + \sum_{j=1}^{n} b_{ij}(s) \int_{-\tau_{ij}}^{0} f_{ij}(\xi)\, x_j(s+\xi)\, d\xi + \sum_{j=1}^{n} c_{ij}(s) \int_{-\sigma_{ij}}^{0} g_{ij}(\xi)\, x_j'(s+\xi)\, d\xi\right] ds \\ &\quad + \sum_{t \le t_k < t+\omega} G_i(t,t_k)\, I_{ik}(x_i(t_k)),\end{aligned} \tag{19}$$

where

$$G_i(t,s) = \frac{e^{-\int_t^s r_i(\theta)d\theta}}{1 - e^{-\int_0^\omega r_i(\theta)d\theta}}, \quad s \in [t, t+\omega], \; i = 1, 2, \ldots, n. \tag{20}$$

It is obvious to see that $G_i(t+\omega, s+\omega) = G_i(t,s)$, $\partial G_i(t,s)/\partial t = r_i(t)G_i(t,s)$, $G_i(t, t+\omega) - G_i(t,t) = -1$, *and*

$$\frac{\delta_i}{1-\delta_i} \le G_i(t,s) \le \frac{1}{1-\delta_i}, \quad s \in [t, t+\omega], \; i = 1, 2, \ldots, n. \tag{21}$$

In what follows, we will give some lemmas concerning E and A defined by (17) and (18), respectively.

Lemma 9. *Assume that* (A_1)*–*(A_3) *hold.*

(i) *If* $\max\{r_i^M, i = 1, 2, \ldots, n\} \le 1$, *then* $T : E \to E$ *is well defined.*

(ii) *If* (A_4) *holds and* $\max\{r_i^M, i = 1, 2, \ldots, n\} > 1$, *then* $T : E \to E$ *is well defined.*

Proof. For any $x \in E$, it is clear that $Tx \in PC^1(R)$. From (18), for $t \in [0,\omega]$, we have

$$\begin{aligned}(T_ix)(t+\omega) &= \int_{t+\omega}^{t+2\omega} G_i(t+\omega, s)\, x_i(s) \times \left[\sum_{j=1}^{n} a_{ij}(s)\, x_j(s) + \sum_{j=1}^{n} b_{ij}(s) \int_{-\tau_{ij}}^{0} f_{ij}(\xi)\, x_j(s+\xi)\, d\xi + \sum_{j=1}^{n} c_{ij}(s) \int_{-\sigma_{ij}}^{0} g_{ij}(\xi)\, x_j'(s+\xi)\, d\xi\right] ds \\ &\quad + \sum_{t+\omega \le t_k < t+2\omega} G_i(t+\omega, t_k)\, I_{ik}(x_i(t_k)) \\ &= \int_t^{t+\omega} G_i(t+\omega, u+\omega)\, x_i(u+\omega) \times \left[\sum_{j=1}^{n} a_{ij}(u+\omega)\, x_j(u+\omega) + \sum_{j=1}^{n} b_{ij}(u+\omega) \times \int_{-\tau_{ij}}^{0} f_{ij}(\xi)\, x_j(u+\omega+\xi)\, d\xi + \sum_{j=1}^{n} c_{ij}(u+\omega) \int_{-\sigma_{ij}}^{0} g_{ij}(\xi)\, x_j'(u+\omega+\xi)\, d\xi\right] ds \\ &\quad + \sum_{t \le t_k < t+\omega} G_i(t, t_k)\, I_{ik}(x_i(t_k)) \\ &= \int_t^{t+\omega} G_i(t,u)\, x_i(u) \times \left[\sum_{j=1}^{n} a_{ij}(u)\, x_j(u) + \sum_{j=1}^{n} b_{ij}(u) \int_{-\tau_{ij}}^{0} f_{ij}(\xi)\, x_j(u+\xi)\, d\xi + \sum_{j=1}^{n} c_{ij}(u) \int_{-\sigma_{ij}}^{0} g_{ij}(\xi)\, x_j'(u+\xi)\, d\xi\right] ds \\ &\quad + \sum_{t \le t_k < t+\omega} G_i(t, t_k)\, I_{ik}(x_i(t_k)) \\ &= (T_ix)(t), \quad i = 1, 2, \ldots, n.\end{aligned} \tag{22}$$

That is, $(T_ix)(t+\omega) = (T_ix)(t)$, $t \in [0,\omega]$. So $Tx \in Y$. In view of (A_2), for $x \in E$, $t \in [0,\omega]$, we have

$$\begin{aligned}
&\sum_{j=1}^{n} a_{ij}(t)\, x_j(t) + \sum_{j=1}^{n} b_{ij}(t) \int_{-\tau_{ij}}^{0} f_{ij}(\xi)\, x_j(t+\xi)\, d\xi \\
&\quad + \sum_{j=1}^{n} c_{ij}(t) \int_{-\sigma_{ij}}^{0} g_{ij}(\xi)\, x_j'(t+\xi)\, d\xi \\
&\quad \geq \sum_{j=1}^{n} a_{ij}(t)\, \delta_j |x_j|_1 - \sum_{j=1}^{n} b_{ij}(t) |x_j|_1 - \sum_{j=1}^{n} c_{ij}(t) |x_j|_1 \\
&\quad = \sum_{j=1}^{n} \left[a_{ij}(t)\,\delta_j - b_{ij}(t) - c_{ij}(t)\right] |x_j'|_1 \geq 0, \quad i = 1, 2, \ldots, n.
\end{aligned} \tag{23}$$

Therefore, for $x \in E,\ t \in [0, \omega]$, we find

$$\begin{aligned}
&|T_i x|_0 \\
&\quad \leq \frac{1}{1-\delta_i} \left\{ \int_0^{\omega} x_i(s) \right. \\
&\qquad \times \left[\sum_{j=1}^{n} a_{ij}(s)\, x_j(s) + \sum_{j=1}^{n} b_{ij}(s) \right. \\
&\qquad \times \int_{-\tau_{ij}}^{0} f_{ij}(\xi)\, x_j(s+\xi)\, d\xi \\
&\qquad \left. + \sum_{j=1}^{n} c_{ij}(s) \int_{-\sigma_{ij}}^{0} g_{ij}(\xi)\, x_j'(s+\xi)\, d\xi \right] ds \\
&\qquad \left. + \sum_{t \leq t_k < t+\omega} I_{ik}(x_i(t_k)) \right\}, \\
&(T_i x)(t) \\
&\quad \geq \frac{\delta_i}{1-\delta_i} \left\{ \int_t^{t+\omega} x_i(s) \right. \\
&\qquad \times \left[\sum_{j=1}^{n} a_{ij}(s)\, x_j(s) \right. \\
&\qquad + \sum_{j=1}^{n} b_{ij}(s) \int_{-\tau_{ij}}^{0} f_{ij}(\xi)\, x_j(s+\xi)\, d\xi \\
&\qquad \left. + \sum_{j=1}^{n} c_{ij}(s) \int_{-\sigma_{ij}}^{0} g_{ij}(\xi)\, x_j'(s+\xi)\, d\xi \right] ds \\
&\qquad \left. + \sum_{t \leq t_k < t+\omega} I_{ik}(x_i(t_k)) \right\}
\end{aligned} \tag{24}$$

$$\begin{aligned}
&\quad = \frac{\delta_i}{1-\delta_i} \\
&\qquad \times \left\{ \int_0^{\omega} x_i(s) \right. \\
&\qquad \times \left[\sum_{j=1}^{n} a_{ij}(s)\, x_j(s) \right. \\
&\qquad + \sum_{j=1}^{n} b_{ij}(s) \int_{-\tau_{ij}}^{0} f_{ij}(\xi)\, x_j(s+\xi)\, d\xi \\
&\qquad \left. + \sum_{j=1}^{n} c_{ij}(s) \int_{-\sigma_{ij}}^{0} g_{ij}(\xi)\, x_j'(s+\xi)\, d\xi \right] ds \\
&\qquad \left. + \sum_{t \leq t_k < t+\omega} I_{ik}(x_i(t_k)) \right\} \\
&\quad \geq \delta_i |T_i x|_0, \quad i = 1, 2, \ldots, n.
\end{aligned} \tag{25}$$

Now, we show that $(T_i x)(t) \geq \delta_i |T_i x|_1$, $i = 1, 2, \ldots, n$, $t \in [0, \omega]$. From (18), we obtain

$$\begin{aligned}
&(T_i x)'(t) \\
&\quad = G_i(t, t+\omega)\, x_i(t+\omega) \\
&\qquad \times \left[\sum_{j=1}^{n} a_{ij}(t+\omega)\, x_j(t+\omega) \right. \\
&\qquad + \sum_{j=1}^{n} b_{ij}(t+\omega) \int_{-\tau_{ij}}^{0} f_{ij}(\xi)\, x_j(t+\omega+\xi)\, d\xi \\
&\qquad \left. + \sum_{j=1}^{n} c_{ij}(t+\omega) \int_{-\sigma_{ij}}^{0} g_{ij}(\xi)\, x_j'(t+\omega+\xi)\, d\xi \right] \\
&\qquad - G_i(t, t)\, x_i(t) \\
&\qquad \times \left[\sum_{j=1}^{n} a_{ij}(t)\, x_j(t) + \sum_{j=1}^{n} b_{ij}(t) \right. \\
&\qquad \times \int_{-\tau_{ij}}^{0} f_{ij}(\xi)\, x_j(t+\xi)\, d\xi \\
&\qquad \left. + \sum_{j=1}^{n} c_{ij}(t) \int_{-\sigma_{ij}}^{0} g_{ij}(\xi)\, x_j'(t+\xi)\, d\xi \right] \\
&\qquad + r_i(t)\, (T_i x)(t) \\
&\quad = r_i(t)\, (T_i x)(t) - x_i(t)
\end{aligned}$$

$$\times\left[\sum_{j=1}^{n}a_{ij}(s)x_j(s)+\sum_{j=1}^{n}b_{ij}(t)\int_{-\tau_{ij}}^{0}f_{ij}(\xi)x_j(t+\xi)\,d\xi\right.$$
$$\left.+\sum_{j=1}^{n}c_{ij}(t)\int_{-\sigma_{ij}}^{0}g_{ij}(\xi)x_j'(t+\xi)\,d\xi\right],$$
$$i=1,2,\ldots,n. \tag{26}$$

It follows from (23) and (25) that if $(T_ix)'(t)\ge 0$, $i=1,2,\ldots,n$, then

$$(T_ix)'(t)\le r_i(t)(T_ix)(t)\le r_i^M(T_ix)(t)\le(T_ix)(t),\quad i=1,2,\ldots,n. \tag{27}$$

On the other hand, from (25) and (A_3), if $(T_ix)'(t)<0$, $i=1,2,\ldots,n$, then

$$\begin{aligned}
&-(T_ix)'(t)\\
&=-r_i(t)(T_ix)(t)+x_i(t)\\
&\quad\times\left[\sum_{j=1}^{n}a_{ij}(t)x_j(t)+\sum_{j=1}^{n}b_{ij}(t)\int_{-\tau_{ij}}^{0}f_{ij}(\xi)x_j(t+\xi)\,d\xi\right.\\
&\quad\left.+\sum_{j=1}^{n}c_{ij}(t)\int_{-\sigma_{ij}}^{0}g_{ij}(\xi)x_j'(t+\xi)\,d\xi\right]\\
&\le|x_i|_1\sum_{j=1}^{n}\left[a_{ij}(t)+b_{ij}(t)+c_{ij}(t)\right]|x_i|_1-r_i^L(T_ix)(t)\\
&\le\left(1+r_i^L\right)\frac{\delta_i^2}{1-\delta_i}|x_i|_1\\
&\quad\times\sum_{j=1}^{n}\left\{\int_0^{\omega}\left[\delta_ja_{ij}(s)-b_{ij}(s)-c_{ij}(s)\right]|x_i|_1ds\right\}\\
&\quad-r_i^L(T_ix)(t)\\
&=\left(1+r_i^L\right)\int_t^{t+\omega}\frac{\delta_i}{1-\delta_i}\delta_i|x_i|_1\\
&\quad\times\sum_{j=1}^{n}\left[\delta_ja_{ij}(s)-b_{ij}(s)-c_{ij}(s)\right]|x_i|_1ds-r_i^L(T_ix)(t)\\
&\le\left(1+r_i^L\right)\left\{\int_t^{t+\omega}\int_t^{t+\omega}G_i(t,s)x_i(s)\right.\\
&\quad\times\left[\sum_{j=1}^{n}a_{ij}(s)x_j(s)+\sum_{j=1}^{n}b_{ij}(s)\right.\\
&\quad\times\int_{-\tau_{ij}}^{0}f_{ij}(\xi)x_j(s+\xi)\,d\xi\\
&\quad\left.+\sum_{j=1}^{n}c_{ij}(s)\int_{-\sigma_{ij}}^{0}g_{ij}(\xi)x_j'(s+\xi)\,d\xi\right]ds\\
&\quad\left.+\sum_{t\le t_k<t+\omega}G_i(t,t_k)I_{ik}(x_i(t_k))\right\}-r_i^L(T_ix)(t)\\
&=\left(1+r_i^L\right)(T_ix)(t)-r_i^L(T_ix)(t)\\
&=(T_ix)(t),\quad i=1,2,\ldots,n.
\end{aligned} \tag{28}$$

It follows from (27) and (28) that $|(T_ix)'|_0\le|T_ix|_0$, $i=1,2,\ldots,n$. So $|T_ix|_1=|T_ix|_0$, $i=1,2,\ldots,n$. By (25), we have $(T_ix)(t)\ge\delta_i|T_ix|_1$, $i=1,2,\ldots,n$. Hence, $Tx\in E$. This completes the proof of (i). In view of the proof of (ii), we only need to prove that $(T_ix)'(t)\ge 0$, $i=1,2,\ldots,n$, implies

$$(T_ix)'(t)\le(T_ix)(t),\quad i=1,2,\ldots,n. \tag{29}$$

From (23), (26), (A_2), and (A_4), we have

$$\begin{aligned}
&(T_ix)'(t)\\
&=r_i(t)(T_ix)(t)-x_i(t)\\
&\quad\times\left[\sum_{j=1}^{n}a_{ij}(t)x_j(t)-\sum_{j=1}^{n}b_{ij}(t)\int_{-\tau_{ij}}^{0}f_{ij}(\xi)x_j(t+\xi)\,d\xi\right.\\
&\quad\left.-\sum_{j=1}^{n}c_{ij}(t)\int_{-\sigma_{ij}}^{0}g_{ij}(\xi)x_j'(t+\xi)\,d\xi\right]\\
&\le r_i(t)(T_ix)(t)-\delta_i|x_i|_1\\
&\quad\times\sum_{j=1}^{n}\left[\delta_ja_{ij}(t)-b_{ij}(t)-c_{ij}(t)\right]|x_j|_1\\
&\le r_i^M(T_ix)(t)-\delta_i|x_i|_1\frac{r_i^M-1}{\delta_i(1-\delta_i)}\\
&\quad\times\sum_{j=1}^{n}\left\{\int_0^{\omega}\left[a_{ij}(s)+b_{ij}(s)+c_{ij}(s)\right]|x_j|_1ds\right\}\\
&=r_i^M(T_ix)(t)-\left(r_i^M-1\right)\int_t^{t+\omega}\frac{1}{1-\delta_i}|x_i|_1\\
&\quad\times\sum_{j=1}^{n}\left[a_{ij}(s)+b_{ij}(s)+c_{ij}(s)\right]|x_j|_1ds\\
&\le r_i^M(T_ix)(t)-\left(r_i^M-1\right)
\end{aligned}$$

$$\times\left\{\int_t^{t+\omega}G_i(t,s)x_i(s)\right.$$
$$\times\left[\sum_{j=1}^{n}a_{ij}(s)x_j(s)+\sum_{j=1}^{n}b_{ij}(s)\right.$$
$$\times\int_{-\tau_{ij}}^{0}f_{ij}(\xi)x_j(s+\xi)\,d\xi$$
$$\left.+\sum_{j=1}^{n}c_{ij}(s)\int_{-\sigma_{ij}}^{0}g_{ij}(\xi)x_j'(s+\xi)\,d\xi\right]ds$$
$$\left.+\sum_{t\le t_k<t+\omega}G_i(t,t_k)I_{ik}(x_i(t_k))\right\}$$
$$=r_i^M(T_ix)(t)-\left(r_i^M-1\right)(T_ix)(t)=(T_ix)(t),\quad i=1,2,\ldots,n. \tag{30}$$

The proof of (ii) is complete. Thus we complete the proof of Lemma 9.

Lemma 10. *Assume that* (A_1)*-*(A_3) *hold and* $R\max_{1\le j\le n}\{\sum_{j=1}^{n}b_{ij}^M\}<1$.

(i) *If* $\max\{r_i^M, i=1,2,\ldots,n\}\le 1$, *then* $T:E\bigcap\overline{\Omega_R}\to E$ *is strict-set-contractive.*

(ii) *If* (A_4) *holds and* $\min\{r_i^M, i=1,2,\ldots,n\}>1$, *then* $T:E\bigcap\overline{\Omega_R}\to E$ *is strict-set-contractive, where* $\Omega_R=\{x\in Y:|x|_1<R\}$.

Proof. We only need to prove (i), since the proof of (ii) is similar. It is easy to see that T is continuous and bounded. Now we prove that a $\alpha_Y(T(S))\le R\max_{1\le j\le n}\{\sum_{j=1}^{n}b_{ij}^M\}\alpha_Y(S)$ for any bounded set $S\in\overline{\Omega}_R$. Let $\eta=\alpha_Y(S)$. Then, for any positive number $\epsilon<R\max_{1\le j\le n}\{\sum_{j=1}^{n}b_{ij}^M\}\eta$, there is a finite family of subsets $\{S_i\}$ satisfying $S=\bigcup_i S_i$ with $\operatorname{diam}(S_i)\le\eta+\epsilon$. Therefore

$$|x-y|_1\le\eta+\epsilon,\quad\text{for any } x,y\in S_i. \tag{31}$$

As S and S_i are precompact in X, it follows that there is a finite family of subsets $\{S_{ij}\}$ of S_i such that $S_i=\bigcup_j S_{ij}$ and

$$|x-y|_0\le\epsilon,\quad\text{for any } x,y\in S_{ij}. \tag{32}$$

In addition, for any $x\in S$ and $t\in[0,\omega]$, we have

$$(T_ix)(t)$$
$$=\int_t^{t+\omega}G_i(t,s)x_i(s)$$
$$\times\left[\sum_{j=1}^{n}a_{ij}(s)x_j(s)+\sum_{j=1}^{n}b_{ij}(s)\int_{-\tau_{ij}}^{0}f_{ij}(\xi)x_j(s+\xi)\,d\xi\right.$$
$$\left.+\sum_{j=1}^{n}c_{ij}(s)\int_{-\sigma_{ij}}^{0}g_{ij}(\xi)x_j'(s+\xi)\,d\xi\right]ds$$
$$+\sum_{t\le t_k<t+\omega}G_i(t,t_k)I_{ik}(x_i(t_k))$$
$$\le\frac{R^2}{1-\delta_i}\int_0^{\omega}\left[\sum_{j=1}^{n}a_{ij}(s)+\sum_{j=1}^{n}b_{ij}(s)+\sum_{j=1}^{n}c_{ij}(s)\right]ds$$
$$+\frac{1}{1-\delta_i}\sum_{t\le t_k<t+\omega}I_{ik}(x_i(t_k)):=B_i,\quad i=1,2,\ldots,n, \tag{33}$$

$$\left|(T_ix)'(t)\right|$$
$$=\left|r_i(t)(T_ix)(t)-x_i(t)\right.$$
$$\times\left[\sum_{j=1}^{n}a_{ij}(t)x_j(t)+\sum_{j=1}^{n}b_{ij}(t)\int_{-\tau_{ij}}^{0}f_{ij}(\xi)x_j(t+\xi)\,d\xi\right.$$
$$\left.\left.+\sum_{j=1}^{n}c_{ij}(t)\int_{-\sigma_{ij}}^{0}g_{ij}(\xi)x_j'(t+\xi)\,d\xi\right]\right|$$
$$\le r_i^MB_i+R^2\sum_{j=1}^{n}\left(a_{ij}^M+b_{ij}^M+c_{ij}^M\right),\quad i=1,2,\ldots,n. \tag{34}$$

Hence,

$$\|(Tx)\|_0\le\sum_{i=1}^{n}B_i,$$
$$\left\|(Tx)'\right\|_0\le\sum_{i=1}^{n}\left[B_ir_i^MB_i+R^2\sum_{j=1}^{n}\left(a_{ij}^M+b_{ij}^M\right)\right]. \tag{35}$$

Applying the Arzela-Ascoli theorem, we know that $T(S)$ is precompact in X. Then, there is a finite family of subsets $\{S_{ijk}\}$ of S_{ij} such that $S_{ij}=\bigcup_k S_{ijk}$ and

$$|Tx-Ty|_0\le\epsilon,\quad\text{for any } x,y\in S_{ijk}. \tag{36}$$

From (23), (26), (31)–(34), and (A_2), for any $x, y \in S_{ijk}$, we have

$$\begin{aligned}
&\left|(T_i x)' - (T_i y)'\right|_0 \\
&= \max_{t\in[0,\omega]} \left\{ \left| r_i(t)(T_i x)(t) - r_i(t)(T_i y)(t) - x_i(t) \right.\right. \\
&\quad \times \left[\sum_{j=1}^n a_{ij}(t) x_j(t) + \sum_{j=1}^n b_{ij}(t) \int_{-\tau_{ij}}^0 f_{ij}(\xi) x_j(t+\xi)\, d\xi + \sum_{j=1}^n c_{ij}(t) \int_{-\sigma_{ij}}^0 g_{ij}(\xi) x_j'(t+\xi)\, d\xi \right] \\
&\quad \left.\left. + y_i(t) \left[\sum_{j=1}^n a_{ij}(t) y_j(t) + \sum_{j=1}^n b_{ij}(t) \int_{-\tau_{ij}}^0 f_{ij}(\xi) y_j(t+\xi)\, d\xi + \sum_{j=1}^n c_{ij}(t) \int_{-\sigma_{ij}}^0 g_{ij}(\xi) y_j'(t+\xi)\, d\xi \right] \right| \right\} \\
&\le \max_{t\in[0,\omega]} \left\{ \left| r_i(t) \left[(T_i x)(t) - (T_i y)(t) \right] \right| \right\} \\
&\quad + \max_{t\in[0,\omega]} \left\{ \left| x_i \left[\left(\sum_{j=1}^n a_{ij}(t) x_j(t) + \sum_{j=1}^n b_{ij}(t) \int_{-\tau_{ij}}^0 f_{ij}(\xi) x_j(t+\xi)\, d\xi + \sum_{j=1}^n c_{ij}(t) \int_{-\sigma_{ij}}^0 g_{ij}(\xi) x_j'(t+\xi)\, d\xi \right) \right.\right.\right. \\
&\quad - \left(\sum_{j=1}^n a_{ij}(t) y_j(t) + \sum_{j=1}^n b_{ij}(t) \times \int_{-\tau_{ij}}^0 f_{ij}(\xi) y_j(t+\xi)\, d\xi + \sum_{j=1}^n c_{ij}(t) \times \int_{-\sigma_{ij}}^0 g_{ij}(\xi) y_j'(t+\xi)\, d\xi \right) \left.\left.\left. \right] \right| \right\} \\
&\quad + \max_{t\in[0,\omega]} \left\{ \left| \left[x_i(t) - y_i(t) \right] \right.\right. \\
&\quad \times \left(\sum_{j=1}^n a_{ij}(t) y_j(t) + \sum_{j=1}^n b_{ij}(t) \times \int_{-\tau_{ij}}^0 f_{ij}(\xi) y_j(t+\xi)\, d\xi + \sum_{j=1}^n c_{ij}(t) \int_{-\sigma_{ij}}^0 g_{ij}(\xi) y_j'(t+\xi)\, d\xi \right) \left.\left. \right| \right\} \\
&\le r_i^M \left| (T_i x) - (T_i y) \right|_0 \\
&\quad + R \max_{t\in[0,\omega]} \left\{ \sum_{j=1}^n \left[a_{ij}(t) \left| x_j(t) - y_j(t) \right| + b_{ij}(t) \times \int_{-\tau_{ij}}^0 f_{ij}(\xi) \times \left| x_i(t+\xi) - y_i(t+\xi) \right| d\xi + c_{ij}(t) \times \int_{-\sigma_{ij}}^0 g_{ij}(\xi) \times \left| x_j'(t+\xi) - y'(t-\xi) \right| d\xi \right] \right\} \\
&\quad + \max_{t\in[0,\omega]} \left\{ \left| \left[x_i(t) - y_i(t) \right] \right.\right. \\
&\quad \times \left(\sum_{j=1}^n a_{ij}(t) y_j(t) + \sum_{j=1}^n b_{ij}(t) \int_{-\tau_{ij}}^0 f_{ij}(\xi) y_j(t+\xi)\, d\xi + \sum_{j=1}^n c_{ij}(t) \times \int_{-\sigma_{ij}}^0 g_{ij}(\xi) y_j'(t+\xi)\, d\xi \right) \left.\left. \right| \right\} \\
&\le r_i^M \epsilon + R\epsilon \sum_{j=1}^n a_{ij}^M + R\epsilon \sum_{j=1}^n b_{ij}^M + |x_i|_0 (\eta + \epsilon) \times \sum_{j=1}^n c_{ij}^M + R\epsilon \sum_{j=1}^n \left(a_{ij}^M + b_{ij}^M + c_{ij}^M \right) \\
&= |x_i|_0 \sum_{j=1}^n c_{ij}^M \eta + \widetilde{B}_i \epsilon,
\end{aligned} \tag{37}$$

where

$$\widetilde{B}_i = r_i^M + 2R \sum_{j=1}^n \left(a_{ij}^M + b_{ij}^M + c_{ij}^M \right), \quad i = 1, 2, \ldots, n. \tag{38}$$

From (36) and (37) we obtain

$$\begin{aligned}\|Tx - Ty\|_1 &\leq \left(\sum_{i=1}^{n} |x_i|_0 \sum_{j=1}^{n} c_{ij}^M\right) \eta + \epsilon \sum_{i=1}^{n} \widetilde{B}_i \\ &\leq R \max_{1\leq i\leq n} \left\{\sum_{j=1}^{n} c_{ij}^M\right\} \eta + \epsilon \sum_{i=1}^{n} \widetilde{B}_i, \quad \text{for any } x, y \in S_{ijk}.\end{aligned} \tag{39}$$

As ϵ is arbitrarily small, it follows that

$$\alpha_Y (T (S)) \leq R \max_{1\leq j\leq n} \left\{\sum_{j=1}^{n} c_{ij}^M\right\} \alpha_Y (S) . \tag{40}$$

Therefore, T is strict-set-contractive. The proof of Lemma 10 is complete.

3. Main Results

Our main result of this paper is as follows.

Theorem 11. *Assume that* (A_1)*–*(A_3)*,* (A_5) *hold.*

(i) *If* $\max\{r_i^M, i = 1, 2, \ldots, n\} \leq 1$*, then system* (2) *has at least one positive* ω*-periodic solution.*

(ii) *If* (A_4) *holds and* $\min\{r_i^M, i = 1, 2, \ldots, n\} > 1$*, then system* (2) *has at least one positive* ω*-periodic solution.*

Proof. We only need to prove (i), since the proof of (ii) is similar. Let

$$\begin{aligned}R &= \left(\min_{i\in[1,n]} \left\{\frac{\delta_i^2}{1-\delta_i} \min_{j\in[1,n]} \{\alpha_{ij}\}\right\}\right)^{-1}, \\ 0 < r &< \min_{i\in[1,n]} \left\{\frac{\delta_i (1-\delta_i) - \gamma_i}{\max_{j\in[1,n]} \{\beta_{ij}\}}\right\}.\end{aligned} \tag{41}$$

Then it is easy to see that $0 < r < R$. From Lemmas 9 and 10, we know that T is strict-set-contractive on $E_{r,R}$. In view of (26), we see that if there exists $x^* \in E$ such that $Tx^* = x^*$, then x^* is one positive ω-periodic solution of system (2). Now, we will prove that condition (ii) of Lemma 7 holds. First, we prove that $Tx \not\geq x$, for all $x \in E$, $\|x\|_1 < r$. Otherwise, there exists $x \in E$, $\|x\|_1 < r$, such that $Tx \geq x$. So $\|x\|_0 > 0$ and $Tx - x \geq 0$, which implies that

$$\begin{aligned}&T_i x (t) - x_i (t) \geq \delta_i |T_i x - x_i|_1 \geq 0, \\ &\text{for any } t \in [0, \omega], \ i = 1, 2, \ldots, n.\end{aligned} \tag{42}$$

Moreover, for $t \in [0, \omega]$, we have

$$\begin{aligned}(T_i x)(t) &= \int_t^{t+\omega} G_i (t, s) x_i (s) \\ &\quad\times \Bigg[\sum_{j=1}^{n} a_{ij} (s) x_j (s) + \sum_{j=1}^{n} b_{ij} (s) \int_{-\tau_{ij}}^{0} f_{ij} (\xi) x_j (s + \xi) d\xi \\ &\quad + \sum_{j=1}^{n} c_{ij} (s) \int_{-\sigma_{ij}}^{0} g_{ij} (\xi) x_j' (s+\xi) d\xi\Bigg] ds \\ &\quad + \sum_{t\leq t_k<t+\omega} G_i (t, t_k) I_{ik} (x_i (t_k)) \\ &\leq \frac{1}{1-\delta_i} |x_i|_0 \left\{\sum_{j=1}^{n} \int_0^{\omega} \left[a_{ij} (s) + b_{ij} (s) + c_{ij} (s)\right] |x_j|_1 ds + \gamma_i\right\} \\ &\leq \frac{\beta_{ij} r + \gamma_i}{1-\delta_i} |x_i|_0 \leq \delta_i |x_i|_0, \quad i = 1, 2, \ldots, n.\end{aligned} \tag{43}$$

In view of (42) and (43), we obtain

$$\|x\|_0 \leq \|Tx\|_0 \leq \max_{1\leq i\leq n} \{\delta_i\} \|x\|_0 < \|x\|_0, \tag{44}$$

which is a contradiction. Finally, we prove that $Tx \not\leq x$, for all $x \in E$, and $\|x\|_1 = R$ also hold. For this case, suppose, for the sake of contradiction, that there exist $x \in E$ and $\|x\|_1 = R$ such that $Tx \leq x$. Furthermore, for any $t \in [0, \omega]$, we have

$$\begin{aligned}&x_i (t) - T_i x (t) \geq \delta_i |x_i - T_i x|_1 \geq 0, \\ &\text{for any } t \in [0, \omega], \ i = 1, 2, \ldots, n.\end{aligned} \tag{45}$$

In addition, for any $t \in [0, \omega]$, we find

$$\begin{aligned}(T_i x)(t) &= \int_t^{t+\omega} G_i (t, s) x_i (s) \\ &\quad\times \Bigg[\sum_{j=1}^{n} a_{ij} (s) x_j (s) + \sum_{j=1}^{n} b_{ij} (s) \int_{-\tau_{ij}}^{0} f_{ij} (\xi) x_j (s + \xi) d\xi \\ &\quad + \sum_{j=1}^{n} c_{ij} (s) \int_{-\sigma_{ij}}^{0} g_{ij} (\xi) x_j' (s + \xi) d\xi\Bigg] ds \\ &\quad + \sum_{t\leq t_k<t+\omega} G_i (t, t_k) I_{ik} (x_i (t_k)) \\ &> \frac{\delta_i^2}{1-\delta_i} |x_i|_1 \sum_{j=1}^{n} \int_0^{\omega} \left[\delta_j a_{ij} (s) - b_{ij} (s) - c_{ij} (s)\right] ds |x_j|_1 \\ &= \frac{\delta_i^2}{1-\delta_i} |x_i|_1 \sum_{j=1}^{n} \left(\alpha_{ij} |x_j|_1\right), \quad i = 1, 2, \ldots, n.\end{aligned} \tag{46}$$

Thus, we have

$$\|Tx\|_0 = \sum_{i=1}^{n} |T_i x|_0 > \sum_{i=1}^{n} \frac{\delta_i^2}{1-\delta_i} |x_i|_1 \sum_{j=1}^{n} \left(\alpha_{ij} |x_j|_1\right)$$
$$\geq \min_{1\leq i\leq n} \left\{ \frac{\delta_i^2 \min_{1\leq j\leq n} \{\alpha_{ij}\}}{1-\delta_i} \right\} \sum_{i=1}^{n} |x_i|_1 \sum_{j=1}^{n} |x_j|_1 \quad (47)$$
$$\geq \min_{1\leq i\leq n} \left\{ \frac{\delta_i^2 \min_{1\leq j\leq n} \{\alpha_{ij}\}}{1-\delta_i} \right\} R^2 = R.$$

From (45) and (47), we obtain

$$\|x\|_1 \geq \|Tx\|_1 \geq \|Tx\|_0 > R, \quad (48)$$

which is a contradiction. Therefore, condition (ii) of Lemma 7 holds. By Lemma 7, we see that T has at least one nonzero fixed point in E. Thus, the system (6) has at least one positive ω-periodic solution. Therefore, it follows from Lemma 6 that system (2) has a positive ω-periodic solution. The proof of Theorem 11 is complete.

Remark 12. If $I_{ik}(x(t_k)) = 0$, we can easily derive the corresponding results in [10]. So we extend the corresponding results in [10].

4. Example

In this section, we give an example to show the effectiveness of our result.

Example 1. Consider the following nonimpulsive system:

$$\begin{aligned}
x_1'(t) &= x_1(t)\left[\frac{1-\sin t}{24} - (10+\sin t)\, x_1(t) - (9-\cos t)\, x_2(t) - (3-2\cos t) \right.\\
&\times \int_{-\tau_{11}}^{0} f_{11}(\xi)\, x_1(t+\xi)\, d\xi - \frac{5+\sin t}{15} \times \int_{-\tau_{12}}^{0} f_{12}(\xi)\, x_2(t+\xi)\, d\xi - (1+\sin t)\\
&\times \int_{-\sigma_{11}}^{0} g_{11}(\xi)\, x_1'(t+\xi)\, d\xi - \frac{4-3\cos t}{6} \left.\times \int_{-\sigma_{12}}^{0} g_{12}(\xi)\, x_2'(t+\xi)\, d\xi\right],\\
x_2'(t) &= x_2(t)\left[\frac{1+\cos t}{8\pi} - (12+3\cos t)\, x_1(t) - (7-\sin t)\, x_2(t) - (2+3\sin t) \right.\\
&\times \int_{-\tau_{21}}^{0} f_{21}(\xi)\, x_2(t+\xi)\, d\xi - (1-\cos t) \times \int_{-\tau_{22}}^{0} f_{22}(\xi)\, x_2(t+\xi)\, d\xi - \frac{4+\sin t}{12}\\
&\times \int_{-\sigma_{21}}^{0} g_{21}(\xi)\, x_2'(t+\xi)\, d\xi - \frac{5-\cos t}{10} \left.\times \int_{-\sigma_{22}}^{0} g_{22}(\xi)\, x_2'(t+\xi)\, d\xi\right],
\end{aligned} \quad (49)$$

where $\tau_{ij}, \sigma_{ij} \in (0, +\infty)$ $(i, j = 1, 2)$, and $f_{ij}, g_{ij} \in C(R, R^+)$ satisfying $\int_{-\tau_{ij}}^{0} f_{ij}(\xi) d\xi = 1$, $\int_{-\sigma_{ij}}^{0} g_{ij}(\xi) d\xi$, $i, j = 1, 2$. Obviously

$$\begin{gathered}
r_1(t) = \frac{1-\sin t}{24}, \qquad r_2 = \frac{1+\cos t}{8\pi},\\
a_{11}(t) = 10 + \sin t,\\
a_{12}(t) = 9 - \cos t, \qquad a_{21}(t) = 12 + 3\cos t,\\
a_{22}(t) = 7 - \sin t, \qquad b_{11}(t) = 3 - 2\cos t,\\
b_{12}(t) = \frac{5+\sin t}{15},\\
b_{21}(t) = 2 + 3\sin t, \qquad b_{22}(t) = 1 - \cos t,\\
c_{11}(t) = 1 + \sin t, \qquad c_{12}(t) = \frac{4-3\cos t}{6},\\
c_{21}(t) = \frac{4+\sin t}{12},\\
c_{22}(t) = \frac{5-\cos t}{10}, \qquad r_1^L = r_2^L = 0.
\end{gathered} \quad (50)$$

Furthermore, we obtain

$$\begin{gathered}
\delta_1 = e^{-(\pi/12)} < 1, \qquad \delta_2 = e^{-(1/4)} < 1,\\
\alpha_{11} = 20\pi\delta_1 - 8\pi \approx 23.8007,\\
\alpha_{12} = 18\pi\delta_2 - 2\pi \approx 37.2423,\\
\alpha_{21} = 24\pi\delta_1 - \frac{14\pi}{3} \approx 44.0594,\\
\alpha_{22} = 14\pi\delta_2 - 3\pi \approx 24.4284,
\end{gathered}$$

$$\min_{1\le j\le 2}\{\alpha_{1j}\} = \alpha_{11} = 20\pi\delta_1 - 8\pi \approx 23.8007,$$

$$\min_{1\le j\le 2}\{\alpha_{2j}\} = \alpha_{22} = 14\pi\delta_2 - 3\pi \approx 24.4284,$$

$$\delta_1 a_{11}(t) - b_{11}(t) - c_{11}(t) > 0.009 > 0,$$

$$\delta_2 a_{12}(t) - b_{12}(t) - c_{12}(t) > 1 > 0,$$

$$\delta_1 a_{21}(t) - b_{21}(t) - c_{21}(t) > 4 > 0,$$

$$\delta_2 a_{22}(t) - b_{22}(t) - c_{22}(t) > 2 > 0,$$

$$\begin{aligned}(1+r_1^L)\frac{\delta_1^2\alpha_{11}}{1-\delta_1} &= \frac{e^{-(\pi/6)}(20\pi\delta_1 - 8\pi)}{1-e^{-(\pi/12)}}\\ &\approx 65.2614 > 18\\ &\ge \max_{0\le t\le 2\pi}\{a_{11}(t)+b_{11}(t)+c_{11}(t)\},\end{aligned}$$

$$\begin{aligned}(1+r_1^L)\frac{\delta_1^2\alpha_{12}}{1-\delta_1} &= \frac{e^{-(\pi/6)}(18\pi\delta_2 - 2\pi)}{1-e^{-(\pi/12)}}\\ &\approx 102.1184 > \frac{347}{30}\\ &\ge \max_{0\le t\le 2\pi}\{a_{12}(t)+b_{12}(t)+c_{12}(t)\},\end{aligned}$$

$$\begin{aligned}(1+r_2^L)\frac{\delta_2^2\alpha_{21}}{1-\delta_2} &= \frac{e^{-(1/2)}(24\pi\delta_1 - (14\pi/3))}{1-e^{-(1/4)}}\\ &\approx 113.3411 > \frac{245}{12}\\ &> \max_{0\le t\le 2\pi}\{a_{21}(t)+b_{21}(t)+c_{21}(t)\},\end{aligned}$$

$$\begin{aligned}(1+r_2^L)\frac{\delta_2^2\alpha_{22}}{1-\delta_2} &= \frac{e^{-(1/2)}(14\pi\delta_2 - 3\pi)}{1-e^{-(1/4)}}\\ &\approx 62.8411 > \frac{53}{5}\\ &> \max_{0\le t\le 2\pi}\{a_{22}(t)+b_{22}(t)+c_{22}(t)\},\end{aligned}$$

$$\frac{\delta_1^2\min_{1\le j\le 2}\{\alpha_{1j}\}}{1-\delta_1} = \frac{e^{-(\pi/6)}(20\pi\delta_1 - 6\pi)}{1-e^{-(\pi/12)}} \approx 65.2614,$$

$$\frac{\delta_2^2\min_{1\le j\le 2}\{\alpha_{2j}\}}{1-\delta_2} = \frac{e^{-(1/2)}(14\pi\delta_2 - 3\pi)}{1-e^{-(1/4)}} \approx 62.8411,$$

$$\min_{1\le i\le 2}\left\{\frac{\delta_i^2\min_{1\le j\le 2}\{\alpha_{ij}\}}{1-\delta_i}\right\} = \frac{\delta_2^2\min_{1\le j\le 2}\{\alpha_{2j}\}}{1-\delta_1} \approx 62.8411,$$

$$c_{11}^M + c_{12}^M = \frac{19}{6}, \qquad c_{21}^M + c_{22}^M = \frac{61}{60},$$

$$\max_{1\le i\le 2}\left\{\sum_{j=1}^{2} d_{ij}^M\right\} = \frac{19}{6}. \tag{51}$$

Therefore,

$$\frac{19}{6} = \max_{1\le i\le 2}\left\{\sum_{i=1}^{2} d_{ij}^M\right\} < \min_{1\le i\le 2}\left\{\frac{\delta_i^2\min_{1\le j\le 2}\{\alpha_{ij}\}}{1-\delta_i}\right\} \approx 62.8411. \tag{52}$$

Hence, (A_1)-(A_3), (A_5) hold and $a_i^M \le 1$, $i = 1, 2$. According to Theorem 11, system (49) has at least one positive 2π-periodic solution.

Acknowledgments

The authors are thankful to the referees and editor for the improvement of the paper. This work was supported by the Construct Program of the Key Discipline in Hunan Province. Research was supported by the National Natural Science Foundation of China (10971229, 11161015), the China Postdoctoral Science Foundation (2012M512162), and Hunan Provincial Natural Science-Hengyang United Foundation of China (11JJ9002).

References

[1] P. Chesson, "Understanding the role of environmental variation in population and community dynamics," *Theoretical Population Biology*, vol. 64, no. 3, pp. 253–254, 2003.

[2] J. Zhen and Z. E. Ma, "Periodic solutions for delay differential equations model of plankton allelopathy," *Computers & Mathematics with Applications*, vol. 44, no. 3-4, pp. 491–500, 2002.

[3] X. Liu and G. Ballinger, "Boundedness for impulsive delay differential equations and applications to population growth models," *Nonlinear Analysis: Theory, Methods & Applications*, vol. 53, no. 7-8, pp. 1041–1062, 2003.

[4] Y. Kuang, *Delay Differential Equations with Applications in Population Dynamics*, Academic Press, New York, NY, USA, 1993.

[5] Y. K. Li and Y. Kuang, "Periodic solutions of periodic delay Lotka-Volterra equations and systems," *Journal of Mathematical Analysis and Applications*, vol. 255, no. 1, pp. 260–280, 2001.

[6] Y. K. Li and L. F. Zhu, "Existence of periodic solutions of discrete Lotka-Volterra systems with delays," *Bulletin of the Institute of Mathematics, Academia Sinica*, vol. 33, no. 4, pp. 369–380, 2005.

[7] H. Zhao and N. Ding, "Existence and global attractivity of positive periodic solution for competition-predator system with variable delays," *Chaos, Solitons & Fractals*, vol. 29, no. 1, pp. 162–170, 2006.

[8] Y. Song, M. Han, and Y. Peng, "Stability and Hopf bifurcations in a competitive Lotka-Volterra system with two delays," *Chaos, Solitons & Fractals*, vol. 22, no. 5, pp. 1139–1148, 2004.

[9] Z. Yang and J. Cao, "Positive periodic solutions of neutral Lotka-Volterra system with periodic delays," *Applied Mathematics and Computation*, vol. 149, no. 3, pp. 661–687, 2004.

[10] Y. K. Li, "Positive periodic solutions of periodic neutral Lotka-Volterra system with distributed delays," *Chaos, Solitons & Fractals*, vol. 37, no. 1, pp. 288–298, 2008.

[11] A. M. Samoikleno and N. A. Perestyuk, *Impulsive Differential Equations*, World Scientific, Singapore, 1995.

[12] S. T. Zavalishchin and A. N. Sesekin, *Dynamic Impulse Systems, Theory and Applications*, Kluwer Academic Publishers, Dordrecht, The Netherlands, 1997.

[13] V. Lakshmikantham, D. D. Bainov, and P. S. Simenov, *Theory of Impulsive Differential Equations*, World Science, Singapore, 1989.

[14] J. R. Yan, A. P. Zhao, and J. J. Nieto, "Existence and global attractivity of positive periodic solution of periodic single-species impulsive Lotka-Volterra systems," *Mathematical and Computer Modelling*, vol. 40, no. 5-6, pp. 509–518, 2004.

[15] X. Liu and G. Ballinger, "Boundedness for impulsive delay differential equations and applications to population growth models," *Nonlinear Analysis: Theory, Methods & Applications*, vol. 53, no. 7-8, pp. 1041–1062, 2003.

[16] W. Zhang and M. Fan, "Periodicity in a generalized ecological competition system governed by impulsive differential equations with delays," *Mathematical and Computer Modelling*, vol. 39, no. 4-5, pp. 479–493, 2004.

[17] S. L. Sun and L. S. Chen, "Existence of positive periodic solution of an impulsive delay logistic model," *Applied Mathematics and Computation*, vol. 184, no. 2, pp. 617–623, 2007.

[18] X. Li, X. Lin, D. Jiang, and X. Zhang, "Existence and multiplicity of positive periodic solutions to functional differential equations with impulse effects," *Nonlinear Analysis: Theory, Methods & Applications*, vol. 62, no. 4, pp. 683–701, 2005.

[19] R. E. Gaines and J. L. Mawhin, *Coincidence Degree Theory and Nonlinear Differential Equations*, Springer, Berlin, Germany, 1977.

[20] N. P. Các and J. A. Gatica, "Fixed point theorems for mappings in ordered Banach spaces," *Journal of Mathematical Analysis and Applications*, vol. 71, no. 2, pp. 547–557, 1979.

[21] D. J. Guo, *Nonlinear Functional Analysis*, in Chinese, ShanDong Science and Technology Press, 2001.

[22] D. Baĭnov and P. Simeonov, *Impulsive Differential Equations: Periodic Solutions and Applications*, vol. 66 of *Pitman Monographs and Surveys in Pure and Applied Mathematics*, 1993.

[23] D. J. Guo, "Positive solutions to nonlinear operator equations and their applications to nonlinear integral equations," *Advances in Mathematics*, vol. 13, no. 4, pp. 294–310, 1984 (Chinese).

18

On the Derivation of a Closed-Form Expression for the Solutions of a Subclass of Generalized Abel Differential Equations

Panayotis E. Nastou,[1] Paul Spirakis,[2,3] Yannis C. Stamatiou,[3,4] and Apostolos Tsiakalos[5]

[1] *Department of Mathematics, University of Aegean, 83200 Samos, Greece*
[2] *Department of Computer Engineering and Informatics, University of Patras Rio, 26504 Patras, Greece*
[3] *Computer Technology Institute and Press "Diophantus" Rio, 26504 Patras, Greece*
[4] *Department of Business Administration, University of Patras Rio, 26504 Patras, Greece*
[5] *Department of Mathematics, University of Ioannina, 45110 Ioannina, Greece*

Correspondence should be addressed to Panayotis E. Nastou; pnastou@aegean.gr

Academic Editor: Elena Braverman

We investigate the properties of a general class of differential equations described by $dy(t)/dt = f_{k+1}(t)y(t)^{k+1} + f_k(t)y(t)^k + \cdots + f_2(t)y(t)^2 + f_1(t)y(t) + f_0(t)$, with $k > 1$ a positive integer and $f_i(t)$, $0 \le i \le k+1$, with $f_i(t)$, real functions of t. For $k = 2$, these equations reduce to the class of *Abel differential equations of the first kind*, for which a standard solution procedure is available. However, for $k > 2$ no general solution methodology exists, to the best of our knowledge, that can lead to their solution. We develop a general solution methodology that for odd values of k connects the closed form solution of the differential equations with the existence of closed-form expressions for the roots of the polynomial that appears on the right-hand side of the differential equation. Moreover, the closed-form expression (when it exists) for the polynomial roots enables the expression of the solution of the differential equation in closed form, based on the class of Hyper-Lambert functions. However, for certain even values of k, we prove that such closed form does not exist in general, and consequently there is no closed-form expression for the solution of the differential equation through this methodology.

1. Introduction

In [1], a differential equation was derived, in the context of the theoretical analysis of a security protocol, that can be written in the following form:

$$(1-t)\frac{dy(t)}{dt} = \left(\frac{k-1}{\left\lceil \frac{k}{2} \right\rceil - 1}\right) y(t)^{k-\lceil k/2\rceil+1}\left(1-y(t)\right)^{\lceil k/2\rceil}. \quad (1)$$

In [2], the differential equations (1) are written as partial fractions in such a way that the form of their solution matches the recursive pattern of the definition of the Hyper-Lambert functions that were proposed in [3] as a generalization of the well-known Lambert W function (see [4] for an introduction to this function and its properties). The differential equations defined by (1) can be seen as belonging to the general differential equation form

$$\frac{dy(t)}{dt} = f_{k+1}(t)\,y(t)^{k+1} + f_k(t)\,y(t)^k + \cdots + f_2(t)\,y(t)^2 + f_1(t)\,y(t) + f_0(t), \quad (2)$$

with $f_i(t)$, $0 \le i \le k+1$ real functions. The class defined by (2) appears to generalize, naturally, the Abel class of differential equations (see [5]). Unfortunately, no general solution strategy exists (to the best of our knowledge) for the solution of this general class for $k > 2$, in analogy with the methodology that exists for the Abel class. Thus, we could not reduce the problem of finding the solution of (1) into the problem of solving (2) for various values of k.

In this paper, we propose a general methodology which for odd values of $k > 2$ provides a closed-form expression of the solution of (1) based on closed-form solutions of the roots of the equation $g(y(t)) - s$ where $g(y(t)) = \binom{k-1}{\lceil k/2\rceil-1} y(t)^{k-\lceil k/2\rceil+1}(1-y(t))^{\lceil k/2\rceil}$ (the polynomial on the right-hand side of (1)) and $s \neq 0$. Moreover, this closed-form solutions are linked with the generalized Hyper-Lambert functions.

However, for certain even values of $k > 2$ (at least for values of k where $k+1$ is prime), we prove that the roots of $g(y(t)) - s$ cannot be expressed by radicals, and, thus, no closed-form solution can be obtained using this methodology. This latter result may be of independent interest.

2. Properties of the Target Class of Differential Equations

The target differential equation class is the following (see [1] for its derivation):

$$\frac{dx(t)}{dt} = -\sum_{j=\lceil k/2\rceil}^{k}\binom{k}{j}\left(1-\frac{x(t)}{(1-t)}\right)^{j}\left(\frac{x(t)}{1-t}\right)^{k-j} \times\left(1-\frac{j}{k}\right)+\left[-\frac{x(t)}{(1-t)}\right] \times\sum_{j=0}^{\lceil k/2\rceil-1}\binom{k}{j}\times\left(1-\frac{x(t)}{(1-t)}\right)^{j}\left(\frac{x(t)}{1-t}\right)^{k-j}. \tag{3}$$

If we set $y(t) = x(t)/(1-t)$, (3) becomes

$$\frac{dy(t)}{dt} = \frac{y(t)}{1-t}\times\left(1-\sum_{j=0}^{\lceil k/2\rceil-1}\binom{k}{j}(1-y(t))^{j}y(t)^{k-j}\right)-\frac{1}{1-t}\times\sum_{j=\lceil k/2\rceil}^{k}\binom{k}{j}\left(1-\frac{j}{k}\right)(1-y(t))^{j}y(t)^{k-j}. \tag{4}$$

After some algebraic manipulations, (4) becomes

$$(1-t)\frac{dy(t)}{dt} = \sum_{j=\lceil k/2\rceil}^{k}\binom{k}{j}(1-y(t))^{j}y(t)^{k-j}\left(1-\frac{j}{k}-y\right). \tag{5}$$

Let $g(y(t))$ be the polynomial, in $y(t)$, at the right-hand side of the differential equation (5). Throughout this paper, when we refer to this polynomial, we will always mean that the polynomial is in $y(t)$, treating $y(t)$ as an independent variable. Based on the following two lemmas, whose proofs follow by straightforward algebraic manipulations, we will present the properties of the class of ordinary differential equations defined by (5).

Lemma 1. *For any l, $0 \le l \le k$, and independently of y, it holds*

$$\sum_{j=l}^{k}\binom{k}{j}y^{l-1-j}(1-y)^{j-l}\left[\left(1-\frac{j}{k}\right)-y\right] = -\binom{k-1}{l-1}. \tag{6}$$

Based on Lemma 1, the following can be proved.

Lemma 2. *The right-hand side of the differential equation (5) can be written as follows:*

$$g(y(t)) = \binom{k-1}{\left\lceil \frac{k}{2}\right\rceil-1}y(t)^{k-\lceil k/2\rceil+1}(1-y(t))^{\lceil k/2\rceil}. \tag{7}$$

From Lemma 2, the following property of the polynomial $g(y(t))$ can be readily derived.

Corollary 3. *The roots of the polynomial $g(y(t))$ are 0 with multiplicity $\lfloor k/2\rfloor + 1$ and 1 with multiplicity $\lceil k/2\rceil$.*

Lemma 4. *The solution $y(t)$ of the differential equation (5) with $y(0) \in [0,1)$ is a function that has the following two properties: (i) all its values are real numbers in the interval $(0,1)$, and (ii) it is monotonically increasing.*

Proof. Since $y(0) \in [0,1)$, because $0 \le x(0) < 1$ and $t = 0$, and the polynomial $g(y(t))$ contains the factors $y(t)$ and $(1-y(t))$ and its roots are 0 and 1, the solution function $y(t)$ of the differential equation (5) cannot get values below 0 or above 1. Thus, statement (i) holds.

Since $dy(t)/dt$ is always positive, as $g(y(t))$ is always positive for $y(t) \in (0,1)$ and $1-t > 0$, then $y(t)$ is a monotonically increasing function, thus proving statement (ii).

Lemma 5. *The roots of the first derivative of $g(y(t))$ as a function of $y(t)$ are $w_1 = 0$ with multiplicity $\lfloor k/2\rfloor$, $w_2 = 1$ with multiplicity $\lceil k/2\rceil$, and $w_3 = (k+1-\lceil k/2\rceil)/(k+1)$, while w_3 is a maximum point of $g(y(t))$.*

Proof. Since $g(y(t))$ is continuous on $[0,1]$, $g(0) = g(1) = 0$, and it is differentiable on $(0,1)$, there is $w \in (0,1)$ where $dg(y(t))/dy(t)|_{y(t)=w} = 0$ and $g(y(t))$ attains minimum and maximum values on $[0,1]$. After some algebraic manipulations we obtain a closed formula for the first derivative of $g(y(t))$:

$$\frac{dg(y(t))}{dy(t)} = \binom{k-1}{\left\lceil \frac{k}{2}\right\rceil-1}\left(k+1-\left\lceil\frac{k}{2}\right\rceil\frac{1}{1-y(t)}\right)\times y(t)^{k-\lceil k/2\rceil}(1-y(t))^{\lceil k/2\rceil}. \tag{8}$$

From (8), it is readily derived that the roots of the first derivative of $g(y(t))$ are $w_1 = 0$ with multiplicity $\lfloor k/2\rfloor$, $w_2 = 1$ with multiplicity $\lceil k/2\rceil$, and $w_3 = (k+1-\lceil k/2\rceil)/(k+1)$.

The second derivative of $g(y(t))$ is given by the following formula:

$$\begin{aligned}\frac{1}{\binom{k-1}{\lceil k/2\rceil-1}}\frac{d^2 g(y(t))}{dy(t)} &= \left(k-\left\lceil\frac{k}{2}\right\rceil\right) w(t)\, y(t)^{k-\lceil k/2\rceil-1} \\ &\quad\times(1-y(t))^{\lceil k/2\rceil} - \left\lceil\frac{k}{2}\right\rceil w(t)\, y(t)^{k-\lceil k/2\rceil} \\ &\quad\times(1-y(t))^{\lceil k/2\rceil-1} - \left\lceil\frac{k}{2}\right\rceil y(t)^{k-\lceil k/2\rceil} \\ &\quad\times(1-y(t))^{\lceil k/2\rceil-2},\end{aligned} \tag{9}$$

where $w(t) = k+1-\lceil k/2\rceil(1/(1-y(t)))$. For $y(t) = w_3$, the factor $w(t)$ of (9) becomes 0, and thus a negative term remains. Thus $d^2 g(y(t))/dy(t) < 0$ on $[0,1]$ which means that w_3 is a maximum point of $g(y(t))$.

Lemma 6. *The function $g(y(t))$ as function of $y(t)$ is not monotonic on $[0,1]$, and it is an increasing function on $(0, w_3)$ and a decreasing function on $(w_3, 1)$.*

Proof. Since $y(t) \in (0,1)$, the sign of the first derivative of $g(y(t))$ in (8) is determined by the factor $(k+1-\lceil k/2\rceil(1/(1-y(t))))$. Specifically, it holds that $(dg(y(t))/dy(t)) > 0$ for $y(t) \in (0, w_3)$ and $(dg(y(t))/dy(t)) < 0$ for $y(t) \in (w_3, 1)$. Since the $dg(y(t))/dy(t)$ changes sign on $[0,1]$, the function $g(y(t))$ is not monotonic, but it is an increasing function on $(0, w_3)$ and a decreasing function on $(w_3, 1)$.

Lemma 7. *For odd k, $w_3 = 1/2$ and $g(w_3) = \binom{k-1}{\lceil k/2\rceil-1}(1/2)^{2\lceil k/2\rceil}$, while for even k, $w_3 = 1-(k/(2(k+1))) > 1/2$ and $\lim_{k\to\infty} w_3 = 1/2$.*

Proof. For an odd k, it holds $\lceil k/2\rceil = (k+1)/2$ and $k-\lceil k/2\rceil+1 = \lceil k/2\rceil$. Thus,

$$w_3 = \frac{k+1-\lceil k/2\rceil}{k+1} = \frac{1}{2}, \tag{10}$$

and by applying $y(t) = w_3$ in (7), we obtain the following:

$$g(w_3) = \binom{k-1}{\left\lceil\frac{k}{2}\right\rceil-1}\left(\frac{1}{2}\right)^{2\lceil k/2\rceil}. \tag{11}$$

For even k, it holds $\lceil k/2\rceil = k/2$. Consequently, we obtain

$$w_3 = 1-\frac{k}{2(k+1)} \tag{12}$$

which means that $w_3 > 1/2$. Applying L'Hospital's rule, for large even k, we have that $\lim_{k\to\infty} w_3 = 1/2$.

The following theorem gives the specific form of the polynomial function $g(y(t))$ of (5).

Table 1: The polynomial of the right-hand side of (13) for various values of k.

k	$g(y(t))$
2	$-y(t)^3 + y(t)^2$
3	$2y(t)^4 - 4y(t)^3 + 2y(t)^2$
4	$3y(t)^5 - 6y(t)^4 + 3y(t)^3$
5	$-6y(t)^6 + 18y(t)^5 - 18y(t)^4 + 6y(t)^3$
6	$-10y(t)^7 + 30y(t)^6 - 30y(t)^5 + 10y(t)^4$
7	$20y(t)^8 - 80y(t)^7 + 120y(t)^6 - 80y(t)^5 + 20y(t)^4$
8	$35y(t)^9 - 140y(t)^8 + 210y(t)^7 - 140y(t)^6 + 35y(t)^5$

Theorem 8. *The right-hand side $g(y(t))$ of differential equation (5) is a polynomial function of one argument of $(k+1)$ degree, and the minimum degree of its terms is $k-\lceil k/2\rceil+1$; that is, it has the form:*

$$(1-t)\frac{dy(t)}{dt} = g(y(t)) = \sum_{j=0}^{\lceil k/2\rceil} a_j \cdot y(t)^{s+j}, \tag{13}$$

where $s = k-\lceil k/2\rceil+1$ and

$$a_j = (-1)^j \binom{\left\lceil\frac{k}{2}\right\rceil}{j}\binom{k-1}{\left\lceil\frac{k}{2}\right\rceil-1}. \tag{14}$$

Proof. Based on (7) and the binomial theorem, the results of the theorem can be readily deduced.

Lemma 9. *The function g is concave on $[0,1]$.*

Proof. The interval $[0,1]$ is convex, since every interval in $\mathbb{R}^1$ is convex. Thus, for $0 \le \lambda \le 1$, the convex combination of any $y(t_1)$ and $y(t_2)$, $y(t_3) = \lambda y(t_1) + (1-\lambda)y(t_2)$ is in the interval $[0,1]$. If λ is close to 0, the convex combination is close to $y(t_2)$. Based on Lemma 6, it is obvious that $g(y(t_3)) \ge \lambda g(y(t_1)) + (1-\lambda)g(y(t_2))$, which means that the function g is concave.

Table 1 presents the polynomial function of the right-hand side of the differential equation using Theorem 8 for various values of k.

Some plots of the polynomial function $g(y(t))$ as a function of $y(t)$ in the same graph for various odd values of k appear in Figure 1. It is easy to verify the result of Lemma 7 that the local maxima of g appear at $y(t) = 0.5$ independently of k value. Moreover, the function g is symmetric around the axis $y(t) = 0.5$; that is, for every value $g(y(t)) \le g(0.5)$, there are $y(t_1) < 0.5$, $y(t_2) > 0.5$, and $|y(t_1) - 0.5| = |y(t_2) - 0.5|$ where $g(y(t_1)) = g(y(t_2))$. By solving the differential equation (5) for a certain odd value of k where $g(y(t)) \in (0, g(0.5))$, we can calculate t that $g(y(t))$ gets its maximum value; that is, the first derivative of $y(t)$ becomes maximum. This value of t along with the fact that $g(y(t))$ is symmetric around the axis $y(t) = 0.5$ can be used to calculate the time value at which the sought function attains a specific target value, which can be of use in determining a particular stopping time for the protocol given in [1].

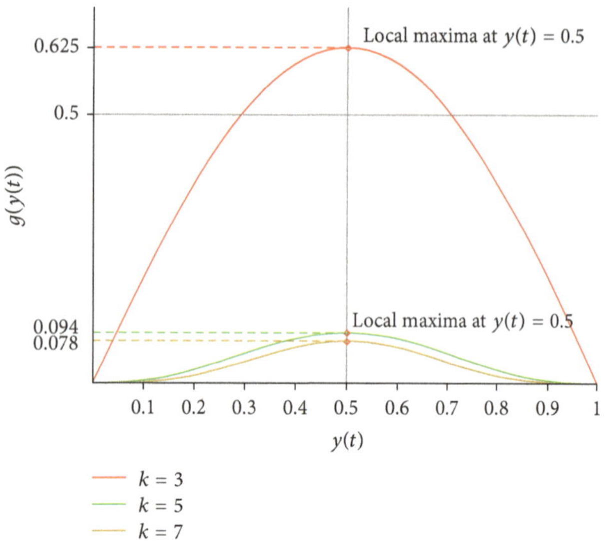

FIGURE 1: A plot of $g(y(t))$ as a function of $y(t)$ for odd values of k.

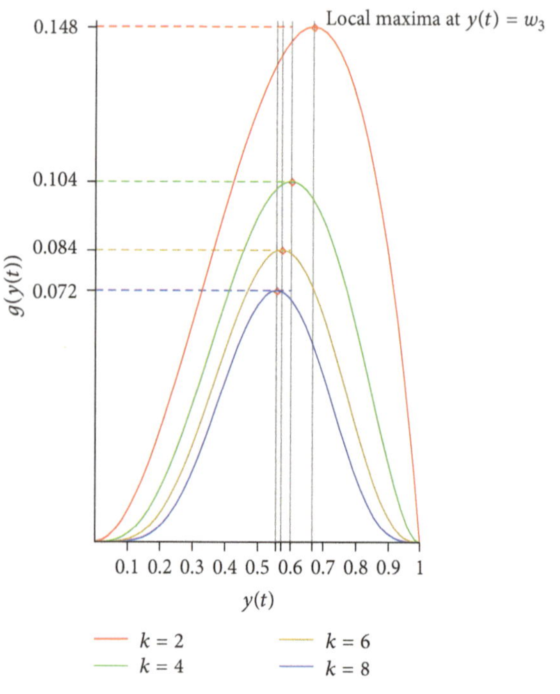

FIGURE 2: A plot of $g(y(t))$ as a function of $y(t)$ for even values of k.

Figure 2 presents some plots of g as a function of $y(t)$ in the same graph for various even values of k. In these plots, the local maxima of g appear at $y(t) = 0.66, 0.6, 0.57$, and 0.55 for $k = 2, 4, 6, 8$ correspondingly. Moreover, as k increases, the local maxima of g are obtained at a $y(t)$ closer to 0.5. But in practice, k takes small values. Moreover, the range $(0, g(w_3))$ where the function g is an increasing function is longer than the range $(g(w_3), 1)$ where the function g is a decreasing function; that is, the values of g are not equally balanced around axis $y(t) = w_3$.

Theorem 10. *For $s \in (0, g(w_3))$, the equation $g(y(t)) = s$ has always one real root $r_1(s)$ within the interval $(0, w_3)$ and one real root $r_2(s)$ within the interval $(w_3, 1)$. In addition, the root $r_1(s)$ is a monotonically increasing function of s, while $r_2(s)$ is a monotonically decreasing function of s.*

Proof. Let $p(y(t)) = g(y(t)) - s$. Since the first derivative of $p(y(t))$ is equal to the first derivative of $g(y(t))$, from Lemma 5, it follows that w_3 is also a root of $dp(y(t))/dt$. By Lemmas 5 and 6 we conclude that the polynomial $p(y(t))$ attains its maximum at w_3. Moreover, it is monotone increasing for $y(t) < w_3$ and monotone decreasing for $y(t) > w_3$.

It holds $p(0) = p(1) = -s < 0$, while $p(w_3) = g(w_3) - s > 0$ since $s \in (0, g(w_3))$. Thus, since $p(y(t))$ as a function of $y(t)$ is continuous on $(0, 1)$ and it has opposite signs at the endpoints of intervals $(0, w_3)$ and $(w_3, 1)$, it has one real solution $r_1(s)$ in the first interval and one real solution $r_2(s)$ in the second one.

Finally, due to the monotonicity of $p(y(t))$ in each of these intervals, the root $r_1(s)$ monotonically increases as s increases, while the root $r_2(s)$ monotonically decreases. □

Lemma 11. *For odd k, it holds $r_1(s) + r_2(s) = 1$.*

Proof. From Lemma 7, $w_3 = 1/2$. Thus $r_1(s) \in (0, 1/2)$ and $r_2(s) \in (1/2, 1)$. Since $p(y(t))$ as a function of $y(t)$ is continuous on $(0, 1)$ and it has opposite signs at the endpoints of intervals $(0, 1/2)$ and $(1/2, 1)$, the roots $r_1(s)$ and $r_2(s)$ can be approximated by the use of the bisection method.

After n iterations of the bisection method on these intervals, the calculated roots $r_1(s)$ and $r_2(s)$ satisfy the identity $r_1(s) + r_2(s) = 1$ since the intervals are balanced, $g(r_1(s)) = g(r_2(s)) = s$, and in each iteration the method operates in one of the two halves of the two intervals.

3. The Solution of the Class of the Differential Equations for Odd k

By the class of differential equations defined by (13), it is easy to see that it has the form $dy(t)/dt = \sum_{i=0}^{k+1} f_i(t)y(t)^i$, with $k > 1$ and $f_i(t), 0 \le i \le k+1$, specific real functions of t. This form appears to generalize the class of *Abel differential equations of the first kind* in that it involves the sought function to powers greater than 3, for $k \ge 3$. To the best of our knowledge, no general solution strategy exists for generalized differential equations of the Abel type that can be exploited to solve (13). In this section we will describe such a solution strategy for odd k that makes use of the result of Lemma 4 and the properties of the polynomial function $g(y(t))$ that have been proved in Section 2.

Definition 12. Let I be an index set, and let $f_i : \mathbb{C} \to \mathbb{C}$ be arbitrary complex functions which do not vanish identically. Assuming $m, n \in \mathbb{N}$ such that $m \ge n$, we define $F_{n,m}(z) : \mathbb{N}^2 \times \mathbb{C} \to \mathbb{C}$ as follows:

$$F_{n,m}(z) = \begin{cases} e^{z} & \text{if } n = 1, \\ e^{f_{m-(n-1)}(z)F_{n-1,m}(z)} & \text{if } n > 1. \end{cases} \quad (15)$$

Definition 13. Let f_i be as in Definition 12 and $z \in \mathbb{C}$. Then we define

$$G\left(f_1(z), f_2(z), \ldots, f_j(z); z\right) = zF_{j+1,j+1}(z). \quad (16)$$

Definition 14. Let f_i be as in Definition 12, G as in Definition 13, and $y \in \mathbb{C}$. Then the function $\mathrm{HW}(\{f_i\}_{i\in I}; y)$ is the function which satisfies the following equation:

$$G\left(\{f_i\}_{i\in I}; \mathrm{HW}\left(\{f_i\}_{i\in I}; y\right)\right) = y. \quad (17)$$

If we set $s \in \mathbb{R}_+$ at the right side of (1), we obtain the following functional equation:

$$(1-t)\frac{dy(t)}{dt} = s. \quad (18)$$

Our approach relies on the following fact for the solution $t(s)$ of the functional equation (18).

Claim 1. Let $r(s)$ be a real function in $\mathbb{R}$, such that each value of t^{-1} belongs to the domain of $r(s)$ for every value $s \in \mathbb{R}_+$. Assuming that $b = t(s)$, then $y(b) = y(t(s)) = r(t^{-1}(b))$.

Lemma 15. *The solution $t(s)$, $s \in \mathbb{R}_+$, of the functional equation*

$$(1-t)\frac{dy(t)}{dt} = s \quad (19)$$

forms a solution of the equation $g(y(t)) = s$.

Proof. Given $s \in \mathbb{R}_+$ and the solution $t(s)$ to (18), it follows from (13) that $y(t(s))$ is a root of the polynomial equation $g(y(t)) = s$.

Theorem 16. *For odd k and $s \in \mathbb{R}_+$, the solution $t(s)$ of the functional equation (18) is equal to*

$$t(s) = 1 + ce^{-\int (1/s)(dr(s)/ds)ds} \quad (20)$$

with $r(s) = (1 \pm \sqrt{1-4h(s)})/2$, where $h(s) = (s/\binom{k-1}{\lceil k/2\rceil-1})^{1/\lceil k/2\rceil}$ and the real positive solution of the equation $g(y(t)) = s$.

Proof. For odd k the exponents of $y(t)$ and $(1-y(t))$ of the polynomial $g(y(t))$ in (7) become equal, that is, $k-\lceil k/2\rceil+1 = \lceil k/2\rceil = l$, and the polynomial $g(y(t))$ could be written as follows:

$$g(y(t)) = \binom{k-1}{\left\lceil \frac{k}{2}\right\rceil - 1} y(t)^l (1-y(t))^l. \quad (21)$$

Based on (21), the polynomial equation $g(y(t)) = s$ becomes

$$y(t)(1-y(t)) = \sqrt[l]{\frac{s}{\binom{k-1}{\lceil k/2\rceil-1}}}. \quad (22)$$

Let $h(s) = \sqrt[l]{s/\binom{k-1}{\lceil k/2\rceil-1}}$, the right-hand side of (22). By Lemmas 5 and 7, we have $s \le \binom{k-1}{\lceil k/2\rceil-1}(1/(2^{2\lceil k/2\rceil}))$ which leads after some algebraic manipulations to $4h(s) \le 1$. This means that the discriminant of the quadratic equation (22) $\Delta = 1-4h(s)$ is positive, and consequently by applying the closed-form formula for the solutions of the quadratic polynomial equation (22), we find that the solutions are the following:

$$r_1(s) = \frac{1+\sqrt{1-4h(s)}}{2}, \qquad r_2(s) = \frac{1-\sqrt{1-4h(s)}}{2}. \quad (23)$$

It is obvious that $r_1(s)+r_2(s) = 1$. Thus, $r_1(s)$ and $r_2(s)$ are real valued functions, where $r_1(s)$ is a monotonically decreasing function with $w_3 \le r_1(s) \le 1$ and $r_2(s)$ is a monotonically increasing function with $0 \le r_2(s) \le w_3$. By Theorem 10 and Lemma 11, the roots $r_1(s)$ and $r_2(s)$ are the appropriate roots, and consequently we set $r(s) = r_2(s)$ for $y(t) \in [0, 0.5]$ and $r(s) = r_1(s)$ for $y(t) \in (0.5, 1]$.

Using the chain differentiation rule, we obtain the following:

$$y(t(s)) = r(s) \Longrightarrow \left(\frac{dy(t)}{dt}\right)_{t=t(s)} \frac{dt(s)}{ds} = \frac{dr(s)}{ds}. \quad (24)$$

However, since $(1-t)(dy(t)/dt) = s$ for $t = t(s)$, it follows that

$$\left(\frac{dy(t)}{dt}\right)_{t=t(s)} = \frac{s}{1-t(s)}. \quad (25)$$

Thus, (24) becomes

$$\left(\frac{dy(t)}{dt}\right)_{t=t(s)} \frac{dt(s)}{ds} = \frac{dr(s)}{ds} \Longrightarrow \frac{s}{1-t(s)}\frac{dt(s)}{ds} = \frac{dr(s)}{ds}. \quad (26)$$

Solving the differential equation in (26), we obtain the solution shown in (20) as required. □

Corollary 17. *Let $a(s) = \sqrt{1-4h(s)}$. Then the solution to (18) is equal to*

$$t(s) = \begin{cases} 1 + c\left(\dfrac{a(s)+1}{a(s)-1}\right)e^{p_1(s)}, & \text{if } r(s) = r_1(s), \\ 1 + c\left(\dfrac{a(s)-1}{a(s)+1}\right)e^{p_2(s)}, & \text{if } r(s) = r_2(s), \end{cases} \quad (27)$$

where $p_1(s)$ and $p_2(s)$ are given by

$$\begin{aligned} p_1(s) &= \sum_{i=2}^{\lceil k/2\rceil} \frac{2^{i-1}\binom{2\lceil k/2\rceil-i-1}{\lceil k/2\rceil-i}}{(i-1)\binom{k-1}{\lceil k/2\rceil-1}} \times \left[\frac{(-1)^{i-1}}{(a(s)-1)^{i-1}} - \frac{1}{(a(s)+1)^{i-1}}\right], \\ p_2(s) &= \sum_{i=2}^{\lceil k/2\rceil} \frac{2^{i-1}\binom{2\lceil k/2\rceil-i-1}{\lceil k/2\rceil-i}}{(i-1)\binom{k-1}{\lceil k/2\rceil-1}} \times \left[\frac{(-1)^{i}}{(a(s)-1)^{i-1}} + \frac{1}{(a(s)+1)^{i-1}}\right]. \end{aligned} \quad (28)$$

Proof. We compute the integral in (20) for $r(s) = r_1(s)$. Let $l = \lceil k/2 \rceil$, $m = \binom{k-1}{\lceil k/2 \rceil - 1}$, and

$$a(s) = \left(1 - 4\left(\frac{s}{m}\right)^{1/l}\right)^{1/2}, \tag{29}$$

$$r(s) = \frac{1}{2} + \frac{1}{2} a(s). \tag{30}$$

After some algebraic manipulations, we obtain the derivative $r'(s)$ of $r(s)$ as follows:

$$r'(s) = \frac{1}{2} a'(s). \tag{31}$$

By (29), we obtain the following expression for s:

$$s = \frac{\left(1 - a(s)^2\right)^l m}{4^l}. \tag{32}$$

Multiplying (31) by $1/s$ and after some algebraic manipulations, we obtain

$$\frac{1}{s} r'(s) = \frac{4^l}{2m} \frac{a'(s)}{\left(1 - a(s)^2\right)^l}. \tag{33}$$

Based on (32) and (33), the integral in (20) becomes

$$\int \frac{1}{s} r'(s) = \frac{(-1)^l 4^l}{2m} \int \frac{a'(s)}{\left(a(s)^2 - 1\right)^l} ds. \tag{34}$$

Setting $a(s) = w$ for convenience, we can form the rational parts expansion of the rational part of the integral in (34) as follows:

$$\begin{aligned} \frac{1}{\left(w^2 - 1\right)^l} &= \frac{1}{(w-1)^l (w+1)^l} \\ &= \sum_{i=1}^{l} \frac{A_i}{(w-1)^i} + \sum_{i=1}^{l} \frac{B_i}{(w+1)^i} \end{aligned} \tag{35}$$

with the coefficients given, by standard calculus techniques, by the following:

$$\begin{aligned} A_i &= \frac{1}{(l-i)!} \left[\frac{d^{(l-i)}}{dw^{(l-i)}} (w-1)^l \frac{1}{(w-1)^l (w+1)^l} \right]_{w=1} \\ &= \frac{1}{(l-i)!} \left[\frac{d^{(l-i)}}{dw^{(l-i)}} \frac{1}{(w+1)^l} \right]_{w=1}, \\ B_i &= \frac{1}{(l-i)!} \left[\frac{d^{(l-i)}}{dw^{(l-i)}} (w+1)^l \frac{1}{(w-1)^l (w+1)^l} \right]_{w=-1} \\ &= \frac{1}{(l-i)!} \left[\frac{d^{(l-i)}}{dw^{(l-i)}} \frac{1}{(w-1)^l} \right]_{w=-1} \end{aligned} \tag{36}$$

for $i = 1, \ldots, l$.

Since

$$\frac{d^{(l-i)}}{dw^{(l-i)}} \frac{1}{(w \pm 1)^l} = (-1)^{l-i} \frac{1}{(w \pm 1)^{2l-i}} \frac{(2l-i-1)!}{(l-1)!}, \tag{37}$$

the coefficients A_i and B_i become

$$\begin{aligned} A_i &= \frac{1}{(l-i)!} (-1)^{l-i} \frac{1}{2^{2l-i}} \frac{(2l-i-1)!}{(l-1)!} \\ &= \binom{2l-i-1}{l-i} (-1)^{l-i} \frac{1}{2^{2l-i}}, \\ B_i &= \frac{1}{(l-i)!} (-1)^{l-i} \frac{1}{(-2)^{2l-i}} \frac{(2l-i-1)!}{(l-1)!} \\ &= \binom{2l-i-1}{l-i} \frac{1}{(-1)^l 2^{2l-i}}. \end{aligned} \tag{38}$$

Based on (34), (35), and (38), the integral in (20) becomes

$$\begin{aligned} &\int \frac{1}{s} r'(s) \\ &= \frac{(-1)^l 4^l}{2m} \int \left[\sum_{i=1}^{l} \frac{a'(s) A_i}{(a(s)-1)^i} + \sum_{i=1}^{l} \frac{a'(s) B_i}{(a(s)+1)^i} \right] \\ &= \frac{\binom{2l-2}{l-1}}{m} \ln\left(\frac{a(s)+1}{a(s)-1}\right) + \frac{(-1)^l 4^l}{2m} \\ &\quad \times \int \left[\sum_{i=2}^{l} \frac{a'(s) A_i}{(a(s)-1)^i} + \sum_{i=2}^{l} \frac{a'(s) B_i}{(a(s)+1)^i} \right] \\ &= \frac{\binom{2l-2}{l-1}}{m} \ln\left(\frac{a(s)+1}{a(s)-1}\right) \\ &\quad + \sum_{i=2}^{l} \left[\frac{2^{i-1} \binom{2l-i-1}{l-i}}{(-1)^i m} \int \frac{a'(s)}{(a(s)-1)^i} \right. \\ &\quad \left. + \frac{2^{i-1} \binom{2l-i-1}{l-i}}{m} \int \frac{a'(s)}{(a(s)+1)^i} \right]. \end{aligned} \tag{39}$$

Since $\binom{2l-2}{l-1} = m$, after some algebraic manipulations (39) is rewritten as follows:

$$\int \frac{1}{s} r'(s) = \ln\left(\frac{a(s)+1}{a(s)-1}\right) + p_1(s), \tag{40}$$

where

$$\begin{aligned} p_1(s) &= \sum_{i=2}^{l} \frac{2^{i-1} \binom{2l-i-1}{l-i}}{(i-1) m} \\ &\quad \times \left[\frac{(-1)^{i-1}}{(a(s)-1)^{i-1}} - \frac{1}{(a(s)+1)^{i-1}} \right]. \end{aligned} \tag{41}$$

For $r(s) = r_2(s)$, the derivative $r'(s)$ of $r(s)$ becomes

$$r'(s) = -\frac{1}{2} a'(s). \tag{42}$$

Applying the previous proof method for $r'(s)$ given by (42), the integral in (20) becomes

$$\int \frac{1}{s} r'(s)\, ds = \ln\left(\frac{a(s)-1}{a(s)+1}\right) + p_4(s), \quad (43)$$

where

$$p_2(s) = \sum_{i=2}^{l} \frac{2^{i-1}\binom{2l-i-1}{l-i}}{(i-1)m} \times \left[\frac{(-1)^i}{(a(s)-1)^{i-1}} + \frac{1}{(a(s)+1)^{i-1}}\right]. \quad (44)$$

Thus, from (20), we obtain the following:

$$t(s) = \begin{cases} 1 + c\left(\dfrac{a(s)+1}{a(s)-1}\right) e^{p_1(s)}, & \text{if } r(s) = r_1(s), \\ 1 + c\left(\dfrac{a(s)-1}{a(s)+1}\right) e^{p_2(s)}, & \text{if } r(s) = r_2(s), \end{cases} \quad (45)$$

completing the proof of the theorem.

We now need to invert the function $t(s)$ or, equivalently, to solve (27), with unknown s. By setting $z = 1 + a(s)$, (27) for $r(s) = r_2(s)$ after some algebraic manipulations becomes

$$z = \frac{c(z-2)}{t(s)-1} e^{Q(z)}, \quad (46)$$

where $Q(z)$ is the rational function of z given by

$$Q(z) = \sum_{i=2}^{l} \frac{2^{i-1}\binom{2l-i-1}{l-i}}{(i-1)m} \times \frac{(-1)^i z^{i-1} + (z-2)^{i-1}}{z^{i-1}(z-2)^{i-1}}. \quad (47)$$

With regard to the constant c, its value may be obtained from $x(0)$ as follows. Since $y(t) = x(t)/(1-t)$, then $y(0) = x(0)$. Setting $t = 0$ in $g(y(t)) = s$, we obtain the value for s. Thus, using this value of s, we have established the relation $t(s) = 0$, and we can compute $a(s)$ and z. Now we can solve (46) for c.

Setting

$$f(z) = \frac{\ln(c(2-z)/(1-t(s)))}{e^z} + \frac{Q(z)}{e^z}, \quad (48)$$

we rewrite (46) as follows:

$$z = e^{f(z)e^z}. \quad (49)$$

Based on Definitions 12, 13, and 14 for $j = 1$, we can rewrite (49) as follows:

$$\begin{aligned} z &= e^{f(z)e^z} \\ &\Longrightarrow z e^{-f(z)e^z} = 1 \\ &\Longrightarrow G(-f(z); z) = 1 \\ &\Longrightarrow z = \mathrm{HW}(-f(z); 1). \end{aligned} \quad (50)$$

Let $l = \lceil k/2 \rceil$ and $m = \binom{k-1}{\lceil k/2 \rceil - 1}$. Since $z = 1 + a(s)$ by substituting $a(s)$ and after some algebraic manipulations, we obtain the following formula for s:

$$\begin{aligned} s &= \frac{m z^l (z-2)^l}{4^l} \\ &= \frac{m \mathrm{HW}(-f(z);1)^l (\mathrm{HW}(-f(z);1) - 2)^l}{4^l}. \end{aligned} \quad (51)$$

Equation (51) gives values of s for specific values of t through Hyper-Lambert functions. Specifically, for a certain value of t, the function $f(z)$ takes a particular form, and through (51), we obtain a value for s. Thus, we have found $t^{-1}(s)$. In [6], (51) has been derived for $k = 3$.

Based on Theorem 16, Lemma 15, (50), and that $r(s) = z/2$, a solution of the differential equation (13) can be obtained by

$$y(t(s)) = r(s) = \frac{\mathrm{HW}(-f(z);1)}{2}. \quad (52)$$

Equivalently, we can obtain the solution by

$$y(t(s)) = r\left(\frac{m \mathrm{HW}(-f(z);1)^l (\mathrm{HW}(-f(z);1) - 2)^l}{4^l}\right) \quad (53)$$

with $r(s) = r_2(s)$.

We will now demonstrate the previous methodology for $k = 3$. Since $l = \lceil k/2 \rceil = 2$ and $m = \binom{k-1}{\lceil k/2 \rceil - 1} = 2$, (21) becomes

$$\begin{aligned} g(y(t)) &= 2y(t)^2(1-y(t))^2 \\ &= 2y(t)^4 - 4y(t)^3 + 2y(t)^2. \end{aligned} \quad (54)$$

From Theorem 16, the real positive solutions $r_1(s)$ and $r_2(s)$ of $g(y(t)) = s$ for $s \le (1/8)$ can be written as follows:

$$r_1(s) = \frac{1 + \sqrt{1 - 2\sqrt{2s}}}{2}, \qquad r_2(s) = \frac{1 - \sqrt{1 - 2\sqrt{2s}}}{2}. \quad (55)$$

Moreover, the rational function $Q(z)$ of z involved in (46) becomes, after some algebraic manipulations, as follows:

$$Q(z) = \frac{2(z-1)}{z(z-2)}. \quad (56)$$

Thus, the function $f(z)$ defined by (48) becomes

$$f(z) = \frac{\ln(c(2-z)/(1-t(s)))}{e^z} + \frac{2(z-1)}{z(z-2)e^z}. \quad (57)$$

Finally, the formula for s defined by (51) and the solution of the differential equation (54) defined by (53) can be written as follows:

$$s = \frac{z^2 (z-2)^2}{8} = \frac{\mathrm{HW}(-f(z);1)^2(\mathrm{HW}(-f(z);1)-2)^2}{8}.$$

$$y(t(s)) = r\left(\frac{\mathrm{HW}(-f(z);1)^2(\mathrm{HW}(-f(z);1)-2)^2}{8}\right), \quad (58)$$

where $r(\cdot) = r_2(s)$.

4. Nonclosed-Form Solvability Results for Even Values of k

In this section, we consider the case where the parameter k is an even integer greater than 2, that is, $k = 2m$, with $m \geq 2$. This case is of distinct interest because it does not seem to be amenable to the methodology we have followed for k odd. In Section 3 where we considered odd values of k, a key step in linking the solution of the differential equation with the class of Hyper-Lambert functions was to find the *closed-form* solution $y(t)$, given s, of the polynomial equation $g(y(t)) = s$, with $g(y(t))$ given by (21). This closed-form solution was needed in Lemma 15 and Theorem 16. When we attempted to apply the same methodology for even values of $k \geq 4$, we were not able to find a closed-form solution of the equation $g(y(t)) = s$, even for fixed value of k (e.g., $k = 4$). Thus, we focused our attention on investigating the possibility that the equation does not have a closed-form solution. As it turned out, this was indeed the case, at least for values of k such that $k+1$ is prime and certain ranges of s. We believe that this result may be of independent interest.

Before proving it, we will state three fundamental theorems of Algebra.

Theorem 18 (Eisenstein's irreducibility criterion). *Let q be a prime number, and let $f(x) = a_n x^n + \cdots + a_1 x^1 + a_0$ be a polynomial of degree n with integer coefficients satisfying the following properties:*

(i) *q does not divide a_n;*

(ii) *q divides $a_0, a_1, \ldots, a_{n-1}$;*

(iii) *q^2 does not divide a_0.*

Then the polynomial f is irreducible over the field of $\mathbb{Q}$ of rational numbers.

Theorem 19. *Let K be a subfield of $\mathbb{R}$, and let $f \in K[X]$ be an irreducible polynomial of prime degree $n > 2$. If the Galois group of f over K is solvable, f has either exactly one root in $\mathbb{R}$ or all its roots in $\mathbb{R}$.*

Theorem 20. *Let f be a polynomial with coefficients in a field K of characteristic zero. Suppose that f is solvable by radicals. Then the Galois group of $\Gamma_K(f)$ of f is a solvable group.*

We now state and prove the main result, concerning the unsolvability, in closed form, of the polynomial equation $g(y(t)) = s$.

Theorem 21. *The equation $g(y(t)) = s$ is not solvable by radicals over $\mathbb{Q}$ for $k = 2m$ with $m \geq 2$, $n = k+1$ a prime, and $s = l/b$ a rational number, given in its lower terms (i.e., $\gcd(l,b) = 1$), such that $s \in (0, g(w_3))$, $lb \neq 0$, and b is a square free integer.*

Proof. Let $s = l/b \in \mathbb{Q}$, with $lb \neq 0$. Then, our target polynomial is $p(y(t)) = g(y(t)) - (l/b)$. We will, first, show that $p(y(t))$ is irreducible over the set of rational numbers $\mathbb{Q}$ using Theorem 18. We will work, in the following, with the polynomial $bp(y(t))$ in order to have a polynomial with integer coefficients, as required by the statement of Theorem 18. The degree of this polynomial according to Theorem 8 is $n = k+1$, which has been assumed to be a prime.

We will show, however, that Theorem 18 is not, directly, applicable to $bp(y(t))$ since the first two conditions of the theorem are not met. The coefficients a_i, $i = 0, \ldots, n$ in Theorem 18 are as follows:

$$\begin{gathered} a_i = b\binom{2m-1}{m-1}\binom{m}{i-(m+1)}(-1)^{i-(m+1)}, \\ i = m+1, \ldots, \quad n = 2m+1, \\ a_i = 0, \quad i = 1, \ldots, m, \\ a_0 = -l. \end{gathered} \quad (59)$$

Since $|a_n| = a_{m+1} = b\binom{2m-1}{m-1}$, the first two conditions of Theorem 18 do not hold. Since 0 is not a root of $p(y(t))$, we can set $y(t) \rightarrow 1/x$, in order to consider the polynomial $f(x) = bp(1/x)$, instead of $bp(y(t))$. Then each nonzero root x_0 of the polynomial $f(x)$ corresponds to a nonzero root y_0 of $p(y(t))$ and vice versa by the correspondence $y_0 \leftrightarrow 1/x_0$. Consequently, $p(y(t))$ is solved by radicals if and only if $f(x)$ is solved by radicals.

Thus, from now on we will consider the polynomial $f(x) = bp(1/x)$. In order to have the inverse of $1/x$, we will use the polynomial $h(x) = x^{2m+1} f(x)$. We will apply Theorem 18 to this polynomial because if it applies to it, then $f(x)$ will be irreducible, as well as, in turn, $p(y(t))$.

After some algebraic manipulations, the polynomial $h(x) = x^{2m+1} f(x)$ can be written as follows:

$$h(x) = b\binom{2m-1}{m-1}\sum_{i=0}^{m}\binom{m}{i}x^{m-i}(-1)^i - lx^{2m+1}. \quad (60)$$

Now the coefficients a_i, $i = 0, \ldots, n = 2m+1$ of the polynomial $h(x)$ in (60) are as follows:

$$\begin{gathered} a_{2m+1} = -l, \\ a_j = 0, \quad j = m+1, \ldots, 2m, \\ a_j = b\binom{2m-1}{m-1}\binom{m}{m-j}(-1)^{m-j}, \quad j = 0, \ldots, m. \end{gathered} \quad (61)$$

In this way, we have decoupled the coefficient $a_n = a_{2m+1}$ from all the rest so as to be feasible to satisfy the first two conditions of Theorem 18. Since, by assumption, $\gcd(l, b) = 1$, in order to satisfy the first two conditions, it suffices to choose as q any prime divisor of b. Since b is, by assumption, a *square free* integer, then the third condition is satisfied too. Thus, Theorem 18 applies to $h(x)$ in (60) and, thus, to $p(y(t))$. Therefore, $p(y(t))$ is irreducible over $\mathbb{Q}$.

We will, now, prove that the number of real roots of $h(x)$ is neither equal to 1 nor its degree, which is $n = k + 1$. From Theorem 10, since we have assumed that $s \in (0, g(w_3))$, it follows that $g(y(t))$ has at least two roots and, thus, $h(x)$ has also at least two roots because of the relationship between their roots. We will prove that not all of its roots are real. Let us assume, towards a contradiction, that $h(x)$ has $2m + 1$ real roots. We can differentiate $h(x)$ m times to obtain its mth order derivative, which is equal to $l((2m+1)!/m!)x^{m+1} + m!b\binom{2m-1}{m-1}$. If $h(x)$ had $2m + 1$ real roots, then its mth order derivative would have, at least, $2m+1-m = m+1 \geq 3$ (since $m \geq 2$, by assumption) real roots.

The roots of $l((2m + 1)!/m!)x^{m+1} + m!b\binom{2m-1}{m-1}$ are the $(m + 1)$th roots of $-B/A$, with $A = l((2m + 1)!/m!)$ and $B = m!b\binom{2m-1}{m-1}$. These roots have the form $|-B/A|^{1/(m+1)}$ $(\cos(\pm\pi + 2d\pi)/(m + 1) + i\sin(\pm\pi + 2d\pi)/(m + 1))$ with $d = 0, \ldots, m$ ($\pm\pi$ is π if $-B/A$ is positive and $-\pi$ if $-B/A$ is negative).

A root is real if and only if the imaginary part is 0; that is, $(\pm\pi + 2d\pi)/(m + 1) = \pi((2d \pm 1)/(m + 1))$ is an integer multiple λ of π. Since $d \leq m$, we only need to consider $\lambda = 1$. Solving $(2d \pm 1)/(m+1) = 1$, we obtain $d = (m+1\mp 1)/2$. It is easy to see that this gives us at most two values for d: (i) two values for even m: $m/2$ (for the "−" sign in $(2d \pm 1)/(m + 1)$) and $(m+2)/2$ (for the "+" sign in $(2d \pm 1)/(m+1)$) and (ii) no such value for odd m. Thus, the mth order derivative of $h(x)$ has at most 2 real roots for even m or no real roots for odd m, while according to the assumption that $h(x)$ has $2m + 1$ real roots, it should have at least $m + 1 \geq 3$ real roots, which is a contradiction.

Therefore, either there are no real roots of $h(x)$, or the number of real roots of $h(x)$ is more than 1 but less than its degree $n = 2m + 1$. Thus, from Theorems 19 and 20, it follows that $h(x)$ is not solvable by radicals over $\mathbb{Q}$. Therefore, $p(y(t))$ is not solvable by radicals over $\mathbb{Q}$ too.

5. Conclusions and Future Work

In this paper we studied the properties of the solution of the differential equation

$$(1-t)\frac{dy(t)}{dt} = \binom{k-1}{\left\lceil \frac{k}{2} \right\rceil - 1} y(t)^{k-\lceil k/2\rceil+1}(1-y(t))^{\lceil k/2\rceil}. \tag{62}$$

This differential equation arose out of the theoretical analysis of a security-related bit-agreement protocol (see [1]) and does not appear to have been documented elsewhere. It can be seen as a subclass of a natural generalization of the well-known class of Abel differential equations of the first kind.

In this paper we showed that, for odd k, the solution of this equation can be expressed in closed form through the roots of the polynomial on its right-hand side. Moreover, the resulting expression can be transformed into a closed-form expression involving the generalized Hyper-Lambert functions that were proposed in [3] as a generalization of the well-known Lambert W function.

For even values of k it was not possible to follow the same approach since, as we showed, no closed-form expression exists for the roots of the polynomial, at least when $k + 1$ is prime. One possible direction for future research is to extend the proof to cover all value of k or show that closed-form solutions exist for nonprime k.

Another, more interesting direction would be to see whether the more general class given in (2) can be attacked using the strategy proposed in this paper, that is, based on deriving closed-form solutions of the polynomial on the right-hand side and linking them with the solution of the differential equation, at least for certain polynomials f_i. There seems to be very scarce research work on the properties of (2), and any result towards the determination of its solution or, at least, the properties of its properties would be of high importance.

References

[1] E. Makri and Y. C. Stamatiou, "An interactive, similarity increasing algorithm for random strings with applications to key agreement in ad hoc networks," *Studies in Applied Mathematics*, vol. 121, no. 2, pp. 141–155, 2008.

[2] P. E. Nastou, Y. C. Stamatiou, and A. Tsiakalos, "Solving a class of ODEs arising in the analysis of a computer security process using generalized Hyper-Lambert Functions," *International Journal of Applied Mathematics and Computation*, vol. 4, no. 3, pp. 285–294, 2012.

[3] I. N. Galidakis, "On some applications of the generalized hyper-Lambert functions," *Complex Variables and Elliptic Equations*, vol. 52, no. 12, pp. 1101–1119, 2007.

[4] R. M. Corless, G. H. Gonnet, D. E. G. Hare, D. J. Jeffrey, and D. E. Knuth, "On the lambert W function," *Advances in Computational Mathematics*, vol. 5, no. 4, pp. 329–359, 1996.

[5] G. M. Murphy, *Ordinary Differential Equations and Their Solutions*, D. Van Nostrand, 1960.

[6] P. E. Nastou, Y. C. Stamatiou, and A. Tsiakalos, "The solution of dierential equation describing the evolution of a key agreement protocol," in *EUROSIAM*, Montreux, Switzerland, 2011.

19

Further Stability Analysis on Neutral Systems with Actuator Saturation and Time-Delays

Xinghua Liu

Department of Automation, School of Information Science and Technology, University of Science and Technology of China, Anhui 230027, China

Correspondence should be addressed to Xinghua Liu; salxh@mail.ustc.edu.cn

Academic Editor: Kartik Ariyur

This paper is concerned with the asymptotic stability analysis for a class of neutral systems with time-delay and saturating actuators, which is further to reduce the conservatism of neutral system. Based on the model transformation and the delay-dividing approach, a new type of augmented Lyapunov functional is constructed, which has fully exploited the information on the lower bound of the delay. Then the delay-dependent conditions for asymptotic stability are derived by applying some integral inequalities and Lyapunov theory. Finally, numerical examples are provided to show that the proposed results significantly improve the allowed upper bounds of the delay size over some existing ones in the literature.

1. Introduction

As is well known, time-delay is often encountered in various mechanical and engineering systems, such as aircraft stabilization, manual control, microwave oscillator, models of lasers, rolling mill systems, and neural network. Since it is shown that the existence of delays may result in instability, oscillations, or poor performances, the problem of stability analysis of time-delay systems has received considerable attention over the decades. Various approaches have been proposed in the literature to obtain stability criteria for delay systems, such as the Lyapunov technique, characteristic equation method, and state solution approaches. In the real world, there are many practical systems containing time-delay of neutral type, that is, the derivative of the past state; we called them neutral systems. Such systems can be found in population ecology [1], processes including steam or water pipes, heat exchanges [2], and other areas [3]. Existing stability criteria of delay systems can be classified into two categories, that is, delay-independent ones and delay-dependent ones. It is generally recognized that delay-independent ones are more conservative than the delay-dependent ones. So, in the literature of these years, the attention was mainly focused in giving delay-dependent stability conditions for neutral systems; see, for example, [4–17] and references therein for more details. However, there is some room for further investigation to reduce the conservatism of neutral systems by novel method.

Besides delays, the saturated controller is apt to cause instability as well. In various practical systems, actuator saturation and time-delays are often observed together in control systems and they are frequently a source of instability. Therefore, much attention has been paid to derive some delay-dependent stability conditions for the time-delay systems with actuator saturation. Based on the Lyapunov-Krasovskii functional, the sufficient conditions have been discussed in [18–23] and the references therein. The controller is constructed in terms of the solution to a set of linear matrix inequalities by employing descriptor model transformation of the systems in [21]. The Lyapunov-Krasovskii functional is employed to investigate the problem of delay-dependent robust stabilization of uncertain neutral systems with saturating actuators in [20], just to name a few. However, to the author's knowledge, it seems that the stability and control synthesis problem of neutral systems in the presence of actuator saturation has not been fully investigated and it is very challenging.

This paper is concerned with the asymptotic stability problem of neutral systems with actuator saturation and the objective is to propose some new delay-dependent conditions, which are less conservative than some existing ones in the literature. The design approach is dependent on the discrete time-delay and neutral time-delay. Applying the delay-dividing approach, the asymptotic stability conditions are formulated as linear matrix inequalities (LMIs) by the model transformation and Lyapunov method. Facing the uncertain case, we obtain the robust asymptotic stability conditions under such hypothesis condition that the system has the norm bounded uncertainties. At last, examples are provided to demonstrate the reduced conservativeness of the proposed criteria by numerically comparing with those reported previously in the literature. The main contributions of this paper can be summarized as follows: (i) model transformation is considered in this paper, but we employ a new stability condition of the transformation operator $\mathscr{L}$ to reduce the conservatism; (ii) the bound of the delay is fully utilized in this paper, that is, improved bounding technique or delay-dividing approach is used to reduce the conservatism; (iii) the constructed Lyapunov functional is novel and very effective in the reduction of conservatism, which has not appeared in the context of neutral systems with time-delay and actuator saturation before; (iv) numerical examples and simulations are provided to verify that the proposed result in this paper is less conservative than some existing ones.

The remainder of this paper is organized as follows: Section 2 contains the problem statement and preliminaries; Section 3 presents the main results; Section 4 provides some numerical examples to verify the effectiveness of the results; Section 5 draws a brief conclusion.

1.1. Notations. In this paper, $\star$ denotes the elements below the main diagonal of a symmetric block matrix, $\mathbb{R}^n$ denotes the n dimensional Euclidean space, and $\mathbb{R}^{m\times n}$ is for the set of all real $m \times n$ matrices. I denotes the identity matrix with proper dimensions. C_0 is the set of all continuous functions from $[-\tau, 0]$ to $\mathbb{R}^n$ and τ is constant neutral time-delay. A^T is the transpose of matrix A. $|A| = [|a_{ij}|]$, with $A = [a_{ij}]$. For real symmetric matrices X and Y, the notation $X \geq Y$ (resp., $X > Y$) means that the matrix X-Y is positive semidefinite (resp., positive definite). $\lambda_{\max(\min)}(A)$ means the eigenvalue of matrix A with maximum (minimum) real part. $\|v\|$ means the Euclidean norm of vector v, $\|v\| = (v^T v)^{1/2}$, while $\|A\|$ is spectral norm of matrix A, $\|A\| = [\lambda_{\max}(A^T A)]^{1/2}$. diag$\{\cdots\}$ denotes a block-diagonal matrix decided by the corresponding elements in the brace. In addition, if not explicitly stated, matrices are assumed to have compatible dimensions.

2. Problem Statement and Preliminaries

In this paper, the investigated neutral systems with time-delay and actuator saturation are described by the following:

$$\dot{x}(t) - D\dot{x}(t-\tau) = Ax(t) + Bx(t-h) + C\mathrm{Sat}(u(t)), \qquad x(s) = \phi(s), \quad s \in [-\rho, 0], \tag{1}$$

where $x(t) \in \mathbb{R}^n$ is the system state and $u(t)$ is the control input. $h > 0$ is a constant discrete time-delay and $\tau > 0$ is a constant neutral time-delay. A, B, C, and D are known real constant parameter matrices of appropriate dimensions. To apply the Lyapunov stability theorem [2] we assume that $\|D\| < 1$. The initial condition $\phi(s)$ is a continuously differentiable vector-valued function. The continuous norm of $\phi(s)$ is defined as

$$\|\phi\|_c = \max_{s\in[-\rho,0]} |\phi(s)|, \quad \rho = \max\{\tau, h\}. \tag{2}$$

Linear state feedback is described by the following:

$$u(t) = Kx(t) \tag{3}$$

the linear state feedback gain K is denoted as

$$K = \left[K_1^T, K_2^T, \ldots, K_m^T\right]^T, \qquad K_i\ (i = 1, 2, \ldots, m) \quad \text{is an } n\text{-dimension row vector.} \tag{4}$$

The notation Sat$(\cdot)$ is to denote the standard saturation function (Figure 1) defined as $u = [u_1, u_2, \ldots, u_m]$:

$$\mathrm{Sat}(u) = \left[\mathrm{Sat}(u_1), \mathrm{Sat}(u_2), \ldots, \mathrm{Sat}(u_m)\right]^T,$$
$$\mathrm{Sat}(u_i) = \begin{cases} u_i^+ & \text{if } u_i > u_i^+, \\ u_i & \text{otherwise}, \\ -u_i^+ & \text{if } u_i < -u_i^+. \end{cases} \tag{5}$$

Here we have slightly abused the notation by using Sat to denote both a scalar valued and a vector valued function. We have also assumed a unity saturation level for the saturation function without loss of generality. Let us decompose the saturation nonlinear function as the sum of a linear part and another nonlinear part.

Definition 1. Nonlinear part $Dz(u(t)) = u(t) - \mathrm{Sat}(u(t))$, where $Dz(u(t)) = [dz(u_1), dz(u_2), \ldots, dz(u_m)]^T$

$$dz(u_i) = \begin{cases} u_i - u_i^+ & \text{if } u_i > u_i^+, \\ 0 & \text{otherwise}, \\ u_i + u_i^+ & \text{if } u_i < -u_i^+. \end{cases} \tag{6}$$

Making use of nonlinear part $Dz(u(t))$, we rewrite the saturated system in the following form:

$$\dot{x}(t) - D\dot{x}(t-\tau) = A_c x(t) + Bx(t-h) - CDz(Kx(t)), \tag{7}$$

where $A_c = A + CK$.

The purpose of this paper is to investigate the sensitivity of stability to variations in the parameters and in the delays. Notice that the stability of a neutral system requires a necessary condition to guarantee the stability of the associated difference operator [24]. Under normal circumstances, we deal with the difference operator given by

$$\mathscr{D}(x(t)) = x(t) - Dx(t-\tau). \tag{8}$$

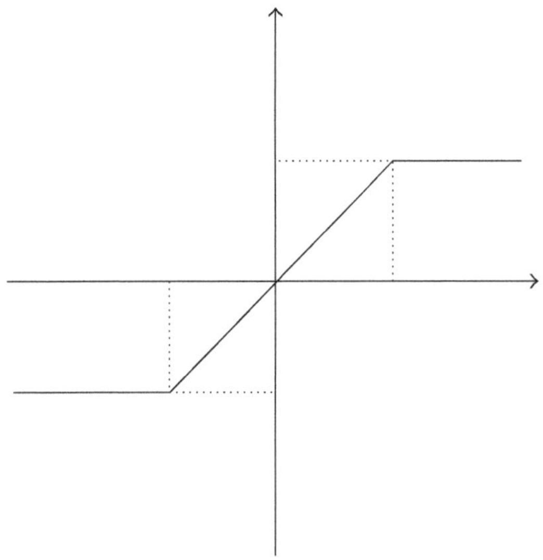

Figure 1: A saturation function $\text{Sat}(u_i(t))$.

Then the stability of operator $\mathscr{D}$ is ensured if $\|D\| < 1$. However, in this paper we introduce a model transformation, which imply the stability of the original model and can be studied by means of Lyapunov-Krasovskii functional.

Utilizing the model transformation, the neutral system (7) can be written as

$$\frac{d}{dt}[\mathscr{L}(x(t))] = \widehat{A}_c x(t) - CDz(Kx), \quad (9)$$

where $\widehat{A}_c = A_c + B = A + B + CK$, $\mathscr{L}(x(t)) = x(t) - Dx(t - \tau) + B\int_{t-h}^{t} x(s)ds$.

Remark 2. The proposed model transformation which uses Leibnitz's rule does not induce any additional dynamics in the characteristic equation for the system without uncertainties (see more details [25, 26]).

It is not difficult to verify that every solution of the neutral system (7) is also solution of (9); then the stability of (9) implies the stability of (7); see [27]. However, the stability of operator $\mathscr{L}$ is required, and it is more difficult to tackle than the untransformed system. In order to give the less conservative stability condition, we introduce Definition 3 and Lemma 4.

Definition 3 (see [24]). The operator $\mathscr{L} : C_0 \rightarrow \mathbb{R}^n$ is said to be stable if the zero solution of the homogeneous difference equation

$$\begin{aligned} &\mathscr{L}(x(t)) = 0, \quad t \geq 0, \\ &x_0 = \psi \in \{\varphi \in C_0 : \mathscr{L}\varphi = 0\} \end{aligned} \quad (10)$$

is uniformly asymptotically stable. The stability of operator $\mathscr{L}$ is necessary for the stability of neutral system (1) with (3), which is always satisfied when $\|h|B| + D\| < 1$.

Lemma 4 (see [28]). *For given positive scalars α_1 and α_2, the operator $\mathscr{L} : C_0 \rightarrow \mathbb{R}^n$ defined by*

$$\mathscr{L}(x(t)) = x(t) + B\int_{t-h}^{t} x(s)\,ds - Dx(t - \tau) \quad (11)$$

is stable if there exists a positive-definite matrix M such that

$$\begin{aligned} &(i) \quad \alpha_1 + \alpha_2 < 1, \\ &(ii) \quad \begin{bmatrix} D^T M D - \alpha_1 M & hD^T M B \\ \star & h^2 B^T M B - \alpha_2 M \end{bmatrix} < 0. \end{aligned} \quad (12)$$

Remark 5. As said in [28], the above criterion (12) is less conservative than the well-known criterion for the stability of the operator $\mathscr{L}$ given in [24]. So the criterion (12) is exploited in this paper to reduce the conservatism of the neutral system with time-delay and actuator saturation.

Then we give the following two lemmas before proceeding main results.

Lemma 6 (see [29]). *Let $U \in R^{n\times n}$ and $V \in R^{n\times n}$, and let $x \in R^n$. Then we have*

$$\begin{aligned} &x^T UVx \leq \|UV\|\, x^T x \leq \|U\|\,\|V\|\, x^T x, \\ &2x^T UVx \leq x^T\left(\frac{1}{\gamma}UU^T + \gamma V^T V\right)x, \quad \forall \gamma > 0. \end{aligned} \quad (13)$$

Considering a matrix $G \in \mathbb{R}^{m\times n}$ and defining the following polyhedral set

$$\mathcal{S} \triangleq \{x \in \mathbb{R}^n; |(K_i - G_i)x| \leq u_i^+, i = 1, 2, \ldots, m\}, \quad (14)$$

where G_i is the row vector of matrix G, the following Lemma 7 concerning the nonlinearity $Dz(Kx(t))$ can be stated.

Lemma 7 (see [30]). *Consider the function $Dz(Kx)$ defined in (6). If $x \in \mathcal{S}$ then the relation*

$$Dz^T(Kx)\Lambda[Dz(Kx) - Gx] \leq 0 \quad (15)$$

is verified for any diagonal and positive definite matrix $\Lambda \in \mathbb{R}^{m\times m}$.

3. Main Results

In this section, we present the stability conditions which guarantee the asymptotic stability of the neutral system with time-delay and actuator saturation. Firstly, we deduce the sufficient asymptotic stability condition for (1) and (3). Then we manage to extend the sufficient conditions to the uncertain case.

3.1. Asymptotic Stability for Certain Neutral System

Theorem 8. *Given scalars α_1 and α_2, a matrix G, and a diagonal matrix Λ of appropriate dimensions, the neutral system with time-delay and actuator saturation which can* *be described by* (1) *and* (3) *is asymptotically stable if there exists a positive definite matrix* M *satisfying* (12) *and there exist* $P > 0$, $Q = [Q_{ij}]_{2\times 2}$, $Q_{ij} > 0$, $R = [R_{ij}]_{2\times 2}$, $R_{ij} > 0$, $T_i > 0$, $(i = 1,2,3,4,5,6)$, $\widetilde{T}_i > 0$, $(i = 3,4)$ *such that the following symmetric linear matrix inequality holds:*

$$\Phi = \begin{bmatrix} \Phi_{11} & \Phi_{12} & 0 & \Phi_{14} & \Phi_{15} & \Phi_{16} & 0 & \Phi_{18} \\ \star & \Phi_{22} & \Phi_{23} & 0 & 0 & 0 & 0 & 0 \\ \star & \star & \Phi_{33} & 0 & 0 & 0 & 0 & 0 \\ \star & \star & \star & \Phi_{44} & \Phi_{45} & 0 & 0 & 0 \\ \star & \star & \star & \star & \Phi_{55} & 0 & 0 & \Phi_{58} \\ \star & \star & \star & \star & \star & \Phi_{66} & 0 & \Phi_{68} \\ \star & \star & \star & \star & \star & \star & \Phi_{77} & 0 \\ \star & \star & \star & \star & \star & \star & \star & \Phi_{88} \end{bmatrix} < 0, \tag{16}$$

where $\widehat{G} = \Lambda G$, $\widetilde{T}_3 = T_3T_5^{-1}T_3$, *and* $\widetilde{T}_4 = T_4T_6^{-1}T_4$

$$\begin{aligned} \Phi_{11} &= \widehat{A}_c^T P + P\widehat{A}_c + Q_{11} + R_{11} + T_1 + T_2 \\ &\quad + \frac{h^2}{2}\left(\widetilde{T}_3 + T_5\right) + \frac{\tau^2}{2}\left(\widetilde{T}_4 + T_6\right), \\ \Phi_{22} &= Q_{22} - Q_{11}, \\ \Phi_{33} &= -Q_{22} - T_1, \\ \Phi_{44} &= R_{22} - R_{11}, \\ \Phi_{55} &= -R_{22} - T_2, \\ \Phi_{66} &= -T_3, \\ \Phi_{77} &= -T_4, \\ \Phi_{88} &= -2\Lambda, \\ \Phi_{12} &= Q_{12}, \\ \Phi_{14} &= R_{12}, \\ \Phi_{15} &= -\widehat{A}_c^T PD, \\ \Phi_{16} &= \widehat{A}_c^T PB, \\ \Phi_{18} &= -PC + \widehat{G}, \\ \Phi_{23} &= -Q_{12}, \\ \Phi_{45} &= -R_{12}, \\ \Phi_{58} &= D^T PC, \\ \Phi_{68} &= -B^T PC. \end{aligned} \tag{17}$$

Proof. Define a legitimate Lyapunov functional candidate as

$$\begin{aligned} V(x(t)) &= V_1(x(t)) + V_2(x(t)) + V_3(x(t)) \\ &\quad + V_4(x(t)) + V_5(x(t)) + V_6(x(t)), \end{aligned} \tag{18}$$

where

$$\begin{aligned} V_1(x(t)) &= \mathscr{L}^T(x(t))\, P\mathscr{L}(x(t)), \\ V_2(x(t)) &= \int_{t-(h/2)}^{t} \begin{bmatrix} x(s) \\ x\left(s-\frac{h}{2}\right) \end{bmatrix}^T \begin{bmatrix} Q_{11} & Q_{12} \\ \star & Q_{22} \end{bmatrix} \times \begin{bmatrix} x(s) \\ x\left(s-\frac{h}{2}\right) \end{bmatrix} ds \\ &\quad + \int_{t-(\tau/2)}^{t} \begin{bmatrix} x(s) \\ x\left(s-\frac{\tau}{2}\right) \end{bmatrix}^T \begin{bmatrix} R_{11} & R_{12} \\ \star & R_{22} \end{bmatrix} \times \begin{bmatrix} x(s) \\ x\left(s-\frac{\tau}{2}\right) \end{bmatrix} ds, \\ V_3(x(t)) &= \int_{t-h}^{t} x^T(s)\, T_1 x(s)\, ds + \int_{t-\tau}^{t} x^T(s)\, T_2 x(s)\, ds, \\ V_4(x(t)) &= \int_{t-h}^{t} \left[\int_s^t x^T(s_1)\, ds_1\right] T_3 \left[\int_s^t x(s_2)\, ds_2\right] ds, \\ V_5(x(t)) &= \int_{t-\tau}^{t} \left[\int_s^t x^T(s_1)\, ds_1\right] T_4 \left[\int_s^t x(s_2)\, ds_2\right] ds, \\ V_6(x(t)) &= \frac{1}{2}\int_{t-h}^{t} (s-t+h)^2 x^T(s)\, T_5 x(s)\, ds \\ &\quad + \frac{1}{2}\int_{t-\tau}^{t} (s-t+\tau)^2 x^T(s)\, T_6 x(s)\, ds, \end{aligned} \tag{19}$$

where $\mathscr{L}(x(t))$ has been defined in (9), $P > 0$, $Q = [Q_{ij}]_{2\times 2}$, $Q_{ij} > 0$, $R = [R_{ij}]_{2\times 2}$, $R_{ij} > 0$, and $T_i > 0$, $i = 1,2,3,4,5,6$.

We solve the time derivative of $V(x(t))$ along the trajectories of the closed-loop system (9) and obtain the following:

$$\begin{aligned} \dot{V}(x(t)) &= \dot{V}_1(x(t)) + \dot{V}_2(x(t)) + \dot{V}_3(x(t)) \\ &\quad + \dot{V}_4(x(t)) + \dot{V}_5(x(t)) + \dot{V}_6(x(t)). \end{aligned} \tag{20}$$

Given that $x(t) \in \mathscr{S}$, which has been discussed in [21]. From Lemma 7, it follows that

$$\begin{aligned} \dot{V}(x(t)) &\le \dot{V}_1(x(t)) + \dot{V}_2(x(t)) + \dot{V}_3(x(t)) + \dot{V}_4(x(t)) \\ &\quad + \dot{V}_5(x(t)) + \dot{V}_6(x(t)) \\ &\quad - 2Dz^T(Kx)\,\Lambda\,[Dz(Kx) - Gx]. \end{aligned} \tag{21}$$

According to (19), we obtain

$$\dot{V}_1(x(t)) = 2\left\{x^T(t) - x^T(t-\tau)D^T + \left[\int_{t-h}^{t} x^T(s)\,ds\right]B^T\right\} \times P\left[\widehat{A}_c x(t) - CDz(Kx)\right]$$
$$= 2x^T P\widehat{A}_c x - 2x^T PCDz(Kx) - 2x^T \widehat{A}_c^T PDx(t-\tau) + 2x^T(t-\tau)D^T PCDz(Kx) + 2x^T(t)\widehat{A}_c^T PB\left[\int_{t-h}^{t} x(s)\,ds\right] - 2\left[\int_{t-h}^{t} x^T(s)\,ds\right]B^T PCDz(Kx)$$
$$= x^T\left[\widehat{A}_c^T P + P\widehat{A}_c\right]x + 2x^T(t-\tau)D^T PCDz(Kx) + 2x^T(t)\widehat{A}_c^T PB\left[\int_{t-h}^{t} x(s)\,ds\right] - 2x^T PCDz(Kx) - 2x^T(t)\widehat{A}_c^T PDx(t-\tau) - 2\left[\int_{t-h}^{t} x^T(s)\,ds\right]B^T PCDz(Kx),$$

$$\dot{V}_2(x(t)) = \begin{bmatrix} x(t) \\ x\left(t-\frac{h}{2}\right)\end{bmatrix}^T \begin{bmatrix} Q_{11} & Q_{12} \\ \star & Q_{22}\end{bmatrix}\begin{bmatrix} x(t) \\ x\left(t-\frac{h}{2}\right)\end{bmatrix} - \begin{bmatrix} x\left(t-\frac{h}{2}\right) \\ x(t-h)\end{bmatrix}^T \begin{bmatrix} Q_{11} & Q_{12} \\ \star & Q_{22}\end{bmatrix}\begin{bmatrix} x\left(t-\frac{h}{2}\right) \\ x(t-h)\end{bmatrix}$$
$$+ \begin{bmatrix} x(t) \\ x\left(t-\frac{\tau}{2}\right)\end{bmatrix}^T \begin{bmatrix} R_{11} & R_{12} \\ \star & R_{22}\end{bmatrix}\begin{bmatrix} x(t) \\ x\left(t-\frac{\tau}{2}\right)\end{bmatrix} - \begin{bmatrix} x\left(t-\frac{\tau}{2}\right) \\ x(t-\tau)\end{bmatrix}^T \begin{bmatrix} R_{11} & R_{12} \\ \star & R_{22}\end{bmatrix}\begin{bmatrix} x\left(t-\frac{\tau}{2}\right) \\ x(t-\tau)\end{bmatrix},$$

$$\dot{V}_3(x(t)) = x^T(t)T_1x(t) - x^T(t-h)T_1x(t-h) + x^T(t)T_2x(t) - x^T(t-\tau)T_2x(t-\tau),$$

$$\dot{V}_4(x(t)) = -\left[\int_{t-h}^{t} x^T(s_1)\,ds_1\right]T_3\left[\int_{t-h}^{t} x(s_2)\,ds_2\right] + \int_{t-h}^{t} x^T(t)T_3\left[\int_{s}^{t} x(s_2)\,ds_2\right]ds + \int_{t-h}^{t}\left[\int_{s}^{t} x(s_2)\,ds_2\right]T_3x(t)\,ds$$
$$= -\left[\int_{t-h}^{t} x^T(s_1)\,ds_1\right]T_3\left[\int_{t-h}^{t} x(s_2)\,ds_2\right] + \int_{t-h}^{t}(s-t+h)x^T(t)T_3x(s)\,ds + \int_{t-h}^{t}(s-t+h)x^T(s)T_3x(t)\,ds$$
$$= -\left[\int_{t-h}^{t} x^T(s_1)\,ds_1\right]T_3\left[\int_{t-h}^{t} x(s_2)\,ds_2\right] - \int_{t-h}^{t}(s-t+h)(T_3x(t) - T_5x(s))^T T_5^{-1} \times (T_3x(t) - T_5x(s))\,ds + \frac{h^2}{2}x^T(t)T_3T_5^{-1}T_3x(t) + \int_{t-h}^{t}(s-t+h)x^T(s)T_5x(s)\,ds$$
$$\le -\left[\int_{t-h}^{t} x^T(s_1)\,ds_1\right]T_3\left[\int_{t-h}^{t} x(s_2)\,ds_2\right] + \frac{h^2}{2}x^T(t)\widetilde{T_3}x(t) + \int_{t-h}^{t}(s-t+h)x^T(s)T_5x(s)\,ds, \quad (22)$$

where $\widetilde{T_3} > 0$, since $\widetilde{T_3} = T_3T_5^{-1}T_3$.

Similarly we have

$$\dot{V}_5(x(t)) \le -\left[\int_{t-\tau}^{t} x^T(s_1)\,ds_1\right]T_4\left[\int_{t-\tau}^{t} x(s_2)\,ds_2\right] + \frac{\tau^2}{2}x^T(t)\widetilde{T_4}x(t) + \int_{t-\tau}^{t}(s-t+\tau)x^T(s)T_6x(s)\,ds, \quad (23)$$

where $\widetilde{T_4} > 0$, since $\widetilde{T_4} = T_4T_6^{-1}T_4$. Considering

$$\dot{V}_6(x(t)) = \frac{1}{2}h^2x^T(t)T_5x(t) - \int_{t-h}^{t}(s-t+h)x^T(s)T_5x(s)\,ds$$

$$+\frac{1}{2}\tau^2 x^T(t) T_6 x(t) - \int_{t-\tau}^{t} (s-t+\tau) x^T(s) T_6 x(s)\, ds. \tag{24}$$

Substituting these into (21) and the time-derivative of V has new upper bound as follows:

$$\dot{V}(x(t)) \le \xi^T(t)\,\Phi\xi(t), \tag{25}$$

where

$$\xi^T(t) = \left[x^T(t)\ \ x^T\left(t-\frac{h}{2}\right)\ \ x^T(t-h)\ \ x^T\left(t-\frac{\tau}{2}\right)\ \ x^T(t-\tau)\ \ \int_{t-h}^{t} x^T(s)\,ds\ \ \int_{t-\tau}^{t} x^T(s)\,ds\ \ Dz(Kx)\right]. \tag{26}$$

Φ is defined as stated in (16).

If linear matrix inequality (16) is feasible, then we can get $\dot{V}(x(t)) < 0, \forall x \in \mathcal{S}$. Therefore, if there exist scalars α_1 and α_2 and a positive definite matrix M satisfying (12) and there exist $P > 0, Q = [Q_{ij}]_{2\times 2}, Q_{ij} > 0, R = [R_{ij}]_{2\times 2}, R_{ij} > 0, T_i > 0, (i = 1,2,3,4,5,6), \widetilde{T}_i > 0, (i = 3,4)$, a matrix $\widehat{G}$, and a diagonal matrix Λ of appropriate dimensions satisfying (16), then we can draw the neutral system which can be described by (1) and (3) is asymptotic stability. This completes the proof. □

Remark 9. Theorem 8 gives a delay-dependent stability criterion for neutral system satisfying (1) and (3) by employing model transformation and delay-dividing approach. Model transformation may be an old technique and can be seen in [7]. But a new stability condition on the operator $\mathcal{L}$ is adopted in this paper to reduce the conservatism of the neutral system. Moreover, the proposed stability condition is not only on the discrete delays but also on the neutral delays, while the conditions in [31, 32] depend only on consideration of the discrete delays.

Remark 10. By dividing the neutral delay interval $[-\tau, 0]$ into $[-\tau, -\tau/2]$ and $[-\tau/2, 0]$, we choose different functional on $[-\tau, -\tau/2]$ and $[-\tau/2, 0]$, respectively. The discrete delay h is to take the same approach. Then the division provides extra freedom for delay terms and reduces the conservation.

We can easily get a delay-dependent stability criterion for system (1) with $\tau \equiv h$. This paper will show the asymptotic stability criterion for this case in Corollary 11.

Corollary 11. *Given scalars α_1 and α_2, a matrix G, and a diagonal matrix Λ of appropriate dimensions, the neutral systems (1) and (3) when $\tau \equiv h$ is asymptotically stable if there exists a positive definite matrix M satisfying (12) and there exist $P > 0, Q = [Q_{ij}]_{2\times 2}, Q_{ij} > 0, T_i > 0, (i = 1,2,3), \widetilde{T}_2 > 0$ such that the following symmetric linear matrix inequality holds:*

$$\overline{\Phi} = \begin{bmatrix} \overline{\Phi}_{11} & \overline{\Phi}_{12} & \overline{\Phi}_{13} & \overline{\Phi}_{14} & \overline{\Phi}_{15} \\ \star & \overline{\Phi}_{22} & \overline{\Phi}_{23} & 0 & 0 \\ \star & \star & \overline{\Phi}_{33} & 0 & \overline{\Phi}_{35} \\ \star & \star & \star & \overline{\Phi}_{44} & \overline{\Phi}_{45} \\ \star & \star & \star & \star & \overline{\Phi}_{55} \end{bmatrix} < 0, \tag{27}$$

where $\widehat{G} = \Lambda G, \widetilde{T}_2 = T_2 T_3^{-1} T_2$

$$\begin{aligned}
\overline{\Phi}_{11} &= \widehat{A}_c^T P + P\widehat{A}_c + Q_{11} + T_1 + \frac{h^2}{2}\left(\widetilde{T}_2 + T_3\right), \\
\overline{\Phi}_{22} &= Q_{22} - Q_{11}, \\
\overline{\Phi}_{33} &= -Q_{22} - T_1, \\
\overline{\Phi}_{44} &= -T_2, \\
\overline{\Phi}_{55} &= -2\Lambda, \\
\overline{\Phi}_{12} &= Q_{12}, \\
\overline{\Phi}_{13} &= -\widehat{A}_c^T PD, \\
\overline{\Phi}_{14} &= \widehat{A}_c^T PB, \\
\overline{\Phi}_{15} &= -PC + \widehat{G}, \\
\overline{\Phi}_{23} &= -Q_{12}, \\
\overline{\Phi}_{35} &= D^T PC, \\
\overline{\Phi}_{45} &= -B^T PC.
\end{aligned} \tag{28}$$

Proof. Define a legitimate Lyapunov functional candidate as

$$V(x(t)) = V_1(x(t)) + V_2(x(t)) + V_3(x(t)) + V_4(x(t)) + V_5(x(t)), \tag{29}$$

where

$$V_1(x(t)) = \mathcal{L}^T(x(t))\, P\mathcal{L}(x(t)),$$

$$V_2(x(t)) = \int_{t-(h/2)}^{t} \begin{bmatrix} x(s) \\ x\left(s-\frac{h}{2}\right) \end{bmatrix}^T \times \begin{bmatrix} Q_{11} & Q_{12} \\ \star & Q_{22} \end{bmatrix} \begin{bmatrix} x(s) \\ x\left(s-\frac{h}{2}\right) \end{bmatrix} ds,$$

$$V_3(x(t)) = \int_{t-h}^{t} x^T(s) T_1 x(s)\, ds,$$

$$V_4(x(t)) = \int_{t-h}^{t} \left[\int_{s}^{t} x^T(s_1)\, ds_1\right] T_2 \left[\int_{s}^{t} x(s_2)\, ds_2\right] ds,$$

$$V_5(x(t)) = \frac{1}{2}\int_{t-h}^{t} (s-t+h)^2 x^T(s) T_3 x(s)\, ds, \tag{30}$$

where $P > 0$, $Q = [Q_{ij}]_{2\times 2}$, $Q_{ij} > 0$ and $T_i > 0$, $i = 1, 2, 3$.

According to (30), we obtain

$$\begin{aligned}\dot{V}_1(x(t)) &= x^T\left[\widehat{A}_c^T P + P\widehat{A}_c\right]x + 2x^T(t-h) D^T PCDz(Kx) \\ &\quad + 2x^T(t)\widehat{A}_c^T PB\left[\int_{t-h}^{t} x(s)\, ds\right] \\ &\quad - 2x^T PCDz(Kx) - 2x^T(t)\widehat{A}_c^T PDx(t-h) \\ &\quad - 2\left[\int_{t-h}^{t} x^T(s)\, ds\right] B^T PCDz(Kx),\end{aligned}$$

$$\begin{aligned}\dot{V}_2(x(t)) &= \begin{bmatrix} x(t) \\ x\left(t-\frac{h}{2}\right)\end{bmatrix}^T \begin{bmatrix} Q_{11} & Q_{12} \\ \star & Q_{22}\end{bmatrix} \begin{bmatrix} x(t) \\ x\left(t-\frac{h}{2}\right)\end{bmatrix} \\ &\quad - \begin{bmatrix} x\left(t-\frac{h}{2}\right) \\ x(t-h)\end{bmatrix}^T \begin{bmatrix} Q_{11} & Q_{12} \\ \star & Q_{22}\end{bmatrix} \begin{bmatrix} x\left(t-\frac{h}{2}\right) \\ x(t-h)\end{bmatrix},\end{aligned}$$

$$\dot{V}_3(x(t)) = x^T(t) T_1 x(t) - x^T(t-h) T_1 x(t-h),$$

$$\begin{aligned}\dot{V}_4(x(t)) &\le -\left[\int_{t-h}^{t} x^T(s_1)\, ds_1\right] T_2 \left[\int_{t-h}^{t} x(s_2)\, ds_2\right] \\ &\quad + \frac{h^2}{2} x^T(t)\widetilde{T_2} x(t) \\ &\quad + \int_{t-h}^{t} (s-t+h) x^T(s) T_3 x(s)\, ds,\end{aligned} \tag{31}$$

where $\widetilde{T_2} > 0$, since $\widetilde{T_2} = T_2 T_3^{-1} T_2$. Considering

$$\begin{aligned}\dot{V}_5(x(t)) &= \frac{1}{2}h^2 x^T(t) T_3 x(t) \\ &\quad - \int_{t-h}^{t} (s-t+h) x^T(s) T_3 x(s)\, ds\end{aligned} \tag{32}$$

given that $x(t) \in \mathcal{S}$, which has been discussed in [21]. From Lemma 7, the time-derivative of V has new upper bound as follows:

$$\begin{aligned}\dot{V}(x(t)) &= \dot{V}_1(x(t)) + \dot{V}_2(x(t)) \\ &\quad + \dot{V}_3(x(t)) + \dot{V}_4(x(t)) + \dot{V}_5(x(t)) \\ &\quad - 2Dz^T(Kx)\Lambda[Dz(Kx) - Gx] \\ &\le \overline{\xi}^T(t)\overline{\Phi}\overline{\xi}(t),\end{aligned} \tag{33}$$

where $\overline{\Phi}$ is defined as stated in (27) and

$$\begin{aligned}&\overline{\xi}^T(t) \\ &= \left[x^T(t)\ \ x^T\left(t-\frac{h}{2}\right)\ \ x^T(t-h)\ \ \int_{t-h}^{t} x^T(s)\, ds\ \ Dz(Kx)\right].\end{aligned} \tag{34}$$

Therefore, if there exist scalars α_1 and α_2 and a positive definite matrix M satisfying (12) and there exist $P > 0$, $\widehat{P} > 0$, $Q = [Q_{ij}]_{2\times 2}$, $Q_{ij} > 0$, $T_i > 0$, $(i = 1, 2, 3)$, $\widetilde{T_2} > 0$ satisfying (27), then we can draw the identical neutral delay and discrete delay system is asymptotic stability. This completes the proof.

Furthermore, since the model transformation which usually may lead to some degree of conservatism of the method, we establish the following theorem based on Theorem 8 instead of using the model transformation.

Theorem 12. *Given a matrix G and a diagonal matrix Λ of appropriate dimensions, the neutral system with time-delay and actuator saturation which can be described by (1) and (3) is asymptotically stable if $\|D\| < 1$ and there exist $P > 0$, $Q = [Q_{ij}]_{2\times 2}$, $Q_{ij} > 0$, $R = [R_{ij}]_{2\times 2}$, $R_{ij} > 0$, $T_i > 0$, $(i = 1, 2, 3, 4, 5, 6)$, $\widetilde{T_i} > 0$, $(i = 3, 4)$ such that the following symmetric linear matrix inequality holds:*

$$\Phi = \begin{bmatrix} \Phi_{11} & \Phi_{12} & \Phi_{13} & \Phi_{14} & \Phi_{15} & 0 & 0 & \Phi_{18} \\ \star & \Phi_{22} & \Phi_{23} & 0 & 0 & 0 & 0 & 0 \\ \star & \star & \Phi_{33} & 0 & \Phi_{35} & 0 & 0 & 0 \\ \star & \star & \star & \Phi_{44} & \Phi_{45} & 0 & 0 & 0 \\ \star & \star & \star & \star & \Phi_{55} & 0 & 0 & \Phi_{58} \\ \star & \star & \star & \star & \star & \Phi_{66} & 0 & 0 \\ \star & \star & \star & \star & \star & \star & \Phi_{77} & 0 \\ \star & \star & \star & \star & \star & \star & \star & \Phi_{88} \end{bmatrix} < 0, \tag{35}$$

where $\Phi_{13} = PB$, $\Phi_{35} = -B^T PD$,

$$\begin{aligned}\Phi_{11} &= A_c^T P + PA_c + Q_{11} + R_{11} + T_1 + T_2 \\ &\quad + \frac{h^2}{2}\left(\widetilde{T_3} + T_5\right) + \frac{\tau^2}{2}\left(\widetilde{T_4} + T_6\right)\end{aligned} \tag{36}$$

and other notations are identical with Theorem 8.

Proof. We replace $\mathcal{L}(x(t))$ with $\mathcal{D}(x(t))$ in $V_1(x(t))$, and then follow the same procedure to prove Theorem 8. Thus it will complete the proof.

Remark 13. It should be pointed out that Theorem 12 is less conservative than Theorem 8, which can be verified by the numerical example. However, if the condition $\|D\| < 1$ cannot be satisfied while (12) can be satisfied, Theorem 12 is no longer applicable and we may apply Theorem 8 in this paper.

3.2. Asymptotic Stability for Uncertain Neutral System. Furthermore, let us consider the following uncertain neutral system with time-delay and actuator saturation:

$$\dot{x}(t) - D\dot{x}(t-\tau) = (A + \Delta A(t))\, x(t) + Bx(t-h) + (C + \Delta C(t))\,\text{Sat}(u(t)), \tag{37}$$

where $\Delta A(t)$ and $\Delta C(t)$ stand for the uncertainties. For simplicity, The constant parameter matrices A, B, C, and D are square matrices. The uncertainties are not known but the spectral norm bound is known and can be described as

$$\|\Delta A(t)\| \le \alpha, \quad \|\Delta C(t)\| \le \beta, \quad \forall t \ge 0. \tag{38}$$

Utilizing the nonlinear function $Dz(\cdot)$, we can also rewrite the uncertain neutral system in the following form:

$$\dot{x}(t) - D\dot{x}(t-\tau) = \left(A_c + \Delta A(t) - \Delta C(t) K\right) x(t) + Bx(t-h) + (C + \Delta C(t))\, Dz(Kx(t)), \tag{39}$$

where A_c is defined as the same with certain neutral system with time-delay and actuator saturation. Specially, when $\|\Delta A(t)\| = 0$ and $\|\Delta C(t)\| = 0$ the uncertain neutral system becomes the certain case.

Similarly, we employ the operator $\mathscr{L} : C_0 \to \mathbb{R}^n$ with

$$\mathscr{L}(x(t)) = x(t) + B\int_{t-h}^{t} x(s)\,ds - Dx(t-\tau). \tag{40}$$

Then we obtain the transformed system

$$\frac{d}{dt}\mathscr{L}(x(t)) = \overline{A_c}x(t) + (C + \Delta C(t))\, Dz(Kx(t)), \tag{41}$$

where $\overline{A_c} = \widehat{A}_c + \Delta A(t) - \Delta C(t)K = A + B + CK + \Delta A(t) - \Delta C(t)K$.

With regard to the uncertain neutral system, we can also prove the following theorem.

Theorem 14. *Given scalars α_1 and α_2, a matrix G, and a diagonal matrix Λ of appropriate dimensions, the uncertain neutral system which can be described as (37) with feedback control (3) is asymptotically stable if there exists a positive definite matrix M satisfying (12) and there exist $P > 0$, $Q = [Q_{ij}]_{2\times 2}$, $Q_{ij} > 0$, $R = [R_{ij}]_{2\times 2}$, $R_{ij} > 0$, $T_i > 0$, $(i = 1,2,3,4,5,6)$, $\widetilde{T}_i > 0$, $(i = 3,4)$ such that the following symmetric linear matrix inequality holds:*

$$\Xi = \begin{bmatrix} \Xi_{11} & \Xi_{12} & 0 & \Xi_{14} & \Xi_{15} & \Xi_{16} & 0 & \Xi_{18} \\ \star & \Xi_{22} & \Xi_{23} & 0 & 0 & 0 & 0 & 0 \\ \star & \star & \Xi_{33} & 0 & 0 & 0 & 0 & 0 \\ \star & \star & \star & \Xi_{44} & \Xi_{45} & 0 & 0 & 0 \\ \star & \star & \star & \star & \Xi_{55} & 0 & 0 & 0 \\ \star & \star & \star & \star & \star & \Xi_{66} & 0 & 0 \\ \star & \star & \star & \star & \star & \star & \Xi_{77} & 0 \\ \star & \star & \star & \star & \star & \star & \star & \Xi_{88} \end{bmatrix} < 0, \tag{42}$$

where $\widehat{G} = \Lambda G$, $\widetilde{T_3} = T_3T_5^{-1}T_3$, and $\widetilde{T_4} = T_4T_6^{-1}T_4$

$$\begin{aligned}
\Xi_{11} &= \widehat{A}_c^T P + P\widehat{A}_c + \frac{1}{\gamma_1}\widehat{P} + \frac{1}{\gamma_2}\widehat{P} \\
&\quad + (\gamma_1 + \gamma_3 + \gamma_5)\left(\alpha^2 + 2\alpha\beta\|K\| + \beta^2\|K\|^2\right) I \\
&\quad + Q_{11} + R_{11} + T_1 + T_2 + \frac{h^2}{2}\left(\widetilde{T_3} + T_5\right) + \frac{\tau^2}{2}\left(\widetilde{T_4} + T_6\right), \\
\Xi_{22} &= Q_{22} - Q_{11}, \\
\Xi_{33} &= -Q_{22} - T_1, \\
\Xi_{44} &= R_{22} - R_{11}, \\
\Xi_{55} &= \left(\frac{1}{\gamma_3} + \frac{1}{\gamma_4}\right) D^T\widehat{P}D - R_{22} - T_2, \\
\Xi_{66} &= \left(\frac{1}{\gamma_5} + \frac{1}{\gamma_6}\right) B^T\widehat{P}B - T_3, \\
\Xi_{77} &= -T_4, \\
\Xi_{88} &= (\gamma_2 + \gamma_4 + \gamma_6)\left(\beta^2 + 2\beta\|C\| + \|C\|^2\right) I - 2\Lambda, \\
\Xi_{12} &= Q_{12}, \\
\Xi_{14} &= R_{12}, \\
\Xi_{15} &= -\widehat{A}_c^T PD, \\
\Xi_{16} &= \widehat{A}_c^T PB, \\
\Xi_{18} &= \widehat{G}, \\
\Xi_{23} &= -Q_{12}, \\
\Xi_{45} &= -R_{12}.
\end{aligned} \tag{43}$$

Proof. In the same way, we define a legitimate Lyapunov functional candidate as

$$V(x(t)) = V_1(x(t)) + V_2(x(t)) + V_3(x(t)) + V_4(x(t)) + V_5(x(t)) + V_6(x(t)), \tag{44}$$

where $V_1(x(t)), V_2(x(t)), V_3(x(t)), V_4(x(t)), V_5(x(t)), V_6(x(t))$ are the same with those in Theorem 8. Given that $x(t) \in \mathcal{S}$, which has been discussed in [21]. From Lemma 7, it also follows that

$$\dot{V}(x(t)) \le \dot{V}_1(x(t)) + \dot{V}_2(x(t)) + \dot{V}_3(x(t)) + \dot{V}_4(x(t)) + \dot{V}_5(x(t)) + \dot{V}_6(x(t)) - 2Dz^T(Kx)\Lambda[Dz(Kx) - Gx]. \tag{45}$$

In fact

$$
\begin{aligned}
\dot{V}_1(x(t)) &= 2\left\{x^T(t) - x^T(t-\tau)D^T + \left[\int_{t-h}^{t} x^T(s)\,ds\right]B^T\right\} \\
&\quad \times P\left[\overline{A_c}x(t) + (C+\Delta C)Dz(Kx)\right] \\
&= 2x^T P\overline{A_c}x + 2x^T P(C+\Delta C)Dz(Kx) \\
&\quad - 2x^T\overline{A_c}^T PDx(t-\tau) - 2x^T(t-\tau)D^T \\
&\quad \times P(C+\Delta C)Dz(Kx) \\
&\quad + 2x^T(t)\overline{A_c}^T PB\left[\int_{t-h}^{t} x(s)\,ds\right] \\
&\quad + 2\left[\int_{t-h}^{t} x^T(s)\,ds\right]B^T P(C+\Delta C)Dz(Kx) \\
&= 2x^T P\widehat{A}_c x + 2x^T P(\Delta A - \Delta CK)x \\
&\quad + 2x^T P(C+\Delta C)Dz(Kx) \\
&\quad - 2x^T(t-\tau)D^T P\widehat{A}_c x(t) \\
&\quad - 2x^T(t-\tau)D^T P(\Delta A - \Delta CK)x \\
&\quad - 2x^T(t-\tau)D^T P(C+\Delta C)Dz(Kx) \\
&\quad + 2\left[\int_{t-h}^{t} x^T(s)\,ds\right]B^T P\widehat{A}_c x \\
&\quad + 2\left[\int_{t-h}^{t} x^T(s)\,ds\right]B^T P(\Delta A - \Delta CK)x \\
&\quad + 2\left[\int_{t-h}^{t} x^T(s)\,ds\right]B^T P(C+\Delta C)Dz(Kx) \\
&\le x^T\left[\widehat{A}_c^T P + P\widehat{A}_c\right]x - 2x^T(t)\widehat{A}_c^T PDx(t-\tau) \\
&\quad + 2x^T(t)\widehat{A}_c^T PB\left[\int_{t-h}^{t} x(s)\,ds\right] \\
&\quad + \frac{1}{\gamma_1}x^T\widehat{P}x + \gamma_1 x^T(\Delta A - \Delta CK)^T \\
&\quad \times (\Delta A - \Delta CK)x \\
&\quad + \frac{1}{\gamma_2}x^T\widehat{P}x + \gamma_2 Dz^T(Kx)(C+\Delta C)^T \\
&\quad \times (C+\Delta C)Dz(Kx) \\
&\quad + \frac{1}{\gamma_3}x^T(t-\tau)D^T\widehat{P}Dx(t-\tau) \\
&\quad + \gamma_3 x^T(\Delta A - \Delta CK)^T(\Delta A - \Delta CK)x \\
&\quad + \frac{1}{\gamma_4}x^T(t-\tau)D^T\widehat{P}Dx(t-\tau) \\
&\quad + \gamma_4 Dz^T(Kx)(C+\Delta C)^T(C+\Delta C)Dz(Kx) \\
&\quad + \frac{1}{\gamma_5}\left[\int_{t-h}^{t} x^T(s)\,ds\right]B^T\widehat{P}B\left[\int_{t-h}^{t} x(s)\,ds\right] \\
&\quad + \gamma_5 x^T(\Delta A - \Delta CK)^T(\Delta A - \Delta CK)x \\
&\quad + \frac{1}{\gamma_6}\left[\int_{t-h}^{t} x^T(s)\,ds\right]B^T\widehat{P}B\left[\int_{t-h}^{t} x(s)\,ds\right] \\
&\quad + \gamma_6 Dz^T(Kx)(C+\Delta C)^T(C+\Delta C)Dz(Kx) \\
&\le x^T\left\{\left(\widehat{A}_c^T P + P\widehat{A}_c\right) + \left(\frac{1}{\gamma_1} + \frac{1}{\gamma_2}\right)\widehat{P} \right. \\
&\qquad + (\gamma_1 + \gamma_3 + \gamma_5) \\
&\qquad \left. \times \left(\alpha^2 + 2\alpha\beta\|K\| + \beta^2\|K\|^2\right)I\right\}x \\
&\quad + (\gamma_2 + \gamma_4 + \gamma_6)\left(\beta^2 + 2\beta\|C\| + \|C\|^2\right) \\
&\quad \times Dz^T(Kx)Dz(Kx) \\
&\quad + x^T(t-\tau)\left(\frac{1}{\gamma_3}D^T\widehat{P}D + \frac{1}{\gamma_4}D^T\widehat{P}D\right)x(t-\tau) \\
&\quad + \left[\int_{t-h}^{t} x^T(s)\,ds\right]\left[\frac{1}{\gamma_5}B^T\widehat{P}B + \frac{1}{\gamma_6}B^T\widehat{P}B\right] \\
&\quad \times \left[\int_{t-h}^{t} x(s)\,ds\right] \\
&\quad - 2x^T(t)\widehat{A}_c^T PDx(t-\tau) \\
&\quad + 2x^T(t)\widehat{A}_c^T PB\left[\int_{t-h}^{t} x(s)\,ds\right],
\end{aligned} \tag{46}
$$

where $\widehat{P} > 0$, since $\widehat{P} = P^2$.

While $\dot{V}_2(x(t))$, $\dot{V}_3(x(t))$, $\dot{V}_4(x(t))$, $\dot{V}_5(x(t))$, and $\dot{V}_6(x(t))$ are obtained by the same method in Theorem 8. Then substituting these into (45), the time-derivative of V has new upper bound as follows:

$$
\begin{aligned}
\dot{V}(x(t)) &\le \dot{V}(x(t)) - 2Dz^T(Kx)\Lambda[Dz(Kx) - Gx] \\
&\le \xi^T(t)\,\Xi\,\xi(t),
\end{aligned} \tag{47}
$$

where $x(t) \in \mathcal{S}$, which has been discussed in [21], and

$$\xi^T(t) = \left[x^T(t) \;\; x^T\left(t - \frac{h}{2}\right) \;\; x^T(t-h) \;\; x^T\left(t - \frac{\tau}{2}\right) \;\; x^T(t-\tau) \;\; \int_{t-h}^{t} x^T(s)\,ds \;\; \int_{t-\tau}^{t} x^T(s)\,ds \;\; Dz(Kx) \right]. \quad (48)$$

Ξ is defined as stated in (42).

If linear matrix inequality (42) is feasible, then we can also get $\dot{V}(x(t)) < 0$, $\forall x \in \mathcal{S}$. Therefore, if there exist scalars α_1 and α_2 and a positive definite matrix M satisfying (12) and there exist $P > 0$, $\widehat{P} > 0$, $Q = [Q_{ij}]_{2\times 2}$, $Q_{ij} > 0$, $R = [R_{ij}]_{2\times 2}$, $R_{ij} > 0$, $T_i > 0$, $(i = 1,2,3,4,5,6)$, $\widetilde{T}_i > 0$, $(i = 3,4)$ satisfying (42) for real scalars $\gamma_i > 0$, $(i = 1,2,3,4,5,6)$, then we can draw the uncertain neutral system which can be described by (37) and (3) is asymptotic stability. This completes the proof.

Theorem 14 provides a new asymptotic stability condition for a class of uncertain neutral system which can be described as (37) and (3). We can also get a Corollary 15 which is a special case of Theorem 14.

Corollary 15. *Given scalars α_1 and α_2, a matrix G, and a diagonal matrix Λ of appropriate dimensions, the uncertain neutral system* (37) *with feedback control* (3) *when $\tau \equiv h$ is asymptotically stable if there exists a positive definite matrix M satisfying* (12) *and there exist $P > 0$, $Q = [Q_{ij}]_{2\times 2}$, $Q_{ij} > 0$, $T_i > 0$, $(i = 1,2,3)$, $\widetilde{T}_2 > 0$ such that the following symmetric linear matrix inequality holds:*

$$\overline{\Xi} = \begin{bmatrix} \overline{\Xi}_{11} & \overline{\Xi}_{12} & \overline{\Xi}_{13} & \overline{\Xi}_{14} & \overline{\Xi}_{15} \\ \star & \overline{\Xi}_{22} & \overline{\Xi}_{23} & 0 & 0 \\ \star & \star & \overline{\Xi}_{33} & 0 & 0 \\ \star & \star & \star & \overline{\Xi}_{44} & 0 \\ \star & \star & \star & \star & \overline{\Xi}_{55} \end{bmatrix} < 0, \quad (49)$$

where $\widehat{G} = \Lambda G$, $\widetilde{T}_2 = T_2 T_3^{-1} T_2$

$$\begin{aligned} \overline{\Xi}_{11} &= \widehat{A}_c^T P + P\widehat{A}_c + \frac{1}{\gamma_1}\widehat{P} + \frac{1}{\gamma_2}\widehat{P} \\ &\quad + (\gamma_1 + \gamma_3 + \gamma_5)\left(\alpha^2 + 2\alpha\beta\|K\| + \beta^2\|K\|^2\right) I \quad (50) \\ &\quad + Q_{11} + T_1 + \frac{h^2}{2}\left(\widetilde{T}_2 + T_3\right), \\ \overline{\Xi}_{22} &= Q_{22} - Q_{11}, \\ \overline{\Xi}_{33} &= \left(\frac{1}{\gamma_3} + \frac{1}{\gamma_4}\right) D^T \widehat{P} D - Q_{22} - T_1, \\ \overline{\Xi}_{44} &= \left(\frac{1}{\gamma_5} + \frac{1}{\gamma_6}\right) B^T \widehat{P} B - T_2, \\ \overline{\Xi}_{55} &= (\gamma_2 + \gamma_4 + \gamma_6)\left(\beta^2 + 2\beta\|C\| + \|C\|^2\right) I - 2\Lambda, \\ \overline{\Xi}_{12} &= Q_{12}, \\ \overline{\Xi}_{13} &= -\widehat{A}_c^T PD, \\ \overline{\Xi}_{14} &= \widehat{A}_c^T PB, \\ \overline{\Xi}_{15} &= \widehat{G}, \\ \overline{\Xi}_{23} &= -Q_{12}. \end{aligned} \quad (51)$$

Proof. Choose a legitimate Lyapunov functional candidate as

$$\begin{aligned} V(x(t)) &= V_1(x(t)) + V_2(x(t)) + V_3(x(t)) \\ &\quad + V_4(x(t)) + V_5(x(t)) \end{aligned} \quad (52)$$

which are the same as (30).

We follow the same processes in Corollary 11 and Theorem 14, then the proof can be easily completed.

Similar to the certain system, we also give the following less conservative condition for the uncertain case.

Theorem 16. *Given a matrix G and a diagonal matrix Λ of appropriate dimensions, the uncertain neutral system which can be described as* (37) *with feedback control* (3) *is asymptotically stable if $\|D\| < 1$ and there exist $P > 0$, $Q = [Q_{ij}]_{2\times 2}$, $Q_{ij} > 0$, $R = [R_{ij}]_{2\times 2}$, $R_{ij} > 0$, $T_i > 0$, $(i = 1,2,3,4,5,6)$, $\widetilde{T}_i > 0$, $(i = 3,4)$ such that the following symmetric linear matrix inequality holds:*

$$\Xi = \begin{bmatrix} \Xi_{11} & \Xi_{12} & \Xi_{13} & \Xi_{14} & \Xi_{15} & 0 & 0 & \Xi_{18} \\ \star & \Xi_{22} & \Xi_{23} & 0 & 0 & 0 & 0 & 0 \\ \star & \star & \Xi_{33} & 0 & \Xi_{35} & 0 & 0 & 0 \\ \star & \star & \star & \Xi_{44} & \Xi_{45} & 0 & 0 & 0 \\ \star & \star & \star & \star & \Xi_{55} & 0 & 0 & 0 \\ \star & \star & \star & \star & \star & \Xi_{66} & 0 & 0 \\ \star & \star & \star & \star & \star & \star & \Xi_{77} & 0 \\ \star & \star & \star & \star & \star & \star & \star & \Xi_{88} \end{bmatrix} < 0, \quad (53)$$

where $\Xi_{13} = PB$, $\Xi_{35} = -B^T PD$,

$$\begin{aligned} \Xi_{11} &= A_c^T P + PA_c + \left(\frac{1}{\gamma_1} + \frac{1}{\gamma_2}\right)\widehat{P} \\ &\quad + (\gamma_1 + \gamma_3)\left(\alpha^2 + 2\alpha\beta\|K\| + \beta^2\|K\|^2\right) I + Q_{11} \\ &\quad + R_{11} + T_1 + T_2 + \frac{h^2}{2}\left(\widetilde{T}_3 + T_5\right) + \frac{\tau^2}{2}\left(\widetilde{T}_4 + T_6\right) \Xi_{66} \\ &= -T_3, \Xi_{88} = (\gamma_2 + \gamma_4)\left(\beta^2 + 2\beta\|C\| + \|C\|^2\right) I - 2\Lambda \end{aligned} \quad (54)$$

and other notations are identical with Theorem 14.

Proof. We replace $\mathcal{L}(x(t))$ with $\mathcal{D}(x(t))$ in $V_1(x(t))$, and then follow the same procedure to prove Theorem 14. Thus it will complete the proof.

4. Numerical Examples

In this section, we present some numerical examples to illustrate the effectiveness of the proposed theoretical results.

4.1. No Uncertainty Case

Example 1. Consider the system (1) without saturation described by the following:

$$\begin{aligned}\dot{x}(t) - \begin{bmatrix} -0.2 & 0 \\ 0.2 & -0.2 \end{bmatrix} \dot{x}(t-h) \\ = \begin{bmatrix} -0.9 & 0.2 \\ 0 & -0.9 \end{bmatrix} x(t) + \begin{bmatrix} -1.1 & -0.2 \\ -0.1 & -1.1 \end{bmatrix} x(t-h).\end{aligned} \tag{55}$$

By Corollary 11 and Theorem 12, the maximum upper bound of h, which satisfies the LMI (27) and (35), can be calculated by solving a quasiconvex optimization problem. This neutral system is considered by the proposed methods in [8, 18, 21, 22, 28]. The results on the maximum upper bound of h are compared in Table 1.

From Table 1, we know that the maximum upper bound of delay $h = 0.7204$ by Corollary 11, and $h = 0.7211$ by and Theorem 12, respectively, while the maximum upper bound of delay $h = 0.6233$ for [18], $h = 0.7000$ for [28], $h = 0.5658$ for [8], $h = 0.7135$ for [21], and $h = 0.7048$ for [22]. It is found that the maximum upper bound in this paper is larger than those in [8, 18, 21, 22, 28]. So it can be demonstrated that the delay-dependent stability condition in this paper yields less conservative results than those previous ones. Additionally, it can be seen that Theorem 12 is less conservative.

Necessarily, let us check the stability of the operator $\mathscr{L}(x(t))$ given in (9) when $h = 0.7204$. By solving inequality (12), we have

$$\alpha_1 = 0.4137, \qquad \alpha_2 = 0.3625, \qquad M = \begin{bmatrix} 0.6157 & 0.0096 \\ 0.0096 & 0.6278 \end{bmatrix} \tag{56}$$

which guarantees the stability of the operator.

Example 2. Consider a time-delay neutral system with an actuator saturated at level ± 1 described as follows:

$$\begin{aligned}\dot{x}(t) - \begin{bmatrix} 0.15 & 0 \\ 0 & 0.15 \end{bmatrix} \dot{x}(t-\tau) \\ = \begin{bmatrix} -2 & 0 \\ 1 & -3 \end{bmatrix} x(t) + \begin{bmatrix} -1 & 0 \\ -0.8 & -1 \end{bmatrix} x(t-0.6) \\ + \begin{bmatrix} 1 & 2 \\ -1 & 4 \end{bmatrix} \text{Sat}(u(t)).\end{aligned} \tag{57}$$

It is found that the nominal system (57) is unstable and it is intended to stabilize the controlled system and find the range of delay time τ by using a state feedback controller K to guarantee that the above system is asymptotically stable. We need to verify the stability of the operator $\mathscr{L} : C_0 \to \mathbb{R}^n$ with

$$\mathscr{L}(x(t)) = x(t) + B\int_{t-h}^{t} x(s)\,ds - Dx(t-\tau). \tag{58}$$

By solving inequality (12), we obtain

$$\alpha_1 = 0.3346, \qquad \alpha_2 = 0.5728, \qquad M = \begin{bmatrix} 0.5364 & 0.0043 \\ 0.0043 & 0.3968 \end{bmatrix} \tag{59}$$

which guarantees the stability of the operator in Example 2.

In this paper, we adopt the feedback control matrix

$$K = \begin{bmatrix} 0.2062 & -0.1304 \\ 0.3755 & 0.4847 \end{bmatrix} \tag{60}$$

in [22]. By using the LMI toolbox in MATLAB, here the solutions of the LMIs given in Theorems 8 and 12 are found to be $\tau < 0.7566$ and $\tau < 0.7572$. So this example guarantees the asymptotic stabilization of system (57) for $0 \le \tau \le 0.7572$. When $D = 0$, we obtain $0 \le \tau \le 0.7828$ while the result of Niculescu guaranteeing the robust stabilization of system (57) is $0 \le \tau \le 0.3819$ in [33]. Hence, our result gives a less conservative result than that obtained by delay-dependent stability in [33].

4.2. Uncertainty Case

Example 3. Consider the system (55) in uncertainty case

$$\begin{aligned}\dot{x}(t) - \begin{bmatrix} -0.2 & 0 \\ 0.2 & -0.2 \end{bmatrix} \dot{x}(t-h) \\ = (A + \Delta A(t))\,x(t) + \begin{bmatrix} -1.1 & -0.2 \\ -0.1 & -1.1 \end{bmatrix} x(t-h),\end{aligned} \tag{61}$$

where A is identical with the system (55). The system (61) has the norm bounded uncertainty $\|\Delta A(t)\| \le 0.01$. The stability of the operator has been verified in Example 1. Then, by Corollary 15 and Theorem 16, the maximum upper bound of h, which satisfies the LMI (49) and (53), can be calculated by solving a quasiconvex optimization problem. This neutral system is considered by the proposed methods in [18], and methods in [8, 21, 22, 28] are no longer applicable to the uncertain case. The results on the maximum upper bound of h are compared in Table 2.

From Table 2, we know that the maximum upper bound of delay $h = 0.6852$ in this paper by Theorem 16, while the maximum upper bound of delay $h = 0.5937$ for [18]. So it also can be demonstrated that the delay-dependent stability condition in this paper yields less conservative results than some previous ones.

Example 4. Consider the system (57) in uncertainty case

$$\begin{aligned}\dot{x}(t) - \begin{bmatrix} 0.15 & 0 \\ 0 & 0.15 \end{bmatrix} \dot{x}(t-\tau) \\ = (A + \Delta A(t))\,x(t) + \begin{bmatrix} -1 & 0 \\ -0.8 & -1 \end{bmatrix} x(t-0.6)\,\text{Sat}(u(t)), \\ + (C + \Delta C(t)),\end{aligned} \tag{62}$$

TABLE 1: Maximum upper bound of h with different methods.

Methods	[8]	[18]	[28]	[22]	[21]	Corollary 11	Theorem 12
h	0.5658	0.6233	0.7000	0.7048	0.7135	0.7204	0.7211

TABLE 2: Maximum upper bound of h with different methods.

Methods	[8]	[18]	[28]	[22]	[21]	Corollary 15	Theorem 16
h	—	0.5937	—	—	—	0.6849	0.6852

where A and C are identical with the system (57). The stability of the operator has been verified in Example 2. The system (62) has the norm bounded uncertainties

$$\|\Delta A(t)\| \leq 0.1, \qquad \|\Delta C(t)\| \leq 0.1. \tag{63}$$

No doubt that this uncertainty will decrease the initial state bound that guarantees the asymptotic stability of the system (62). In this paper, we obtain $0 \leq \tau \leq 0.4369$ by Theorem 16 which guarantees the robust stabilization of system (62).

5. Conclusions

The asymptotic stability of neutral delay-differential systems with saturating actuators in terms of the LMI approach is developed in this paper. Applying the model transformation and the delay-dividing approach, a new augmented Lyapunov functional is constructed and the delay-dependent stability conditions are derived by some integral inequalities. Then the conditions are extended to the uncertain system, which has the norm bounded uncertainties. Examples have been provided to compare with recent papers, which show our results are more effective than those in earlier reports.

Conflict of Interests

The authors declare that there is no conflict of interests regarding the publication of this paper.

Acknowledgments

The authors would like to thank the editor and the anonymous reviewers for their constructive comments and suggestions to improve the quality of the paper. This work is partly supported by the Fundamental Research Funds for the Central Universities.

References

[1] Y. Kuang, *Delay Differential Equations with Applications in Population Dynamics*, Academic Press, San Diego, Calif, USA, 1993.

[2] V. B. Kolmanovskii and A. Myshkis, *Applied Theory of Functional Differential Equations*, Kluwer Academic, Dordrecht, The Netherlands, 1999.

[3] S.-I. Niculescu, *Delay Effects on Stability: A Robust Control Approach*, vol. 269 of *Lecture Notes in Control and Information Sciences*, Springer, London, UK, 2001.

[4] C.-H. Lien, "New stability criterion for a class of uncertain nonlinear neutral time-delay systems," *International Journal of Systems Science*, vol. 32, no. 2, pp. 215–219, 2001.

[5] C.-H. Lien, K.-W. Yu, and J.-G. Hsieh, "Stability conditions for a class of neutral systems with multiple time delays," *Journal of Mathematical Analysis and Applications*, vol. 245, no. 1, pp. 20–27, 2000.

[6] D. Ivănescu, S.-I. Niculescu, L. Dugard, J.-M. Dion, and E. I. Verriest, "On delay-dependent stability for linear neutral systems," *Automatica*, vol. 39, no. 2, pp. 255–261, 2003.

[7] E. Fridman, "New Lyapunov-Krasovskii functionals for stability of linear retarded and neutral type systems," *Systems & Control Letters*, vol. 43, no. 4, pp. 309–319, 2001.

[8] J. D. Chen, C.-H. Lien, K. K. Fan, and J. H. Chou, "Criteria for asymptotic stability of a class of neutral systems via a LMI approach," *IEE Proceedings: Control Theory and Applications*, vol. 148, no. 6, pp. 442–447, 2001.

[9] J.-H. Park, "A new delay-dependent criterion for neutral systems with multiple delays," *Journal of Computational and Applied Mathematics*, vol. 136, no. 1-2, pp. 177–184, 2001.

[10] J.-H. Park, "Stability criterion for neutral differential systems with mixed multiple time-varying delay arguments," *Mathematics and Computers in Simulation*, vol. 59, no. 5, pp. 401–412, 2002.

[11] J.-H. Park, "Convex optimization approach to dynamic output feedback control for delay differential systems of neutral type," *Journal of Optimization Theory and Applications*, vol. 127, no. 2, pp. 411–423, 2005.

[12] J.-H. Park and S. Won, "Asymptotic stability of neutral systems with multiple delays," *Journal of Optimization Theory and Applications*, vol. 103, no. 1, pp. 183–200, 1999.

[13] J.-H. Park and S. Won, "A note on stability of neutral delay-differential systems," *Journal of the Franklin Institute*, vol. 336, no. 3, pp. 543–548, 1999.

[14] O. M. Kwon, M. J. Park, J. H. Park, S. M. Lee, and E. J. Cha, "New delay-partitioning approaches to stability criteria for uncertain neutral systems with time-varying delays," *Journal of the Franklin Institute*, vol. 349, no. 9, pp. 2799–2823, 2012.

[15] Q.-L. Han, "On delay-dependent stability for neutral delay-differential systems," *International Journal of Applied Mathematics and Computer Science*, vol. 11, no. 4, pp. 965–976, 2001.

[16] Q.-L. Han, "Robust stability of uncertain delay-differential systems of neutral type," *Automatica*, vol. 38, no. 4, pp. 719–723, 2002.

[17] S.-I. Niculescu, "On delay-dependent stability under model transformations of some neutral linear systems," *International Journal of Control*, vol. 74, no. 6, pp. 609–617, 2001.

[18] A. Haurani, H. H. Michalska, and B. Boulet, "Delay-dependent robust stabilization of uncertain neutral systems with saturating actuators," in *Proceedings of the American Control Conference*, pp. 509–514, Denver, Colo, USA, June 2003.

[19] A. H. Glattfelder and W. Schaufelberger, "Stability analysis of single loop control systems with saturation and antireset-windup circuits," *IEEE Transactions on Automatic Control*, vol. 28, no. 12, pp. 1074–1081, 2003.

[20] J.-J. Yan, J.-S. Lin, and T.-L. Liao, "Robust dynamic compensator for a class of time delay systems containing saturating control input," *Chaos, Solitons & Fractals*, vol. 31, no. 5, pp. 1223–1231, 2007.

[21] J. M. G. da Silva Jr., A. Seuret, E. Fridman, and J. P. Richard, "Stabilisation of neutral systems with saturating control inputs," *International Journal of Systems Science*, vol. 42, no. 7, pp. 1093–1103, 2011.

[22] P.-L. Liu, "Stabilization criteria for neutral time delay systems with saturating actuators," *Journal of the Franklin Institute*, vol. 347, no. 8, pp. 1577–1588, 2010.

[23] P.-L. Liu and H.-L. Hung, "Stability of bilinear time-delay systems with saturating actuators," in *Proceedings of the IEEE International Symposium on Industrial Electronics (ISIE '99)*, pp. 1082–1086, Bled, Slovenia, July 1999.

[24] J. K. Hale and S. M. V. Lunel, *Introduction to Functional-Differential Equations*, vol. 99 of *Applied Mathematical Sciences*, Springer, New York, NY, USA, 1993.

[25] S. A. Rodrìguez, J.-M. Dion, and L. Dugard, "Stability of neutral time delay systems: a survey of some results," in *Advances in Automatic Control*, vol. 754, pp. 315–335, Kluwer Academic, Boston, Mass, USA, 2004.

[26] V. B. Kolmanovskii and J.-P. Richard, "Stability of some linear systems with delays," *IEEE Transactions on Automatic Control*, vol. 44, no. 5, pp. 984–989, 1999.

[27] V. L. Kharitonov, "Robust stability analysis of time-delay systems: a survey," in *Proceedings of the IFAC Conference on System Structure and Control*, pp. 1–12, Nantes, France, 1998.

[28] D. Yue, S. Won, and O. Kwon, "Delay dependent stability of neutral systems with time delay: an LMI approach," *IEE Proceedings: Control Theory and Applications*, vol. 150, no. 1, pp. 23–27, 2003.

[29] H. B. Ji, *Algebra Foundation of Control Theory*, University of Science and Technology of China Press, Hefei, China, 2008.

[30] S. Tarbouriech, J. M. G. da Silva Jr., and G. Garcia, "Delay-dependent anti-windup strategy for linear systems with saturating inputs and delayed outputs," *International Journal of Robust and Nonlinear Control*, vol. 14, no. 7, pp. 665–682, 2004.

[31] B. Yang, J. C. Wang, X. J. Pan, and C. Q. Zhong, "Delay-dependent criteria for robust stability of linear neutral systems with time-varying delay and nonlinear perturbations," *International Journal of Systems Science*, vol. 38, no. 6, pp. 511–518, 2007.

[32] Y.-Y. Cao and J. Lam, "Computation of robust stability bounds for time-delay systems with nonlinear time-varying perturbations," *International Journal of Systems Science*, vol. 31, no. 3, pp. 359–365, 2000.

[33] S.-I. Niculescu, J.-M. Dion, and L. Dugard, "Robust stabilization for uncertain time-delay systems containing saturating actuators," *IEEE Transactions on Automatic Control*, vol. 41, no. 5, pp. 742–747, 1996.

Multiscale Splitting Method for the Boltzmann-Poisson Equation: Application to the Dynamics of Electrons

Jürgen Geiser

Department of Physics, Ernst-Moritz-Arntz University of Greifswald, Domstraße 14, 17487 Greifswald, Germany

Correspondence should be addressed to Jürgen Geiser; geiser@mathematik.hu-berlin.de

Academic Editor: Shuyu Sun

We present a model based on dynamics of electrons in a plasma using a simplified Boltzmann equation coupled with Poisson's equation. The motivation arose from simulating active plasma resonance spectroscopy, which is used for plasma diagnostic techniques; see Braithwaite and Franklin (2009), Lapke et al. (2010), and Oberrath et al. (2011). Mathematically, we are interested in designing splitting methods for the model problem. While the full Boltzmann equation is delicate to solve, we decouple it into a transport and collision part, which are then solved in different ways. First we reduce it to a simplified transport-collision equation and start to analyse the abstract Cauchy problem using semigroup methods. Second, we pass to the coupled transport and collision model and apply the splitting ideas, resecting the different discretization schemes. The results are discussed first with numerical experiments and then we verify the underlying theoretical novelties.

1. Introduction

Our motive arose from studying the simulation of active plasma resonance spectroscopy, a well-established plasma diagnostic technique. To study this technique with simulation models, we concentrate on an abstract kinetic model that describes the dynamics of the electrons in plasma by using a Boltzmann equation. The Boltzmann equation is coupled with the electric field and we obtain coupled partial differential equations. The full Boltzmann equation is a very delicate equation to solve. Therefore, we decouple this into transport and collision contributions, which are solved in different ways. While finite difference schemes are applied to the transport parts, the collision part is solved with numerical integration schemes. The underlying splitting scheme is theoretically discussed as an abstract Cauchy problem. Here, we could address the Cauchy problem with functional analytical tools. First, we will discuss the description of a positive semigroup, which helps carry out the numerical estimates in the splitting schemes. Second, a numerical method is discussed with respect to separate differential and integral parts of the equations. The numerical approximation of the abstract splitting scheme is made by applying an iterative splitting method of the second order.

The paper is outlined as follows. In Section 2, we present our mathematical model and a possible reduced model for further approximations. The functional analytical setting with semigroups is discussed in Section 3. The splitting schemes are presented in Section 4. The results of some numerical experiments are exhibited in Section 6. In the conclusions that are given in Section 7, we summarize our results.

2. Mathematical Model

In the following a model is presented whose physical motivation is explained in [1–3].

The kinetic model considers a fluid dynamical approach to treat the natural ability of plasma to resonate near the electron plasma frequency ω_{pe}.

Here we specialise in an abstract kinetic model to describe the dynamics of the electrons in the plasma which allows for this resonance.

The Boltzmann equation for the electrons is

$$\frac{\partial f(x,v,t)}{\partial t} = \mathscr{A}\left[f(x,v,t)\right] + \mathscr{B}\left[f(x,v,t)\right],$$
$$(x,v) \in \Omega_x \times \Omega_v, \quad t \in [0,T],$$

$$\mathscr{A}[f] = -v \cdot \nabla_x f(x,v,t) - \frac{e}{m_e} \nabla_x \phi \cdot \nabla_v f(x,v,t),$$

$$\mathscr{B}[f] = -\sigma(x,v,t) f(x,v,t) + \int_V \kappa(x,v,v') f(x,v',t)\, dv',$$

$$f(x,v,0) = f_0(x,v), \quad (x,v) \in \Omega_{P,x} \times \Omega_{P,v},$$

$$f(x,v,t) = f_1(x,v), \quad (x,v) \in \partial\Omega_{P,x} \times \partial\Omega_{P,v}, \tag{1}$$

where $\Omega_{P,x} \times \Omega_{P,v} \subset \mathbb{R}^3 \times \mathbb{R}^3$ is the six-dimensional phase space. $f : \Omega_{P,x} \times \Omega_{P,v} \times [0,T] \to \mathbf{X}_c$ is the density function and $\mathbf{X}_c$ is an appropriate smooth space; for example, C_2. f_0 is the initial function and the f_1 is the boundary condition in the plasma phase space $\Omega_{P,x} \times \Omega_{P,v}$.

We assume based on the different materials (a plasma and a dielectricum) a complete reflection of the electrons due to the sheath $f(v_{||} + v_{\perp})$ with $v_{||}$, which is the parallel, and $v_{\perp}$ which is perpendicular to the surface normal vector. Further the electric force field is given as $F = e\nabla_x \phi$, where $\phi : \Omega_{P,x} \to \mathbf{X}_c$ is the potential and $\mathbf{X}_c$ is an appropriate smooth space, for example, C_2.

Boltzmann's equation is coupled with the electric field $\nabla\phi$, while the potential is approximated via Poisson's equation. The electrostatic approximation of the field is represented by a potential that is valid on the complete velocity volume $S \subset \Omega_{P,v}$, and apply Poisson's equation:

$$-\nabla_x \cdot (\epsilon \nabla \phi) = \begin{cases} e\left(n_i(x) - \int f\, dS\right) & \text{in } x \in \Omega_{P,x} \\ 0 & \text{in } x \in \Omega_{D,x}, \end{cases} \tag{2}$$

where the full space domain is $\Omega_x = \Omega_{P,x} \cup \Omega_{D,x} \subset \mathbb{R}^3$ and the full velocity domain is $\Omega_v = \Omega_{P,v} \cup \Omega_{D,v}$. The permittivity is equal to ϵ_0 in the plasma $\Omega_{P,x}$ and $\epsilon_0 \epsilon_D$ in the dielectric $\Omega_{D,x}$. ϕ fulfills the boundary conditions U_n at any electrode E_n and $-\mathbf{n} \cdot \nabla\phi = 0$ at isolator I, whereas $\mathbf{n}$ is the normal vector of the isolator surface.

On the surface of the dielectric, a surface charge σ may accumulate, which leads to a transition condition

$$\Delta(\epsilon \nabla \phi) = -\sigma. \tag{3}$$

We will make the following assumption to modify the model into a simpler and more tractable system of equations, which allows linearizing and simplifying the Boltzmann equations; see [4].

First we assume that we have an analytical solution of the potential ϕ, such that we could replace the convective term of the velocity-variable to a diffusive term of the space-variable as

$$\frac{F}{m_e} \cdot \nabla_v \approx \nabla_x D \nabla_x, \tag{4}$$

where D is the diffusion tensor and $F = e\nabla\phi$ is the external force. This modified model allows employing the following semigroup theory and deriving splitting schemes for the model.

The modified and splittable model equation is

$$\frac{\partial f(x,v,t)}{\partial t} = \mathscr{A}[f(x,v,t)] + \mathscr{B}[f(x,v,t)],$$

$$(x,v) \in \Omega_x \times \Omega_v, \quad t \in [0,T],$$

$$\mathscr{A}[f] = -v \cdot \nabla_x f(x,v,t) - \nabla_x D \nabla_x f(x,v,t),$$

$$\mathscr{B}[f] = -\sigma(x,v,t) f(x,v,t) + \int_V \kappa(x,v,v') f(x,v',t)\, dv',$$

$$f(x,v,0) = f_0(x,v), \quad (x,v) \in \Omega_{P,x} \times \Omega_{P,v},$$

$$f(x,v,t) = f_1(x,v), \quad (x,v) \in \partial\Omega_{P,x} \times \partial\Omega_{P,v}. \tag{5}$$

Next, we will discuss the relevant semigroup theory.

3. Semigroups for Transport Equations

In the following, we derive the exponential growth of the transport semigroups that are used in the section on the numerical methods.

We discuss two aspects:

(i) neutron transport,

(ii) electron transport.

The differences of the two transport schemes are as follows.

(i) Neutron transport is not influenced by an electric field.

(ii) Electron transport is influenced by an electric field.

3.1. Transport Model for the Neutrons. For this model we can assume that $f(x,v,t)$ describes the density distribution of the particles at position $x \in S$ with speed $v \in V$ at time $t \in [0,T]$; see also [5, 6].

The space S is assumed to be a compact and convex subset of $\mathbb{R}^3$ with nonempty interior, and the velocity space V is

$$V := \left\{ v \in \mathbb{R}^3 : v_{\min} \le \|\ \|_2 \le v_{\max} \right\}, \tag{6}$$

for $v_{\min} > 0$ and $v_{\max} < \infty$.

Assumption 1. We make the following assumptions.

(i) Particles move according to their speed v.

(ii) Particles are absorbed according to the probability given by the function σ which depends on x and v.

(iii) Particles are scattered according to a scattering kernel κ depending on x, the incoming speed v', and the outgoing speed v.

Then the neutron transport is

$$\frac{\partial f(x,v,t)}{\partial t} = -v \cdot \nabla f(x,v,t) - \sigma(x,v,t) f(x,v,t) + \int_V \kappa(x,v,v') f(x,v',t)\, dv', \tag{7}$$

$$f(x,v,0) = f_0(x,v),$$

and boundary conditions are included in the transport operator A_0; see, in the following, the abstract Cauchy problem.

We next treat the abstract Cauchy problem for this simplified model.

Abstract Cauchy Problem: Transport Model for the Neutrons. We have a Banach space $\mathbf{X} := L^1(S \times V)$ with Lebesgue measure on $S \times V \subset \mathbb{R}^6$ and define the abstract Cauchy problem as

$$\begin{gathered}\frac{du(t)}{dt} = Bu(t), \\ \frac{du(t)}{dt} = (A_0 - M_\sigma + K_\kappa)\,u(t), \\ u(0) = u_0,\end{gathered} \tag{8}$$

where $u \in \mathbf{X}$.

We have the following operators:

(1) collisionless transport operator,

(2) absorption operator,

(3) scattering Operator.

Important results for the further numerical analysis is the fact that the transport semigroup can be estimated by an exponential growth; see [5].

Corollary 2. *One assumes that $s(B) > -\infty$ is a dominant eigenvalue and $(S(t))_{t\geq 0}$ is irreducible. Then the transport semigroup $(S(t))_{t\geq 0}$ has balanced exponential growth. There exists a one-dimensional projection P satisfying $0 < Pf$ whenever $0 < f$ such that*

$$\left\|\exp(-s(B)\,t)\,S(t) - P\right\| \leq M\exp(-\epsilon t), \tag{9}$$

for all $t \geq 0$ and appropriate $M \geq 1$ and $\epsilon > 0$.

3.2. Transport Model for the Electrons or Ions. For this model we can assume that $f(x, v, t)$ describes the density distribution of particles at position $x \in S$ with speed $v \in V$ at time $t \in [0, T]$; see also [5, 6].

The space S is assumed to be a compact and convex subset of $\mathbb{R}^3$ with nonempty interior, and the velocity space V is

$$V := \left\{v \in \mathbb{R}^3 : v_{\min} \leq \|\ \|_2 \leq v_{\max}\right\}, \tag{10}$$

for $v_{\min} > 0$ and $v_{\max} < \infty$.

Assumption 3. We make the following assumptions.

(i) Particles move according to their speed v.

(ii) Particles are absorbed according to the probability given by the function σ which depends on x and v.

(iii) Particles are scattered according to a scattering kernel κ depending on x, the incoming speed v', and the outgoing speed v.

(iv) Particles are influenced by the static electric field ϕ, which can be derived by the kinetic theory.

The electron transport is

$$\begin{aligned}\frac{\partial f(x, v, t)}{\partial t} &= -v \cdot \nabla_x f(x, v, t) - \frac{e}{m_e}\nabla_x\phi \cdot \nabla_v f(x, v, t) \\ &\quad - \sigma(x, v, t)\,f(x, v, t) + \int_V \kappa(x, v, v')\,f(x, v', t)\,dv', \\ f(x, v, 0) &= f_0(x, v),\end{aligned} \tag{11}$$

and boundary conditions are included in the transport operators. ϕ is the electric field.

Furthermore, we have Poisson's equation,

$$-\nabla_x \cdot (\epsilon\nabla\phi) = \begin{cases} e\left(n_i - \int f\,dS\right) & \text{in } P \\ 0 & \text{in } D,\end{cases} \tag{12}$$

and the permittivity is equal to ϵ_0 in the plasma P and $\epsilon_0\epsilon_D$ in the dielectric D.

For the simplification, we assume to solve Poisson's equation analytically with

$$\begin{aligned}\phi &= \int_\Omega \mathcal{D}(f)\,dx \\ &= \begin{cases}\int_{\Omega_P} \epsilon^{-1}\int_{\Omega_P} e\left(n_i - \int f\,dS\right)dx\,dx & \text{in } P \\ 0 & \text{in } D,\end{cases}\end{aligned} \tag{13}$$

where $\mathcal{D}$ is the analytical function of the electric field $\nabla_x\phi$.

We embed the electric field analytically into the transport equation

$$\begin{aligned}\frac{\partial f(x, v, t)}{\partial t} &= -v \cdot \nabla_x f(x, v, t) - \nabla_x\mathcal{D}(f) \cdot \nabla_x f(x, v, t) \\ &\quad - \sigma(x, v, t)\,f(x, v, t) + \int_V \kappa(x, v, v')\,f(x, v', t)\,dv', \\ f(x, v, 0) &= f_0(x, v),\end{aligned} \tag{14}$$

and boundary conditions are included in the transport operators A_0 and A_1; see, in the following, the treatment of the abstract Cauchy problem. $\mathcal{D}(f)$ is the diffusion parameter that includes the electric field and we assume to approximate it via a constant operator $\mathcal{D} \approx \mathcal{D}(f)$.

Next we treat the abstract Cauchy problem for a transport model for the electrons or ions.

Abstract Cauchy Problem: Transport Model for the Electrons or Ions. We have a Banach space $\mathbf{X} := L^1(S \times V)$ with Lebesgue measure on $S \times V \subset \mathbb{R}^6$ and define the abstract Cauchy problem as

$$\frac{du(t)}{dt} = Bu(t),$$

$$\frac{du(t)}{dt} = (A_0 + A_1 - M_\sigma + K_\kappa)u(t), \quad (15)$$

$$u(0) = u_0,$$

where $u \in \mathbf{X}$.

We have the following operators:

(1) collisionless transport operator,

(2) diffusion operator,

(3) absorption operator,

(4) scattering Operator.

An important result for the further numerical analysis is the fact that the transport semigroup can be estimated by an exponential growth. Due to the analytical embedding of the electric field, we could also estimate such operator.

Corollary 4. *One assumes that* $s(\widetilde{B}) > -\infty$ *is a dominant eigenvalue and* $(S(t))_{t\geq 0}$ *is irreducible. Then the transport semigroup* $(S(t))_{t\geq 0}$ *has balanced exponential growth. There exists a one-dimensional projection* $\widetilde{P}$ *satisfying* $0 < \widetilde{P}f$ *whenever* $0 < f$ *such that*

$$\left\| \exp\left(-s\left(\widetilde{B}\right)t\right)S(t) - \widetilde{P} \right\| \leq \widetilde{M}\exp(-\epsilon t), \quad (16)$$

for all $t \geq 0$ *and appropriate* $\widetilde{M} \geq 1$ *and* $\epsilon > 0$.

In the next section we discuss the splitting schemes.

4. Splitting Schemes

In general, operator splitting methods are used to solve complex models in geophysical and environmental physics. They have been developed and applied in [7–9].

4.1. Sequential Splitting Method for Nonlinear Problems. For our problems, nonlinear splitting schemes are necessary; see [10]. We could use the result for the general formulation of nonlinear ordinary differential equations, which are given by

$$c'(t) = F_1(t, c(t)) + F_2(t, c(t)), \quad (17)$$

where the initial conditions are $c^n = c(t^n)$.

As before, we can decouple the above problem into two (usually simpler) subproblems, namely,

$$\frac{\partial c^*(t)}{\partial t} = F_1(t, c^*(t))$$

$$\text{with } t^n \leq t \leq t^{n+1}, \quad c^*(t^n) = c^n, \quad (18)$$

$$\frac{\partial c^{**}(t)}{\partial t} = F_2(t, c^{**}(t))$$

$$\text{with } t^n \leq t \leq t^{n+1}, \quad c^{**}(t^n) = c^*(t^{n+1}),$$

where the initial values are given by $c^n = c(t^n)$ and the split approximation on the next time level is defined as $c^{n+1} = c^{**}(t^{n+1})$.

For this case, the splitting error can be determined by using the Jacobians of the nonlinear mappings F_1 and F_2:

$$\rho_n = \frac{1}{2}\tau\left[\frac{\partial F_1}{\partial c}F_2, \frac{\partial F_2}{\partial c}F_1\right](t^n, c(t^n)) + \mathcal{O}(\tau_n^2). \quad (19)$$

Hence, for the general case, the splitting error is of first order, that is, $O(\tau_n)$.

Remark 5. Higher order splitting methods are given in [11]. Based on the Strang splitting, higher order nonlinear splitting methods are also possible; see [12].

In the next subsection we present the iterative splitting method.

4.2. Iterative Splitting Method. Alternatives are iterative approaches to nonlinear splitting schemes.

We concentrate again on nonlinear differential equations of the form

$$\frac{du}{dt} = A(u(t))u(t) + B(u(t))u(t), \quad \text{with } u(t^n) = u^n, \quad (20)$$

where $A(u)$, $B(u)$ are matrices with nonlinear entries and densely defined, where we assume that the entries involve the spatial derivatives of c; see [13]. In the following we discuss the standard iterative operator splitting method as a fixed-point iteration method to linearize the operators.

We split our nonlinear differential equation (20) by applying

$$\frac{du_i(t)}{dt} = A(u_{i-1}(t))u_i(t) + B(u_{i-1}(t))u_{i-1}(t), \quad \text{with } u_i(t^n) = c^n,$$

$$\frac{du_{i+1}(t)}{dt} = A(u_{i-1}(t))u_i(t) + B(u_{i-1}(t))u_{i+1}(t), \quad \text{with } u_{i+1}(t^n) = c^n, \quad (21)$$

where the time step is $\tau = t^{n+1} - t^n$. The iterations are $i = 1, 3, \ldots, 2m+1$. $u_0(t) = c_n$ is the starting solution, where we assume that the solution c^{n+1} is near c^n or $u_0(t) = 0$. Thus we have to solve the local fixed-point problem. c^n is the known split approximation at time level $t = t^n$.

The split approximation at time level $t = t^{n+1}$ is defined as $c^{n+1} = u_{2m+2}(t^{n+1})$. We assume that the operators $A(u_{i-1}(t^{n+1}))$, $B(u_{i-1}(t^{n+1}))$ are constant for $i = 1, 3, \ldots, 2m+1$. Here the linearization is done with respect to the iterations, such that $A(u_{i-1})$, $B(u_{i-1})$ are at least nondependent operators in the iterative equations, and we can apply the linear theory. For the linearization we assume that at least in the first equation $A(u_{i-1}(t)) \approx A(u_i(t))$ and in the second equation $B(u_{i-1}(t)) \approx B(u_{i+1}(t))$, for small t.

We have

$$\left\| A\left(u_{i-1}\left(t^{n+1}\right)\right) u_i\left(t^{n+1}\right) - A\left(u^{n+1}\right) u\left(t^{n+1}\right) \right\| \leq \epsilon, \quad (22)$$

for sufficient iterations $i \in \{1, 3, \ldots, 2m+1\}$.

Remark 6. The linearization with the fixed-point scheme can be used for smooth or weakly nonlinear operators, but otherwise we lose the convergence behavior, not converging to a local fixed point; see [14].

5. Numerical Integration of the Integrodifferential Part

We treat the following integrodifferential equation:

$$\frac{\partial u}{\partial t} = \int_0^t u(s)\, ds, \qquad u(0) = u_0. \quad (23)$$

The integration part is done numerically with trapezoidal rule:

$$\int_a^b f(x)\, dx \approx \frac{b-a}{n}\left(\frac{f(a)+f(b)}{2} + \sum_{k=1}^{n-1} f\left(a + k\frac{b-a}{n}\right)\right), \quad (24)$$

where the subintervals have the form $[kh, (k+1)h]$, with $h = (ba)/n$ and $k = 0, 1, 2, \ldots, n1$.

The higher order formulas are given as closed Newton-Cotes formulas, where f_i is a shorthand for $f(x_i)$, with $x_i = a + i(b-a)/n$, and n is the degree. We obtain the following formula for the trapezoid rule:

$$\frac{\partial u}{\partial t} = \frac{t}{2(u(0) + u(t))} ds, \qquad u(0) = u_0, \quad (25)$$

and obtain the analytical result

$$u(t) = \frac{2}{2}\exp\left(\frac{t^2}{4}\right) u(0) - \frac{1}{2} u(0). \quad (26)$$

For a higher order formula like Simpson's rule, we have

$$\frac{\partial u}{\partial t} = \frac{t}{6(u(0) + 4u(t/2) + u(t))} ds, \qquad u(0) = u_0. \quad (27)$$

We apply the idea of a polynomial solution:

$$u(t) = a_0 + a_1 t + a_2 t^2 + a_3 t^3 + \cdots, \quad (28)$$

and we obtain, after differentiating the coefficients,

$$\begin{aligned} a_1 &+ 2a_2 t + 3a_3 t^2 + \cdots \\ &= t\left(6\left(a_0 + 4\left(a_0 + \frac{a_1 t}{2} + \frac{a_2 t^2}{4} + \frac{a_3 t^3}{8} + \cdots\right)\right.\right. \\ &\qquad \left.\left. + a_0 + a_1 t + a_2 t^2 + a_3 t^3 + \cdots\right)\right)^{-1} \\ a_0 &= u_0. \end{aligned} \quad (29)$$

Now, comparing coefficients yields

$$\begin{aligned} a_0 &= u_0, \\ a_1 &= a_3 = a_5 = \cdots = 0, \\ a_2 &= 3a_0, \\ a_4 &= \frac{1}{12} a_2, \ldots. \end{aligned} \quad (30)$$

6. Numerical Experiments

We present the results of our numerical experiments based on the neutron transport. A simplified one-dimensional model is given by

$$\partial_t c + v\partial_x c - D\partial_{xx} c + \sigma c = \int_\Omega \kappa\left(x, v, v'\right) c\left(x, v', t\right) dv'. \quad (31)$$

The velocity v and the diffusion D are given by the plasma model. The initial conditions are given by $c(x, 0) = c_0(x)$ and the boundary conditions are trivial; $\partial_n c(x, t) = 0$.

A first integral operator is

$$\int_\Omega \kappa\left(x, v, v'\right) c\left(x, v', t\right) dv' = \int_0^T c(x, t)\, dt. \quad (32)$$

A second integral operator is as follows.

We assume a simple collision operator $\kappa(x, v, v') = q(v')(1 + v'^2)$, where $q(v')$ is the potential, for example, v'^2.

We deal with the first integral operator and define the following operators:

$$\begin{aligned} A &= v\frac{1}{\Delta x}\left[-1 \;\; 1 \;\; 0\right] I - D\frac{1}{\Delta x^2}\left[1 \;\; -2 \;\; 1\right] I, \\ B &= (-\sigma + t) I, \end{aligned} \quad (33)$$

while

$$\exp(Bt) = \exp\left(\left(-\sigma t + \frac{t^2}{2}\right) I\right), \quad (34)$$

where I is the identity matrix.

In the following, the simplified real-life problem for a neutron transport equation, which includes the gain and loss of a neutron, will be presented.

We concentrate on the computational benefits of a fast computation of the iterative scheme, given with matrix exponentials.

The equation is

$$\partial_t c + \nabla \cdot \mathbf{F} c = -\lambda_1 c + \int_0^t \lambda_2 c(x,t)\,dt, \quad \text{in } \Omega \times [0,t],$$
$$\mathbf{F} = \mathbf{v} - D\nabla,$$
$$c(\mathbf{x},t) = c_0(\mathbf{x}), \quad \text{on } \Omega,$$
$$c(\mathbf{x},t) = c_1(\mathbf{x},t), \quad \text{on } \partial\Omega \times [0,t]. \tag{35}$$

In the following, we deal with the semidiscretized equation given by the matrices

$$\partial_t \mathbf{C} = (A - \Lambda_1 + \Lambda_2)\,\mathbf{C}, \tag{36}$$

where $\mathbf{C} = (c_1, \ldots, c_I)^T$ is the solution of the species in the mobile phase in each spatial discretization point ($i = 1, \ldots, I$).

We have the following two operators for the splitting method:

$$A = \frac{D}{\Delta x^2} \cdot \begin{pmatrix} -2 & 1 & & & \\ 1 & -2 & 1 & & \\ & \ddots & \ddots & \ddots & \\ & & 1 & -2 & 1 \\ & & & 1 & -2 \end{pmatrix} + \frac{v}{\Delta x} \cdot \begin{pmatrix} 1 & & & & \\ -1 & 1 & & & \\ & \ddots & \ddots & & \\ & & -1 & 1 & \\ & & & -1 & 1 \end{pmatrix} \in \mathbb{R}^{I \times I}, \tag{37}$$

where I is the number of spatial points. Consider the following:

$$\Lambda_1 = \begin{pmatrix} \lambda_1 & 0 & & & \\ 0 & \lambda_1 & 0 & & \\ & \ddots & \ddots & \ddots & \\ & & 0 & \lambda_1 & 0 \\ & & & 0 & \lambda_1 \end{pmatrix} \in \mathbb{R}^{I \times I}. \tag{38}$$

For the integral term we have the following ideas.

Case 1. Consider the following:

$$\int_0^t \lambda_2 c(x,t)\,dt \approx \lambda_2 t c(x,t), \tag{39}$$

and we obtain the matrix

$$\Lambda_2 = \begin{pmatrix} \frac{\lambda_2 t^2}{2} & 0 & & & \\ 0 & \frac{\lambda_2 t^2}{2} & 0 & & \\ & \ddots & \ddots & \ddots & \\ & & 0 & \frac{\lambda_2 t^2}{2} & 0 \\ & & & 0 & \frac{\lambda_2 t^2}{2} \end{pmatrix} \in \mathbb{R}^{I \times I}. \tag{40}$$

For the operator splitting scheme, we apply A and $B = -\Lambda_1 + \Lambda_2$ and we apply the iterative splitting method, given in (21).

Case 2. We integrate the operator B with respect to the previous solutions $\mathbf{C}_{i-1}$, $\Lambda_2(\mathbf{C}_{i-1})$ and we obtain the matrix

$$\Lambda_2(\mathbf{C}_{i-1}) = \begin{pmatrix} \int_0^t \lambda_2 c_{1,i-1}(x,s)\,ds & 0 & \cdots & 0 \\ 0 & \int_0^t \lambda_2 c_{2,i-1}(x,s)\,ds & 0 & \\ \vdots & \ddots & \ddots & \vdots \\ 0 & \cdots & 0 & \int_0^t \lambda_2 c_{I,i-1}(x,s)\,ds \end{pmatrix} \in \mathbb{R}^{I \times I}. \tag{41}$$

We obtain $B(\mathbf{C}) = \Lambda_2(\mathbf{C}_{i-1}) + \Lambda_1 \mathbf{C}$.

The iterative scheme is given by the following.

For $i = 1, 2, \ldots$

$$\mathbf{C}_i(t) = \exp\left(A\left(t - t^n\right)\right)\mathbf{C}\left(t^n\right) + \int_{t^n}^t \exp\left(\left(t-s\right)A\right) B\left(\mathbf{C}_{i-1}(s)\right)ds, \quad t \in \left(t^n, t^{n+1}\right]. \tag{42}$$

For the reference solution, we apply a fine time and spatial scale without decoupling the equations. Figure 1 presents the numerical errors between the exact and the numerical solution. Here we obtain the optimal results for one-sided iterative schemes on the operator B, meaning that we iterate with respect to B and use A as the right-hand side.

Remark 7. For all iterative schemes, we can reach results faster than with the standard schemes, due to the fact that the iterative schemes benefit from their fast computations of the exponential matrices. With from four to five iterative steps, we obtain more accurate results than we did with

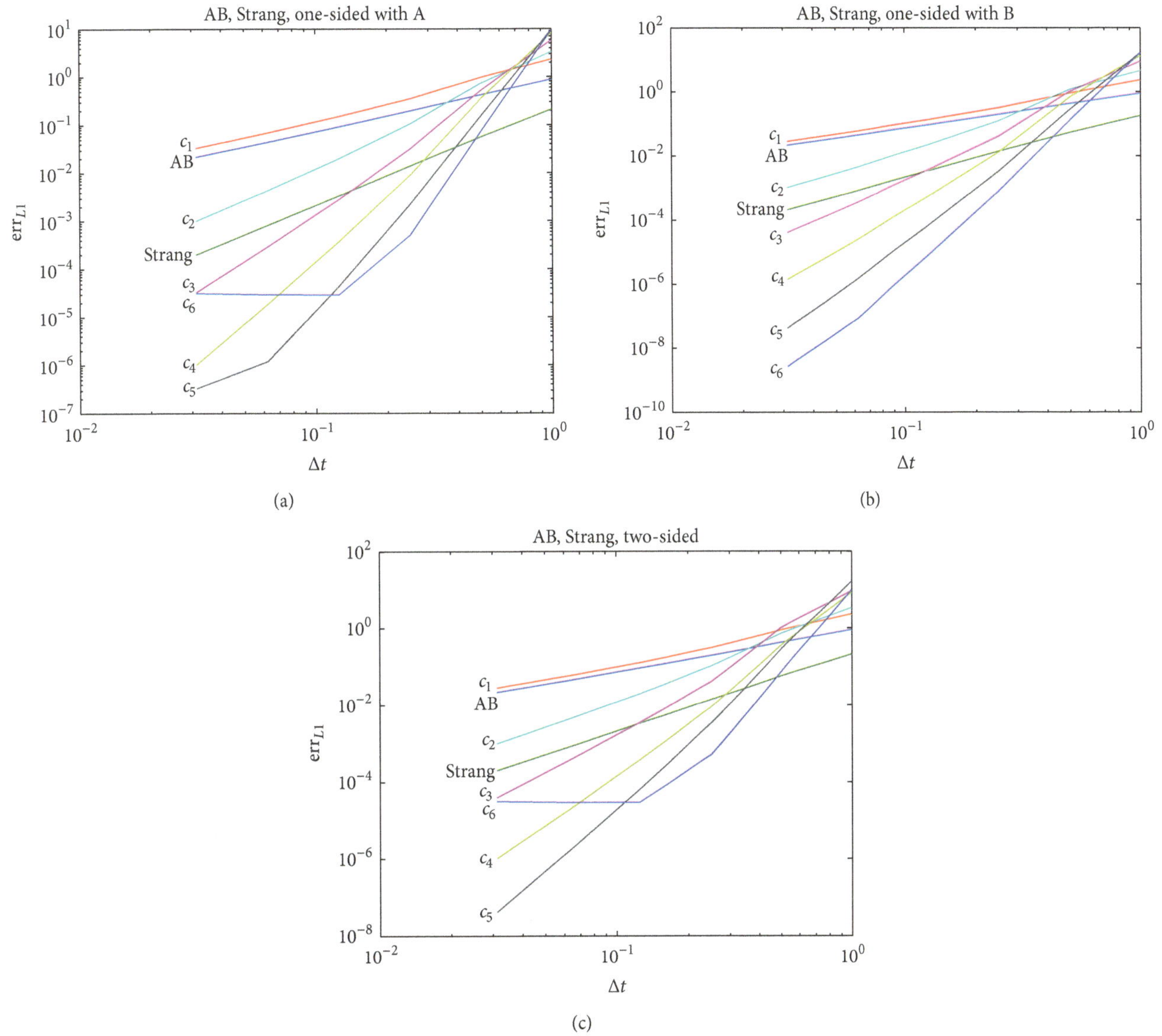

FIGURE 1: Numerical errors of the one-sided splitting scheme with A (a), the one-sided splitting scheme with B (b), and the iterative schemes with $1, \ldots, 6$ iterative steps (c).

the expensive standard schemes. With one-sided iterative schemes, we obtain the best convergence results.

7. Conclusions and Discussion

We presented the coupled model for the transport of deposition species in a plasma environment. We assumed that the flow field could be computed by the plasma model and the transport of the deposition species by a transport reaction model.

Such a first model can help understand the important modeling of the plasma environment in a CVD reactor.

Conflict of Interests

The author declares that there is no conflict of interests regarding the publication of this paper.

References

[1] N. S. J. Braithwaite and R. N. Franklin, "Reflections on electrical probes," *Plasma Sources Science and Technology*, vol. 18, no. 1, Article ID 014008, 2009.

[2] M. Lapke, T. Mussenbrock, and R. P. Brinkmann, "Modelling of volume- and surface wave based plasma resonance spectroscopy," in *Proceedings of the Abstracts IEEE International Conference on Plasma Sciences*, pp. 8–9, 2010.

[3] J. Oberrath, M. Lapke, T. Mussenbrock, and R. P. Brinkmann, "A functional analytical description of active plasma resonance spectroscopy in terms of kinetic theory," in *Proceedings of the 30th International Conference on Phenomena in Ionized Gases (ICPIG '11)*, Belfast, Ireland, August 2011.

[4] S. Ukai and T. Yang, *Mathematical Theory of Boltzmann Equation*, Lecture Notes Series No 8, Liu Bie Ju Center of Mathematical Sciences, City University of Hong Kong, 2006.

[5] K.-J. Engel and R. Nagel, *One-Parameter Semigroups for Linear Evolution Equations*, Springer, New York, NY, USA, 2000.

[6] A. Rhandi, *Spectral Theory for Positive Semigroups and Applications*, Quaderno Q. 1-2002, University of Lecce, Lecce, Italy, 2002.

[7] G. Strang, "On the construction and comparison of difference schemes," *SIAM Journal on Numerical Analysis*, vol. 5, pp. 506–517, 1968.

[8] J. G. Verwer and B. Sportisse, "A note on operator splitting in a stiff linear case," Tech. Rep. MAS-R9830, CWI, 1998.

[9] Z. Zlatev, *Computer Treatment of Large Air Pollution Models*, Kluwer, 1995.

[10] J. Geiser, *Splitting Methods for Differential Equations*, Chapman & Hall/CRC, 2011.

[11] E. Hansen and A. Ostermann, "Exponential splitting for unbounded operators," *Mathematics of Computation*, vol. 78, no. 267, pp. 1485–1496, 2009.

[12] S. Descombes and M. Thalhammer, "The LieTrotter splitting method for nonlinear evolutionary problems involving critical parameters. A compact local error representation and application to nonlinear Schrdinger equations in the semi-classical regime," *IMA Journal of Numerical Analysis*, vol. 33, no. 2, pp. 722–745, 2012.

[13] E. Zeidler, *Nonlinear Functional Analysis and Its Applications. II/B Nonlinear Montone Operators*, Springer, Berlin, Germany, 1990.

[14] J. F. Kanney, C. T. Miller, and C. T. Kelley, "Convergence of iterative split-operator approaches for approximating nonlinear reactive problems," *Advances in Water Resources*, vol. 26, no. 3, pp. 247–261, 2003.

Permissions

The contributors of this book come from diverse backgrounds, making this book a truly international effort. This book will bring forth new frontiers with its revolutionizing research information and detailed analysis of the nascent developments around the world.

We would like to thank all the contributing authors for lending their expertise to make the book truly unique. They have played a crucial role in the development of this book. Without their invaluable contributions this book wouldn't have been possible. They have made vital efforts to compile up to date information on the varied aspects of this subject to make this book a valuable addition to the collection of many professionals and students.

This book was conceptualized with the vision of imparting up-to-date information and advanced data in this field. To ensure the same, a matchless editorial board was set up. Every individual on the board went through rigorous rounds of assessment to prove their worth. After which they invested a large part of their time researching and compiling the most relevant data for our readers. Conferences and sessions were held from time to time between the editorial board and the contributing authors to present the data in the most comprehensible form. The editorial team has worked tirelessly to provide valuable and valid information to help people across the globe.

Every chapter published in this book has been scrutinized by our experts. Their significance has been extensively debated. The topics covered herein carry significant findings which will fuel the growth of the discipline. They may even be implemented as practical applications or may be referred to as a beginning point for another development.

The editorial board has been involved in producing this book since its inception. They have spent rigorous hours researching and exploring the diverse topics which have resulted in the successful publishing of this book. They have passed on their knowledge of decades through this book. To expedite this challenging task, the publisher supported the team at every step. A small team of assistant editors was also appointed to further simplify the editing procedure and attain best results for the readers.

Our editorial team has been hand-picked from every corner of the world. Their multi-ethnicity adds dynamic inputs to the discussions which result in innovative outcomes. These outcomes are then further discussed with the researchers and contributors who give their valuable feedback and opinion regarding the same. The feedback is then collaborated with the researches and they are edited in a comprehensive manner to aid the understanding of the subject.

Apart from the editorial board, the designing team has also invested a significant amount of their time in understanding the subject and creating the most relevant covers. They scrutinized every image to scout for the most suitable representation of the subject and create an appropriate cover for the book.

The publishing team has been involved in this book since its early stages. They were actively engaged in every process, be it collecting the data, connecting with the contributors or procuring relevant information. The team has been an ardent support to the editorial, designing and production team. Their endless efforts to recruit the best for this project, has resulted in the accomplishment of this book. They are a veteran in the field of academics and their pool of knowledge is as vast as their experience in printing. Their expertise and guidance has proved useful at every step. Their uncompromising quality standards have made this book an exceptional effort. Their encouragement from time to time has been an inspiration for everyone.

The publisher and the editorial board hope that this book will prove to be a valuable piece of knowledge for researchers, students, practitioners and scholars across the globe.

List of Contributors

Aftab Khan
Shaheed Benazir Bhutto University Sheringal, Dir Upper 18000, Pakistan
Government College University, Abdus Salam School of Mathematical Sciences (ASSMS), Lahore 54600, Pakistan

Gul Rahmat
Government College University, Abdus Salam School of Mathematical Sciences (ASSMS), Lahore 54600, Pakistan

Akbar Zada
Department of Mathematics, University of Peshawar, Peshawar 25000, Pakistan

Siniša Miličić, Mervan Pašić and Darko Žubrinić
Department of Mathematics, Faculty of Electrical Engineering and Computing, University of Zagreb, 10000 Zagreb, Croatia

Kusuo Kobayashi and Norio Yoshida
Department of Mathematics, University of Toyama, Toyama 930-8555, Japan

B. K. Bonzi, S. Ouaro and F. D. Y. Zongo
Laboratoire d'Analyse Mathématique des Equations (LAME), UFR, Sciences Exactes et Appliquées, Université de Ouagadougou, 03 BP 7021 Ouaga 03, Ouagadougou, Burkina Faso

Zhenbin Fan
Department of Mathematics, Changshu Institute of Technology, Suzhou, Jiangsu 215500, China

Gisèle Mophou
Laboratoire CEREGMIA, Université des Antilles et de la Guyane, Campus Fouillole, Guadeloupe (FWI) 97159 Pointe-à-Pitre, France

Tursuneli Niyaz and Ahmadjan Muhammadhaji
College of Applied Mathematics, Xinjiang University of Finance and Economics, Ürümqi 830012, China
College of Mathematical Sciences, Xinjiang Normal University, Urumqi 830054, China

Gung-Min Gie
Department of Mathematics, University of California, Riverside, 900 University Avenue, Riverside, CA 92521, USA

Chang-Yeol Jung
Ulsan National Institute of Science and Technology, San 194, Banyeon-ri, Eonyang-eup, Ulju Gun, Ulsan 689-798, Republic of Korea

Roger Temam
The Institute for Scientific Computing and Applied Mathematics, Indiana University, 831 EastThird Street, Bloomington, IN 47405, USA

Davide Catania
Dipartimento di Matematica, Facoltà di Ingegneria,Università degli Studi di Brescia, Via Valotti 9, 25133 Brescia, Italy

Leonid Pankratov
Laboratoire de Mathèmatiques et de leurs Applications, CNRS-UMR 5142, UNIV Pau & Pays Adour,
Avenue de lÚniversitè, 64000 Pau, France
Department of Mathematics, B. Verkin Institute for Low Temperature Physics and Engineering, 47, Avenue Lenin, Kharkov 61103, Ukraine

Brahim Amaziane
Laboratoire de Mathématiques et de leurs Applications, CNRS-UMR 5142, UNIV Pau & Pays Adour, Avenue de l'Université, 64000 Pau, France

Lakshman Mahto and Syed Abbas
School of Basic Sciences, Indian Institute of Technology Mandi, Mandi, Himachal Pradesh 175001, India

Angelo Favini
Dipartimento di Matematica, Universitá di Bologna, Pizza di Porta S. Donato 5, 40126 Bologna, Italy

Mandeep Singh and Amit K. Verma
Department of Mathematics, BITS Pilani, Pilani, Rajasthan 333031, India

Yūki Naito
Ehime University, Matsuyama 790-8577, Japan

Mervan Pašić
Department of Mathematics, University of Zagreb, FER, 10000 Zagreb, Croatia

A. Guezane-Lakoud and R. Khaldi
Laboratory of Advanced Materials, Faculty of Sciences, University Badji Mokhtar-Annaba, P.O. Box 12, 23000 Annaba, Algeria

Mervan Pašić
Department of Applied Mathematics, Faculty of Science, University of Zagreb, 10000 Zagreb, Croatia

Satoshi Tanaka
Okayama University of Science, Okayama 700-0005, Japan

Hassan A. El-Arabawy
Department of Mathematics, Faculty of Science, University of Tabuk, P.O. Box 741, Tabuk 71491, Saudi Arabia
Department of Mathematics, Faculty of Science, Ain Shams University, Egypt

Abdelhalim Ebaid and Nader Y. Abd Elazem
Department of Mathematics, Faculty of Science, University of Tabuk, P.O. Box 741, Tabuk 71491, Saudi Arabia

Renato Colucci, Gerardo R. Chacón and Andrés Vargas
Departamento de Matemáticas, Pontificia Universidad Javeriana, Carrera 7 No. 40-62, Bogotá, Colombia

Zhenguo Luo
Department of Mathematics, National University of Defense Technology, Changsha 410073, China
Department of Mathematics, Hengyang Normal University, Hengyang, Hunan 421008, China

Liping Luo
Department of Mathematics, Hengyang Normal University, Hengyang, Hunan 421008, China

Panayotis E. Nastou
Department of Mathematics, University of Aegean, 83200 Samos, Greece

Paul Spirakis
Department of Computer Engineering and Informatics, University of Patras Rio, 26504 Patras, Greece
Computer Technology Institute and Press "Diophantus" Rio, 26504 Patras, Greece

C. Stamatiou
Computer Technology Institute and Press "Diophantus" Rio, 26504 Patras, Greece
Department of Business Administration, University of Patras Rio, 26504 Patras, Greece

Apostolos Tsiakalos
Department of Mathematics, University of Ioannina, 45110 Ioannina, Greece

Xinghua Liu
Department of Automation, School of Information Science and Technology, University of Science and Technology of China, Anhui 230027, China

Jürgen Geiser
Department of Physics, Ernst-Moritz-Arntz University of Greifswald, Domstraße 14, 17487 Greifswald, Germany